FUNDAMENTALS OF GROUND WATER

FRANKLIN W. SCHWARTZ
The Ohio State University

HUBAO ZHANG
The Ohio State University

JOHN WILEY & SONS, INC.

This book is dedicated to our families:
Diane, Cynthia (Schwartz)
Ping, Dayu, and Michael (Zhang)

Acquisitions Editor   *Ryan Flahive*
Assistant Editor   *Denise Powell*
Marketing Manager   *Kevin Molloy*
Senior Production Editor   *Norine M. Pigliucci*
Design Director   *Madelyn Lesure*
Production Management Services   *Hermitage Publishing Services*

This book was set in Times Roman by Hermitage Publishing Services and printed and bound by Donnelley Willard.
The cover was printed by Phoenix Color Corporation.

This book is printed on acid-free paper. ∞

Library of Congress Cataloging in Publication Data:
Schwartz, F. W. (Franklin W.)
    Fundamentals of Ground Water / Franklin W. Schwartz and Hubao Zhang.
        p. cm.
    Includes bibliographical references.
    ISBN 0-471-13785-5 (cloth; alk. paper)
    1. Groundwater. I. Zhang, Hubao. II. Title.

GB1003.2.S39 2002
551.49–dc21   2002028811

PREFACE

Fundamentals of Ground Water is written at an introductory level to facilitate learning and teaching, while maintaining appropriate rigor. To provide an introductory look, feel and substance, we have included: numerous worked examples and exercises in each chapter; a section at the end of each chapter emphasizing highlights in those chapters; several windows-based computer programs for simulating ground-water processes; several Excel spreadsheet programs to demonstrate the calculation of mass transport processes; Power Point shows that illustrate field methods and technology, and take you on virtual tours of interesting study areas and field problems. These materials can be downloaded from http://www.wiley.com/college/schwartz. The publisher, John Wiley & Sons, Inc., has also contributed to this effort by providing an open, reader-oriented layout for the book and clearly drafted figures.

This text covers what we consider the basic areas of ground water. Chapter 1 introduces hydrologic cycle and especially the ground-water component. Chapter 2 focuses on the hydrologic processes at the earth's surface and the connection between these processes and ground-water flow. Chapters 3 through 5 provide the basic concepts of ground-water/ surface-water interactions, geology and ground water, and the theory of ground-water flow. Chapter 6 extends the discussions of flow to the unsaturated zone and fractured media. Chapter 7 details basic methods for geologic and hydrogeologic investigations in the field. Chapter 8 concentrates on regional ground-water flow and conceptual/mathematical models. Chapters 9 through 14 discuss well hydraulics and present common methods used to interpret results from aquifer tests. The last chapter is concerned specifically with ground-water flow. Chapter 15 deals with the management of ground-water resources and ground-water models.

The last part of the book deals with concepts of mass transport, chemistry, and contamination. Chapters 16 through 18 provide the basic concepts of water chemistry necessary to tackle both natural and contaminated systems. Chapter 19 develops conceptual and mathematical models for mass transport, so essential to frame problems and issues in water chemistry. Chapter 20 discusses isotopes as tracers and age dating techniques used in ground water. Chapter 21 explains how chemical concepts are used in the study of natural systems. Chapter 22 covers contaminated systems and investigation techniques for contamination problems. A bonus chapter, Chapter 23, introduces basic models and techniques in contaminant transport modeling. The bonus chapter can be downloaded from the book's web site, http://www.wiley.com/college/schwartz. Several Excel programs and a Windows based computer program have been developed to assist users of this chapter in simulating mass transport problems using analytical solutions.

This book is designed as an introductory ground-water text for upper-division undergraduate students or lower-division graduate students, or a reference book for ground-water professionals. Elementary college calculus and chemistry will help to understand the derivation of equations in the book. However, this book doesn't require readers to have this background because it provides step-by-step solutions for the application of mathematical equations. We have made each chapter as self sufficient as possible to make the teaching of the material and concepts easier. We have also deliberately made the text longer than one course can cover to provide instructors with choice of materials. In some cases, instructors might use the text for two courses, one for ground-water flow and one for contaminant hydrology.

As with most book writing efforts, we benefited from the support and assistance of our colleagues and, especially, our families. Cynthia Schwartz helped in the drafting of figures and production of graphic material for the PowerPoint shows. Dr. P. LaMoreaux kindly provided photographs of the Western Desert of Egypt. A. Crowe, E.A. Sudicky, and T. Davis helped by providing photos or other materials used in the preparation of illustrations. Colleagues of Hubao Zhang at Duke Engineering Services, D. Peterson, J. Studer, D. Walker, and G. Freeze provided references from their library and helpful advice. The authors also appreciate comments on the draft manuscript of the book by three anonymous reviewers and many useful suggestions by our publisher.

Our friend and colleague, Pat Domenico, followed our efforts in producing this book with great interest. Regretfully, he passed away in the late summer of 2001 before we had completed this text. We think that he would approve of this book, especially the process-based organization, the strong linkages between theory and practice, and the use of quantitative approaches.

Franklin W. Schwartz
Hubao Zhang

CONTENTS

1

INTRODUCTION TO GROUND WATER

▶ 1.1 WHY STUDY GROUND WATER?

▶ 1.2 GROUND WATER AND THE HYDROLOGIC CYCLE

This book is concerned with the theory and practice of *ground-water hydrology,* the science of water in subsurface environments. It is divided into two main parts: (1) basic concepts dealing with the origin, movement of ground water, and its recovery from wells, and (2) a qualitative and quantitative assessment of the chemistry of natural and contaminated systems. The book is developed around a process-oriented theme that organizes hydrologic phenomena based on physical and mathematical principles. It emphasizes the application of knowledge to the solution of practical problems encompassing areas of site characterization, hydraulic testing, and ground-water contamination.

One of the important ways to learn about ground-water problems is to experience them first hand. Thus, the book relies on case studies and on demonstrations of techniques through worked problems. In spite of the fact that most of the hydrogeological world is hidden from view, there are exciting things to see in the field. We have attempted to bring some of these to you through the colored photographs and illustrations that are included on the book's website, www.wiley.com/college/schwartz.

▶ 1.1 WHY STUDY GROUND WATER?

Scientists and engineers study ground water for a variety of reasons. First and foremost, ground water is a key source of drinking water that is essential to life on Earth, as we know it. The Earth has an estimated 330 million cubic miles (mi^3) of water, with most of it occurring as nonpotable seawater (Table 1.1). Interestingly, ground water makes up only a tiny fraction, just 0.06%, of the Earth's available water. However, this relatively small volume is critically important because it represents 98% of the freshwater readily available to humans (Zaporozec and Miller, 2000; Table 1.2). Abundant freshwater is tied up in glaciers and is essentially unavailable. Surprisingly, the other reservoirs (for example, streams and lakes) have local importance but are much less significant on a global scale (Table 1.2).

Ground water is found in aquifers, which have the capability of both storing and transmitting ground water. An *aquifer* is defined formally as a geologic unit that is sufficiently permeable to supply water to a well. Commonly, the large volumes of water stored in aquifers provide a reliable source during periods of drought lasting months or years. Moreover, ground water inflows to rivers, termed *baseflow*, can sustain flow over days and months with little rain. As Figure 1.1 illustrates, ground water can remain in the subsurface from days to millennia depending on the length of the flow path. In very extensive aquifers, the water presently being withdrawn by wells may have entered the subsurface thousands of years ago. Given the long time periods involved, the dynamics of ground-water flow can be influenced significantly by long-term changes in climate.

TABLE 1.1   Distribution of water stored on the earth

Pools of Water	Stored Volume (cubic miles)
Oceans	322,600,000
Glacial ice	6,000,000
Ground water	2,000,000
Atmosphere	3,000

Source: From W. H. Schlesinger, 1991, *Biochemistry—an analysis of global change*. Reprinted by permission of Academic Press.

TABLE 1.2   Distribution of the world's unfrozen freshwater supply

Source	Volume (Million mi^3)
Ground water	2.0
Lakes	0.031
Soil moisture	0.012
Streams	0.002
Swamps	<0.002

Source: Data from Zaporozec and Miller (2000).

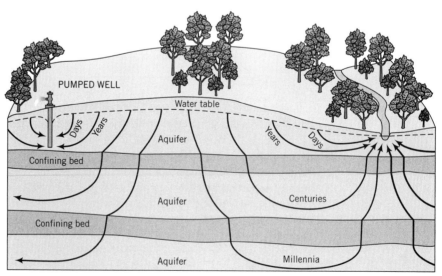

Figure 1.1   Water infiltrating the subsurface flows through the ground-water system and eventually discharges in streams, lakes, or oceans. The residence time in the subsurface can vary from days to thousands of years (from Winter et al., 1998).

The Kharga Oases in the Western Desert of Egypt exemplify the importance of ground water as a resource in a desert setting with no other (Case Study 1–1) surface-water sources.

Another feature of ground water that makes it valuable as a resource is its physical chemical quality. Unlike many surface-water supplies, natural ground water has few suspended solids, small concentrations of bacteria and viruses, and often only minimal concentrations of dissolved mineral salts. These characteristics make ground water an ideal source of water to support human life. Unfortunately, however, the connection of the ground-water pathway in the hydrologic cycle to the land surface provides the opportunity for humans to pollute natural ground waters and devalue the resource.

Not surprisingly, then, issues of ground-water pollution and the protection of ground-water resources provide another important reason to study ground water. Over the last 20 years, ground-water scientists and engineers have become aware of the health threat posed by contamination and the daunting technical challenges involved in cleaning up contaminated sites.

The human attack on ground water comes from many different directions. Significant contamination is added to ground water around the world through the careless disposal of human and animal wastes; haphazard disposal of wastes from industrial mining and oil operations; leaks from storage tanks, pipelines, or disposal ponds; and everyday activities such as farming and solid waste disposal. The most developed countries of the world have recognized these problems and are moving to clean up historical problems of contamination and to prevent new problems from developing. Other countries will require more time and money to confront the problems of contamination. Case Study 1–2 illustrates the too common problem of contamination that can develop in shallow aquifers.

Ground water also influences the design and construction of engineered facilities. The long-term viability of dams depends on controlling ground-water flow under or around the actual dam. The stability of excavations (e.g., trenches, or open-pit mines) often requires that ground-water levels in adjacent geologic units be controlled through an engineering dewatering program. Similarly, tunnels in rock often rely on extensive grouting programs to control ground-water inflow, while mining proceeds. Knowledge of ground-water

---

CASE STUDY 1-1

KHARGA OASES

The Kharga Oases are found in a large topographic depression in western Egypt (Figure 1.2). What makes the Kharga Oases interesting from a water resource perspective is that ground waters have been exploited at this site since Paleolithic man migrated there as early as 25,000 years B.P. Ground water originally discharged as a series of mound springs in the valley floor. Some of the first artesian wells were drilled to depths of 75 m during times of Persian rule. Later, Persian, Greco/Roman, Roman, and Arab peoples occupied the oases. One can understand the importance of the ground water at this location given that the mean annual rainfall is less than 1 mm year.

The ground water that is provided to wells and springs in Kharga and other oases in the Western Desert of Egypt comes from the Nubian Sandstone, an aquifer that underlies a large area of North Africa (Figure 1.3). The map of the flow paths (Figure 1.3) shows how rainfall, having infiltrated the subsurface, flows through the aquifer some 800 km to the Kharga Oases. LaMoreaux and others (1985) cite various studies indicating that some 50,000 years is required for water to flow from the nearest intake bed for rainfall (recharge areas) to the oases. Over these 50,000 years there have been significant changes in climate—influencing both the hydrology and human settlements. The website (www.wiley.com/college/schwartz) has photos of the Western Desert of Egypt courtesy of Dr. Philip LaMoreaux (show/WEgypt.ppt).

After LaMoreaux and others, 1985

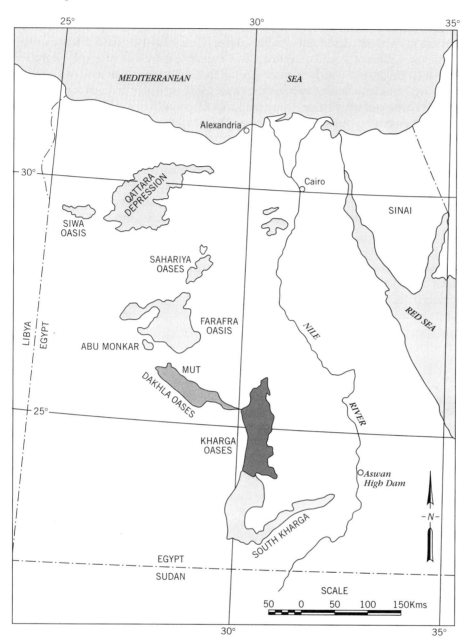

Figure 1.2 Location of oases in the Western Desert of Egypt (with permission from LaMoreaux et al., 1985).

hydrology is key to assessing the viability of geologic storage of nuclear wastes in many countries around the world.

Ground water is also important because of the geologic work that it can do. Valuable economic deposits owe their origins to ground-water systems. For example, ground water moving along flow systems can transport metals to zones of accumulation in roll-front uranium deposits or Mississippi Valley-type lead and zinc deposits. Oil formed by the maturation of organic matter deep in sedimentary basins can migrate through a three-dimensional ground-water system and accumulate in petroleum reservoirs. To an important extent,

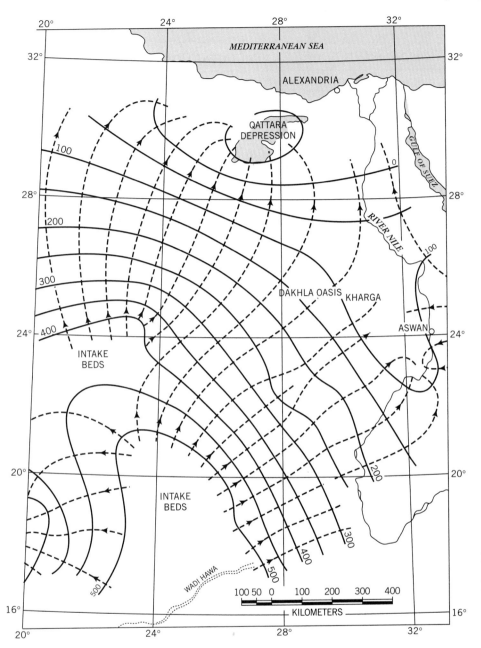

Figure 1.3 Pattern of ground-water flow in the Nubian Sandstone aquifer of Egypt and Libya (modified from Hellstrom, 1940).

the migration and accumulation of hydrocarbons depends on the existence of permeable pathways and, ultimately, on conditions capable of trapping hydrocarbons.

Not all of the geological processes in which ground water participates have dollar signs so prominently displayed. The development of surface landscapes is influenced by ground water. Higgins and Coates (1990) discuss the role of subsurface waters in Earth-surface processes and landforms. Ground water contributes to the development of landslides, rock falls, and stream channel networks. The dissolution and channelization within near surface carbonate rocks or salt deposits lead to the formation of *karst*, a landscape typified by sinkholes, sinking streams, and subsurface drainage. Karst landscapes occur in many countries

PHOSPHORUS TRANSPORT IN SEWAGE-CONTAMINATED GROUND WATER

The land disposal of treated sewage to a shallow sandy aquifer on Cape Cod, Massachusetts, has created a contaminant plume more than 3.5 miles long (LeBlanc and others, 1999). The site has been extensively studied as part of the Toxic Substances Program of the U.S. Geological Survey. Although this plume contains a variety of contaminants, phosphorus is of particular concern because water with concentrations in excess of 2 mg/L has been discharging into nearby Ashumet Pond. This phosphorus loading to the pond could promote changes in the ecology of this beautiful recreational pond.

Phosphorus was found to be strongly sorbed to the aquifer solids and to move more slowly than other contaminants. Nevertheless, 60 years of disposal of treated sewage between 1936 and 1995 promoted the development of a phosphorus plume about 3500 feet long (Figure 1.4). Phosphorus concentrations in the center of the plume were typically 3 to 4 mg/L, with a maximum concentration of 11 mg/L. Cleanup of this problem is complicated because the 90 to 99% of the phosphorus that is sorbed onto the aquifer materials will be slow to be flushed from the system.

After Walter and others, 1999

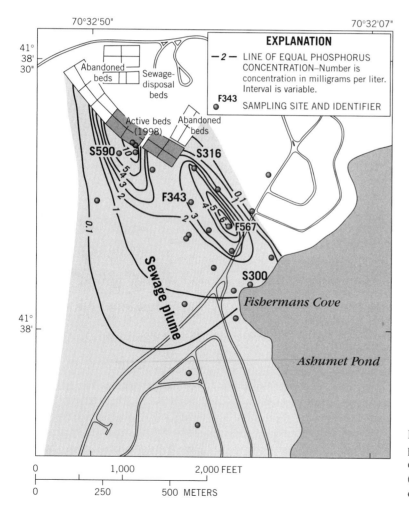

Figure 1.4 Distribution of dissolved phosphorus concentrations between sewage disposal beds and Ashumet Pond (1993–1994). Ground water is flowing south or southeast (from Walter et al., 1999).

SOIL SALINIZATION AT BLACKSPRING RIDGE, ALBERTA

The development of saline soils is a problem affecting dryland agriculture in arid lands around the world. Soil salinization occurs at sites where ground water carrying a load of dissolved mineral matter discharges at the ground surface or becomes prone to evaporation. Recent increases in the areas of saline soils have been attributed to farming and other land-use changes. Problems of soil salinity were studied at Blackspring Ridge, located southeast of Calgary, Alberta, Canada. Blackspring

Ridge is a till-covered bedrock ridge trending north-south that rises approximately 50 m above a surrounding lacustrine plain. Field studies show how local flow systems are responsible for severe soil salinization (Figure 1.5). Salts are leached from bedrock and deposited in topographically low areas along the flank of Blackspring Ridge (Figure 1.6), where the water table occurs less than 1.5 m from the ground surface.

After Stein, 1987, and Stein and Schwartz, 1991

Figure 1.5 Area of soil salinity showing severe salt crusting and salt-tolerant vegetation (red samphire) (with permission from Stein, 1987).

around the world and provide the most spectacular evidence of the geologic work of ground water.

Ground water plays an important role in the formation of soils and in the alteration of soils through salinization. *Saline soils* contain unusually high concentrations of soluble salts, which can be deleterious to the growth of many plants. Saline soils can form naturally or as the result of farming practices in arid lands. Case Study 1–3, based on a study in western Canada, illustrates how farming has led to the formation of saline soils.

We have identified many of the important issues that bear on ground water because these are what this book is all about. Although the water resource and engineering aspects of ground-water hydrology remain the focus of practice, the field has a rich relationship with the other earth sciences. *Ground-water hydrology* or *hydrogeology* is the study of the laws governing the movement of subterranean water, the mechanical, chemical, and thermal interaction of this water with the porous solid, and the transport of energy, chemical constituents, and particulate matter by flow (Domenico and Schwartz, 1998). Historically, the term *ground-water hydrology* has implied an emphasis on engineering applications, and the term *hydrogeology* has implied an emphasis on geologic problems.

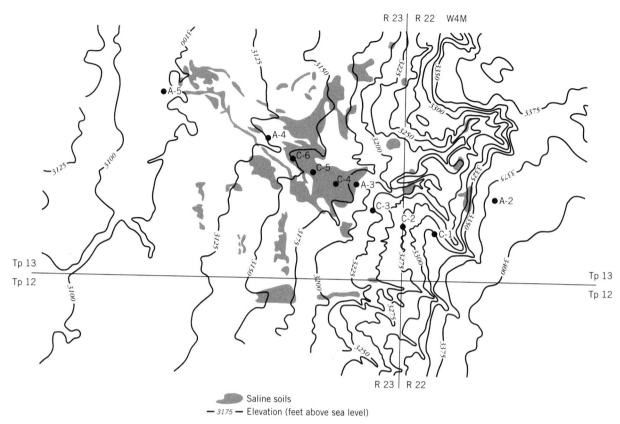

Figure 1.6 Topographic map showing Blackspring Ridge (east side). Saline soils are developed on the flank and toward the bottom of the ridge (with permission from Stein, 1987).

▶ 1.2 GROUND WATER AND THE HYDROLOGIC CYCLE

Water circulates on Earth from the oceans to the atmosphere to land and back to the oceans in what is called the hydrologic cycle. The main pathways in the hydrologic cycle are shown schematically in Figure 1.7. Water evaporates from oceans, lakes, and rivers into the atmosphere. This water vapor is transported with the atmospheric circulation and eventually falls as rain or snow onto the land, lakes, rivers, and oceans. Of the water falling on land, a proportion quickly evaporates, some flows into streams or lakes as overland flow, and another proportion infiltrates into the subsurface. Of the water entering the soil, some is transpired back into the atmosphere by plants. The remaining water follows a subsurface pathway back to surface (Figure 1.7).

Although this book touches on these various components of the hydrologic cycle, our main emphasis is on the water that occurs in the subsurface. As Figure 1.8 shows, subsurface water can occur in the *soil-water zone*, the *vadose zone*, and the *phreatic zone*. The soil-water zone contains voids caused by cracking and the decay of plant roots. This feature enhances the infiltration of precipitation. Pores in both the soil-water zone and the vadose or unsaturated zone contain both air and water. In the phreatic or saturated zone, water completely fills the pore space. The saturated and unsaturated zones are separated by an imaginary surface called the *water table*. The actual definition of the water table is somewhat more complicated, and we will defer further discussion to a later chapter. Classically, ground water refers to water that occurs in the zone of saturation, below the water table.

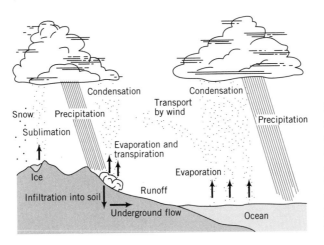

Figure 1.7 Schematic representation of the hydrologic cycle (from Domenico and Schwartz, 1998, *Physical and chemical hydrogeology*). Copyright ©1998 by John Wiley & Sons, Inc. Reprinted by permission of John Wiley & Sons, Inc.

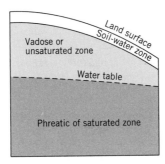

Figure 1.8 Occurrence of water in the surface (modified from Winter et al., 1998).

The hydrologic cycle, as shown in Figure 1.7, describes how water moves into and out of various domains, for example, the atmosphere or the subsurface. Within each of these domains, one can account for the water that is present with the following equation:

$$\text{input} - \text{output} = \text{change of storage} \tag{1.1}$$

Equation 1.1 is a conservation equation that accounts for all of the water moving into and out of the various domains along the hydrologic cycle. The term *conservation* implies that water moving in the hydrologic cycle is neither gained nor lost. This equation is used on both a global scale (Figure 1.7) and a basin scale (Figure 1.9). On a global scale, the water balance equation can be applied to one of three domains: atmosphere, land (surface and underground), and oceans. For example, water balance for land is written as

$$P - E - T - R_o = \Delta S \tag{1.2}$$

where P is precipitation on land, E is evaporation from land, T is transpiration from land, R_o is total outflow from land to oceans, and ΔS is the change of water storage on and under land.

We will examine the various components of the water balance equation in detail in Chapter 2. For the purposes of our discussion here, precipitation refers to water reaching the Earth's surface from rain and snow, *evaporation* is the loss of water from soils or free water by a phase change from liquid to water vapor, and *transpiration* is the loss of water through the plant system to the atmosphere. The runoff term accounts for all the water carried back to the ocean from the channel network of streams. The unit of all the variables

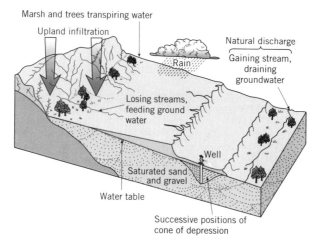

Figure 1.9 Key components of the basin hydrologic cycle. Modified from Theis, 1940, The source of water derived from wells—essential factors controlling the response of an aquifer to development. Copyright 1940 by ASCE and reproduced with permission of the publisher, ASCE.

is volume per unit time $[L^3/T]$. Similar equations can be written for atmosphere and oceans. Example 1.1 illustrates the application of the global water balance for land.

▶ **EXAMPLE 1.1**

On a global scale, it is known that $P = 4200$ bgd (billion gallons per day), $E + T = 2900$ bgd, and $R_o = 1300$ bgpd. Calculate the water storage change on the land area.

SOLUTION   The change in storage is

$$\Delta S = P - E - T - R_o = \text{(4200 bgd) - (2900 bgd) - (1300 bgd)} = 0$$

This result is interesting in that it tells us that over the long term the sources of water to the land (that is, precipitation) are exactly balanced by the losses (that is, E, T, R_o).

Basin-Scale Water Balances

In most instances, people are interested in water balances on a scale smaller than the global cycle. The basin scale is often a useful choice in this respect. As before, rainfall and evapotranspiration remain as key inputs and outputs to the ground-water system (Figure 1.9). Reducing the scale, however, adds more detail and complexity. The figure shows several different ways in which water is added or removed from the ground-water system. For example, in upland mountainous areas (Figure 1.9), ground-water recharge comes as melting snow, and spring rains infiltrate the unsaturated zone. Commonly, much of this water is discharged to nearby streams (Figure 1.10a). Water is also added to the ground-water system from *losing streams*. These streams leak water into the ground-water system (Figure 1.10b). Alternatively, water can be discharged from the ground water to surface waters (Figure 1.10c). The *gaining stream* in this case receives discharge from the downstream end of the aquifer.

In many areas, ground-water withdrawals from wells provide another significant component of the ground-water balance. As Figure 1.10d shows, wells can extract water from both the ground-water system and the surface-water system. On a basin scale, the water balance equation for ground-water storage may be expressed as

$$R_N + Q_i - T - Q_o = \Delta S \tag{1.3}$$

where R_N is the recharge to ground water, Q_i is surface water inflow to ground-water storage, T is transpiration, and Q_o is outflow from ground-water storage to surface water.

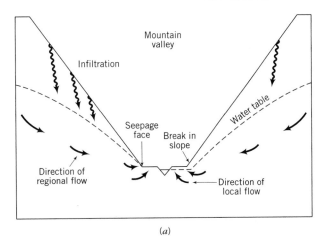

(a)

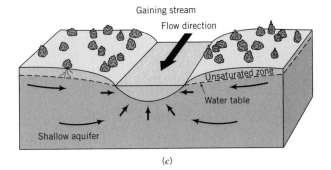

(b)

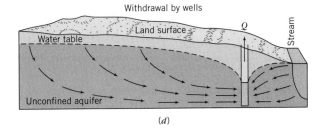

(c)

(d)

Figure 1.10 Examples of outflow from a ground-water flow system (Panels *a*, *c*, and *d*) and inflow to the ground water from a losing stream (Panel *b*) (modified from Winter et al., 1998).

If withdrawals from wells are included, Eq. (1.3) becomes

$$R_N + Q_i - T - Q_o - Q_p = \Delta S \tag{1.4}$$

where Q_p is the total pumping rate in the basin. With the help of the following example, let us consider how this equation is used.

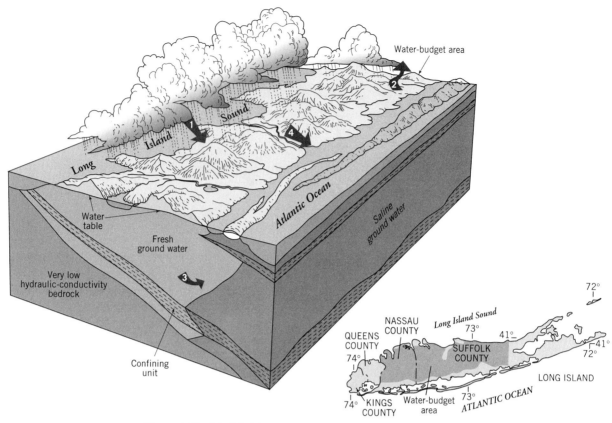

Figure 1.11 Modified from Cohen et al. (1968).

► **EXAMPLE 1.2**
Ground-water Changes in Response to Pumping (adapted from Alley et al., 1999)

This example illustrates how water-budget calculations might be used to assess the impact of ground-water withdrawals on Long Island, New York. The area of interest includes parts of Nassau and Suffolk counties (Figure 1.11). As Figure 1.11 implies, the ground-water system is one of Long Island's important sources of freshwater. Cohen and others (1968) estimated the major contributions to water balance (Table 1.1) before extensive development of the ground water (that is, the mid-seventeenth century). For reference purposes, the numbers in the table are keyed to the arrows in Figure 1.11.

1. Write an equation to describe the predevelopment water balance for the two counties. Those items on the left side of Table 1.3 are inflows, and those on the right side are outflows.

SOLUTION The water balance equation can be written in words as:

water input from precipitation − evapotranspiration of precipitation − evapotranspiration of

ground water − streamflow discharging to the sea − ground water discharging to the sea

− spring flow = change in storage.

$$P - ET_p - ET_{gw} - Q_{swo} - Q_{gwo} - Q_{so} = \Delta S$$

where P is precipitation to the land, ET is evaporation of precipitation and ground water, and Q is outflow to the ocean from surface water, ground water, and springs.

2. Given the water-budget information in Table 1.3 determine whether or not the system is at steady state.

TABLE 1.3   Overall Water Budget before Ground Water Was Extensively Developed

Inflow to Long Island Hydrologic System	(ft³/s)	Outflow from Long Island Hydrologic System	(ft³/s)
1. Precipitation	2475	2. Evapotranspiration of precipitation	1175
		3. Ground-water discharge to sea	725
		4. Streamflow to sea	525
		5. Evapotranspiration of ground water	25
		6. Spring flow	25

Source: Modified from Cohen et al. (1968).

TABLE 1.4   Water Budget for the Ground-Water System before Development

Inflow to Long Island Ground-Water System	(ft³/s)	Outflow from Long Island Ground-Water System	(ft³/s)
7. Recharge	1275	8. Ground-water discharge to streams	500
		9. Ground-water discharge to sea	725
		10. Evapotranspiration of ground water	25
		11. Spring flow	25

Source: Modified from Cohen et al. (1968)

SOLUTION   Substitute the appropriate values into the water balance equation to calculate ΔS.

$$2475 - 1175 - 25 - 725 - 525 - 25 = \Delta S = 0$$

Storage change in the system (ΔS) is zero, which means the system is at steady state. Outflows from the hydrologic system are exactly matched to the inflows.

3. Cohen et al. (1968) also created a water balance for the ground-water system before the resource was developed. This budget is described by the following equation:

$$R_N - Q_{gws} - Q_{gwo} - ET_{gw} - Q_s = \Delta S$$

where R_N is ground-water recharge, Q_{gws} is ground-water discharge to streams, and the other terms as described previously. Estimates for these components of the ground-water budget are provided in Table 1.4.

Substitution of the inflow and outflow values from Table 1.4 to the water-budget equation yields

$$1275 - 500 - 725 - 25 - 25 = 0$$

Given this information, assess what is likely to happen as the ground-water system is developed by wells. More specifically, by 1985, wells were withdrawing some 200 cfs that was used and turned into wastewater. This wastewater was sent directly into the ocean.

SOLUTION   To start, we need to modify the water balance equation to include the effects of ground-water withdrawals

$$R_N - Q_{gws} - Q_{gwo} - ET_{gw} - Q_s - Q_p = \Delta S$$

If we assume that the system finds a new steady state with recharge (R_N) remaining at 1275 cfs, then natural outflows from the system need to be reduced to accommodate the 200 cfs of loss due to ground-water pumping. Given the relatively large size of the pumping demands, the reductions would be felt mostly in Q_{gws}, and Q_{gwo}. The water balance could look something like this

$$1275 - 410 - 625 - 25 - 15 - 200 = 0$$

As flow to streams (Q_{gws}) is decreased from 500 cfs to 410 cfs, there will be a tendency for headwaters to dry up and streams to become shorter (Alley et al., 1999). As outflow from the ground-water system to the ocean (Q_{gwo}) is decreased from 725 cfs to 625 cfs, the interface between salt- and freshwater (Figure 1.11) will move landward, developing the problem known as saltwater intrusion.

► EXERCISE

1.1. Ten-year averages (1981–1991) for a ground-water inventory for the San Fernando Basin are as follows: Recharge from precipitation and other sources, $79.3 \times 10^6 \text{m}^3$, recharge from injection wells, $38 \times 10^6 \text{m}^3$, ground-water inflow, $1.01 \times 10^6 \text{m}^3$, ground-water pumping, $120.55 \times 10^6 \text{m}^3$, ground-water outflow, $0.521 \times 10^6 \text{m}^3$, and baseflow to streams $3.8 \times 10^6 \text{m}^3$. Calculate ground-water storage change based on the 10-year averages.

► CHAPTER
 HIGHLIGHTS

1. Water is an important topic for study because it is an essential requirement for life on Earth as we know it. Although there are about 330 million mi^3 of water on Earth, most of it is either in oceans and therefore not suitable for human or animal consumption, or else locked in glaciers and ice caps. Ground water comprises 98% of the world's unfrozen supply of freshwater.

2. Most of the work of hydrogeologists is concerned with developing this important resource and protecting the chemical and biological quality of water. Significant contamination of ground water comes from inappropriate disposal of waste into the ground, widespread use of fertilizers, herbicides, and pesticides, and accidental spills from pipelines or storage tanks.

3. Knowledge of hydrogeology is also essential for the construction of dams and underground facilities. The geologic work of ground water is important in shaping the landscape, especially in karst regions, in forming some types of uranium and lead-zinc deposits, and in contributing to the migration of oil.

4. The hydrologic cycle is the circulation of water from the oceans, to the atmosphere, to the land, and back to the ocean. Water circulates among the major reservoirs (that is, oceans, atmosphere, ice, and ground water) through key hydrogeological processes such as atmospheric transport, precipitation, evapotranspiration, river flow, and ground-water flow.

5. Our main interest in this book is with the subsurface component of the hydrologic cycle that begins as some small quantity of the precipitation falling on land infiltrates to the subsurface. Some of this water is transpired; the remainder follows a ground-water flow path through the subsurface and back to the surface. The residence time of this water varies from days to thousands of years.

6. The vadose or unsaturated zone is found above the water table and is an environment where the pore space is filled with both soil gas and water. In the phreatic or saturated zone, below the water table, the pores are filled completely with water.

7. The water balance equation (input − output = change in storage) describes the response of the major reservoirs or domains in the hydrologic cycle. Because water is neither created nor lost from the hydrologic cycle, this is a conservation equation. More detailed forms of these equations are written for ground-water systems to account for the inputs due to recharge and infiltration from surface waters and losses due to transpiration and pumping.

2

HYDROLOGIC PROCESSES AT THE EARTH'S SURFACE

Chapter 1 provided a brief introduction to the hydrologic cycle. In this chapter, we look in more detail at hydrologic processes occurring at the Earth's surface and introduce concepts and methods used to elucidate these processes.

▶ 2.1 PRECIPITATION

Precipitation includes the various forms of water, liquid or solid, that fall to the ground from the atmosphere. Precipitation can occur as rain, drizzle, snow, hail, sleet, and ice crystals. It can be measured at a location using a variety of devices. The simplest instruments are containers that collect precipitation through the storm event. The quantity of precipitation is measured by an observer following the storm. The standard U.S. precipitation gage has an 8-in. (20.32-cm) diameter mouth and height of about 30 in. (76.2 cm) (Figure 2.1*a*). More sophisticated rain gages (e.g., Figure 2.1*b*) have the capability of recording the time, duration, and intensity of precipitation. Typically, they store this information or transmit results electronically.

The accuracy of precipitation measurements is affected by the physical setting and by disturbances due to the presence of the gage itself. For example, strong winds during precipitation events can cause considerable difference between measured and actual precipitation. Measurement errors can also result from small amounts of dew, frost, and rime accidentally included in the total measured precipitation. With snow measurements, all gages typically underestimate the actual precipitation, particularly at high wind speeds.

Precipitation data are often available from networks of government weather stations, such as those operated by the National Weather Service. Historical data can be purchased for specified geographic areas in the United States in computer-compatible formats.

Many applications require that amounts of precipitation be estimated over relatively large areas. A few simple estimation approaches provide areal estimates of precipitation

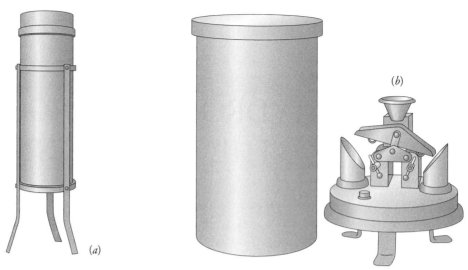

Figure 2.1 Examples of gages used for measuring precipitation. Panel (*a*) shows an 8-inch standard U.S. rain gage. The collection funnel inside directs the rain to a plastic measurement tube. Panel (*b*) shows a tipping bucket recording rain gage. The buckets are calibrated to tip after 0.01 in. of rainfall and are recorded by a data system.

from a distributed network of weather stations. A simple *arithmetic average* of precipitation for an area is given as

$$P_a = \frac{1}{N} \sum_{i=1}^{N} P_i \tag{2.1}$$

where P_a is the average precipitation for the area, N is the number of weather stations in the area, and P_i is the precipitation at station i.

The *Thiessen-weighted average* method is a more sophisticated approach. It is based on the idea that precipitation data for a station are most representative of an area immediately surrounding it. A Thiessen network is constructed by plotting the stations on a map and connecting adjacent stations by straight lines and bisecting each connecting line perpendicularly. The two sets of perpendicular lines define a polygon around each station. Precipitation at a station is applied to each polygon closest to it. The area-weighted average precipitation is given by the following equation:

$$P_a = \frac{\sum_{i=1}^{N} P_i a_i}{\sum_{i=1}^{N} a_i} \tag{2.2}$$

where a_i is the area of the polygon around station i.

The *isohyetal method* is based on areas computed from a contoured precipitation map. This average can be calculated by

$$P_a = \frac{\sum_{i=1}^{N_c-1} P_{ai} A_i}{\sum_{i=1}^{N_c-1} A_i} \tag{2.3}$$

where N_c is the number of contour lines, A_i is the area between two contour lines, and P_{ai} is the average precipitation of adjacent two contour values.

► **EXAMPLE 2.1**

Precipitation (in inches) was recorded at several stations (Figure 2.2*a*). Characterize the precipitation over the area as the arithmetic average, the Thiessen-weighted average, and the isohyetal average.

SOLUTION   Arithmetic Average:

$$P_a = \frac{1}{9}(4.5 + 4.4 + 4.4 + 4.1 + 4.0 + 3.7 + 3.5 + 2.5 + 2.7) = 3.76 \text{ in.}$$

Thiessen-Weighted Average:
    Adjacent stations are connected by a straight line, and the connecting line is bisected perpendicularly (Figure 2.2*b*). The area of each polygon around each station is measured and tabulated in Table 2.1. The table also calculates and lists the total area and the sum of the product of the polygon

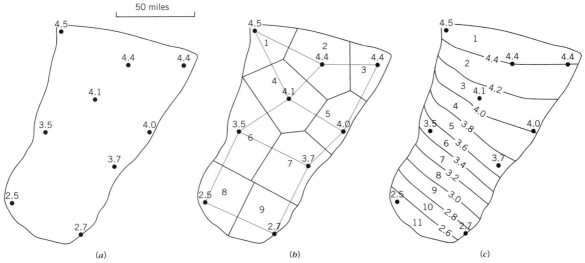

Figure 2.2   Calculation of average precipitation using: (*a*) arithmetic average method; (*b*) Thiessen-weighted average method; (*c*) isohyetal method.

TABLE 2.1   Calculation of average precipitation by Thiessen-weighted average method

Polygon Index	a_i(mi.^{2})	P_i(in.)	a_iP_i
1	781	4.50	3516
2	1262	4.40	5552
3	625	4.40	2750
4	1563	4.10	6406
5	888	4.00	3553
6	1575	3.50	5513
7	1375	3.70	5088
8	1164	2.50	2910
9	952	2.70	2570
Total	10,185		37,857

area and the precipitation. The average precipitation by the Thiessen-weighted average is:

$$P_a = \frac{\sum_{i=1}^{N} P_i a_i}{\sum_{i=1}^{N} a_i} = \frac{(37,857 \text{ mi.}^2 \cdot \text{in.})}{(10,185 \text{ mi.}^2)} = 3.72 \text{ in.}$$

Isohyetal Average:

Contour lines of precipitation are drawn using an interval of 0.2 in. (Figure 2.2c). We measure the area between each pair of contour lines and find the average of the pair of contour values to apply to the area (Table 2.2). Other calculations are shown in the table. The average precipitation by the isohyetal method is

$$P_a = \frac{\sum_{i=1}^{N_c-1} P_{ai} A_i}{\sum_{i=1}^{N_c-1} A_i} = \frac{(37,609 \text{ mi.}^2 \cdot \text{in.})}{(10,185 \text{ mi.}^2)} = 3.69 \text{ in.}$$

In the SI unit system, the precipitation is in centimeters (cm) (1 in. $= 2.54$ cm, or 1 cm $= 0.3937$ in.).

New technological advances have provided a fundamentally new approach to measuring precipitation on an areal basis. Data from Doppler weather radar (NEXRAD; Model WSR-88D) can be processed to yield almost continuous space/time estimates of many meteorological variables, including precipitation. Each NEXRAD station monitors many thousands of square kilometers on a continuous basis. For a given storm, the strength of the reflected radar signal is proportional to the precipitation intensity. Commonly, intensities are presented in qualitative terms (that is, light or heavy). When calibrated, Doppler radar can provide highly resolved estimates of precipitation amounts, especially for complex rainstorms or storms in areas where the coverage by conventional gages is limited. The United States is far along in deploying these systems.

We can show the advantages of the radar-based techniques by looking at data for a severe precipitation and associated flooding event in the Appalachian Mountains on January

TABLE 2.2 Calculation of average precipitation by isohyetal method

Contour Index	A_i (mi.^{2})	P_{ai} (in.)	$A_i P_{ai}$
1	1303	4.45	5799
2	1528	4.3	6571
3	1453	4.1	5958
4	967	3.9	3772
5	967	3.7	3578
6	661	3.5	2314
7	661	3.3	2182
8	661	3.1	2049
9	661	2.9	1917
10	661	2.7	1785
11	661	2.55	1686
Total	10,185		37,609

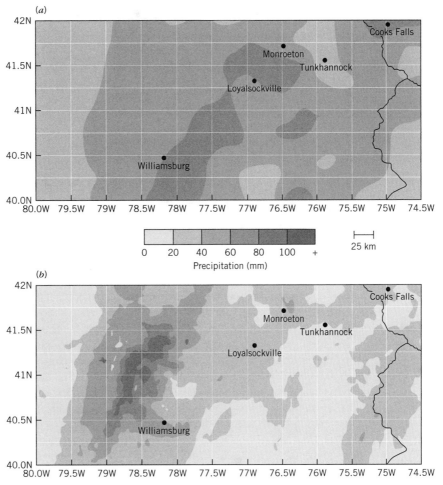

Figure 2.3 Interpretations of precipitation distribution in the Appalachian Mountains from (*a*) a network of standard rain gages and (*b*) WSR-88D weather radar (from Barros and Kuligowski, 1998).

19, 1996 (Figure 2.3). The radar data (Figure 2.3*b*) provided the basis for correlating the distribution of the precipitation with topographic features. In addition, these data provided an estimated rainfall distribution that was consistent with the observed patterns of flooding. The conventional monitoring system (Figure 2.3*a*) was much less useful in characterizing the spatial distribution of the precipitation.

▶ 2.2 EVAPORATION, EVAPOTRANSPIRATION, AND POTENTIAL EVAPOTRANSPIRATION

Evaporation is the physical process by which a liquid is transformed to a gas. In hydrologic applications, the term *evaporation* refers to the quantity of water lost from soils, rivers, and lakes. A variety of methods can be used to estimate or measure the amount of evaporation. Since it is beyond the scope of this book to discuss these approaches in detail, interested readers can refer to standard references such as Gray et al. (1970). Measurements with an evaporation pan, a shallow tank that contains water, are straightforward. The U.S. Weather

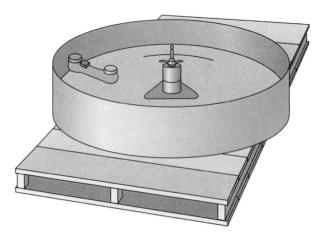

Figure 2.4 Sketch showing the U.S. Weather Service Class A evaporation pan. It has a diameter of 47.5 in. and is 10 in. deep. Mounted on the interior are a stilling basin, a hook gage for accurate measurements, and a min/max thermometer.

Service Class A pan (Figure 2.4) is 4 ft (1.22 m) in diameter and 10 in. (25.4 cm) deep. The water depth is measured daily, along with data on maximum and minimum temperatures and the volume of water added or removed from the pan to adjust for losses from evaporation or gains in precipitation. Although the calculation of the daily evaporation from the pan is straightforward, interpreting the value requires additional work. In general, evaporation from a pan is higher than a nearby lake. The water in the pan can heat up considerably and is able to evaporate more rapidly than a lake. Lake evaporation is determined as the product of the pan value and a *pan coefficient*—an empirical correction factor that varies between about 0.58 and 0.78 depending on the month (Roberts and Stall, 1967). Evaporation from plants is called *transpiration*, whereas the combined process of evaporation and transpiration is *evapotranspiration*. The concept of potential evapotranspiration recognizes that evapotranspiration can't occur if there is no water to evaporate. *Potential evapotranspiration* is the amount of water that would evaporate or transpire if sufficient water was available in the soil to meet the demand. In other words, the potential evapotranspiration is the maximum evapotranspiration assuming the water is there to evaporate. In arid areas, the actual evapotranspiration would be less than the potential rate. For example, the actual evapotranspiration in a desert is essentially zero because there is no water available in the soil.

As was the case with evaporation, both empirical and measurement-based techniques are available for estimating evapotranspiration. A lysimeter has proven useful in such an application. A *lysimeter* is a soil-filled tank on which plants are grown. By carefully accounting for the quantity of water added through precipitation, lost through deep percolation, and storage changes in the tank, the rate of evaporation can be calculated. In effect, the conservation equations that we introduced in Chapter 1 are applied on a tank scale.

Lysimeter data are usually unavailable, and one must therefore turn to one of several empirical approaches. The Thornthwaite (1948) method is commonly used for estimating the potential evapotranspiration from an area. The Thornthwaite equation (see Gray et al., 1970) is:

$$E_T = 1.62 \left(\frac{10 T_{ai}}{I} \right)^a \tag{2.4}$$

$$I = \sum_{i=1}^{12} \left(\frac{T_{ai}}{5} \right)^{1.5} \tag{2.5}$$

$$a = 0.492 + 0.0179 I - 0.0000771 I^2 + 0.0000006751 I^3 \tag{2.6}$$

where E_T is the potential evapotranspiration in cm/month, T_{ai} is the mean monthly air temperature in $C°$ for month i, I is the annual heat index, and a is constant. It is assumed that each day has 12 hours of sunshine with 30 days in a month. Table 2.3 lists the correction factors to adjust the hours of sunshine in each day for various latitudes. Temperature may also be in $F°(C° = [F° - 32] \times 5/9$, or $F° = C° \times 9/5 + 32)$.

▶ **EXAMPLE 2.2**
(from Gray et al., 1970)

The second column of Table 2.4 lists the monthly temperatures between April and October in 1961 at Saskatoon, Canada. Calculate the potential evapotranspiration for an alfalfa crop using the Thornthwaite method.

SOLUTION Temperature needs to be converted from $F°$ to $C°$. Next, the heat index is calculated using Eq. (2.5). Each term is calculated and listed in the fourth column of Table 2.4. The heat index is the summation of all the terms.

$$I = \sum_{i=1}^{12} \left(\frac{T_{ai}}{5} \right)^{1.5} = 0.21 + 3.56 + 7.92 + 7.68 + 8.90 + 2.32 + 0.89 = 31.48$$

TABLE 2.3 **NRC mean hours of bright sunshine expressed in units of 30 days of 12 hours each day**

North Lat.	J	F	M	A	M	J	J	A	S	O	N	D
0E	1.04	0.94	1.04	1.01	1.04	1.01	1.04	1.04	1.01	1.04	1.01	1.04
10E	1.00	0.91	1.03	1.03	1.08	1.06	1.08	1.07	1.02	1.02	0.98	0.99
20E	0.95	0.90	1.03	1.05	1.13	1.11	1.14	1.11	1.02	1.00	0.93	0.94
30E	0.90	0.87	1.03	1.08	1.18	1.17	1.20	1.14	1.03	0.98	0.89	0.88
35E	0.87	0.85	1.03	1.09	1.21	1.21	1.23	1.16	1.03	0.97	0.86	0.85
40E	0.84	0.83	1.03	1.11	1.24	1.25	1.27	1.18	1.04	0.96	0.83	0.81
45E	0.80	0.81	1.02	1.13	1.28	1.29	1.31	1.21	1.04	0.94	0.79	0.75
50E	0.74	0.78	1.02	1.15	1.33	1.36	1.37	1.25	1.06	0.92	0.76	0.70

Source: Gray et al., 1970. *Handbook on the principles of hydrology.* Reprinted by permission of the National Research Council of Canada.

TABLE 2.4 **Potential evapotranspiration for alfalfa crop at Saskatoon, 1961**

Month	T_m (EF)	T (EC)	i	ET(cm)	Daylight Factor	Adjusted ET (cm)	Adjusted ET(in.)
Jan.							
Feb.							
Mar.							
Apr.	35.2	1.8	0.21	0.93	1.15	1.06	0.42
May	52.8	11.6	3.56	5.97	1.33	7.93	3.12
June	67.4	19.7	7.92	10.13	1.36	13.78	5.42
July	66.8	19.3	7.68	9.92	1.37	13.60	5.35
Aug.	70.3	21.3	8.90	10.95	1.25	13.69	5.39
Sept.	47.7	8.7	2.32	4.47	1.06	4.74	1.87
Oct.	40.2	4.6	0.89	2.37	0.92	2.18	0.86
Nov.							
Dec.							

Source: Modified from Gray et al., 1970. *Handbook on the principles of hydrology.* Reprinted by permission of the National Research Council of Canada.

The exponential constant a is calculated as

$$a = 0.492 + 0.0179 \cdot 31.48 - 0.0000771 \cdot 31.48^2 + 0.000000675 \cdot 31.48^3 = 1.0$$

Calculations of potential evapotranspiration for each month are listed in the fifth column of the table. With Saskatoon located at 50 degrees north latitude, one can determine the daylight correction factors using Table 2.3. The corrected potential evapotranspiration for each month is listed in the last two columns of the table. To apply the results to an alfalfa field, it is important to include only the growing season for the crop. For example, if the crop only grew from May 16 to September 24, 1961, the total potential evapotranspiration would be

$$ET_{\text{Alfalfa}} = \frac{16}{31}(3.12) + 5.42 + 5.35 + 5.39 + \frac{24}{31}(1.87) = 19.22 \text{ in.}$$

▶ 2.3 INFILTRATION, OVERLAND FLOW, AND INTERFLOW

The relationship between precipitation, ground-water flow, and streamflow was introduced in Chapter 1 as the basin hydrological cycle (see Figure 2.5). Here, we examine the key

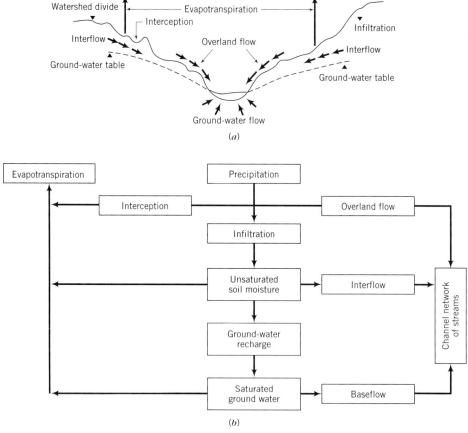

Figure 2.5 The basin hydrologic cycle (from Domenico and Schwartz, 1998. *Physical and chemical hydrogeology*). Copyright ©1998 by John Wiley & Sons, Inc. Reprinted by permission of John Wiley & Sons, Inc.

features of the basin cycle in more detail. *Interception* is a process whereby some precipitation is captured on the leaves and branches of plants and evaporates back to the atmosphere. The total energy available for evaporation of water from leaf surface is the same whether evaporation is supplied from intercepted water or from within the leaves of plants.

Infiltration is the process of downward water entry into the soil. The rate of infiltration is usually sensitive to near-surface conditions as well as the antecedent water content of the soil. Hence, infiltration rates are subject to significant change with soil use and management, and time. Horton (1933, 1940) pointed out that the maximum permissible infiltration rate decreases with increasing time. When water is initially applied to a soil surface, the infiltration rate is equal to the rate of application. All of the water infiltrates, and there is no overland flow. As water continues to be added, a time will be reached when free water accumulates or ponds on the ground surface. Initially, such ponding will be restricted to local depressions and will be essentially absent from the rest of the surface.

Once ponding begins, control on infiltration shifts from the rate of water addition to the characteristics of the soil. Surface-connected pores and cracks become effective in conducting water downward. Infiltration with free water present on the ground surface is referred to as *ponded infiltration*. With the initiation of ponding, the infiltration rate usually decreases appreciably with time because of the deeper wetting of the soil, which results in a reduction in the energy available for flow, and in the closing of cracks and other surface-connected macropores. After long, continued wetting under ponded conditions, the rate of infiltration reaches a constant or steady-state infiltration rate. This stage is referred to as steady ponded infiltration. Under ideal conditions, the saturated hydraulic conductivity of the soil (this term will be discussed in Chapter 3) within a depth of 0.5 to 1 m is a useful predictor of the steady ponded infiltration rate. Figure 2.6*a, b* illustrates an infiltration rate and accumulated infiltration through time. Total depth of infiltration increases linearly with time, although the infiltration rate is reduced to a steady state.

Once ponding begins, overland flow is possible. *Overland flow* or *surface runoff* is flow across the land surface to a nearby channel. Another pathway for near-surface flow to streams is *interflow*—a lateral flow of water above the water table during storms. Interflow can proceed directly to a lake or stream, or it can reemerge at the ground surface. Overland flow and interflow are sometimes grouped together as *direct runoff*.

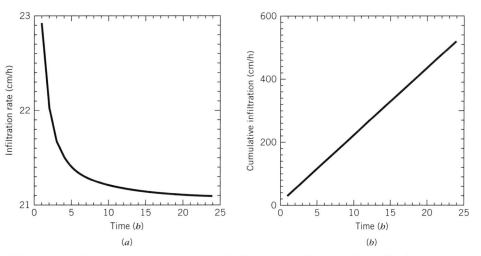

Figure 2.6 Infiltration change with time: (*a*) infiltration rate; (*b*) cumulative infiltration.

A *watershed* or *basin* defines the area that contributes surface runoff to some given system of stream channels. The surface area of a watershed can be determined using topographic maps or digital elevation data. The boundary between adjacent watersheds or basins is termed a *surface water divide,* which is often, but not always, a boundary between ground-water basins. In some cases, ground water flows beneath a surface-water divide.

▶ 2.4 A SIMPLE APPROACH TO RUNOFF ESTIMATION

The peak discharge from surface runoff can be estimated by the rational, peak discharge, tabular, and unit hydrograph methods (Virginia Soil and Water Conservation Commission, 1980). Here, we demonstrate the rational method, which applies to small basins of up to 200 acres. The peak discharge, Q, in cubic feet per second is calculated as

$$Q = CiA \tag{2.7}$$

where C is the runoff coefficient (Table 2.5); i is the average intensity of rainfall (inches per hour) for a selected frequency of occurrence or return period for the time of concentration

TABLE 2.5 Soil water runoff coefficient in rational formula

Land Use	C	Land Use	C
Business		Lawns	
Downtown areas	0.70–0.95	Sandy soil, flat, 2%	0.05–0.10
Neighborhood areas	0.50–0.70	Sandy soil, average, 2–7%	0.10–0.15
		Sandy soil, steep, 7%	0.15–0.20
Residential		Heavy soil, flat, 2%	0.13–0.17
Single-family areas	0.30–0.50	Heavy soil, average, 2–7%	0.18–0.22
Multi-units, detached	0.40–0.60	Heavy soil, steep, 7%	0.25–0.35
Multi-units, attached	0.60–0.75		
Suburban	0.25–0.40	Agricultural land	
		Bare packed soil	
Industrial		Smooth	0.30–0.60
Light areas	0.50–0.80	Rough	0.20–0.50
Heavy areas	0.60–0.90	Cultivated rows	
		Heavy soil no crop	0.30–0.60
Parks, cemeteries	0.10–0.25	Heavy soil with crop	0.20–0.50
		Sandy soil no crop	0.20–0.40
Playgrounds	0.20–0.35	Sandy soil with crop	0.10–0.25
		Pasture	
Railroad yard areas	0.20–0.40	Heavy soil	0.15–0.45
		Sandy soil	0.05–0.25
Unimproved areas	0.10–0.30	Woodlands	0.05–0.25
Streets			
Asphaltic	0.70–0.95		
Concrete	0.80–0.95		
Brick	0.70–0.85		
Drives and walks	0.75–0.85		
Roofs	0.75–0.95		

Note: The designer must use judgment in selecting the appropriate C value within the range. Generally, larger areas with permeable soils, flat slopes, and dense vegetation should have lowest (C) values. Smaller areas with dense soils, moderate to steep slopes, and sparse vegetation should be assigned highest (C) values.
Source: From Virginia Soil and Water Conservation Commission (1980).

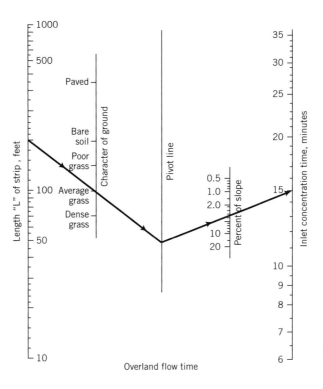

Figure 2.7 Nomograph for calculating the overland flow time (from Virginia Soil and Water Conservation Commission, 1980).

(T_c) (min) (the *time of concentration* being the estimated time required for runoff to flow from the most remote part of the area under consideration to the point under consideration); and A is the drainage area (acres). The time of concentration may include overland flow time and channel flow time. Figure 2.7 is the nomograph for calculating the overland flow time. The channel flow time is equal to the length of channel divided by average water-flow velocity in the channel. For a small drainage area, the time of concentration can be estimated from Figure 2.8.

▶ **EXAMPLE 2.3**

For a drainage area of 80 acres (30% of rooftops, 10% of streets and driveways, 20% of lawn on sandy soil, and 40% of woodland) in Lynchburg, Virginia, the height of the most remote point is 100 ft above the flow outlet, with a maximum travel length of 3000 ft. Calculate the peak runoff from a 10-year frequency storm.

SOLUTION   To calculate the peak runoff using Eq. (2.7), we need to determine runoff coefficient C, time of concentration, and rainfall intensity i.

Runoff coefficient: The runoff coefficient of the basin is the weighted average of different land uses. The second column of Table 2.6 is the percentage of land use in the basin; the third column lists the runoff coefficients for different land uses; and the fourth column is the contribution from each land use to the runoff coefficient of the basin. The runoff coefficient is 0.43 for the basin.

Time of concentration: Given the height difference (100 ft) and distance (3000 ft) between the flow outlet and the remote point, the time of concentration is obtained from Figure 2.8 (T_c=14 min).

Rainfall intensity: The rainfall intensity can be read from a rainfall frequency-intensity-duration chart (Figure 2.9) for a given return period (10 years) and duration (14 min) ($i = 4.9$ in./h). Peak discharge is:

$$Q = CiA = 0.43(4.9)(80) = 169 cfs$$

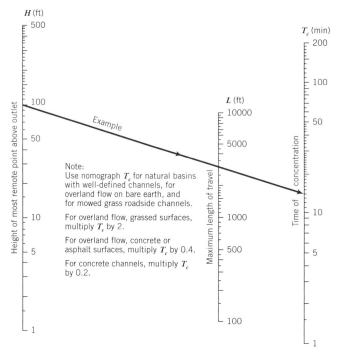

Figure 2.8 Nomograph for estimating time of concentration for a small drainage basin (from Virginia Soil and Water Conservation Commission, 1980).

TABLE 2.6 Calculation of runoff coefficient

Land Use	% of Area	C	% of Area $\times C$
Rooftops	30	0.9	0.27
Streets	10	0.9	0.09
Lawns	20	0.15	0.03
Woodland	40	0.10	0.04
Total	100		0.43

Source: From Virginia Soil and Water Conservation Commission (1980).

▶ 2.5 STREAMFLOW AND THE BASIN HYDROLOGIC CYCLE

Streams and rivers operate in the hydrologic cycle by returning surface runoff, interflow, ground-water discharge, and channel precipitation back to the oceans. From a ground-water perspective, understanding features of the surface-water hydrology directly and indirectly lets us learn something about the subsurface hydrologic properties. *Discharge* of a stream or river is the volume of water flowing per unit of time (Manning, 1987). Thus, discharge has units of $[L^3/T]$, for example, m³/s or ft³/s. Commonly, hydrologists present discharge data graphically as a *discharge hydrograph*, which displays the discharge as a function of time (Figure 2.10).

Measuring Stream Discharge

Stream discharge is a basic parameter that needs to be measured to support a variety of hydrologic assessments. Because discharge is so difficult to measure, it is more convenient

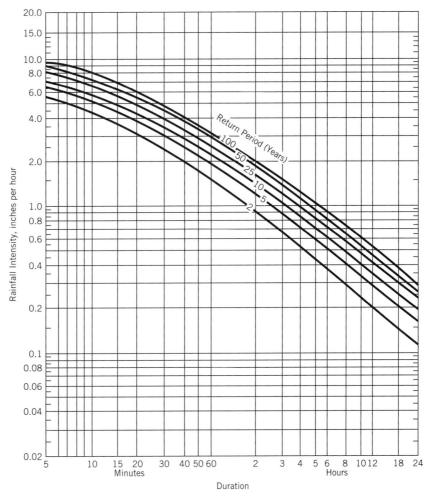

Figure 2.9 Rainfall frequency-intensity-duration chart for Lynchburg. Virginia (from Virginia Soil and Water Conservation Commission, 1980).

to provide velocity measurements instead. Discharge is easily calculated as the product of streamflow velocity and area of flow (Figure 2.11). In real streams, this approach for estimating discharge is complicated by the fact that velocity varies within the stream. However, you can get around this complication by subdividing the stream cross section into smaller vertical sections and estimating the discharge through each piece, as the product of the velocity in that section and its area (Figure 2.12). The total discharge is the sum of discharges through all the segments. The average velocity in each vertical section is determined by averaging values at depths of two-tenths and eight-tenths. In shallow water near shore, a single measurement is made at a six-tenths depth. The vertical sections are selected in such a way that no more than 10% of the flow comes through each.

A Price current meter measures the flow velocities (Figure 2.13). The cups on the meter are turned by the flowing water, with the speed of rotation proportional to velocity. Calibration information provided with the meter permits the number of revolutions to be converted over some interval of time to velocity. One keeps track of where the measurements are made by using a long tape stretched across the stream and the scale on the wading rod that holds the current meter (Figure 2.13).

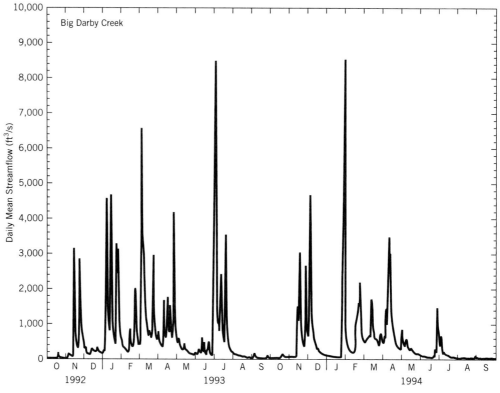

Figure 2.10 This figure is an example of a stream hydrograph. Daily mean discharge is plotted from 1992 to 1994 for Big Darby Creek, which is located in central Ohio (modified from Hambrook et al., 1997).

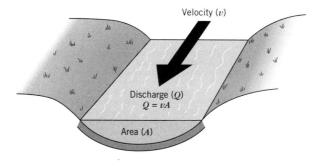

Figure 2.11 Ideally, discharge is calculated for a stream as the product of the flow velocity and the area of the channel cross section.

▶ **EXAMPLE 2.4**   Shown in Table 2.7 are a series of measurements collected on Big Darby Creek in central Ohio. Estimate the stream discharge.

SOLUTION   Proceed by estimating the area and the mean velocity in each vertical section. Discharge for each vertical section is calculated as the product of velocity and discharge. The summation in Table 2.7 provides a total discharge of 10,317 ft^3/s. Using Figure 2.10 for comparison shows that Big Darby Creek was gaged during a major flood event.

Monitoring stream discharge by gaging on a routine basis is not practical because the approach is so labor intensive. However, methods are available to provide estimates

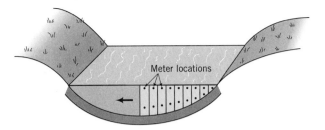

Figure 2.12 The cross section of the stream is divided into vertical sections. The streamflow velocity is measured at two-tenths and eight-tenths depths.

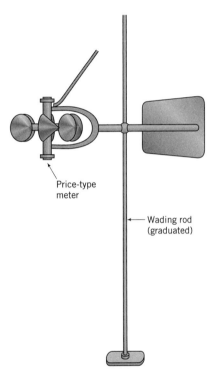

Figure 2.13 The Price current meter attached to a wading rod is one of the most common streamflow meters used for stream gaging.

of stream discharge from the stream stage. As Figure 2.14 shows, *stream stage* is the elevation of the stream above some datum. During storm flows, the stream stage increases (Figure 2.14).

The simplest way to measure stream stage is to use a staff gage (Figure 2.14). An observer visits the site and records the stream stage using the staff gage. A more automated approach uses a stilling well together with a continuous water-level recorder (Figure 2.15). This automated approach requires that the gaging site be visited much less frequently. With the most modern systems, stage data are collected automatically and sent via satellite to ground stations for further processing. Within a matter of a few hours, discharge data are posted on the Web for interested users. Take a look at the near-real-time data for Big Darby Creek at Darbyville and see what the discharge is today (http://water.usgs.gov/realtime.html).

Turning stage data into discharge estimates requires a *rating curve.* A rating curve relates stream stage to stream discharge at a gaging site (Figure 2.16). A rating curve is created over a number of years by measuring discharge and stage together during periodic

TABLE 2.7 USGS stream-gaging data for Big Darby Creek at Darbyville, Ohio, April 30, 1996

Section Number	Width (ft)	Stream Depth (ft)	Observation Depth (ft)	Velocity at Point (ft/s)	Mean Velocity (ft s^{-1})	Area (ft^2)	Discharge (ft^3 s^{-1})
1	13	2.4	0.6	0.242	0.242	31	8
2	20	4.5	0.6	0.402	0.402	90	36
3	20	7.0	0.2	2.24	1.81	140	253
			0.8	1.38			
4	50	7.9	0.2	1.66	0.982	395	388
			0.8	0.303			
5	47.5	10.1	0.2	1.29	1.51	480	653
			0.8	1.73			
6	12.5	11.0	0.2	3.80	3.89	138	537
			0.8	3.98			
7	8.5	11.0	0.2	4.85	4.85	94	456
			0.8	4.85			
8	7.0	11.7	0.2	6.41	6.24	82	512
			0.8	6.06			
9	7.0	12.9	0.2	7.26	7.04	90	634
			0.8	6.81			
10	6.0	14.3	0.2	8.17	7.56	86	650
			0.8	6.96			
11	5.0	14.7	0.2	8.38	7.60	74	562
			0.8	6.81			
12	5.0	15.2	0.2	8.38	7.52	76	572
			0.8	6.67			
13	5.0	15.7	0.2	8.54	7.82	78	610
			0.8	7.11			
14	6.0	15.7	0.2	7.78	6.98	94	656
			0.8	6.17			
15	7.0	15.8	0.2	6.96	6.62	111	735
			0.8	6.29			
16	7.0	16.0	0.2	6.54	6.14	112	688
			0.8	5.74			
17	8.5	15.9	0.2	5.32	5.32	135	718
			0.8	5.32			
18	10	15.3	0.2	4.26	3.84	153	588
			0.8	3.43			
19	10	12.7	0.2	3.37	3.58	127	455
			0.8	3.80			
20	12.5	11.6	0.2	2.10	2.36	145	342
			0.8	2.61			
21	15	10.7	0.2	1.41	1.52	160	243
			0.8	1.62			
22	17	4.0	0.6	0.308	0.308	68	21
						$\sum 2959$	$\sum 10,317$

visits to the gaging site. In constructing a rating curve, it is important to make measurements across a complete spectrum of river conditions, particularly high flows.

Gaging sites are usually established at places where the stream channel is relatively stable. Erosion or deposition in the channel can cause the rating curve to change over time. In small watersheds, structures called *weirs* are sometimes installed to provide stable cross

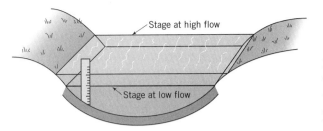

Figure 2.14 Stream stage is the elevation of the water surface above some datum. The staff gage provides the simplest way to measure the stage.

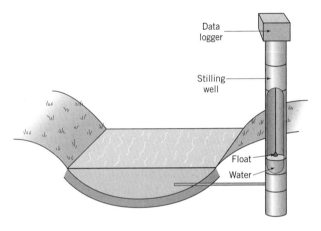

Figure 2.15 A gaging station equipped with a water-level recorder provides a continuous record of stream stage. The stilling well protects the fragile float from debris and damps out waves that might be generated during floods. Data are recorded digitally and downloaded to a PC or sent via satellite.

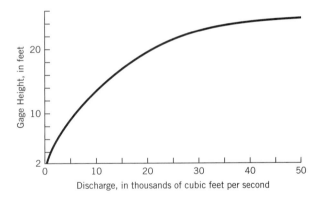

Figure 2.16 A rating curve is constructed by measuring stream stage and discharge during a variety of different flow conditions (from Kennedy, 1984).

sections at gaging stations. Specialized weirs such as triangular weirs permit discharge estimates from stage measurements alone. In other words, one calculates the discharge from equations rather than empirical relationships with stage. We have included several photographs of weirs and streams in a PowerPoint Show (show/GagingSites.ppt) on the website.

Hydrograph Shape

A look at a typical hydrograph (Figure 2.10) reveals that the discharge of a stream varies with time. During a storm flow event, the total flow increases because in addition to baseflow, contributions come from overland flow, interflow, and direct precipitation on the stream

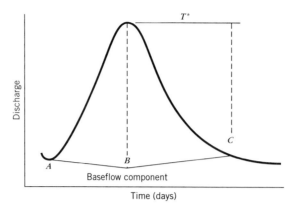

Figure 2.17 Determining baseflow component from streamflow hydrograph (from Domenico and Schwartz, 1998. *Physical and chemical hydrogeology*). Copyright ©1998 by John Wiley & Sons, Inc. Reprinted by permission of John Wiley & Sons, Inc.

(Figure 2.17). A storm hydrograph has three distinct parts: a rising limb, a peak, and a recession. The *rising limb,* at the beginning of a storm, reflects significant new contributions from overland flow and interflow, and enhanced ground-water discharge. Once the storm ends, the stream will reach a peak and begin to recede. The *recession* of the hydrograph represents a period where contributions from surface runoff end and enhanced ground-water contributions begin to fall off. The following case study from Wilson Creek Watershed in Manitoba Canada illustrates some typical features of streamflow behavior.

Estimating Ground-Water Recharge from Hydrographs

Chapter 1 illustrated the complexities of the interactions between ground and surface-water systems. We saw how a gaining stream received ground-water inflow. During periods of precipitation, recharge to the ground-water system can lead to increased ground-water flow to the stream. Analysis of a stream hydrograph following a storm can provide an estimate.

CASE STUDY 2-1

WILSON CREEK WATERSHED

Wilson Creek Watershed is located on the Manitoba Escarpment (Figure 2.18) in western Canada. Its main tributaries, Packhorse and Bald Hill creeks, rise at an elevation of approximately 2450 ft above sea level and flow down the escarpment to its base at approximately 1100 ft a.s.l. Thus, while Wilson Creek Watershed is relatively small in area, 8.5 mi^2, it has high relief. The main interest in studying this watershed came from an attempt to understand and control the erosion of shale bedrock during large floods. Valuable farmland at the base of the escarpment ended up covered by pieces of rock as Wilson Creek fanned out onto the prairie. You can become more familiar with Wilson Creek by looking at the Wilson Creek PowerPoint show on the website (show/WilsonCr.ppt).

In late June 1968, the watershed experienced the 100-year storm, a rare storm that statistically might be expected to return only once every 100 years. Over a three-day period, approximately 10 in. of rain fell on the higher elevation parts of the watershed. At lower elevations, rainfall amounts were about 4.9 in. (Figure 2.19). The hydrograph on Packhorse Creek related to this storm exhibited a steep rising limb. Although not obvious on the hydrograph, the peak flow occurred approximately three to five hours following the midpoint of the precipitation event. Some of the photographs from this event (show/WilsonCr.ppt) hint at the surprising discharges observed during this storm event. Recharge from the precipitation raised the water table several feet across the watershed. Recession from this and smaller storms is relatively steep.

Figure 2.18 The Wilson Creek watershed is located in western Manitoba, Canada. Two tributaries, Packhorse Creek and Bald Hill Creek, join to form Wilson Creek.

Linsley and others (1958) proposed a simple equation to determine when the direct runoff component ends at some time (T^*) after the streamflow peak. T^* in days is expressed as

$$T^* = A^{0.2} \tag{2.8}$$

where A is the drainage area in square miles and 0.2 is some empirical constant.

The first step in interpreting ground-water information is to separate the hydrograph into its component parts. Through the years, a variety of empirical and measurement-based approaches have been developed for this purpose. One simple procedure for separating

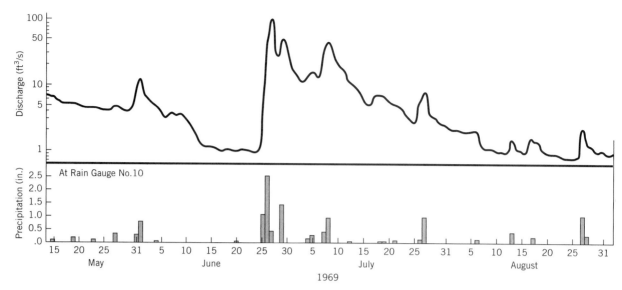

Figure 2.19 Discharge hydrograph for Packhorse Creek from 1969 (from Schwartz, 1970).

ground-water discharge from the total streamflow is summarized, as follows, with reference to Figure 2.17.

1. Extend the recession of the hydrograph prior to the storm (point A; Figure 2.17) to point B under the peak of the hydrograph.
2. Calculate the time in days when the direct runoff contribution stops and mark this time point as C.
3. Connect B and C.
4. Ground-water discharge through the storm event is the area below the curve ABC.

In this example, the ground-water contribution during the storm is reduced because the high stage in the stream reduces the ground-water discharge into the stream. This behavior is not universal—ground-water discharge can increase through storm events as well. From point C, the ground-water discharge can be approximated by

$$Q = Q_0 e^{-K_0 t} \tag{2.9}$$

where Q_0 is the ground-water discharge at point C, Q is the discharge at time t when point C is defined as time zero, and K_0 is a recession constant.

This equation for the ground-water recession provides the basis for a useful approach to estimating the quantities of recharge and discharge to a ground-water system. Meyboom (1961) defines the total potential ground-water discharge as

$$V = \frac{Q_0 K}{2.306} \tag{2.10}$$

where V is the total potential ground-water discharge, defined as the total volume of water that will discharge from ground-water storage if allowed to do so for infinite time without recharge, and K is the recession index. Glove (1964) shows that the total potential ground-water discharge at critical time (T_c) after the peak in a streamflow is equal to approximately one-half of the total volume of water that recharged the ground-water system (Figure 2.20):

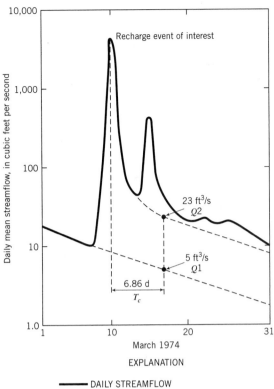

Figure 2.20 Illustration for calculating ground-water recharge (from Rutledge and Daniel, 1994). Reprinted by permission of *Ground Water.* Copyright ©1994. All rights reserved.

Within the figure:

10,000 — Recharge event of interest

Daily mean streamflow, in cubic feet per second

23 ft³/s Q2

5 ft³/s Q1

6.86 d
T_c

March 1974

EXPLANATION

── DAILY STREAMFLOW
─ ─ ─ EXTRAPOLATED GROUND-WATER DISCHARGE

PROCEDURE
1. Compute recession index, K (32 d/log cycle)
2. Compute critical time, T_c (0.2144 × K, or 6.86 d)
3. Locate time that is 6.86 days after peak
4. Extrapolate pre-event recession to $Q1$ (5 ft³/s)
5. Extrapolate post-event recession to $Q2$ (23 ft³/s)
6. Compute total recharge by equation 6,

$$\frac{2 \times (18 \text{ft}^3/\text{s}) \times 32 \text{ d}}{2.3026} \times \frac{86400 \text{ s}}{1 \text{ d}} = 4.32 \times 10^7 \text{ ft}^3$$

where R is total volume of recharge (ft³), Q_1 is ground-water discharge (ft³/sec, or cfs) at critical time as extrapolated from the hydrograph recession preceding the peak, and Q_2 is ground-water discharge (cfs) at the critical time, as extrapolated from the hydrograph recession after the peak. A procedure, the so-called recession-curve-displacement method, is developed using Eq. (2.11). Rutledge and Daniel (1994) summarized the procedure as follows.

$$R = \frac{2(Q_2 - Q_1)K}{2.3026} \tag{2.11}$$

1. Determine the recession index K (days/log cycle), which is the time required for the ground-water discharge to decline through one log cycle in the recession curve.

2. Calculate the critical time (days) using the following equation

$$T_c = 0.2144K \tag{2.12}$$

3. Determine the ground-water discharge Q_1 at the critical time by extending the recession curve preceding the storm event.

4. Determine the ground-water discharge Q_2 at the critical time by extending the recession curve after the storm event.

5. Calculate recharge volume using Eq. (2.11).

▶ **EXAMPLE 2.5**
(from Rutledge and
Daniel, 1994)

A hydrograph for streamflow on Big Alamance Creek near Elon, North Carolina, is shown in Figure 2.21. Calculate the ground-water recharge related to all storm events.

SOLUTION   To determine the recession index K, traces of recession curves are superimposed (Figure 2.22a) and the average of the recession curves is determined as a template of the master recession curve (MRC) (Figure 2.22b). Extending the ground-water discharge over one log cycle, we see that the recession index is the time difference in days.

$$K = T_2 - T_1 = 45 \text{ days}$$

The critical time is calculated as

$$T_c = 0.2144\, K = 0.2144(45 \text{ days}) = 9.6 \text{ days}$$

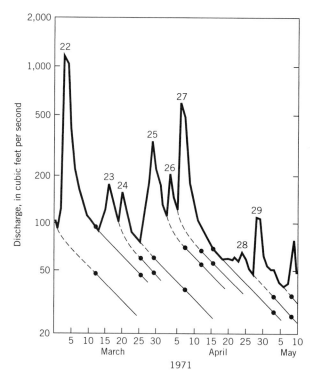

1971 water year storm-event number	Q2 (ft³/s)	Q1 (ft³/s)	Ground-water recharge ((ft³/s) × d)
22	93.0	47.0	1,800
23	59.0	47.0	470
24	60.0	49.0	430
25	72.0	38.5	1,311
26	68.0	56.0	470
27	70.0	57.0	509
28	34.5	28.0	254
29	35.0	26.0	352

Figure 2.21 A hydrograph of streamflow in Big Alamance Creek near Elon, North Carolina (from Rutledge and Daniel, 1994). Reprinted by permission of *Ground Water.* Copyright ©1994. All rights reserved.

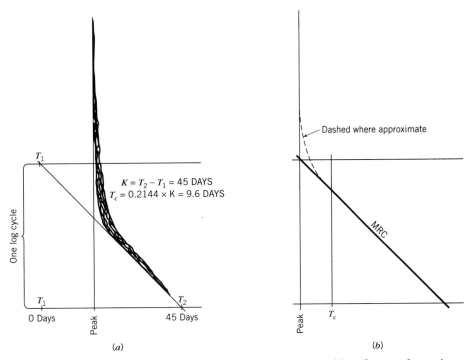

Figure 2.22 Determination of the recession index K: (a) superimposition of traces of recession curves; (b) a template of master recession curve (MRC) (from Rutledge and Daniel, 1994). Reprinted by permission of *Ground Water*. Copyright ©1994. All rights reserved.

The template is characterized by the peak time, critical time, and nonlinear and linear segments of the master recession curve. The template is fixed on a light table, and the hydrograph is extended across the template to achieve a best fit for each storm event. The linear segments of the recession curves and ground-water discharge (Q_1 and Q_2) are indicated in Figure 2.21. The estimated ground-water recharge attributed to each storm is also shown below the hydrograph in the figure.

The streamflow is in cubic meters per second in the SI unit system ($1\ \text{ft}^3/\text{sec} = 0.028317\ \text{m}^3/\text{sec}$, or $1\ \text{m}^3/\text{sec} = 35.315\ \text{ft}^3/\text{sec}$).

▶ 2.6 FLOOD PREDICTIONS

Historical flood discharge records in a stream can be used to determine the probability or recurrence interval of floods in the future. The *recurrence interval* is defined as

$$T = \frac{n+1}{m} \tag{2.13}$$

(see Riggs, 1989) where T is recurrence interval in years, n is number of items in a sample, and m is order number of the individual in the sample array. The probability that a given flow will be exceeded in any one year is

$$P = \frac{1}{T} \tag{2.14}$$

The graphical procedure for creating a plot of the annual flood discharge versus recurrence interval is as follows.

1. Determine the total number of records in the sample (n).
2. Arrange the historical flood data in order of magnitude, beginning with the largest flood first. This rank value is m in Eq. (2.13).
3. Calculate the recurrence interval for each flood using Eq. (2.13).
4. Plot the annual flood on a linear scale against recurrence interval on a log scale using semi-log graph paper.

▶ **EXAMPLE 2.6**   Annual discharges for the years 1915–1950 are listed in Table 2.8. Calculate the recurrence interval and the probability that a given flow will be exceeded in any one year.

TABLE 2.8   Computation of recurrence interval, 1915–1950

Water Year	Q (cfs)	Order Number (m)	T	P
1915	264	34	1.09	0.92
1916	374	11	3.36	0.30
1917	332	19	1.95	0.51
1918	346	16	2.31	0.43
1919	359	13	2.85	0.35
1920	333	18	2.06	0.49
1921	483	3	12.33	0.08
1922	417	5	7.40	0.14
1923	346	17	2.18	0.46
1924	320	21	1.76	0.57
1925	271	31	1.19	0.84
1926	214	36	1.03	0.97
1927	530	2	18.50	0.05
1928	304	25	1.48	0.68
1929	271	32	1.16	0.86
1930	271	33	1.12	0.89
1931	304	26	1.42	0.70
1932	400	9	4.11	0.24
1933	327	20	1.85	0.54
1934	415	6	6.17	0.16
1935	402	8	4.63	0.22
1936	362	12	3.08	0.32
1937	320	22	1.68	0.59
1938	272	30	1.23	0.81
1939	244	35	1.06	0.95
1940	279	28	1.32	0.76
1941	303	27	1.37	0.73
1942	310	24	1.54	0.65
1943	275	29	1.28	0.78
1944	317	23	1.61	0.62
1945	350	15	2.47	0.41
1946	387	10	3.70	0.27
1947	359	14	2.64	0.38
1948	449	4	9.25	0.11
1949	406	7	5.29	0.19
1950	570	1	37.00	0.03

Source: Modified from Riggs (1989).

SOLUTION   The order of data in the sample (m) is numbered in column 3 of Table 2.8. The total number of records in the sample is 36. Thus, $n + 1 = 37$. The recurrence interval (T) is calculated as $(n + 1)/m$ and listed in column 4. The probability (P) of exceeding a given discharge in any one year is represented as $1/T$ in column 5.

▶ EXERCISES

2.1. Calculate the average precipitation in Figure 2.23 using the arithmetic average, Thiessen-weighted average, and isohyetal methods. Note that precipitation is given in inches.

2.2. Here is a summary of the average monthly temperature (°F) at Minneapolis-St. Paul, Minnesota, for 1964 and 1993.

Year	Jan	Feb	Mar	Apr	May	Jun	Jul	Aug	Sep	Oct	Nov	Dec
1964	20.0	23.9	25.8	46.8	61.5	68.7	76.0	68.5	58.9	48.2	35.0	14.8
1993	14.6	17.2	29.5	44.2	57.2	64.5	70.3	70.4	55.0	46.5	30.6	22.2

Calculate the potential evapotranspiration for each month of the two years using the Thornthwaite method and compare the results.

2.3. The streamflow at a gaging station (Muskingum River, North Coshocton, Ohio) is shown for the first 180 days of 1995 in Figure 2.24. The drainage area for the stream at the station is 4859 mi^2. There are four major storm events in the period. Calculate recharge for each storm.

2.4. The length from a remote point on a strip of land to a stream is 800 ft. The slope of the bare soil land is 10%. Calculate overland flow time using Figure 2.10. If the length of the stream from the point where the overland flow enters the stream to the exit point in the basin is 3000 ft and average streamflow velocity is 10 ft/min, what is the time of concentration at the exit?

2.5. For a drainage area of 150 acres in a single-family residential area, calculate the peak discharge due to surface runoff for a rainfall intensity of 2.5 in./hour using the rational method.

Year	Flow (cfs)	Year	Flow (cfs)	Year	Flow (cfs)
1905	13006	1945	4526	1966	248
1906	10906	1946	4706	1967	709
1907	12506	1947	694	1968	746
1908	5006	1948	5686	1969	639
1909	16206	1949	956	1970	762
1913	7406	1950	4186	1971	900
1914	17006	1951	689	1972	5976
1915	5656	1952	12906	1973	794
1916	6666	1953	6306	1974	6056
1917	16206	1954	2926	1975	4946
1918	18406	1955	2676	1976	3516
1935	9146	1956	7026	1977	2026
1936	6666	1957	1310	1978	9366
1937	4056	1958	942	1979	7746
1938	7766	1959	5156	1980	7266
1939	5576	1960	5646	1981	3126
1940	4066	1961	617	1982	3626
1941	6026	1962	747	1983	12906
1942	4696	1963	368	1984	13106
1943	5536	1964	660	1985	6156
1944	403	1965	1070	1986	7106

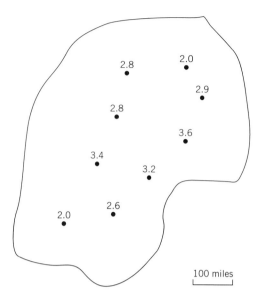

Figure 2.23 Precipitation in inches for a basin.

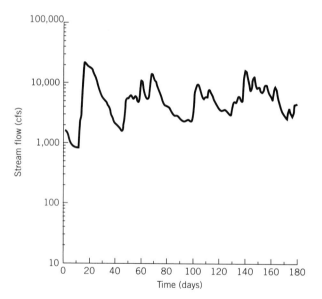

Figure 2.24 Discharge hydrograph for the Muskingum River in Ohio.

2.6. In the accompanying table (pg. 39), you will find the annual peak discharges for a gaging station on the Colorado River near Grand Lake, Colorado from 1905 to 1986. Calculate the recurrence interval and the probability of exceedence in any one year.

2.7. (a) Delineate a watershed and calculate its area using a topographic map provided by your instructor. (b) For a storm in the basin, calculate how many days will be needed for all of the direct runoff to reach a stream in the basin.

2.8. Streamflow data for most gaging stations in the United States can be retrieved from the USGS Web site: http://water.usgs.gov/public/realtime.html. Download the average daily streamflow data from the USGS Web site for a gaging station in the basin in Exercies 2.7 and calculate the ground-water recharge in the basin in a year.

▶ CHAPTER
HIGHLIGHTS

1. Precipitation can occur in a variety of different forms—rain, drizzle, snow, and sleet. The standard U.S. precipitation gage and various types of recording gages (for example, tipping bucket) form the basis for regional, station-based measurements. New NEXRAD radars provide a capability of continuous precipitation measurements over thousands of square miles. This coverage is particularly well suited for terrain that has variable relief and complex storms.

2. Various statistical techniques are available to estimate average precipitation from several stations scattered across an area. The simplest technique is the arithmetic average. More sophisticated Thiessen polygon and isohyetal methods provide different ways of weighting the individual data points according to their area of influence.

3. Evaporation and evapotranspiration lead to water losses from the surface and subsurface. Evaporation from surface-water bodies is estimated from pan measurements. Evapotranspiration can be measured using a lysimeter—a soil-filled tank with plants that has instrumentation capable of measuring small gains and losses of water. Empirical methods (for example, Thornthwaite, 1948) provide a useful alternative for estimating potential evapotranspiration.

4. Infiltration, overland flow, and interflow processes occur at or close to the ground surface. Of the precipitation falling to the ground, some fraction moves downward and enters the soil as infiltration. Sometimes a small proportion of infiltrated water flows in the unsaturated zone to a nearby stream as interflow. As a rainstorm continues, overland flow moves some of the water on the ground surface to nearby streams. All of these processes vary with time and space and depend on rainfall rate, soil characteristics, vegetation, and topography.

5. A key parameter of interest to hydrologists is the stream discharge, which is defined as the volume of water flowing past some location per unit time (units like ft^3/s). Measuring stream discharge involves making a series of velocity measurements with a current meter in vertical sections across the stream. Discharge in a segment is the mean velocity times the area of the segment. Add up the segment discharges to get the total discharge.

6. As the discharge increases during a storm, so does the stage—the elevation of the stream surface. Once a correlation between these two parameters is defined by a rating curve, the stage can be measured to provide an estimate of the stream discharge. In the United States and Canada, stilling wells with automated recording systems are provided at gaging sites to monitor stages on a continuous basis. Periodic stream-discharge measurements are, however, required to keep the rating curve current.

7. Discharge records that are collected over many years at a gaging site provide useful information on the likely response of the stream during droughts and storms. Flood frequency analysis uses data on the maximum discharge in each year of a record to determine the probability that a given flood discharge will be exceeded in a given year. This information is useful in designing hydraulic structures like dams and canals.

3

BASIC PRINCIPLES OF GROUND-WATER FLOW

Water flows in sediments or rocks through open spaces, which range from tiny imperfections along crystal boundaries in igneous rocks to huge caverns in limestone. This chapter introduces the basic concepts and principles of ground-water flow in sediments and rock.

▷ 3.1 POROSITY OF A SOIL OR ROCK

Total porosity of a rock or soil is defined as the ratio of the void volume to the total volume of material:

$$n_T = \frac{V_v}{V_T} = \frac{V_T - V_s}{V_T} \tag{3.1}$$

where n_T is the total porosity, V_v is the volume of voids, V_s is the volume of solids, and V_T is the total volume. In some cases, porosity is expressed as a percentage.

Primary porosity refers to the original interstices (or voids) created when some rock or soil was formed. These interstices include pores in soil or sedimentary rocks, and vesicles, lava tubes, and cooling fractures in basalt (Figure 3.1*a, b*; Heath 1988). *Secondary porosity* refers to joints, faults in igneous, metamorphic, and consolidated sedimentary rocks, and solution-enlarged openings in carbonate and other soluble rocks (e.g., Figure 3.1*c–f*; Heath, 1988). Porosity may also be defined in terms of grain density and bulk density.

$$n_T = 1 - \frac{\rho_b}{\rho_s} \tag{3.2}$$

where ρ_b is the *bulk density* (density of dry soil or rock sample) and ρ_s is the density of solids.

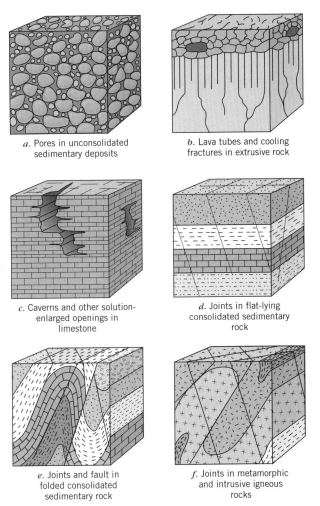

Figure 3.1 Types of openings in selected water-bearing rocks. Block (*a*) is a few millimeters to a few tens of centimeters wide depending on the medium. The remaining blocks are a few tens of meters wide. Openings in Panels (*a*) and (*b*) are primary; those in the others are secondary (from *Hydrogeology*, Heath, 1988). Reproduced with permission of the publisher, the Geological Society of America, Boulder, Colorado USA. Copyright ©1988 by the Geological Society of America, Inc.

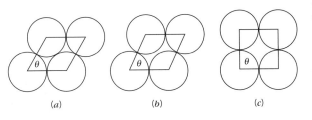

Figure 3.2 Sections of four contiguous spheres of equal size: (*a*) the most compact arrangement, lowest porosity; (*b*) less compact arrangement, higher porosity; (*c*) least compact arrangement, highest porosity (from Slichter, 1899).

The porosity of a soil or rock depends on the degree of compaction of grains, the shape of grains, and the particle-size distribution. If a material consists of spheres of equal size, the greater the compaction, the lower the porosity (Figure 3.2). The shape of the grains can cause the porosity to be larger or smaller than average, depending on how the grains are

TABLE 3.1   Range in values of porosity

Material	Porosity (%)
Sedimentary	
Gravel, coarse	24–36
Gravel, fine	25–38
Sand, coarse	31–46
Sand, fine	26–53
Silt	34–61
Clay	34–60
Sedimentary Rocks	
Sandstone	5–30
Siltstone	21–41
Limestone, dolomite	0–40
Karst limestone	0–40
Shale	0–10
Crystalline Rocks	
Fractured crystalline rocks	0–10
Dense crystalline rocks	0–5
Basalt	3–35
Weathered granite	34–57
Weathered gabbro	42–45

Source: In part from Davis (1969) and Johnson and Morris (1962).

arranged and connected. A strongly sorted medium (soil particles of relatively equal sizes) possesses a higher porosity than a poorly sorted medium because particles of a smaller size tend to occupy the void spaces between larger ones.

The porosity of a soil and rock can range from zero to more than 60% (Table 3.1). On average, the porosity is much higher for unlithified materials than for lithified materials. For unlithified sediments, the smaller the grain size, the higher the porosity.

The concept of effective porosity recognizes that not all of the pores participate meaningfully in the flow of water. A case in point is flow through fractured shale. Although the unfractured shale contains water-filled pores, almost all of the flow takes place through the fractures. In other words, the fracture system provides the effective pathway for flow through the rock. The *effective porosity* of a sediment or rock is the ratio of volume of the interconnected interstices to the total volume of the soil or rock. Effective porosity is more closely related to the flow of ground water in a medium than total porosity. In some cases, the effective porosity can differ significantly from the total porosity (Table 3.2).

▶ 3.2 DARCY'S EXPERIMENTAL LAW

Henry Darcy was a civil engineer in the mid-1800s concerned with the public water supply of Dijon, France. He was interested in acquiring data that would improve the design of filter sands for water purification. In his search for this information (Darcy, 1856), Darcy conducted experiments using an apparatus similar to that shown in Figure 3.3. His testing system consisted of a cylinder having a known cross-sectional area A (L^2), which was filled with various filter sands. Appropriate plumbing was provided to flow water through the column. The cylinder contained two manometers whose intakes were separated by a distance Δl (L). *Manometers* are nothing more than small open tubes that provide measurements of the energy available for flow at their open end in the medium. Water was flowed into

TABLE 3.2   Range in values of total porosity

Material	Total Porosity (%)	Effective Porosity (%)
Anhydrite[a]	0.5–5	0.05–0.5
Chalk[a]	5–40	0.05–2
Limestone, dolomite[a]	0–40	0.1–5
Sandstone[a]	5–15	0.5–10
Shale[a]	1–10	0.5–5
Salt[a]	0.5	0.1
Granite[b]	0.1	0.0005
Fractured crystalline rock[b]	—	0.00005 – 0.01

[a] Data from Croff et al. (1985).
[b] Data from Norton and Knapp (1977).

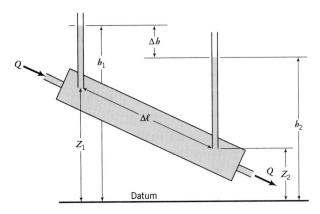

Figure 3.3 Laboratory apparatus to demonstrate Darcy's law (from Domenico and Schwartz, 1998. *Physical and chemical hydrogeology*). Copyright ©1998 by John Wiley & Sons, Inc. Reprinted by permission of John Wiley & Sons, Inc.

(and out of) the cylinder at a known rate Q (L^3/T), and the elevation of water levels in the manometers, h_1 and h_2 (L), was measured relative to a local datum.

Darcy conducted a variety of experiments in which the flow rate Q or the types of filter medium were changed. He derived the following relationship, now known as *Darcy's equation*

$$\frac{Q}{A} = K\frac{(h_1 - h_2)}{\Delta l} \tag{3.3}$$

where K is a constant of proportionality termed *hydraulic conductivity*. The term $(h_1 - h_2)/\Delta l$ is known as the *hydraulic gradient*. The term Q/A, representing the volumetric flow rate per unit cross-sectional area of the cylinder, is the *Darcy velocity* (q), or *specific discharge*. In words, Eq. (3.3) states that the velocity of flow is proportional to the hydraulic gradient. If the hydraulic gradient is denoted as i, then

$$i = \frac{(h_1 - h_2)}{\Delta l} = -\frac{dh}{dl} \tag{3.4}$$

and Darcy's equation is written as

$$q = K i \tag{3.5}$$

or

$$Q = K i A \tag{3.6}$$

Darcy's equation is valid for flow through most granular material as long as the flow is laminar. Under conditions of turbulent flow, the water particles take more tortuous paths. At the other extreme, in very-low-permeability materials, a minimum threshold gradient could be required before flow takes place (Bolt and Groenevelt, 1969).

Linear Ground-Water Velocity or Pore Velocity

Built into the Darcy velocity is an assumption that flow occurs over the entire surface area of the soil column. Because water only flows in the pore space, the actual flow velocity or pore velocity is greater than the Darcy velocity. The *pore velocity* (v), also termed the *linear velocity,* is defined as the volumetric flow rate per unit interconnected pore space.

$$v = \frac{q}{n_e} \tag{3.7}$$

where n_e is the effective porosity. The pore velocity is the true velocity of water flow in a porous medium, whereas the Darcy velocity or specific discharge is the apparent velocity. For example, if a hypothetical tracer particle is placed in the water, the particle will travel at the pore velocity. As will be evident in later chapters, contaminants are transported with the linear ground-water velocity.

Hydraulic Head

In Darcy-type experiments (Figure 3.3), the fact that water moves implies the existence of some kind of energy gradient. In effect, water has more energy available for flow at one point in the column than at another point. The measure of energy available for flow is reflected by the height of water in the manometer above some datum—the higher the water level, the greater the energy available for flow. Thus, in Figure 3.3, the fact that the water level is higher at h_1 than at h_2 implies that flow is moving from left to right in the column.

In field settings, the direction of ground-water flow is determined by looking at how water levels change from place to place. Measurements are made with a *piezometer,* which is essentially the field counterpart to the laboratory manometer. In its simplest form, a piezometer is a standpipe installed to some depth in the saturated system (Figure 3.4). The water elevation with sea level as the datum is a measure of the energy for flow at the intake or opening at the bottom of the piezometer.

This energy available for flow is given the name *hydraulic head.* It consists of three components, related to elevation, pressure, and velocity. The relationship among hydraulic head, elevation, pressure, and velocity is expressed by the Bernoulli equation.

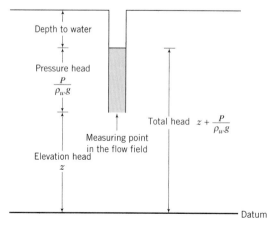

Figure 3.4 Diagram showing elevation, pressure, and total head for a point in the flow field (from Domenico and Schwartz, 1998. *Physical and chemical hydrogeology*). Copyright ©1998 by John Wiley & Sons, Inc. Reprinted by permission of John Wiley & Sons, Inc.

$$h = z + \frac{P}{\rho_w g} + \frac{v^2}{2g} \tag{3.8}$$

where h is the hydraulic head [L], z is the elevation [L], P is the pressure exerted by water column [M/LT^2], ρ_w is the fluid density [M/L], g is the gravitation acceleration [L/T], and v is the velocity [L/T]. In ground-water settings, the flow velocity is so low that the energy contained in velocity can be neglected when computing the total energy. Thus, the hydraulic head is written as

$$h = z + \frac{P}{\rho_w g} \tag{3.9}$$

This relationship is illustrated in Figure 3.4. The hydraulic head is the sum of elevation head and pressure head. In SI units, h is in meters (m), z is in meters (m) above the datum (usually sea level), P is in Pascal (Pa), ρ_w is in kg/m^3, and g is in m/s^2. The density ρ_w varies as a function of temperature and chemical composition, with fresh water at 15.5°C having a density of 1000 kg/m^3. The gravitational constant, g, is 9.81 m/s^2. The Pascal is defined as

$$1 \text{ Pascal } = 1 \, kg/m/s^2 \tag{3.10}$$

In English engineering units, the pressure is in psi (pounds per square inches). Table 3.3 lists the unit conversion factors between different units of pressure.

► **EXAMPLE 3.1**

With reference to Figure 3.4, assume that the elevation of the ground surface is 1000 m above sea level, the depth to water is 25 m, the total length of the piezometer is 50 m, and the water has density of 1000 kg/m^3. What are (a) the total hydraulic head at the measurement point, (b) the pressure head, and (c) the pressure?

SOLUTION
(a) Total hydraulic head at the bottom of the piezometer

$$h = 1000 - 25 = 975 \text{ m}$$

(b) Pressure head

$$\frac{P}{\rho_w g} = h - z = 975 - 950 = 25 \text{ m}$$

(c) Pressure

$$P = \rho_w g(h - z) = (1000 \text{ kg/m}^3)(9.81 \text{ m/s}^2)(25 \text{ m}) = 2.45 \times 10^5 \text{ kg/m/s}^2 = 0.245 \text{ MPa}$$

TABLE 3.3 Unit conversion factors between different units of pressure

	Psi	Kg/cm^2	Pascal	Atmospheres	Inches of Hg	Millibar	Ft of H$_2$O
Psi	1	7.031×10^{-2}	6.895×10^3	6.807×10^{-2}	2.036	6.895×10^1	2.307
kg/cm^2	1.422×10^1	1	9.807×10^4	9.681×10^{-1}	2.896×10^1	9.807×10^2	3.281×10^1
Pascal	1.45×10^{-4}	1.02×10^{-5}	1	9.872×10^{-6}	2.953×10^{-4}	$1. \times 10^{-2}$	3.346×10^{-4}
Atmosphere	1.469×10^1	1.033	1.013×10^5	1	2.992×10^1	1.013×10^3	3.389×10^1
Inches of Hg	4.911×10^{-1}	3.453×10^{-2}	3.386×10^3	3.343×10^{-2}	1	3.386×10^1	1.133
Millibar	1.45×10^{-2}	1.02×10^{-3}	1×10^2	9.872×10^{-4}	2.953×10^{-2}	1	3.346×10^{-2}
ft of H$_2$O	4.335×10^{-1}	3.048×10^{-2}	2.989×10^3	2.951×10^{-2}	8.827×10^{-1}	2.989×10^1	1

Note: H$_2$O at 4°C and Hg at 0°C.

We can also calculate pressure in Pa using Table 3.3.

$$P = 25 \text{ m of water} = 82 \text{ ft of water} = 82 \times 2989 \text{ Pa} = 0.245 \text{ MPa}$$

▶ 3.3 HYDRAULIC GRADIENT AND GROUND-WATER-FLOW DIRECTION

Darcy's experiments showed that for flow to occur there must be differences in hydraulic head creating a hydraulic gradient. The *hydraulic gradient* is formally defined as the change in hydraulic head in a given direction

$$i = -\frac{dh}{dl} = \frac{h_1 - h_2}{\Delta l} \tag{3.11}$$

where h_1 and h_2 is the hydraulic head at points 1 and 2, respectively, and Δl is the distance between points 1 and 2. The hydraulic gradient i in Eq. (3.11) is the hydraulic gradient from point 1 to 2.

In a field setting, it is possible to install a large number of piezometers in a unit and to contour the resulting hydraulic head values (Figure 3.5). Starting at one of the piezometers, for example, a in Figure 3.5, it is likely that the head will decrease in some directions and increase in others. The gradient is oriented in the direction of maximum head decrease. For the simple example, the maximum gradient is perpendicular to the lines of equal hydraulic head—called *equipotential lines*. The hydraulic gradient essentially defines the direction of ground-water flow from regions of high hydraulic head to regions of lower hydraulic head.

At a minimum, it takes three hydraulic-head measurements to determine the hydraulic gradient and the direction of ground-water flow. Let's demonstrate how this calculation is done. Consider three piezometers, located as shown in Figure 3.6 at the corners of a triangle. Here are the steps.

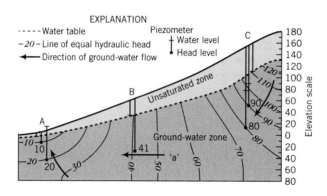

Figure 3.5 Example of the hydraulic-head distribution defined along a hypothetical cross section using piezometers and water-table observation wells (from Winter et al., 1998).

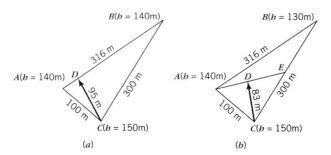

Figure 3.6 Determination of flow direction.

1. Make sure that the three hydraulic heads in three wells (A, B, and C) are not equal.

2. If the hydraulic heads at two of the wells (A and B) are equal, draw a line through C and perpendicular to $\overline{AB}$ (with the intercept point D). If the hydraulic head at well C is greater than the hydraulic head at wells A and B, ground water will flow from C to D. The hydraulic gradient from C to D is

$$i_{CD} = \frac{h_C - h_D}{\overline{CD}}$$

where h_C and h_B are the hydraulic heads at wells C and B, and $\overline{CD}$ is the distance between C and D.

3. If the hydraulic heads are different at the three wells, find a location (E) on the line connecting the well with the highest head (C) to the well with the lowest head (B), at which the head is the same as that in the intermediate well (A). Draw a line between A and E and another line through C, intercepting the line $\overline{AE}$. The flow direction is from C to D as the arrow shows. Again, the hydraulic gradient from C to D will be

$$i_{CD} = \frac{h_C - h_D}{\overline{CD}}$$

▶ **EXAMPLE 3.2**   With reference to Figure 3.6, find the flow direction and calculate the hydraulic gradient.

SOLUTION   In Figure 3.6a, two of the hydraulic heads are equal. According to the procedure outlined here, an intercept D is found on line $\overline{AB}$ so that $\overline{CD}$ will be perpendicular to $\overline{AB}$. The flow direction from C to D is marked in Figure 3.6a. The hydraulic gradient from C to D is

$$i_{CD} = \frac{150\,\text{m} - 140\,\text{m}}{95\,\text{m}} = 0.105$$

In Figure 3.6b, the hydraulic heads at three wells are different. First, find the location E with an average hydraulic head of the wells B and C on line $\overline{CB}$. Then, draw a line $\overline{CD}$ perpendicular to line $\overline{AE}$. Ground water will flow from C to D. The hydraulic gradient is calculated as

$$i_{CD} = \frac{150\,\text{m} - 140\,\text{m}}{83\,\text{m}} = 0.12$$

▶ 3.4 HYDRAULIC CONDUCTIVITY AND PERMEABILITY

Hydraulic conductivity is introduced in Darcy's law as a constant of proportionality relating the specific discharge to the hydraulic gradient. Qualitatively, *hydraulic conductivity* is a parameter describing the ease with which flow takes place through a porous medium. It has relatively large values for permeable units like sand and gravel and relatively small values for poorly permeable materials like clay or shale.

Hydraulic conductivity has units of velocity [L/T]. Thus, with a flow rate in m^3/day and both hydraulic head and lengths in meters, the hydraulic conductivity has units of m/day. In the English Engineering Unit System with a flow rate in gallons per day, the hydraulic conductivity has units of gallons per day per foot squared (gpd/ft^2). The following equations describe conversions among these units.

$$1\,\text{m/day} = 3.28\,\text{ft/day} = 24.5\,\text{gpd/ft}^2 \tag{3.12}$$

$$1\,\text{ft/day} = 0.3048\,\text{m/day} = 7.48\,\text{gpd/ft}^2 \tag{3.13}$$

$$1\,\text{gpd/ft}^2 = 4.075 \times 10^{-2}\,\text{m/day} = 0.1337\,\text{ft/day} \qquad (3.14)$$

▶ **EXAMPLE 3.3**

Ground water flows through a buried-valley aquifer with a cross-sectional area of 1.0×10^6 ft² and a length of 2×10^4 ft. Hydraulic heads at the ground-water entry and exit points in the aquifer are 1000 and 960 ft, respectively. At the downstream end of the aquifer, ground water discharges into a stream at a rate of 1.0×10^5 ft³/day. What is the hydraulic conductivity of the buried-valley aquifer in ft/day, m/day, and gpd/ft²? If the effective porosity of the material is 0.3, what is the linear ground-water velocity?

SOLUTION

1. Calculation of the Darcy velocity:

$$q = \frac{Q}{A} = \frac{1.0 \times 10^5\,\text{ft}^3/\text{day}}{1.0 \times 10^6\,\text{ft}^2} = 0.1\,\text{ft/day}$$

2. Calculation of the hydraulic gradient:

$$-dh/dl = \frac{(1000\,\text{ft}) - (960\,\text{ft})}{2 \times 10^4\,\text{ft}} = 2.0 \times 10^{-3}$$

3. Calculation of the hydraulic conductivity:

$$K = -\frac{q}{dh/dl} = \frac{0.1\,\text{ft/day}}{2.0 \times 10^{-3}} = 50\,\text{ft/day} = 15.24\,\text{m/day} = 373\,\text{gpd/ft}^2$$

4. Calculation of the linear ground-water velocity:

$$v = \frac{q}{n_e} = \frac{0.1\,\text{ft/day}}{0.3} = 0.33\,\text{ft/day}$$

Many tens of thousands of hydraulic conductivity values have been measured for geological materials of all kinds. Subsequent chapters will describe hydraulic conductivity in a variety of geological materials and structural settings, and the variety of field and laboratory approaches designed to provide estimates.

In the absence of actual data, it is often necessary to estimate hydraulic-conductivity values from knowledge of the rock type. This method is only slightly better than an educated guess, but occasionally it is the only available approach. Hydraulic-conductivity values for common rocks and sediments are listed in Table 3.4.

Intrinsic Permeability

Experiments have shown that hydraulic conductivity depends on both properties of the porous medium and the fluid (for example, density and viscosity). For many ground-water studies, water is the fluid of interest, providing more or less constant values of density and viscosity (neglecting temperature dependencies). Thus, measurements of hydraulic conductivity are useful in comparing differences in hydraulic behavior of the actual materials. In looking more generally at systems where the fluids other than water are present (such as, air, oil, and gasoline), hydraulic conductivity becomes an awkward parameter because the density and viscosity of the fluid vary together with the medium properties.

A convenient alternative is to write Darcy's equation in a form where the properties of the medium and the fluid are represented explicitly

TABLE 3.4 Representative values of hydraulic conductivity for various rock types

Material	Hydraulic Conductivity (m/s)
Sedimentary	
Gravel	$3 \times 10^{-4} - 3 \times 10^{-2}$
Coarse sand	$9 \times 10^{-7} - 6 \times 10^{-3}$
Medium sand	$9 \times 10^{-7} - 5 \times 10^{-4}$
Fine sand	$2 \times 10^{-7} - 2 \times 10^{-4}$
Silt, loess	$1 \times 10^{-9} - 2 \times 10^{-5}$
Till	$1 \times 10^{-12} - 2 \times 10^{-6}$
Clay	$1 \times 10^{-11} - 4.7 \times 10^{-9}$
Unweathered marine clay	$8 \times 10^{-13} - 2 \times 10^{-9}$
Sedimentary Rocks	
Karst and reef limestone	$1 \times 10^{-6} - 2 \times 10^{-2}$
Limestone, dolomite	$1 \times 10^{-9} - 6 \times 10^{-6}$
Sandstone	$3 \times 10^{-10} - 6 \times 10^{-6}$
Siltstone	$1 \times 10^{-11} - 1.4 \times 10^{-8}$
Salt	$1 \times 10^{-12} - 1 \times 10^{-10}$
Anhydrite	$4 \times 10^{-13} - 2 \times 10^{-8}$
Shale	$1 \times 10^{-13} - 2 \times 10^{-9}$
Crystalline Rocks	
Permeable basalt	$4 \times 10^{-7} - 2 \times 10^{-2}$
Fractured igneous and metamorphic rock	$8 \times 10^{-9} - 3 \times 10^{-4}$
Weathered granite	$3.3 \times 10^{-6} - 5.2 \times 10^{-5}$
Weathered gabbro	$5.5 \times 10^{-7} - 3.8 \times 10^{-6}$
Basalt	$2 \times 10^{-11} - 4.2 \times 10^{-7}$
Unfractured igneous and metamorphic rocks	$3 \times 10^{-14} - 2 \times 10^{-10}$

Source: From Domenico and Schwartz, 1998. *Physical and chemical hydrogeology.* Copyright ©1998 by John Wiley & Sons, Inc. Reprinted by permission of John Wiley & Sons, Inc.

$$q = -\frac{k \rho_w g}{\mu} \frac{dh}{dl} \tag{3.15}$$

where q is the rate of flow per unit area, k is the intrinsic permeability, ρ_w is the density of water, g is the acceleration due to gravity, μ is the dynamic viscosity of water, and dh/dl is the unit change in hydraulic head per unit length of flow. The *intrinsic permeability* of a rock or soil is a measure of its ability to transmit fluid as the fluid moves through it. The permeability is independent of the fluid moving through the medium. If q is measured in m/sec, μ is in kg/(m.sec), ρ_w is in kg/m^3, g in m/sec^2, and dh/dl in m/m, the unit for k is m^2.

$$[k] = \frac{(\text{m/sec})(\text{kg/(m.sec)})}{(\text{kg/m}^3)(\text{m/sec}^2)(\text{m/m})} = m^2$$

Intrinsic permeability also has units like cm^2 or the darcy. Equations are available for conversion between the units.

$$1\,\text{m}^2 = 10^4\,\text{cm}^2 = 1.013 \times 10^{12}\text{darcy} \tag{3.16}$$

$$1\,\text{cm}^2 = 10^{-4}\,\text{m}^2 = 1.013 \times 10^8\text{darcy} \tag{3.17}$$

$$1\,\text{darcy} = 9.87 \times 10^{-13}\,\text{m}^2 = 9.87 \times 10^{-9}\,\text{cm}^2 \tag{3.18}$$

Comparing Eqs. (3.3) and (3.15), we see that hydraulic conductivity is a function of both the porous medium (permeability) and the fluid (density and viscosity) such that

$$K = \frac{k\rho_w g}{\mu} \tag{3.19}$$

For water, viscosity and density are functions of pressure and temperature. At $20°$ C and 1 atm pressure, the density and dynamic viscosity of water are $998.2\,\text{kg/m}^3$ and 1.002×10^{-3} kg/(m·sec), respectively. Under these conditions, Eq. (3.19) can be simplified to

$$K = \frac{k\rho_w g}{\mu} = \frac{(998.2\text{kg/m}^3)(9.81\text{m/sec}^2)}{1.002 \times 10^{-3}\text{kg/(m}\cdot\text{sec})}k = \left(9.77 \times 10^6 \frac{1}{m \cdot sec}\right) \cdot k \tag{3.20}$$

where K is in m/sec and k is in m^2. To convert from hydraulic conductivity to intrinsic permeability, Eq. (3.20) can be written as

$$k = (1.023 \times 10^{-7}\text{m} \cdot \text{sec}) \cdot K \tag{3.21}$$

▶ **EXAMPLE 3.4**

What is the intrinsic permeability of a water-saturated medium that has a hydraulic conductivity of 15.24 m/day?
Assume the ground water is at $20°$C and 1 atm pressure.

SOLUTION The hydraulic conductivity in m/sec is calculated as

$$K = 15.24\text{m/day} = (15.24\text{m/day})\left(\frac{1}{24 \times 60 \times 60}\frac{\text{day}}{\text{sec}}\right) = 1.764 \times 10^{-4}\text{m/sec}$$

Intrinsic permeability is calculated from Eq. (3.21) as

$$k = (1.023 \times 10^{-7}\text{m} \cdot \text{sec})(1.764 \times 10^{-4}\text{m/sec}) = 1.8 \times 10^{-11}\text{m}^2$$

Empirical Approaches for Estimation

The intrinsic permeability of a rock or soil is related to the diameter of the grains and porosity. A number of empirical equations have been derived to estimate the intrinsic permeability and hydraulic conductivity from grain-size properties (Table 3.5). A grain-size analysis of the material is necessary to use these equations. The *effective grain diameter* is the grain size of the smallest 10% of the grains (90% larger). The grain-size distribution of coarse-grained soils is determined directly by sieve analysis. The particle distribution for fine-grained soils (or the fine fraction of a coarse soil) is indirectly determined by hydrometer analysis. A sieve analysis involves passing a sample through a set of sieves and weighting the amount of material retained in each sieve. Sieves are constructed of wire screens, with square openings of standard sizes from 3 in. to 0.074 mm (No. 200). Hydrometer analysis should be performed on only the fraction passing the No. 200 sieve.

TABLE 3.5   Examples of empirical relationships for estimating hydraulic conductivity or permeability values

Source	Equation	Parameters
Hazen (1911)	$K = Cd_{10}^2$	K = hydraulic conductivity (cm/s)
		C = constant 100 to 150 $(\text{cm/s})^{-1}$ for loose sand
		d_{10} = effective grain size cm (10% particles are finer, 90% coarser)
Harleman et al. (1963)	$k = (6.54 \times 10^{-4})d_{10}^2$	k = permeability (cm^2)
Krumbein and Monk (1943)	$k = 760d^2 e^{-1.31\sigma}$	k = permeability (darcys)
		d = geometric mean grain diameter (mm)
		σ = log standard deviation of the size distribution
Kozeny (1927)	$k = Cn^3/S^{*2}$	C = dimensionless constant: 0.5, 0.562, and 0.597 for circular, square, and equilateral triangle pore openings
		k = permeability (L^2)
		n = porosity
		S^* = specific surface-interstitial surface areas of pores per unit bulk volume of the medium
Kozeny–Carmen Bear (1972)	$K = \left(\dfrac{\rho_w g}{\mu}\right) \dfrac{n^3}{(1-n)^2}\left(\dfrac{d_m^2}{180}\right)$	K = hydraulic conductivity
		ρ_w = fluid density
		μ = fluid viscosity
		g = gravitational constant
		d_m = any representative grain size
		n = porosity

Source: From Domenico and Schwartz, 1998. *Physical and chemical hydrogeology.* Copyright ©1998 by John Wiley & Sons, Inc. Reprinted by permission of John Wiley & Sons, Inc.

► **EXAMPLE 3.5**   Figure 3.7 shows the results of a grain-size analysis. Estimate the permeability and hydraulic conductivity.

SOLUTION   The lower horizontal axis shows grain size, while the vertical axis is the percent passing by weight. The effective grain diameter is 0.009 mm. The permeability can be estimated from Harleman's equation in Table 3.5.

$$k = (6.54 \times 10^{-4})d_{10}^2 = (6.54 \times 10^{-4}) \cdot 0.0009^2 = 5.3 \times 10^{-10} \text{ cm}^2$$

The hydraulic conductivity may be calculated from Eq. (3.24):

$$K = (9.77 \times 10^6 \frac{1}{\text{m/sec}}) \cdot k = (9.77 \times 10^6 \frac{1}{\text{m/sec}}) \cdot 5.3 \times 10^{-10} \text{ cm}^2 = 5.2 \times 10^{-5} \text{ cm/sec}$$

Or the hydraulic conductivity may be estimated from Hazen's equation:

$$K = Cd_{10}^2 = \left(100 \frac{1}{\text{sec/cm}}\right)(0.0009 \text{ cm})^2 = 8.1 \times 10^{-5} \text{ cm/sec}$$

The two results are in the same order of magnitude.

► **3.5 LABORATORY MEASUREMENT OF HYDRAULIC CONDUCTIVITY**

A number of methods can be used to determine the hydraulic conductivity of geological materials, including field hydraulic tests, laboratory measurements, and empirical methods based on correlation with grain sizes. Usually, field hydraulic tests (see Chapters 9 through 12) are thought to provide the most reliable estimates, for they permit the testing of a

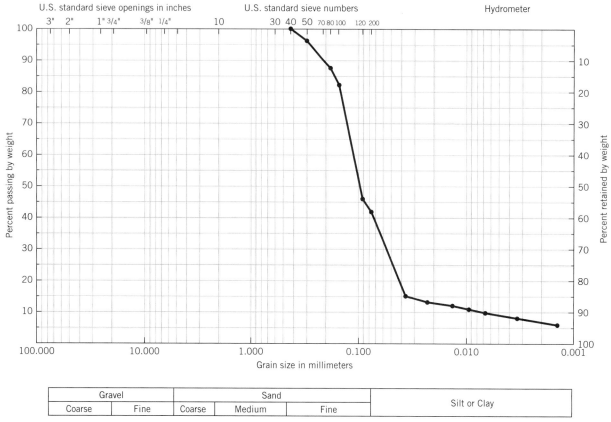

Figure 3.7 Grain-size analysis results (data from Krumbein and Pettijohn, 1938).

large volume of rock with one pumping well and one or more observation wells. In this section, we introduce two laboratory methods for testing cores: the constant-head and the falling-head tests (Figure 3.8). In the constant-head test, a valve at the base of the sample is opened and water starts to flow through the core. Once a steady flow has developed, the volumetric flow rate is determined using data on hydraulic heads and the volume of outflow as a function of time. Hydraulic conductivity is determined with Darcy's equation written in the following form:

$$K = \frac{Q\,L}{A\,h} \tag{3.22}$$

where L is the length of the sample, A is the cross-sectional area of the sample, and h is the constant head shown in Figure 3.8a.

In the falling-head test, hydraulic head is measured in the column (Figure 3.8b) as a function of time. For a sample of length L and a cross-sectional area A, hydraulic conductivity is given by

$$K = 2.3 \frac{aL}{A(t_1 - t_0)} \log_{10} \frac{h_0}{h_1} \tag{3.23}$$

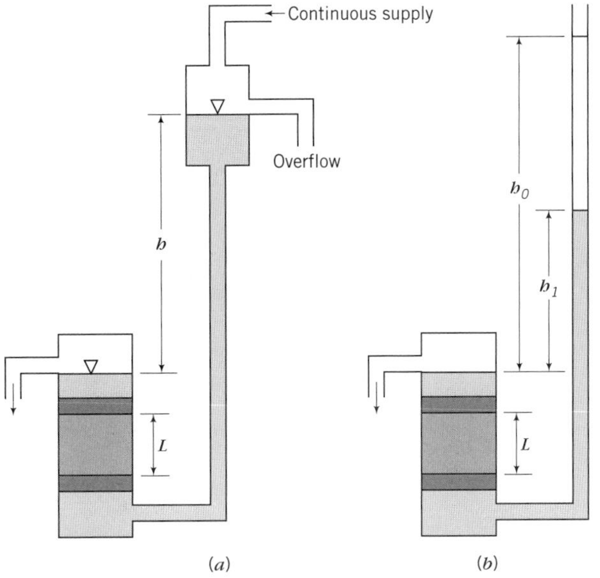

Figure 3.8 Laboratory equipment for measuring hydraulic conductivity of porous media. Panel (*a*) shows a constant-head permeameter. Panel (*b*) shows a falling-head permeameter (from Domenico and Schwartz, 1998. *Physical and chemical hydrogeology*). Copyright ©1998 by John Wiley & Sons, Inc. Reprinted by permission of John Wiley & Sons, Inc.

where a is the cross-sectional area of the standpipe and $(t_1 - t_0)$ is the time required for the head to fall from h_0 to h_1.

For fine-grained soils, Q is small and is difficult to measure accurately. Thus, the constant-head test is used principally for testing coarse-grained media (clean sands and gravels) with K values greater than about 10×10^{-4} cm/sec. The falling-head test is generally used for less permeable media (fine sands to fat clays) with K value less than 10×10^{-4} cm/sec.

► 3.6 DARCY'S EQUATION FOR ANISOTROPIC MATERIALS

In this section, we examine cases where the hydraulic conductivity changes as a function of the direction of flow. The term *anisotropic* is used to describe the permeability or hydraulic conductivity of materials, which at a point exhibit directional dependency. When the permeability is the same in all directions, the material at that point is said to be *isotropic*. Davis (1969) cites several cases for bedded sediments in which permeability is greater in the direction of stratification and smaller perpendicular to the stratification. Darcy's law in anisotropic materials is expressed as

$$q = -K\nabla h \tag{3.24}$$

where q is the Darcy velocity vector, K is the hydraulic-conductivity matrix tensor, and ∇h is the gradient of hydraulic head. In Cartesian coordinates, q, K, and ∇h are written as

$$q = q_x i + q_y j + q_z k \tag{3.25}$$

$$-\nabla h = \frac{\partial h}{\partial x} i + \frac{\partial h}{\partial y} j + \frac{\partial h}{\partial z} k \tag{3.26}$$

$$K = \begin{bmatrix} K_{xx} & K_{xy} & K_{xz} \\ K_{yx} & K_{yy} & K_{yz} \\ K_{zx} & K_{zy} & K_{zz} \end{bmatrix} \tag{3.27}$$

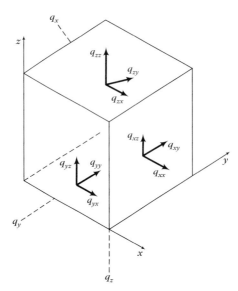

Figure 3.9 Array of nine components of three velocity vectors acting at a point and the three faces of a cube.

where i, j, and k are unit vectors along the x-, y-, and z-directions, respectively. Equation (3.24) is Darcy's equation for fluid flow in anisotropic media. When Eqs. (3.24) through (3.27) are combined, the Darcy velocity vector may be broken down into components in x-, y-, and z-directions (Figure 3.9).

$$q_x = -K_{xx}\frac{\partial h}{\partial x} - K_{xy}\frac{\partial h}{\partial y} - K_{xz}\frac{\partial h}{\partial z} \tag{3.28}$$

$$q_y = -K_{yx}\frac{\partial h}{\partial x} - K_{yy}\frac{\partial h}{\partial y} - K_{yz}\frac{\partial h}{\partial z} \tag{3.29}$$

$$q_z = -K_{zx}\frac{\partial h}{\partial x} - K_{zy}\frac{\partial h}{\partial y} - K_{zz}\frac{\partial h}{\partial z} \tag{3.30}$$

where

$$K_{xy} = K_{yx}, \quad K_{xz} = K_{zx}, \quad K_{yz} = K_{zy} \tag{3.31}$$

If the principal directions of anisotropy coincide with the x-, y-, and z-directions, the nondiagonal components of hydraulic conductivity tensor in Eq. (3.27) are zero.

$$K_{xy} = K_{yx} = K_{xz} = K_{zx} = K_{yz} = K_{zy} = 0 \tag{3.32}$$

And the Darcy velocities in the x-, y-, and z-directions are simplified as

$$q_x = -K_{xx}\frac{\partial h}{\partial x} \tag{3.33}$$

$$q_y = -K_{yy}\frac{\partial h}{\partial y} \tag{3.34}$$

$$q_z = -K_{zz}\frac{\partial h}{\partial z} \tag{3.35}$$

TABLE 3.6   The anisotropic character of some rocks

Material	Horizontal Conductivity (m/s)	Vertical Conductivity (m/s)
Anhydrite	10^{14}–10^{-12}	10^{-15}–10^{-13}
Chalk	10^{-10}–10^{-8}	5×10^{-11}–5×10^{-9}
Limestone, dolomite	10^{-9}–10^{-7}	5×10^{-10}–5×10^{-8}
Sandstone	5×10^{-13}–10^{-10}	2.5×10^{-13}–5×10^{-11}
Shale	10^{-14}–10^{-12}	10^{-15}–10^{-13}
Salt	10^{-14}	10^{-14}

Source: From Domenico and Schwartz, 1998. *Physical and chemical hydrogeology.* Copyright ©1998 by John Wiley & Sons, Inc. Reprinted by permission of John Wiley & Sons, Inc.

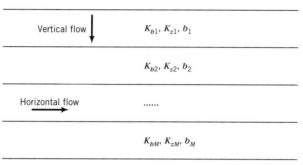

Figure 3.10   Horizontal hydraulic units.

Table 3.6 provides a summary of information on the anisotropic nature of some sedimentary materials as determined from core samples. The ratio between the horizontal and vertical hydraulic conductivities for these rocks is between one and several thousands.

► 3.7 HYDRAULIC CONDUCTIVITY IN HETEROGENEOUS MEDIA

The classical definition considers a unit to be *homogeneous* if the permeability in a given direction is the same from point to point in a geological unit. Materials that do not conform with this condition are *heterogeneous*. A simple example of heterogeneity in permeability is represented by a layered geological unit with variable hydraulic conductivities (Figure 3.10). According to Leonards (1962), an equivalent vertical horizontal hydraulic conductivity in the horizontal direction is

$$K_h = \frac{\sum\limits_{i=1}^{M} b_i K_{hi}}{\sum\limits_{i=1}^{M} b_i} \tag{3.36}$$

where K_h is the equivalent horizontal hydraulic conductivity, K_{hi} is the horizontal hydraulic conductivity of layer i, and b_i is the thickness of layer i. An equivalent vertical hydraulic conductivity is

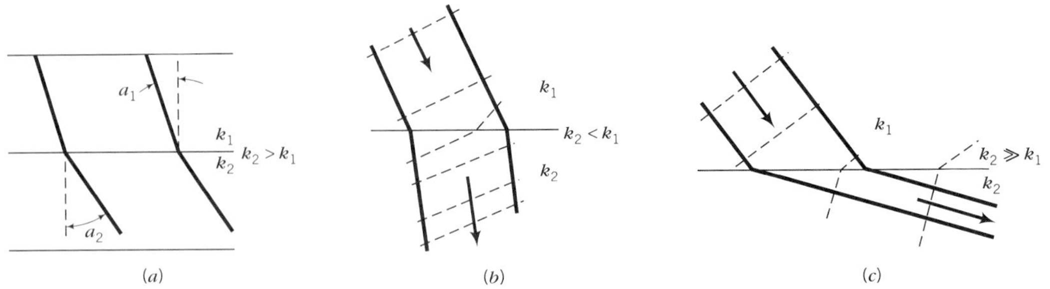

Figure 3.11 Diagrams of flow line refraction and conditions at the boundaries between materials of differing permeability (from Domenico and Schwartz, 1998. *Physical and chemical hydrogeology*). Copyright ©1998 by John Wiley & Sons, Inc. Reprinted by permission of John Wiley & Sons, Inc.

$$K_v = \frac{\displaystyle\sum_{i=1}^{M} b_i}{\displaystyle\sum_{i=1}^{M} b_i/K_{vi}} \tag{3.37}$$

where K_v is equivalent vertical hydraulic conductivity and K_{vi} is the vertical hydraulic conductivity of layer i.

In cases where flow is at some angle to a geological boundary between layers, ground-water flow obeys the tangent refraction law (Figure 3.11a). The extent of refraction of the flow lines will depend on the hydraulic conductivity ratio of the two layers. Mathematically, the ratio of the tangents of the angles that the flow lines make with the normal to the boundary is equal to the ratio K_1/K_2, or

$$\frac{K_1}{K_2} = \frac{\tan(\alpha_1)}{\tan(\alpha_2)} \tag{3.38}$$

where ground water flows from unit 1 into unit 2, K is hydraulic conductivity, and α is the refraction angle. In Figure 3.11b, the hydraulic gradient in the lower hydrological unit is steepened to accommodate the flow crossing the boundary from the unit of higher hydraulic conductivity. In Figure 3.11c, both the hydraulic gradient and the cross-sectional flow area in the high-conductivity unit are decreased to accommodate the flow crossing the boundary from the low-conductivity unit.

▶ **EXAMPLE 3.6**

Consider a 300-m sequence of interbedded sandstone and shale that has 75% sandstone. The sandstone has a horizontal and vertical hydraulic conductivity of 10^{-5} m/s, and shale has a horizontal and vertical hydraulic conductivity of 1.92×10^{-12} m/s. Calculate the equivalent conductivities for a system of layers. If the ground water flows in the sandstone with an angle of 45° away from the vertical direction toward the sandstone–shale boundary, what is the flow direction in the shale?

SOLUTION The equivalent horizontal hydraulic conductivity is

$$K_x = \frac{(225\,\text{m})(1 \times 10^{-5}\,\text{m/s}) + (75\,\text{m})(1.92 \times 10^{-12}\,\text{m/s})}{(300\,\text{m})} = 7.5 \times 10^{-6}\,\text{m/s}$$

The equivalent vertical hydraulic conductivity is

$$K_z = \frac{(300\,\text{m})}{\dfrac{(225\,\text{m})}{(1 \times 10^{-5}\,\text{m/s})} + \dfrac{(75\,\text{m})}{(1.92 \times 10^{-12}\,\text{m/s})}} = 7.7 \times 10^{-12}\,\text{m/s}$$

Thus, for horizontal flow the most permeable units dominate the system; for vertical flow the least permeable units dominate the system. The flow direction in the shale is calculated by Eq. (3.38).

$$\alpha_2 = \tan^{-1}\left[\frac{K_2}{K_1}\tan(\alpha_1)\right] = \tan^{-1}\left[\frac{(1.92 \times 10^{-12} \text{ m/s })}{(10^{-5} \text{ m/s })}(1)\right] = 0°$$

The result shows that the flow in the shale is normal to the boundary.

For a uniform flow, arithmetic, harmonic, and geometric means are commonly used to scale measured hydraulic conductivities in heterogeneous fields. The arithmetic mean (η) is the summation of measurements divided by the number of measurements.

$$\eta = \frac{1}{N}\sum_{i=1}^{N}X_i \tag{3.39}$$

where N is the number of measurements in the sample and X_i is the individual measurement. The harmonic mean (H) of a set of numbers is given as

$$H = \frac{N}{\sum_{i=1}^{N}\frac{1}{X_i}} \tag{3.40}$$

Geometric mean (G) is the Nth root of a product of N numbers.

$$G = (X_1 X_2 X_3 \ldots X_N)^{1/N} \tag{3.41}$$

▷ **EXAMPLE 3.7**

The second column of Table 3.7 lists hydraulic-conductivity values measured for a hypothetical geological unit. Calculate arithmetic, harmonic, and geometric means.

SOLUTION It is easy to use a spreadsheet program to calculate the mean values as shown in Table 3.7. Columns 3 and 4 are log(K) and 1/K for measured values, respectively. The last row in the table is the summation of K, log(K), and 1/K. The arithmetic mean is the sum in column 2 divided by the number of measurements.

$$\eta = \frac{8.45 \times 10^{-6}}{32} = 2.64 \times 10^{-7} \text{ (m/s)}$$

The logarithm of geometric mean is the sum in column 3 divided by the number of measurements. Thus,

$$G = 10^{-224.099/32} = 9.93 \times 10^{-8} \text{ (cm/s)}$$

The harmonic mean is the number of measurements divided by the sum in column 4 of the table.

$$H = \frac{32}{5.59 \times 10^9} = 5.72 \times 10^{-9} \text{ (cm/s)}$$

Matheron (1967) shows that, for a uniform flow in a statistically isotropic hydraulic-conductivity field, the average hydraulic conductivity for a system of any dimension is between the harmonic and arithmetic means.

In many studies, hydraulic conductivity has been shown to be a lognormally distributed parameter. Like a typical normal distribution, this distribution is characterized

TABLE 3.7 Calculation of arithmetic, harmonic, and geometric means

No.	K	Log (K)	1/K	$[\log(K)-\eta]^2$
1	4.00×10^{-8}	-7.40	2.50×10^7	0.16
2	1.00×10^{-7}	-7.00	1.00×10^7	0.00
3	2.00×10^{-8}	-7.70	5.00×10^7	0.49
4	4.00×10^{-7}	-6.40	2.50×10^6	0.36
5	1.00×10^{-6}	-6.00	1.00×10^6	1.00
6	1.50×10^{-7}	-6.82	6.67×10^6	0.03
7	6.00×10^{-8}	-7.22	1.67×10^7	0.05
8	1.50×10^{-6}	-5.82	6.67×10^5	1.38
9	8.00×10^{-9}	-8.10	1.25×10^8	1.20
10	1.00×10^{-7}	-7.00	1.00×10^7	0.00
11	1.00×10^{-7}	-7.00	1.00×10^7	0.00
12	1.00×10^{-8}	-8.00	1.00×10^8	1.00
13	8.00×10^{-7}	-6.10	1.25×10^6	0.82
14	5.00×10^{-8}	-7.30	2.00×10^7	0.09
15	4.00×10^{-7}	-6.40	2.50×10^6	0.36
16	5.00×10^{-7}	-6.30	2.00×10^6	0.49
17	2.50×10^{-8}	-7.60	4.00×10^7	0.36
18	2.00×10^{-7}	-6.70	5.00×10^6	0.09
19	2.00×10^{-7}	-6.70	5.00×10^6	0.09
20	5.00×10^{-8}	-7.30	2.00×10^7	0.09
21	2.00×10^{-7}	-6.70	5.00×10^6	0.09
22	2.50×10^{-8}	-7.60	4.00×10^7	0.36
23	1.00×10^{-7}	-7.00	1.00×10^7	0.00
24	1.00×10^{-6}	-6.00	1.00×10^6	1.00
25	2.00×10^{-10}	-9.70	5.00×10^9	7.28
26	1.00×10^{-7}	-7.00	1.00×10^7	0.00
27	8.00×10^{-8}	-7.10	1.25×10^7	0.01
28	4.00×10^{-8}	-7.40	2.50×10^7	0.16
29	3.00×10^{-7}	-6.52	3.33×10^6	0.23
30	6.00×10^{-7}	-6.22	1.67×10^6	0.61
31	2.50×10^{-7}	-6.60	4.00×10^6	0.16
32	4.00×10^{-8}	-7.40	2.50×10^7	0.16
Sum	8.45×10^{-6}	-224.10	5.59×10^9	18.12

by an arithmetic mean (or mean) and standard deviation. The standard deviation (σ) is defined as

$$\sigma = \left[\frac{1}{N-1} \sum_{i=1}^{N} (X_i - \eta)^2 \right]^{1/2} \tag{3.42}$$

The equation for a normal (Gaussian) probability density curve is

$$f(X) = \frac{1}{\sigma \sqrt{2\pi}} e^{\frac{(X-\eta)^2}{2\sigma^2}} \tag{3.43}$$

where $f(X)$ is the normal probability density function. For $\eta = 0.5$ and $\sigma = -8.0$, the $f(X)$ versus X is shown in Figure 3.12. A plot for the frequency of observed values versus the observed values is called a *frequency histogram*. In constructing the histogram, a frequency table records how often observed values fall within certain intervals. The frequency

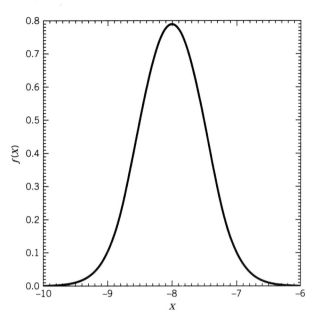

Figure 3.12 Normal probability distribution function.

histogram is often constructed to see how well its shape matches certain distributions, like a normal distribution.

► **EXAMPLE 3.9**   Prepare a frequency distribution histogram for each of the two sets of hydraulic conductivity data in Table 3.7. Determine by inspection whether one or both of these distributions is normally distributed. If the hydraulic conductivity appears to be normal, calculate the mean and the standard deviation.

SOLUTION   The histograms for hydraulic-conductivity and log-transformed hydraulic- conductivity data are shown in Figures 3.13a and b, respectively. Notice how the histogram of the log-transformed hydraulic-conductivity data is similar to a normal distribution. The mean for the transformed hydraulic conductivity is the summation of log(K) listed in column 3 of Table 3.7 divided by the number of measurements.

$$\eta = \frac{-224.1}{32} = -7.0$$

The standard deviation is the square root of summation of $[\log(K) - \eta]^2/(N-1)$ listed in column 5 of Table 3.7.

$$\sigma = \sqrt{\frac{1}{32-1} 18.12} = 0.76$$

So far, this chapter has presented a rather simple model of hydraulic conductivity for individual geological units. In reality, hydraulic-conductivity values vary from point to point in a systematic manner. Hydraulic conductivity in a geological unit is actually a correlated random variable. The extent of correlation changes as a function of direction.

► 3.8 MAPPING FLOW IN GEOLOGICAL SYSTEMS

Section 3.6 showed how the spatial variation in a hydraulic head determined the hydraulic gradient and, ultimately, the direction of flow. Hydraulic heads are determined by measuring

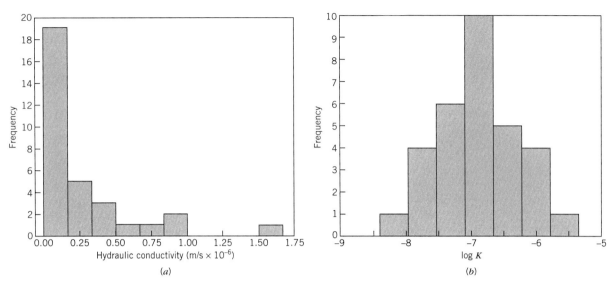

Figure 3.13 (*a*) Frequency distribution histogram of the original 32 values of hydraulic conductivity, (*b*) histogram of the log-transformed hydraulic-conductivity data (from Domenico and Schwartz, 1998. *Physical and chemical hydrogeology*). Copyright ©1998 by John Wiley & Sons, Inc. Reprinted by permission of John Wiley & Sons, Inc.

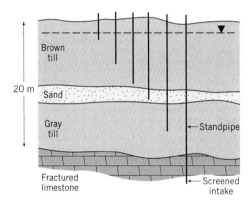

Figure 3.14 Illustration of a piezometer nest including a water-table observation well. Piezometers are emplaced in both high- and low-permeability units (from Domenico and Schwartz, 1998. *Physical and chemical hydrogeology*). Copyright ©1998 by John Wiley & Sons, Inc. Reprinted by permission of John Wiley & Sons, Inc.

the water levels in wells or piezometers installed at different locations and depths. Normally, piezometers are located within a study area to sample different topographical regimes in the area of interest. Often, at each drilling location, several piezometers are completed to provide hydraulic-head measurements at different depths in the same unit or in various units (Figure 3.14). Such a collection of piezometers is referred to as a *piezometer nest*.

Hydraulic-head distributions vary in three spatial directions and time. The time element can be removed by assuming that all the head measurements are made at the same time (i.e., a "snapshot" of hydraulic-head conditions at one time). The remaining difficulty is representing a three-dimensional field where the flow can be interpreted as largely two-dimensional. In other words, one can construct two-dimensional planes (e.g., a cross section or a map view) where the flow is essentially two-dimensional. However, the resulting cross section or map must be viewed as a two-dimensional projection of a three-dimensional field (Hubbert, 1940). Some projections will contain most of the variations in hydraulic head that have to be addressed.

Hydrogeological cross sections are vertical sections through a three-dimensional flow region. By aligning the section parallel to the direction of mean ground-water flow, flow conditions can be represented accurately in terms of a two-dimensional cross section. Figure 3.15a illustrates a simple topographic map encompassing an upland area and an adjacent valley. The hydrogeological cross section A-A' is located parallel to the direction of mean flow. Normally, the section includes basic information about the stratigraphy and variations in hydraulic conductivity, as well as hydraulic-head data from nests of piezometers located along the section (Figure 3.15b). A convenient way of presenting these data is to represent the measurement point (i.e., intake of the piezometer or elevation of water table) by a dot and noting the measured value of the hydraulic head. Contouring these head data defines the spatial variation in the hydraulic head. For this simple example, ground-water flow, as represented by the arrows, is normal to the equipotential lines (contour lines of the same head). In a more permeable lower unit, the equipotential lines are vertical, and flow in the aquifer is essentially horizontal. Within this unit, there is only a tiny vertical hydraulic gradient.

Another way of presenting hydraulic-head measurements is to contour them on a map in a horizontal plane. Such a map of hydraulic heads is referred to as a *potentiometric*

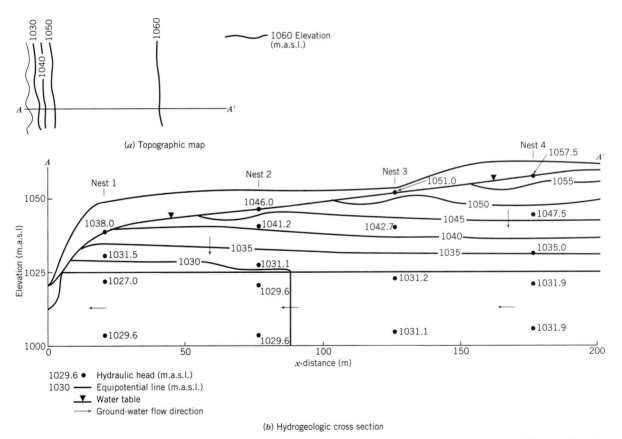

Figure 3.15 Panel (a) shows the orientation of the cross section in relation to a hill slope. Panel (b) is an example of a hydrogeologic cross section describing the pattern of ground-water flow (from Domenico and Schwartz, 1998. *Physical and chemical hydrogeology*). Copyright ©1998 by John Wiley & Sons, Inc. Reprinted by permission of John Wiley & Sons, Inc.

surface, defined by Meinzer (1923) as an imaginary surface that everywhere coincides with level of water in the aquifer. If the aquifer is unconfined, the contoured surface is a good approximation of the water table.

Figure 3.16*a* shows the potentiometric surface for the Dakota Sandstone in the Dakota aquifer system (Figure 3.16*b*). This artesian aquifer crops out along the eastern flanks of the Black Hills in South Dakota and dips eastward. Water enters this unit in its elevated intake areas and moves downdip in an easterly direction. The water moves from where the head is high to where it is low, with each of the lines presumably connecting points of equal head. In working with these maps, be aware of three important points. First, a potentiometric map must be related to a single aquifer. Other aquifers deeper or shallower in the section will

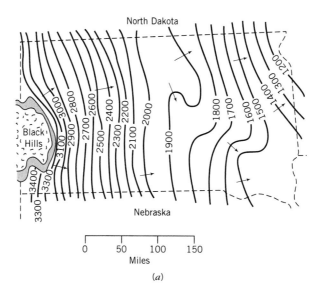

(a)

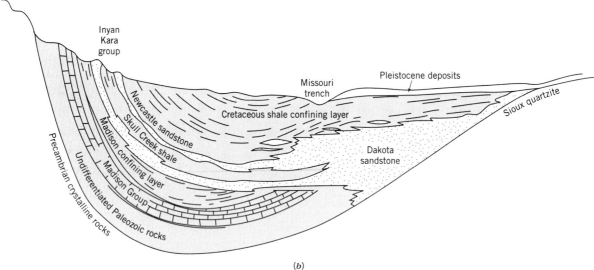

(b)

Figure 3.16 Panel (*a*) illustrates the potentiometric surface of the Dakota Sandstone, contour interval 100 ft (from Darton, 1909); Panel (*b*) illustrates the Dakota aquifer system (from Bredehoeft et al., 1983).

have different potentiometric surfaces with hydraulic heads that are higher or lower than the one of immediate concern. For example, the Madison Formation underlies the Dakota Sandstone throughout much of South Dakota and has its own potentiometric surface. The slope of this surface is about the same, but hydraulic heads are higher than those encountered in the Dakota Sandstone.

Second, it is assumed that flow in the aquifer is horizontal, that is, parallel to upper and lower confining layers. Thus, if a piezometer is placed in such an aquifer, the hydraulic head is presumed not to change as a function of depth in the aquifer. Hence, the potentiometric surface is in reality a projection of vertical equipotential lines into horizontal plane. Lastly, if head losses between adjacent pairs of equipotential lines are equal, the hydraulic gradient varies inversely with distance between lines of equal head.

▶ **EXAMPLE 3.10**

A waste disposal facility is to be constructed in glacial till (Figure 3.17). However, a sand aquifer occurs immediately beneath the lowermost till unit. The facility will occupy a region of 278 m long (in the direction of ground-water flow) by 200 m wide. The trenches will extend about 13 m into the oxidized till and penetrate the water table, which is found about 5 m below land surface. This upper till has a measured hydraulic conductivity of 10^{-7} m/s. The underlying unoxidized till has a vertical hydraulic conductivity of 10^{-8} m/s. A water-table map has been prepared and shows an average gradient of about 0.072 across the site. The materials underlying the small stream on the right-hand side of the diagram are to be excavated to the top of the unoxidized zone. This trench will be backfilled with gravel along with some sort of collection system. Assume an effective porosity of 0.1 for the entire till layer. Calculate the ground-water flow velocity in the tills, the volumetric flow rate into the trench, and the travel time for waste to reach the sand aquifer.

SOLUTION The horizontal velocity in the oxidized till layer is:

$$v_x = -\frac{K_{xx}}{n_e}\frac{\partial h}{\partial x} = \frac{(10^{-7}\,\text{cm/s})}{(0.1)}(7.2\times10^{-2}) = 7.2\times10^{-8}\,\text{m/s} = 2.3\,\text{m/yr}$$

The volumetric flow into the trench is:

$$Q_x = K_{xx}iA = (10^{-7}\,\text{m/s})(7.2\times10^{-2})(200\times8\,\text{m}^2) = 1.15\times10^{-5}\,\text{m}^3/\text{s}$$

The vertical velocity through the unoxidized till layer is:

$$v_z = -\frac{K_{zz}}{n_e}\frac{\partial h}{\partial x} = \frac{(10^{-8}\,\text{cm/s})}{(0.1)}\left(\frac{9}{13}\right) = 6.9\times10^{-8}\,\text{m/s} = 2.18\,\text{m/yr}$$

The travel time for waste to reach the sand aquifer is:

$$t = \frac{L}{v_z} = \frac{(13\,\text{m})}{(2.18\,\text{m/yr})} = 6\,\text{yr}$$

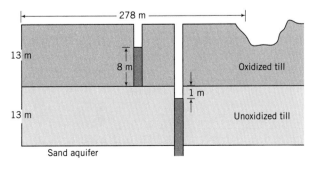

Figure 3.17 A diagram showing an aquifer system where waste will be placed (from Domenico and Schwartz, 1998. *Physical and chemical hydrogeology*). Copyright ©1998 by John Wiley & Sons, Inc. Reprinted by permission of John Wiley & Sons, Inc.

▶ EXERCISES

3.1. With reference to Figure 3.15, calculate the average horizontal and vertical hydraulic gradients in the two hydrogeological units (the upper unit having a lower hydraulic conductivity than the lower unit). Explain why the vertical hydraulic gradient is higher in the lower-conductivity unit than the horizontal hydraulic gradient in the higher conductivity unit using Darcy's law.

3.2. Discuss the similarity and difference of hydraulic conductivity and intrinsic permeability. What is the fundamental difference between them?

3.3. In an experiment, ground water flows through a sediment column (43.5 cm in length and 5 cm in diameter) at a rate of 4.4 ml/min. The porosity of the sediment is 0.4. Calculate the Darcy velocity and linear ground-water velocity. If the hydraulic conductivity is 0.11 cm/s, calculate the hydraulic gradient.

3.4. Figure 3.18 shows a grain-size analysis result. Estimate the hydraulic conductivity and intrinsic permeability using the equations in Table 3.5.

3.5. In a constant-head test, the Darcy flux is 5 cm^3/s for a column of 5 cm in diameter and 50 cm in length. The constant head is 60 cm. Calculate the hydraulic conductivity.

3.6. In a falling-head test, the initial head at $t = 0$ is 60 cm. At $t = 30$ min, the head is 57 cm. The diameters of the standpipe and the specimen are 1 cm and 20 cm, respectively. The length of the specimen is 20 cm. Calculate the hydraulic conductivity and intrinsic permeability of the specimen.

3.7. A hydrological system consists of five horizontal formations. The hydraulic conductivities of the formations are 20, 10, 15, 50, and 1 m/day, respectively. Calculate equivalent horizontal and vertical

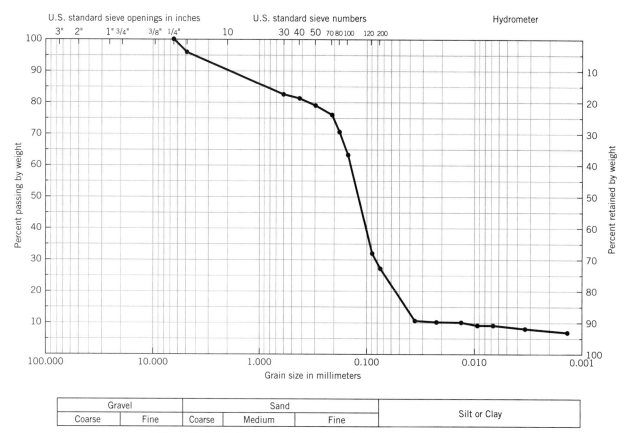

Figure 3.18 A diagram showing the results of a grain-size analysis.

TABLE 3.8   Measured hydraulic conductivities in a formation (cm/s)

4.17×10^{-7}	7.59×10^{-3}	2.63×10^{-7}	1.78×10^{-8}	2.29×10^{-6}
1.00×10^{-4}	5.25×10^{-4}	1.62×10^{-6}	6.17×10^{-10}	2.04×10^{-7}
3.16×10^{-16}	3.16×10^{-13}	3.31×10^{-6}	2.09×10^{-2}	2.42×10^{-1}
6.03×10^{-9}	5.50×10^{-9}	8.71×10^{-7}	5.01×10^{-12}	5.01×10^{-5}
4.57×10^{-8}	2.14×10^{-5}	4.90×10^{-7}	8.32×10^{-8}	7.41×10^{-8}
4.27×10^{-10}	1.12×10^{-3}	2.24×10^{-3}	1.02×10^{-6}	1.91×10^{-8}
1.74×10^{-5}	2.40×10^{-4}	2.82×10^{-5}	2.51×10^{-9}	1.74×10^{-9}
1.26×10^{-11}	1.26×10^{-9}	5.01×10^{-11}	2.14×10^{-10}	9.12×10^{-4}
2.69×10^{-8}	4.79×10^{-6}	1.00×10^{-5}	3.98×10^{-11}	9.77×10^{-9}
1.26×10^{-4}	1.26×10^{-7}	7.41×10^{-1}	1.02×10^{-10}	6.17×10^{-6}

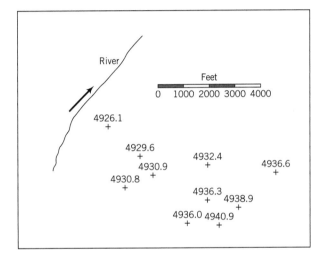

Figure 3.19 Measured hydraulic heads for a confined aquifer.

hydraulic conductivities. If the flow in the uppermost layer is at an angle of 30 away from the normal direction relative to the boundary, calculate flow directions in all of the formations.

3.8. Table 3.8 lists measured hydraulic conductivity values for a formation. (a) Calculate the arithmetic, geometric, and harmonic means of K; (b) Construct histograms of K and log(K) and determine the probability distribution of the sample. If the distribution is normal, or log-normal, calculate the standard deviation of the sample.

3.9. Figure 3.19 is a diagram showing measured heads for a confined aquifer. Hydraulic tests determined that the transmissivity of the aquifer is 2000 m/day. Draw contour lines of measured heads, and calculate the average hydraulic gradient in the aquifer and water discharge rate from the aquifer to the stream

▶ CHAPTER
HIGHLIGHTS

1. Natural geological materials contain voids or pores. The porosity is a measure of the volume of the voids to the total volume of a sample (V_v/V_T). Sands and gravels have porosities close to 40%. Rocks typically have smaller porosities, in some cases approaching zero.

2. In some media like fractured shale, flow only occurs through the secondary or fracture porosity. The term *effective porosity* describes specifically the porosity along the major pathways for water flow. In sands and gravels, the effective porosity will be much the same as the total porosity. In fractured shale, the effective porosity is often small—1×10^{-3} (or even smaller)—because the volume of the fractures is small relative to the total volume of any sample.

3. The fact that water moves from place to place implies that there is a gradient in the driving force that makes this movement possible. In ground-water systems, the energy for flow is given the name *hydraulic head*. It is measured simply by sticking a hollow standpipe in the ground (called a piezometer) below the water table and recording the elevation to which water rises. By convention, sea level is the datum for such measurements.

4. The Darcy equation is one of the most important equations that hydrologists use to study ground water. This equation is written as

$$\frac{Q}{A} = Ki$$

where K is a constant of proportionality, termed the *hydraulic conductivity*, and i is the hydraulic gradient. The hydraulic gradient measures the change in hydraulic head (h) as a function of distance. Darcy's equation explains that the quantity of flow through a porous medium is proportional to the hydraulic conductivity and the hydraulic gradient.

5. The hydraulic-conductivity value in the Darcy equation has important physical significance. It is a measure of the ease with which water can pass through a geologic material. Permeable materials like sand and gravel have relatively high hydraulic-conductivity values. Clay and shale have relatively low values. Hydraulic conductivity has the same units as velocity, such as m/sec or ft/day.

6. A variety of laboratory and field tests, as well as empirical approaches, are available to estimate hydraulic conductivity values. This chapter provides equations to estimate hydraulic conductivity from characteristics of a grain-size distribution. It also discusses both constant-head and falling-head permeameters. These devices permit estimation of hydraulic conductivity by measuring hydraulic heads and the rates of water flow through a rock core or an unlithified sediment sample. Details of the field approaches will be presented in later chapters.

7. If the permeability or hydraulic conductivity at a point is the same in all directions, the material is considered to be isotropic at that point. If the medium exhibits directional dependencies in hydraulic conductivity, the medium is considered to be anisotropic at that point.

8. A porous medium is homogeneous if the permeability or hydraulic conductivity in a given direction is the same from point to point. Where permeability changes from point to point, the medium is heterogeneous.

9. In a ground-water system, one maps flow by defining how the pattern of hydraulic head varies as a function of space. Normal practice is to take measured hydraulic-head data, plot them on a two-dimensional section, contour the hydraulic heads to define equipotential lines, and draw flow lines. A hydrogeological cross section is a vertical section oriented parallel to the direction of mean ground-water flow. A potentiometric surface map is a map of hydraulic heads measured in an aquifer.

4

GEOLOGY AND GROUND WATER

Life as a hydrogeologist would be downright boring if field settings looked anything like the soil-filled pipes or simple media we have considered so far. The reality is that the complexities of the geology are magnificently manifested in the hydrogeological world that we are starting to explore. The geologic setting provides a context for hydrogeological investigations. This chapter will link elements of geology and hydrogeology.

The discussion begins with aquifers and confining beds, which are manifestations of the geological setting. Next, we examine how key hydrologic variables, like hydraulic conductivity and porosity, are influenced by the geology and geologic processes.

▶ 4.1 AQUIFERS AND CONFINING BEDS

From a resource perspective, the primary unit in ground-water investigations is the *aquifer*, a lithologic unit or combination of lithologic units capable of yielding water to pumped wells or springs (Domenico, 1972). An aquifer can be co-extensive with geologic formations, a group of formations, or part of a formation. It may cut across formations in a way that makes it independent of any geologic unit. Units of low permeability that bound an aquifer are called *confining beds*.

Linking the definition of an aquifer to features of water supply can create confusion. In areas with prolific aquifers, a low-permeability unit might be considered a confining bed. However, in ground-water-poor regions, the same deposit could be considered an aquifer. In actual field studies, this ambiguity in the definition of an aquifer turns out not to be much of a problem because hydraulic conductivity or porosity values explicitly define the hydraulic character of the unit.

Aquifers and confining beds come in flavors. The terms *water table* or *unconfined* are applied to aquifers where the water table forms the upper boundary (Figure 4.1). When shallow wells or piezometers are installed into such an aquifer, the water levels in these wells approximately define the position of the water table.

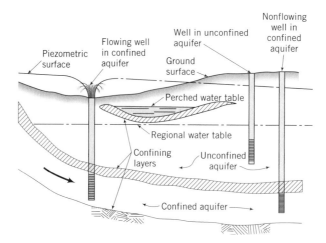

Figure 4.1 Conceptual model of aquifers developed in a field setting (modified from Bureau of Reclamation, 1995).

A *confined* (or *artesian*) *aquifer* has its upper and lower boundaries marked by confining beds (Figure 4.1). Stated another way, an aquifer is confined by overlying and underlying low-permeability beds. The water level of a well or piezometer installed in a confined aquifer occurs somewhere above its upper boundary. Occasionally, the water level of a well occurs above the ground surface. This condition can produce a flowing artesian well (Figure 4.1). As noted in Section 3.8, a contoured map of hydraulic heads for a large number of wells installed in the same aquifer is a potentiometric surface.

A *perched aquifer* is an unconfined aquifer that develops above the regional water table. In effect, there is an unsaturated zone below the low-hydraulic conductivity layer on which the perched zone develops.

Occasionally, the terms *aquifuge, aquitard,* and *aquiclude* are applied to various types of confining beds. The use of these terms has fallen out of favor, but sometimes readers might encounter them. An *aquifuge* is the ultimate low-hydraulic conductivity unit, which is a poor conductor of ground water and is essentially impermeable. An *aquitard* is a low-permeability unit that is capable of storing water and transmitting water between adjacent aquifers. This stored and transmitted water is available to wells being pumped in nearby aquifers. The term *aquiclude* is essentially a synonym for confining bed.

CASE STUDY 4-1

AQUIFERS OF LONG ISLAND, NEW YORK

Long Island, located on the East Coast of the United States (Figure 4.2), has important ground-water resources. A sequence of Cretaceous aquifers (Lloyd Aquifer, Magothy Aquifer) and confining beds (Raritan Confining unit) dip from north to south. Above the Cretaceous deposits is a series of marine clays and clayey sands (Gardiners Clay and Monmouth Greensand) that act as a confining unit above the Magothy Aquifer. The cross section (Figure 4.2) shows that the Lloyd Aquifer and some of the Magothy Aquifer can be classified as confined aquifers, given the presence of confining units above and below.

The uppermost aquifer consists of a thick sequence of outwash deposits, related to the most recent glaciation. These deposits consist mainly of stratified sand and gravel with little or no clay and silt. The water table is found at shallow depth in this unit, which makes the uppermost aquifer unconfined. Later in the book, we will revisit the unconfined glacial aquifer on Long Island because of its susceptibility to ground-water contamination.

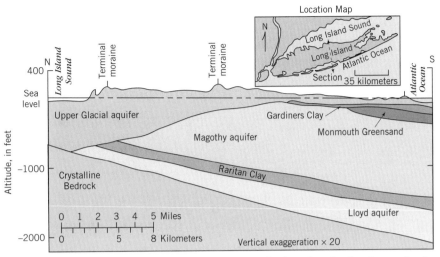

Figure 4.2 Generalized cross section showing the distribution of confined and unconfined aquifers, and confining beds on Long Island, New York (insert) (from Wexler, 1988).

▶ 4.2 TRANSMISSIVE AND STORAGE PROPERTIES OF AQUIFERS

Aquifers play a key role in supplying water to wells. When a pump is turned on in a well, the water level in the well casing (and the hydraulic head) is reduced, causing ground water to flow from the aquifer into the well. Of the water that is pumped from the well, much of it initially comes from "storage" in the aquifer. Thus, aquifers have at least two important characteristics—some ability to store ground water and to transmit this water to a nearby well. These properties depend to an important extent on the geologic setting.

Transmissivity

The term *transmissivity* describes the ease with which water can move through an aquifer. More explicitly, it is the rate at which water of prevailing kinematic viscosity is transmitted through a unit width of the aquifer under a unit hydraulic gradient (Figure 4.3). The concept of transmissivity is similar to hydraulic conductivity. The main difference is that transmissivity is a measurement that applies across the vertical thickness of an aquifer. If the thickness of the aquifer is b, the transmissivity (T) is

$$T = bK \tag{4.1}$$

where K is the hydraulic conductivity of the aquifer. Transmissivity has units of $[L^2/T]$ (for example, ft^2/day, m^2/day). In English engineering units, transmissivity has units of gpd/ft. The following set of factors provides a basis for unit conversions

$$1\,\text{m}^2/\text{day} = 10.76\,\text{ft}^2/\text{day} = 80.52\ \text{gpd/ft} \tag{4.2}$$

$$1\,\text{ft}^2/\text{day} = 0.0929\,\text{m}^2/\text{day} = 7.48\ \text{gpd/ft} \tag{4.3}$$

$$1\ \text{gpd/ft} = 0.01242\,\text{m}^2/\text{day} = 0.1337\,\text{ft}^2/\text{day} \tag{4.4}$$

Not surprisingly, one can develop a form of the Darcy equation that applies to an aquifer. This equation, which applies to a homogeneous confined aquifer (Figure 4.4),

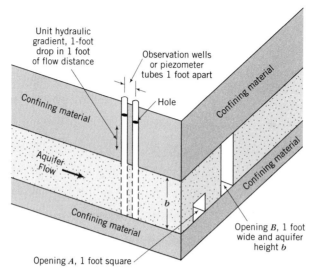

Figure 4.3 Diagram illustrating hydraulic conductivity and transmissivity in an aquifer (from Ferris et al., 1962).

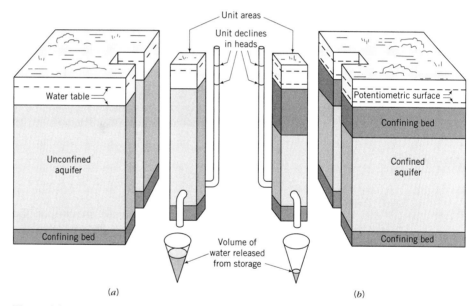

Figure 4.4 Diagrams illustrating the concept of storativity in (a) an unconfined aquifer and (b) a confined aquifer (from Heath, 1982).

is written as

$$Q = -WT\frac{dh}{dl} \qquad (4.5)$$

where W is the width of the aquifer [L], T is the transmissivity of the aquifer [L^2/T], and Q is discharge rate [L^3/T].

▶ **EXAMPLE 4.1**   The hydraulic conductivity of a confined aquifer with a thickness of 10 ft is 374 gpd/ft^2. Calculate the transmissivity of the aquifer in gpd/ft, ft^2/day, and m^2/day.

SOLUTION

$$T = K b = (374 \text{ gpd/ft}^2)(10 \text{ ft}) = 3740 \text{ gpd/ft}$$

Conversion to ft^2/day and m^2/day

$$T = 46.45 \text{ m}^2/\text{day} = 500.0 \text{ ft}^2/\text{day}$$

Storativity (or Coefficient of Storage) and Specific Storage

Aquifers have the ability to store water. How this storage is accomplished differs depending on whether the aquifer is confined or unconfined. When a well is pumped in a confined aquifer, the declining hydraulic head in the vicinity of the well enables the pressurized water to expand slightly, adding a small volume of additional water. In addition, the decline in hydraulic head lets the aquifer collapse slightly, thereby compensating for the volume of water that flows to the well.

In an unconfined aquifer, the main source of water is the drainage of water from pores as the water table declines in response to pumping. For a comparable unit decline in hydraulic head, an unconfined aquifer releases much more water from storage than a confined aquifer (Figure 4.4).

The *storativity* of an aquifer is defined as the volume of water that an aquifer releases from or takes into storage per unit surface area of the aquifer per unit change in head (Figure 4.4).

$$S = \frac{\text{volume of water}}{(\text{unit area})(\text{unit head change})} = \frac{\text{m}^3}{\text{m}^3} \tag{4.6}$$

where S is storativity (dimensionless). In a confined aquifer, values of storativity range from 10^{-3} to 10^{-5}. A related measure of the water stored in an aquifer is specific storage. *Specific storage* is defined as the volume of water that an aquifer releases from or takes into storage per unit surface area of the aquifer per unit aquifer thickness per unit change in head.

$$S_s = \frac{\text{volume for water}}{(\text{unit area})(\text{unit aquifer thickness})(\text{unit head change})} = \frac{1}{m} \tag{4.7}$$

where S_s is the specific storage of an aquifer [1/L]. Equation (4.7) indicates that the unit of specific storage in the SI unit system is 1/m. Specific storage is related to storativity by

$$S = S_s b \tag{4.8}$$

where b is the thickness of an aquifer.

Storage in Confined Aquifers

For a confined aquifer, the mathematical definition of specific storage reflects the storage coming from compression of the granular matrix and the expansion of water. The specific storage in a confined aquifer (see Domenico and Schwartz, 1998) is

$$S_s = \rho_w \, g(\beta_p + n \, \beta_w) \tag{4.9}$$

where ρ_w is the density of water [ML^{-3}], g is gravitational constant (9.81 m/s^2) [LT^{-2}], n is the porosity of the aquifer, β_p is the vertical compressibility of rock matrix, and β_w

is the compressibility of water. The unit for compressibility is the inverse of pressure. The compressibility of ground water (β_w) is 4.8×10^{-10} m²/N (or 2.3×10^{-8} ft²/lb) at 25°C. Table 4.1 lists the compressibility of some common geological materials. To use the compressibility, it is important to review the units of force.

$$1 \text{ Newton(N)} = 1 \text{ kg} \cdot \text{m/s}^2 \tag{4.10}$$

$$1 \text{ kg of force} = 9.80665 \text{ N} = 2.02046 \text{ lb of force} \tag{4.11}$$

Equation (4.9) can be used to estimate the range of specific storage and storativity of some aquifer types.

▶ **EXAMPLE 4.2**

A confined aquifer is composed of dense, sandy gravel with a thickness of 100 m and a porosity of 20%. Estimate the likely range for specific storage and storativity. For a total head drop of 100 m in an area of 1×10^9 m², how much water is released from the storage?

SOLUTION From Table 4.1, the compressibility of dense, sandy gravel ranges from 5.2×10^{-9} to 1.0×10^{-8} m²/N. To calculate the specific storage, it is computationally convenient to replace unit N with kg · m/s².

The specific storage due to the compressibility of water is

$$S_s^w = \rho_w g n \beta_w = (1000 \text{ kg/m}^3)(9.81 \text{ m/s}^2)(0.2)\left(4.8 \times 10^{-10} \frac{\text{m}^2}{\text{kg} \cdot \text{m/s}^2}\right)$$

$$= 9.4 \times 10^{-7} \frac{1}{\text{m}}$$

The specific storage due to the compressibility of granular matrix is

$$S_s^M = \rho_w g \beta_p = (1000 \text{ kg/m}^3)(9.81 \text{ m/s}^2)[(0.52 \sim 1.0) \times 10^{-8} \frac{\text{m}^2}{\text{kg} \cdot \text{m/s}^2}]$$

$$= (5.1 \sim 9.81) \times 10^{-5} \frac{1}{\text{m}}$$

The specific storage for the aquifer combines these

$$S_s = \rho_w g(\beta_p + n\beta_w) = 9.4 \times 10^{-7} + (5.1 \sim 9.81) \times 10^{-5} = (5.2 \sim 9.9) \times 10^{-5} \frac{1}{\text{m}}$$

TABLE 4.1 Vertical compressibility

Material	Coefficient of Vertical Compressibility		
	ft²/lb	m²/N	bars⁻¹
Plastic clay	$1 \times 10^{-4} - 1.25 \times 10^{-5}$	$2 \times 10^{-6} - 2.6 \times 10^{-7}$	$2.12 \times 10^{-1} - 2.65 \times 10^{-2}$
Stiff clay	$1.25 \times 10^{-5} - 6.25 \times 10^{-6}$	$2.6 \times 10^{-7} - 1.3 \times 10^{-7}$	$2.65 \times 10^{-2} - 1.29 \times 10^{-2}$
Medium-hard clay	$6.25 \times 10^{-6} - 3.3 \times 10^{-6}$	$1.3 \times 10^{-7} - 6.9 \times 10^{-8}$	$1.29 \times 10^{-2} - 7.05 \times 10^{-3}$
Loose sand	$5 \times 10^{-6} - 2.5 \times 10^{-6}$	$1 \times 10^{-7} - 5.2 \times 10^{-8}$	$1.06 \times 10^{-2} - 5.3 \times 10^{-3}$
Dense sand	$1 \times 10^{-6} - 6.25 \times 10^{-7}$	$2 \times 10^{-8} - 1.3 \times 10^{-8}$	$2.12 \times 10^{-3} - 1.32 \times 10^{-3}$
Dense, sandy gravel	$5 \times 10^{-7} - 2.5 \times 10^{-7}$	$1 \times 10^{-8} - 5.2 \times 10^{-9}$	$1.06 \times 10^{-3} - 5.3 \times 10^{-4}$
Rock, fissured	$3.3 \times 10^{-7} - 1.6 \times 10^{-8}$	$6.9 \times 10^{-10} - 3.3 \times 10^{-10}$	$7.05 \times 10^{-4} - 3.24 \times 10^{-5}$
Rock, sound	less than 1.6×10^{-8}	less than 3.3×10^{-10}	less than 3.24×10^{-5}
Water at 25° C	2.3×10^{-8}	4.8×10^{-10}	5×10^{-5}

Source: Modified from Domenico and Mifflin (1965), *Water resources res.* 4. p. 563–576. Copyright by American Geophysical Union.

The storativity due to the compressibility of water is

$$S^W = b S_s^W = (100\,\text{m}) \left(9.4 \times 10^{-7} \frac{1}{\text{m}} \right) = 9.4 \times 10^{-5}$$

The storativity due to the compressibility of the matrix is

$$S^M = b S_s^W = (100\,\text{m})(5.1 \sim 9.81) \times 10^{-5} \frac{1}{\text{m}} = (5.1 \sim 9.81) \times 10^{-3}$$

The overall storativity of the aquifer is

$$S = S^M + S^W = 9.4 \times 10^{-5} + (5.1 \sim 9.8) \times 10^{-3} = (5.2 \sim 9.9) \times 10^{-3}$$

The volume of water, which is withdrawn from the storage due to a drop in hydraulic head of 100 m in an area of 10^9 m^2, is

$$V = S \, \Delta h \, A = (5.2 \sim 9.9) \times 10^{-3} (100\,\text{m})(1 \times 10^9\,\text{m}^2) = (5.2 \sim 9.9) \times 10^8\,\text{m}^3$$

In this example, most of the water comes from the compression of the matrix. In some areas of California and Texas, overpumping of ground water leads to land subsidence.

Storage in Unconfined Aquifers

In an unconfined aquifer, the ground-water response to pumping is different from a that in a confined aquifer. At an early time, when there is no significant change of water level, water comes from expansion of the water and compression of the grains. Later on, water comes mainly from the gravity drainage of pores in the aquifer through which the water table is falling. The storativity of an unconfined aquifer is expressed as

$$S = S_y + b S_s \tag{4.12}$$

where S_y is the specific yield of the aquifer. The specific yield ranges from 0.1 to 0.3, while the product of aquifer thickness and specific storage is in the range of 10^{-3} to 10^{-5}. Thus, specific yield is the storage term for an unconfined aquifer.

In some cases, an aquifer may be confined at an early stage of pumping, only to become unconfined at a late time. Water levels that initially started out above the aquifer end up falling below the top of the aquifer as it dewaters. As the aquifer changes from a confined to an unconfined aquifer, storativity values change accordingly.

Specific Yield and Specific Retention

Specific yield is the water released from a water-bearing material by gravity drainage. The specific yield is expressed as the ratio of the volume of water yielded from soil or rock by gravity drainage, after being saturated, to the total volume of the soil or rock (Meinzer, 1923).

$$S_y = \frac{V_d}{V_T} \tag{4.13}$$

where S_y is the specific yield and V_d is the volume of water that drains from a total volume of V_T.

Not all of the water initially present in the rock or sediment is released from storage. The term *specific retention* describes the water that is retained as a film on the surface of grains or held in small openings by molecular attraction. The specific retention is expressed as the ratio of volume of water that is retained, after being saturated, to the total volume of the soil or rock (Meinzer, 1923).

$$S_r = \frac{V_r}{V_T} \qquad (4.14)$$

where S_r is the specific retention and V_r is the volume water retained against gravity. The porosity defined in Section 3.1 is related to specific yield and specific retention by

$$n = S_y + S_r \qquad (4.15)$$

That is, the sum of specific yield and specific retention equals porosity. The specific retention increases with decrease of grain size and pore size of a soil or rock (Table 4.2).

▶ **EXAMPLE 4.3**   After a soil sample is drained by gravity, the weight of the soil sample is 85 g. After the sample is oven-dried, the sample weighs 80 g. The bulk density of the wet soil is 1.65 g/cm^3, and the density of water is 1 g/cm^3. Calculate the specific yield, specific retention, and porosity of the sample. Assume water that was drained by gravity is 20 g.

SOLUTION   The total volume of the sample is

$$V_T = \frac{(85\,\text{g} + 20\,\text{g})}{(1.65\,\text{g/cm}^3)} = 63.6\,\text{g/cm}^3$$

The volume of water retained in the sample after it was drained by gravity is

$$V_r = \frac{(85 - 80)\text{g}}{(1\,\text{g/cm}^3)} = 5\,\text{cm}^3$$

The volume of water that was drained by gravity is

$$V_d = \frac{(20\text{g})}{(1\,\text{g/cm}^3)} = 20\,\text{cm}^3$$

The specific retention is

$$S_r = \frac{V_r}{V_T} = \frac{(5\,\text{cm}^3)}{(63.636\,\text{cm}^3)} = 0.079 = 7.9\%$$

TABLE 4.2   Selected values of porosity, specific yield, and specific retention (values in percent by volume)

Material	Porosity	Specific Yield	Specific Retention
Soil	55	40	15
Clay	50	2	48
Sand	25	22	3
Gravel	20	19	1
Limestone	20	18	2
Sandstone (semiconsolidated)	11	6	5
Granite	0.1	0.09	0.01
Basalt (young)	11	8	3

Source: From Heath (1989).

while the specific yield is

$$S_y = \frac{V_d}{V_T} = \frac{(20\,\text{cm}^3)}{(63.636\,\text{cm}^3)} = 0.314 = 31.4\%$$

The porosity of the sample is now calculated as

$$n = S_y + S_r = 7.9\% + 31.4\% = 39.3\%$$

▶ 4.3 GEOLOGY AND HYDRAULIC PROPERTIES

The hydraulic properties of geological materials are closely related to sediment or rock type, and the historical fingerprint is impressed on the materials by geological events. For example, the hydraulic conductivity of sedimentary rocks depends mainly on the porosity of the original sediments. Subsequent erosion, tectonic, and other geological activities often cause fractures that can increase their hydraulic conductivity. This section discusses how geology determines the development of key aquifers. Following the lead of the U.S Geological Survey (Figure 4.5), we will examine aquifers related to five types of materials: unconsolidated sediments, semi-unconsolidated sediments, carbonate rocks, sandstone rocks, and volcanic and other crystalline rocks. To learn more about aquifers in the United States, visit http://capp.water.usgs.gov/gwa, which is the site for the Ground Water Atlas of the United States.

Aquifers in Unconsolidated Sediments

Various types of aquifers are composed of unconsolidated sediments. Examples include blanket sand and gravel aquifers, basin-fill aquifers, and glacial-deposit aquifers. *Basin-fill aquifers* consist of sand and gravel deposits that partly fill depressions that were formed by faulting or erosion, or both. These aquifers are also commonly called valley-fill aquifers because the basins that they occupy are topographic valleys. *Blanket sand* and *gravel aquifers* consist mostly of medium to coarse sand and gravel. These aquifers mostly contain water under unconfined, or water-table, conditions but, locally, confined conditions exist where the aquifers contain beds of low-permeability silt, clay, or marl. The majority of blanket sand and gravel aquifers form from alluvial deposits. Occasionally, some of these aquifers, such as the High Plains aquifer, contain windblown sand, whereas others, such as the surficial aquifer system of the southeastern United States, form as a complex assemblage of alluvium, beach deposits, and shallow marine sands.

Glacial-deposit aquifers are sediments that were deposited during cycles of continental glaciation across the northern United States and Canada. Glacial sediments incorporated in the ice were redistributed as ice-contact or meltwater deposits, or both, during retreats. Ice-laid or meltwater deposits are often called *glacial drift. Till* is an unsorted and unstratified material that ranges in size from boulders to clay, which is deposited directly by the ice. *Outwash*, which is mostly stratified sand and gravel, and glacial-lake deposits consisting mostly of stratified clay, silt, and fine sand are related to meltwater processes. In the Midwest, outwash is commonly deposited along river valleys cut into bedrock. *Ice-contact deposits* consisting of local bodies of sand and gravel were deposited at the face of the ice sheet or in cracks in the ice.

The hydraulic conductivity of unconsolidated sediments depends on the grain size, mineral composition, and sorting (Table 4.3). For example, the hydraulic conductivity of

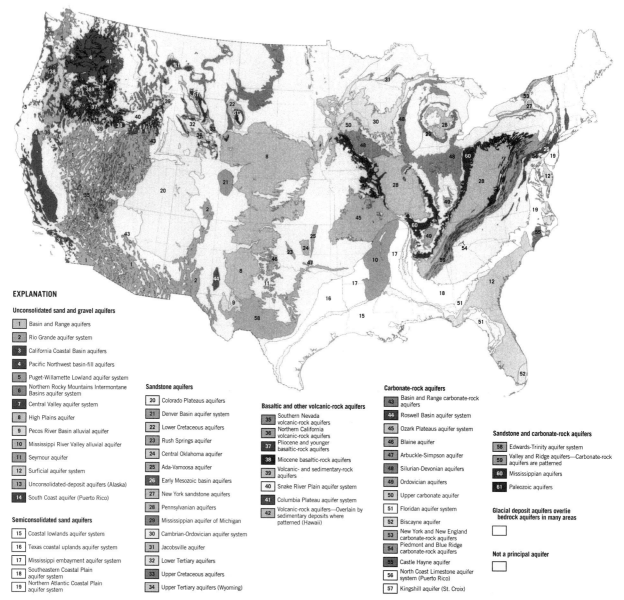

EXPLANATION

Unconsolidated sand and gravel aquifers

1	Basin and Range aquifers
2	Rio Grande aquifer system
3	California Coastal Basin aquifers
4	Pacific Northwest basin-fill aquifers
5	Puget-Willamette Lowland aquifer system
6	Northern Rocky Mountains Intermontane Basins aquifer system
7	Central Valley aquifer system
8	High Plains aquifer
9	Pecos River Basin alluvial aquifer
10	Mississippi River Valley alluvial aquifer
11	Seymour aquifer
12	Surficial aquifer system
13	Unconsolidated-deposit aquifers (Alaska)
14	South Coast aquifer (Puerto Rico)

Semiconsolidated sand aquifers

15	Coastal lowlands aquifer system
16	Texas coastal uplands aquifer system
17	Mississippi embayment aquifer system
18	Southeastern Coastal Plain aquifer system
19	Northern Atlantic Coastal Plain aquifer system

Sandstone aquifers

20	Colorado Plateaus aquifers
21	Denver Basin aquifer system
22	Lower Cretaceous aquifers
23	Rush Springs aquifer
24	Central Oklahoma aquifer
25	Ada-Vamoosa aquifer
26	Early Mesozoic basin aquifers
27	New York sandstone aquifers
28	Pennsylvanian aquifers
29	Mississippian aquifer of Michigan
30	Cambrian-Ordovician aquifer system
31	Jacobsville aquifer
32	Lower Cretaceous aquifers
33	Upper Cretaceous aquifers
34	Upper Tertiary aquifers (Wyoming)

Basaltic and other volcanic-rock aquifers

35	Southern Nevada volcanic-rock aquifers
36	Northern California volcanic-rock aquifers
37	Pliocene and younger basaltic-rock aquifers
38	Miocene basaltic-rock aquifers
39	Volcanic- and sedimentary-rock aquifers
40	Snake River Plain aquifer system
41	Columbia Plateau aquifer system
42	Volcanic-rock aquifers—Overlain by sedimentary deposits where patterned (Hawaii)

Carbonate-rock aquifers

43	Basin and Range carbonate-rock aquifers
44	Roswell Basin aquifer system
45	Ozark Plateaus aquifer system
46	Blaine aquifer
47	Arbuckle-Simpson aquifer
48	Silurian-Devonian aquifers
49	Ordovician aquifers
50	Upper carbonate aquifer
51	Floridan aquifer system
52	Biscayne aquifer
53	New York and New England carbonate-rock aquifers
54	Piedmont and Blue Ridge carbonate-rock aquifers
55	Castle Hayne aquifer
56	North Coast Limestone aquifer system (Puerto Rico)
57	Kingshill aquifer (St. Croix)

Sandstone and carbonate-rock aquifers

58	Edwards-Trinity aquifer system
59	Valley and Ridge aquifers—Carbonate-rock aquifers are patterned
60	Mississippian aquifers
61	Paleozoic aquifers

Glacial deposit aquifers overlie bedrock aquifers in many areas

Not a principal aquifer

Figure 4.5 Principal aquifers of the United States are in six types of rocks and deposits. The colored areas show the extent of each principal aquifer at or near the land surface (from Miller, 1990).

clay is less than 3×10^{-4} m/day, whereas the hydraulic conductivity of coarse gravel is over 100 m/day. Well-sorted sediments are more permeable than poorly sorted materials. With glacial deposits, hydraulic-conductivity values vary from 10^{-6} m/day for till to 100 m/day for outwash (Table 4.4; Stephenson and others, 1988). Table 4.4 also indicates that weathering and fracturing will increase the hydraulic conductivity of the fine-grained deposits.

In the United States, most of the highest yield aquifers are deposits of unconsolidated materials. These aquifers (Figure 4.5) include the Basin and Range aquifers in southwestern United States, Rio Grande aquifer system, California Coastal Basin aquifers, Pacific

TABLE 4.3   Hydraulic conductivity of unconsolidated materials

| | Hydraulic Conductivity (m/day) | | | | | |
| | Degree of Sorting | | | Silt Content | | |
Grain-Size Class	Poor	Moderate	Well	Slight	Moderate	High
1. Fine-grained Materials						
Clay			0.0003			
Silt, clayey			0.3–1.2			
Silt, slightly sandy			1.5			
Silt, moderately sandy			2.1–2.5			
Silt, very sandy			2.7–3.5			
Sandy silt			3.4			
Silty sand			4			
2. Sands and Gravels						
Very fine sand	4	6	8	7	6	4
Very fine to fine sand	8	8	—	7	6	4
Very fine to medium sand	11	12–14	—	10	8	6
Very fine to coarse sand	15	—	—	12	9	7
Very fine to very coarse sand	18	—	—	16	12	9
Very fine sand to fine gravel	23	—	—	20	16	12
Very fine sand to medium gravel	30	—	—	24	20	15
Very fine sand to coarse gravel	39	—	—	33	26	20
Fine sand	8	12	16	10	8	6
Fine to medium sand	16	20	—	15	12	9
Fine to coarse sand	17	20 – 22	—	16	13	10
Fine to very coarse sand	21	—	—	18	14	11
Fine sand to fine gravel	27	—	—	23	18	13
Fine sand to medium gravel	35	—	—	29	23	17
Fine sand to coarse gravel	44	—	—	33	27	22
Medium sand	20	24	29	20	16	12
Medium to coarse sand	23	29	—	22	17	13
Medium to very coarse sand	26	30–34	—	22	19	15
Medium sand to fine gravel	31	—	—	26	21	16
Medium sand to medium gravel	40	—	—	35	25	20
Medium sand to coarse gravel	50	—	—	41	33	25
Coarse sand	24	33	41	29	23	16
Coarse to very coarse sand	29	41	—	29	23	17
Coarse sand to fine gravel	35	41–48	—	33	27	21
Coarse sand to medium gravel	45	—	—	35	29	23
Coarse sand to coarse gravel	56	—	—	41	30	28
Very coarse sand	33	45	57	35	29	23
Very coarse sand to fine gravel	41	65	—	37	32	27
Very coarse sand to medium gravel	52	61–69	—	45	37	30
Very coarse sand to coarse gravel	63	—	—	49	40	32
Fine gravel	49	65	81	69	43	33
Fine to medium gravel	61	102	—	61	51	41
Fine to coarse gravel	75	88–102	—	71	58	44
Medium gravel	73	70	122	73	61	49
Medium to coarse gravel	90	143	—	90	74	58
Coarse gravel	102	143	183	102	87	71

Source: Lappala (1978).

TABLE 4.4   Hydraulic conductivity values for glacial deposits

Lithology	Location	Hydraulic Conductivity (m/s)	Source
Till			
Compact	—	1.2×10^{-7}–1.2×10^{-11}	a
Compact	NE Appalachians	2×10^{-7}	b
Compact	S Saskatchewan	$1 \times 10^{-11} - 1 \times 10^{-12}$	c
	Saskatchewan	1.7×10^{-6}–4.5×10^{-9}	d
Nonweathered	S Alberta	9.1×10^{-10}	e
Weathered	S Alberta	1.9×10^{-8}	e
Weathered	—	1.2×10^{-6}–1.2×10^{-9}	a
Fractured	—	1.2×10^{-5}–1.2×10^{-9}	a
Lacustrine Silt and Clay			
Unweathered	—	1.2×10^{-9}–1.2×10^{13}	a
Unweathered	NE Appalachians	2×10^{-9}	b
	Manitoba	3.6×10^{-11}	d
Weathered	S Alberta	2.1×10^{-7} (hor.)	e
		3.2×10^{-8} (vert.)	
Loess			
Unweathered	—	1.2×10^{-5}–1.2×10^{-11}	a
Weathered	—	1.2×10^{-7}–1.2×10^{-10}	a
Outwash			
	—	1.2×10^{-3}–1.2×10^{-7}	a

[a] Stephenson et al. (1988).
[b] Randall et al. (1988).
[c] Hendry and Wassenaar (1999).
[d] Grisak and Jackson (1978).
[e] Stein (1987).

Northwest basin-fill aquifers, Puget-Willamette Lowland aquifer system, Northern Rocky Mountains Intermontane Basins aquifer system, Central Valley aquifer system in California, High Plains aquifer, Pecos River Basin alluvial aquifer in Texas and New Mexico, Mississippi River Valley alluvial aquifer, Seymour aquifer in Texas, Surficial aquifer system in eastern United States, Unconsolidated-deposit aquifers in Alaska, and South Coast aquifer in Puerto Rico.

Blanket Sand and Gravel Aquifers

With blanket sand and gravel aquifers, variability in the primary pore structure is related to spatial or temporal variability in the processes operating to form the sediment. This point can be demonstrated for clastic deposits (for example, sand, sandstone, and gravel aquifers, and shale-confining beds), which represent some of the most important aquifers in the world. Table 4.5 describes the general geometry deposits resulting from deposition in various types of sedimentary environments. In many countries, it is possible to identify aquifers and confining beds that at a large scale have the geometries described in the table.

We will examine several of the key environments listed in Table 4.5 to illustrate the inherent complexities of natural systems and the inherent difficulties in describing the variability. The terms *fluvial deposit* or *alluvial aquifer* describe deposits formed by the action of streams and rivers. As a general rule, these deposits take on the shape of the river valleys in which they form and produce aquifers that are long, narrow, and thin (Table 4.5). Within

TABLE 4.5 Geometry of clastic sedimentary rocks related to important environments of deposition

Environment of Deposition	Geometry of Resulting Hydrogeologic Units
	Marine
Shoreline (beach, bars, lagoons, and deltas)	Shale tends to be more or less continuous over areas of 10 s to 100 s km². Thicknesses are in meters or 10 s of meters. Sand beds more elongated unless in prograded sheet. Ancient bar and channel deposits, 100 s m wide, 1 to 10 s m thick, 10 s km long.
Shelf and slope	Sand and shale beds continuous over 100 s to 1000 s km², 10 s to few 100 s m thick.
Floor	Extensive shale over 1000 s to 10,000 s km², thicknesses 100 s m. Cyclic turbidite sands, individual beds, 10 s cm to few m thick. Lateral extent measured in km.
	Continental
Fluvial (floodplains, channel deposits, alluvial fans)	Less continuous than marine units. Large rivers produce sands a few km long, 10 s m to few km wide and few m to few 10 s m thick. Shales more continuous. Ancient alluvial fans produce thick, poorly sorted units with local well-sorted units, few m thick, 10 s m wide, and 100 s m long.
Aeolian (sand dunes, loess deposits)	Extensive sandstones up to a few 100 m thick and 1000 s km² area.
Lacustrine (shoreline, deltas, lake bottom)	Lake-bottom shale, extensive, few 1000 km² to 10,000 km², and up to a few 100 m thick. Deltaic and shoreline sand bodies are elongate and more local than shale.
Glacial (till plains, moraines, ice-contact deposits)	Tillite forms continuous sheets 10 s of m thick over 1000 s of km².

Source: From Hydrogeology, Davis (1988). Reproduced with permission of the publisher, the Geological Society of America, Boulder, Colorado USA. Copyright ©1988 by the Geological Society of America, Inc.

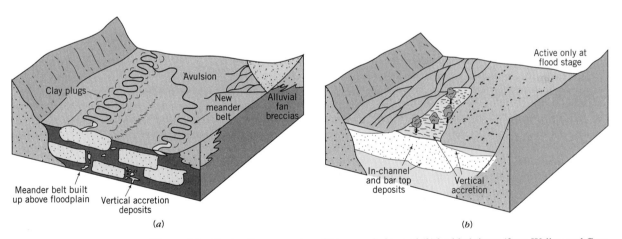

Figure 4.6 Contrasting geometry of (*a*) meandering and (*b*) braided rivers (from Walker and Cant, 1984, with permission of the Geological Association of Canada).

a fluvial deposit, there are marked spatial and temporal changes in grain size reflecting the complexities of the depositional environment.

Boggs (1987) recognizes three broad environmental settings for fluvial systems: braided rivers, meandering rivers, and alluvial fans. Figure 4.6 contrasts the geometry of meandering and braided streams. According to Cant (1982), meandering streams generally produce linear shoestring sand bodies that are aligned parallel to the river course, and these are normally bounded below and on both sides by finer materials. The shoestring sands are many times wider than they are thick. The braided river, on the other hand, frequently produces sheetlike sands that contain beds of clay enclosed within them. The reason postulated by Cant (1982) lies in the meander width, with braided streams capable of extensive lateral migration. Meandering streams are confined more rigorously in narrow channels. Braided rivers are characterized by many channels, separated by islands or bars. Meandering streams have a greater sinuosity and finer sediment load. Many river systems developed at the margins of glaciers (e.g., Ohio River) often start out as a braided stream and change to a meandering stream. Sharp (1988) and Rosenshein (1988) discuss alluvial aquifers and their hydraulic properties in detail.

Alluvial fans form at the base of mountains where erosion provides a supply of sediment (Figure 4.7). Fans can occur in both arid areas, such as Death Valley, California, and humid

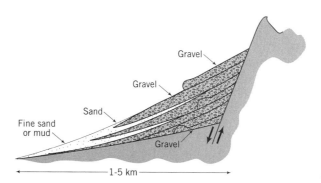

Figure 4.7 Diagrammatic cross section of an alluvial fan (from Rust and Koster, 1984, with permission of the Geological Association of Canada.)

CASE STUDY 4-2

HIGH PLAINS AQUIFER

The High Plains aquifer underlies an area of about 174,000 square miles in parts of Colorado, Kansas, Nebraska, New Mexico, Oklahoma, South Dakota, Texas, and Wyoming. It consists of all or parts of several geologic units of Quaternary and Tertiary age. Water in the High Plains aquifer generally is under unconfined, or water-table, conditions (Figure 4.8). Locally, water levels in wells completed in some parts of the aquifer may rise slightly above the regional water table because of artesian pressure created by local confining beds. The saturated thickness of the High Plains aquifer in 1980 ranged from 0 (where the sediments that compose the aquifer were unsaturated) to about 1000 feet. The entire High Plains aquifer contained about 21.7 billion acre-feet of saturated material in 1992. The quantity of drainable water in storage in the aquifer can be estimated by multiplying the volume of saturated material by the average specific yield (15 percent). Therefore, about 3.25 billion acre-feet of drainable water were in storage in the High Plains aquifer in 1992.

The Ogallala Formation of Miocene age is the principal geologic unit included in the High Plains aquifer and is found commonly at the land surface. The Ogallala Formation consists of unconsolidated gravel, sand, silt, and clay. Locally, it also includes *caliche*, which is a hard deposit of calcium carbonate that precipitated when part of the ground water that moved through the formation evaporated. The Ogallala Formation was deposited by an extensive eastward-flowing system of braided streams that drained the eastern slopes of the Rocky Mountains during late Tertiary time. This stream system migrated during a long period of time and deposited the Ogallala Formation over about 134,000 square miles in eastern Colorado, Kansas, Nebraska, New Mexico, Oklahoma, South Dakota, Texas, and Wyoming.

Miller and Appeal, 1997

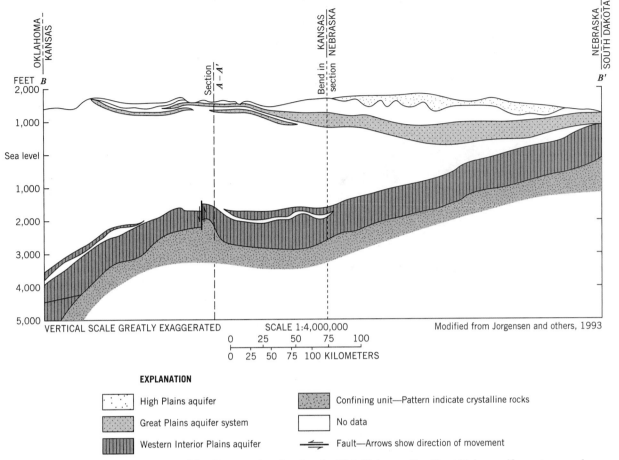

Figure 4.8 Cross section showing the High Plains aquifer, Great Plains aquifer systems, and confining units (from Jorgensen et al., 1993).

areas. The upper part of a fan is characterized by coarse sediment because flow is confined to one or a few channels. The toe of the fan has the finest sediment, where more than one channel persisted over long periods of time. Case Study 4-2 provides a description of the High Plains aquifer, which is a well-known blanket sand and gravel aquifer.

Basin-Fill Aquifers

Before the most recent period of tectonic activity in the southwestern United States, which began in the middle Miocene time (about 17 million years before present), the Basin and Range region had moderate relief. Streams were unable to transport large volumes of sediment. As the mountains were uplifted, stream gradients increased and the transporting power of the streams greatly increased. Steep, narrow canyons and gulches were incised into the sharp escarpments that bounded the mountain ranges, and enormous volumes of materials were eroded from mountains. The sediments eroded, transported, and deposited by streams form the basin-fill aquifers. Because regional aquifers are not continuous within the Basin and Range province, individual basins, which are encircled by topographic drainage divides, are classified as one of four types based on similar recharge–discharge relations (Figure 4.9). The simplest type is the *undrained, closed basin*, a single valley in which

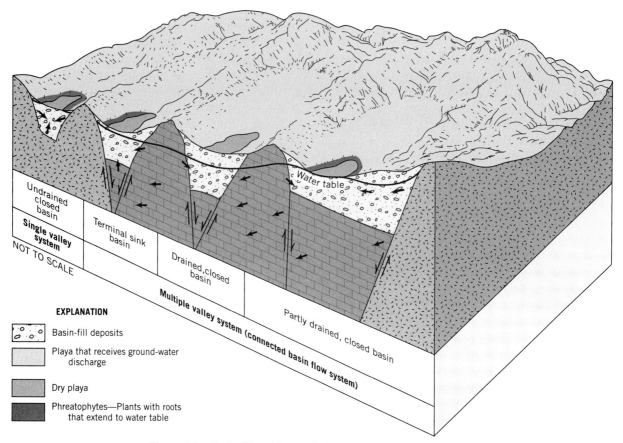

Figure 4.9 Basin-fill aquifers: undrained closed basin, terminal sink basin, drained, closed basin, and partly drained, closed basin (from Eakin et al., 1976).

the underlying and surrounding bedrock is practically impermeable and does not allow interbasin flow. All recharge is discharged at a sink, a playa near the center of the basin. Basins underlain by permeable bedrock commonly are hydraulically connected as multiple valley systems. The *partly drained, closed basin* is underlain or surrounded by bedrock that is moderately permeable and allows for some ground-water underflow. In this type of basin, some water is evaporated or transpired at the upgradient side of a playa, but most of the water continues to flow past the downgradient side of the playa and leave the basin. The *drained, closed basin* has a deep water table that prevents evapotranspiration. The bedrock is sufficiently permeable to allow all recharge to flow through it and out of the basin. The *terminal sink basin* is underlain or surrounded by bedrock that is sufficiently permeable to conduct flow into the basin, and the playa in the basin is the discharge point for recharge from several connected basins. The Great Basin aquifer (Case Study 4–3) is a classic example of basin-fill aquifers.

Glacial-Deposit Aquifers

The glacial environment is another setting that affords some opportunities for generalization. This setting is a composite one and incorporates fluvial, lacustrine, and eolian environments (Figure 4.10 and Table 4.5). The high permeability resides in the glaciofluvial depositional environments, including alluvial valleys, alluvial deposits in buried bedrock

CASE STUDY 4-3

GREAT BASIN AQUIFER

The ground-water flow systems of the Great Basin aquifer area are located in individual basins or in two or more hydraulically connected basins through which ground water flows to a terminal discharge point or sink. Except for relatively small areas that drain to the Colorado River, water is not discharged to major surface-water bodies but is lost solely through evapotranspiration. Each basin has essentially the same characteristics—the impermeable rocks of the mountain ranges serve as boundaries to the flow system, and the majority of the ground water flows through basin-fill deposits. In the area where carbonate rocks underlie the basins, substantial quantities of water can flow between basins through the carbonate rocks and into the basin-fill deposits, but this water also is ultimately discharged

by evapotranspiration. Most recharge to the basin-fill deposits originates in the mountains as snowmelt, and, where the mountain streams emerge from bedrock channels, the water infiltrates into the alluvial fans and replenishes the basin-fill aquifer. Intense thunderstorms may provide some direct recharge to the basin-fill deposits, but, in most cases, any rainfall that infiltrates the soil is either immediately evaporated or taken up as soil moisture; little water percolates downward through the unsaturated zone to reach the water table in the valleys. In mountain areas underlain by permeable carbonate rocks, most of the recharge may enter the carbonate rocks, and little water remains to supply runoff.

Planert and Williams, 1995

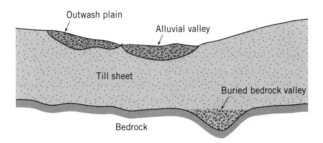

Figure 4.10 Glaciofluvial deposits in glaciated terrain (from Domenico and Schwartz, 1998. *Physical and chemical hydrogeology).* Copyright ©1998 by John Wiley & Sons, Inc. Reprinted by permission of John Wiley & Sons, Inc.

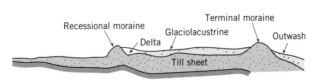

Figure 4.11 Deposits associated with a hypothetical advance and retreat of a glacier (from Domenico and Schwartz, 1998. *Physical and chemical hydrogeology).* Copyright ©1998 by John Wiley & Sons, Inc. Reprinted by permission of John Wiley & Sons, Inc.

valleys, and the well-sorted sand and gravel resulting from glacial melt water (outwash plains). *Buried bedrock valleys* are valleys that are no longer occupied by the streams that cut them. Glacial till is the most extensive deposit and consists of an assortment of grain sizes ranging from boulders to clay. Tills usually have a low hydraulic conductivity and are considered confining beds. Frequently, the till is fractured, leading to a higher permeability than was present originally.

The advance and retreat of continental glaciers generally produces a complex vertical stratigraphy. This complexity near an ice margin is illustrated in Figure 4.11. The basal

MAHOMET BEDROCK VALLEY AQUIFER

The Mahomet Valley aquifer (Figure 4.12) provides an interesting example of a fluvial aquifer system buried by subsequent glacial advances. This aquifer is related to the Teays River drainage system that extended from Illinois eastward to West Virginia. The Teays system was an ancestral version of the present-day Mississippi system that developed in late Tertiary time. This aquifer was buried as a consequence of Pleistocene glaciation.

The Mahomet sand is mainly outwash that was deposited in the bedrock valley system. In this part of Illinois, the Mahomet

Panno et al., 1994

sands range in thickness from 0 to 60 m. The mean thickness is about 30 m (Figure 4.12). Transmissivities of the Mahomet Valley aquifer range from 7×10^{-4} m²/s to 8×10^{-2} m²/s, with a mean hydraulic conductivity of 1.4×10^{-3} m/s. Figure 4.13 is a cross section along the bottom of the buried valley. Besides showing how the Mahomet sand occurs, it depicts other glacial sands developed within shallower glacial deposits. Overall, the pattern of occurrence of the Mahomet Valley aquifer and overlying aquifers is spatially complex—a point that is made repeatedly in this section.

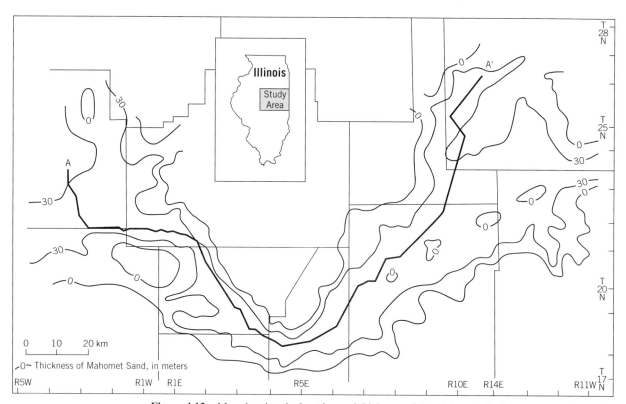

Figure 4.12 Map showing the location and thickness of the Mahomet Sand in east-central Illinois (modified from Panno et al., 1994). Reprinted by permission of *Ground Water*. Copyright ©1994. All rights reserved.

till sheet marks the advance of a glacier across a region. The terminal moraine marks the furthest ice advance, with subsequent melting forming glaciofluvial deposits in advance of the moraine. Lacustrine deposits are associated with a glacial lake that was trapped between the retreating glacier and the end moraine complex. The recessional moraine marks

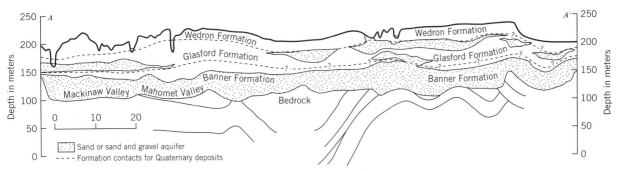

Figure 4.13 Longitudinal cross section A-A' showing the stratigraphy of a segment of the Mahomet Valley aquifer in east-central Illinois (modified from Panno et al., 1994). Reprinted by permission of *Ground Water*. Copyright ©1994. All rights reserved.

a position where the retreating glacier temporarily halted, with its meltwater forming the lake and associated delta deposits.

Aquifers in Semiconsolidated Sediments

This next family of aquifers is formed of semiconsolidated sand interbedded with silt, clay, and minor carbonate rocks. Porosity is intergranular, and the hydraulic conductivity of the aquifers is moderate to high. The aquifers of the Coastal Plains of the eastern and southern United States are an example. They are sediments of fluvial, deltaic, and shallow marine origin, which occur as a thick wedge of sediments that dip and thicken coastward. In places, these aquifers are more than 650 m thick. The varied depositional environments produce complex interbedding of fine- and coarse-grained materials. Accordingly, some aquifers are local in extent, whereas others extend over hundreds of square kilometers.

In topographically high-recharge areas, these aquifers are typically unconfined. Moving seaward, they become confined with upward leakage to shallower aquifers or discharge to saltwater bodies. Because flow is sluggish near the ends of long regional flow systems, the aquifers commonly contain unflushed saline water in their deeply buried, downdip parts. When shallow aquifers are pumped near coasts, saltwater intrusion can lead to salinization of the ground water. During 1985 in the United States, more than 30 million cubic meters per day were withdrawn from these types of aquifers.

Semiconsolidated aquifers in the United States include the Coastal lowlands aquifer system (Gulf Coast), Texas coastal uplands aquifer system, Mississippi embayment aquifer system, Southeastern Coastal Plain aquifer system, and Northern Atlantic Coastal Plain aquifer system. The Coastal lowlands aquifer system is a good example of aquifers of semiconsolidated sediments (Case Study 4-5).

Sandstone Aquifers

Sandstone retains only a small part of the primary sand porosity; compaction and cementation greatly reduce the pore space. Secondary openings, such as joints and fractures, along with bedding-plane parings, facilitate the flow of ground water in sandstone. Typically, the hydraulic conductivity of sandstone aquifers is low to moderate, but because they extend over large areas, these aquifers provide large amounts of water. Sandstone aquifers in the United States (Figure 4.5) include aquifers of the Colorado Plateau, Denver Basin aquifer system, Lower Cretaceous aquifers (Northern Great Plains aquifer system), Rush Springs aquifer (Oklahoma), Central Oklahoma aquifer, Ada-Vamoosa aquifer

CASE STUDY 4-5

COASTAL LOWLANDS AQUIFER SYSTEM

The Coastal lowlands aquifer system consists of mostly Miocene and younger unconsolidated deposits that lie above and eastward of the Vicksburg–Jackson confining unit and extends to land surface (Figure 4.14). The lithology is generally sand, silt, and clay and reflects three depositional environments—continental (alluvial plain), transitional (delta, lagoon, and beach), and marine (continental shelf). The gradual subsidence of the depositional basin and rise of the land surface caused the deposits to thicken gulfward, which resulted in a wedge-shaped configuration of

Ryder, 1996

the hydrological unit (Figure 4.14). Most of the ground water is confined with storage coefficients between 10^{-4} and 10^{-3}. Where the aquifer is shallow and unconfined, the specific yield is from 10 to 3%. A model analysis suggests that hydraulic-conductivity values are 17 to 21 ft/day (or 5 to 7 m/day). The transmissivity of the aquifer system ranges from 5000 to 35,000 ft^2/day (or 500 to 4000 m^2/day). Fifty percent of total withdrawals from the aquifer system (1090 million gallons per day in 1985) is concentrated in the Houston area for public water supply.

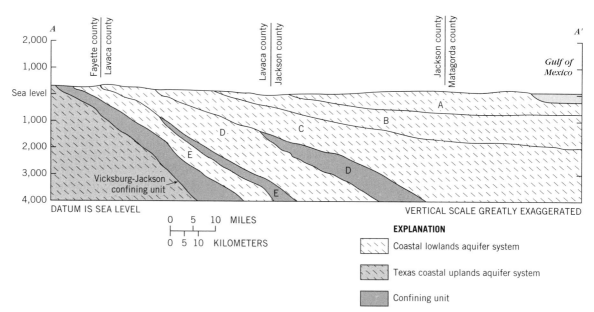

Figure 4.14 Cross section showing the Coastal lowlands aquifer system and confining units (from Ryder, 1996).

(Oklahoma), Early Mesozoic Basin aquifers (eastern U.S.), New York sandstone aquifers, Pennsylvanian aquifers (central and eastern U.S.), Mississippian aquifer of Michigan, Cambrian-Ordovician aquifer system (northern Midwest), Jacobsville aquifer (Michigan), Lower Tertiary aquifers (northern Great Plains), Upper Cretaceous aquifers (northern Great Plains), and Wyoming Tertiary aquifers. A good example of sandstone aquifers is provided by the Dakota aquifer system.

Davis (1988) provides representative porosities and permeabilities of sandstone and shale (Table 4.6).

Carbonate-Rock Aquifers

Aquifers in carbonate rocks are most extensive in the eastern United States. Most carbonate-rock aquifers are limestone, but locally, dolomite and marble yield water. The water-yielding

TABLE 4.6 Representative values of porosity and hydraulic conductivity for sandstone and shale. Other sedimentary rocks are provided for comparison

Lithology	Formation or Geologic Age	Location	Porosity (%)	Hydraulic Conductivity (m/s)	Source
Sandstone	Paskapoo	Alberta	—	4.11×10^{-5}	a
Sandstone	Paskapoo	Alberta	—	2.53×10^{-5}	a
Sandstone	Wilcox	—	—	4.55×10^{-5}	a
Sandstone	Bradford	—	14.8	2.59×10^{-8}	a
Sandstone	Berea	—	19	3.67×10^{-6}	a
Sandstone	Pennsylvanian	Illinois	19	1.7×10^{-6}	b
Sandstone	Chesterian	Illinois	17	1.3×10^{-6}	b
Sandstone	Ancell	Illinois	16	4.8×10^{-6}	b
Sandstone	Mt. Simon	Illinois	12	7.4×10^{-6}	b
Mudstone	Cenozoic	North Dakota	—	4.41×10^{-10}	a
Shale (fractured)	Paskapoo	Alberta		3.10×10^{-5}	a
Shale	Graneros	Kansas	11.6	4.51×10^{-13}	a
Shale	Wolfcamp	Texas	—	9.21×10^{-13} (hor.) 9.21×10^{-14} (vert.)	a
Shale	New Albany	Illinois	1.5	1.9×10^{-8}	c
Limestone	Mammoth Cave	Illinois	14	3.7×10^{-6}	b
Limestone and dolostone	Hunton	Illinois	14	1.3×10^{-6}	b
Dolostone	Ottawa	Illinois	8	1.3×10^{-7}	b

[a] Davis (1988).
[b] van Den Berg (1980).
[c] Kalyoncu et al. (1978).

CASE STUDY 4-6

DAKOTA AQUIFER SYSTEM

The aquifer system stretches from the Black Hills of western South Dakota eastward under the entire state (Figure 4.15). There are three major aquifers present in the basin—Mississippian carbonate rocks collectively referred to as the Madison Group, the Inyan Kara Group Sandstones, and the Newcastle Sandstone. All of these aquifers crop out on the eastern flank of the Black Hills (Figure 4.15). The confining beds are largely shale and include the Madison Group confining layer, the Skull Creek Shale, and the Cretaceous Shale confining layer. The Skull Creek Shale thins eastward and eventually pinches out, permitting the Inyan Kara and New Castle Sandstones to merge and form the Dakota Sandstone of eastern South Dakota.

The Madison confining layer also pinches out to the east permitting waters from the Madison Group to enter the basal sands of the Dakota Sandstone. Swenson (1968) thinks that the Madison is thus able to provide recharge to the Dakota Sandstone. In summary, this example shows at a large scale how patterns of sedimentation on a basinal scale control the distribution of units and the development of aquifers and confining beds. Porosity and permeability are variable in the Dakota aquifer system. Yields of most wells range from 5 to 60 gallons per minute. Yields of some wells in the aquifer system may exceed 500 to 1000 gallons per minute. Some wells completed in these aquifers are 5000 feet deep or more.

Bredehoeft et al., 1983

---

properties of carbonate rocks vary widely; some yield almost no water and are considered to be confining units, whereas others are among the most productive aquifers known. According to the *Ground Water Atlas of the United States,* the carbonate-rock aquifers may

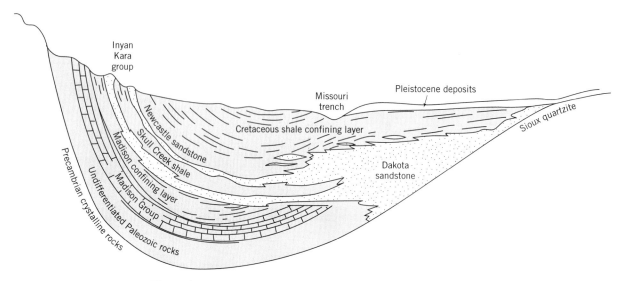

Figure 4.15 West-East cross section showing the most important aquifers and confining beds comprising the Dakota aquifer system (from Bredehoeft et al., 1983).

include Basin and Range carbonate-rock aquifers (southwestern United States), Roswell Basin aquifer system (New Mexico), Ozark Plateaus aquifer system (Missouri), Blaine aquifer (Oklahoma and Texas), Arbuckle-Simpson aquifer (Oklahoma), Silurian-Devonian aquifers (northern Midwest), Ordovician aquifers (Tennessee and Kentucky), Upper Carbonate aquifer (Minnesota and Iowa), Floridan aquifer system, Biscayne aquifer (Florida), New York and New England carbonate-rock aquifers, Piedmont and Blue Ridge carbonate-rock aquifers (eastern United States), Castle Hayne aquifer (North Carolina), North Coast Limestone aquifer system (Puerto Rico), and Kingshill aquifer (Virgin Islands).

Enhancement of Permeability and Porosity by Dissolution

Most carbonate rocks originate as sedimentary deposits in marine environments. Compaction, cementation, and dolomitization processes might act on the deposits as they lithify and greatly change their porosity and permeability. However, the principal postdepositional change in carbonate rocks is the dissolution of part of the rock by circulating, slightly acidic ground water. Solution openings in carbonate rocks range from small tubes and widened joints to caverns that may be tens of meters wide and hundreds to thousands of meters in length. Where they are saturated, carbonate rocks with well-connected networks of solution openings yield large amounts of water to wells that penetrate the openings, although the intact rock between the secondary openings is poorly permeable.

Some rocks, like limestone, dolostone, and halite, can be dissolved by flowing ground water. Such dissolution can enlarge pore spaces, joints, fractures, and bedding-plane partings, leading to increased hydraulic conductivity and porosity. In some cases, solution conduits or caves can develop that connect recharge areas to discharge areas at the downstream end of the flow system. The term *solution conduit* refers to long openings enlarged by ground water. Such openings are generally considered to be smaller than caves (Palmer, 1990).

Major solutional enhancement of permeability in carbonate aquifers is dependent on the circulation of ground water. Naturally, chemically aggressive ground water must be

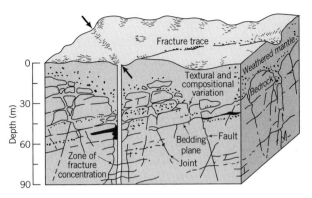

Figure 4.16 Solution enhancement of original fracture porosity of carbonate rock. Highest well yields occur in zones where multiple fractures intersect (from Lattman and Parizek, 1964). Reprinted from *J. Hydrol.* v. 2, Relationship between fracture traces and the occurrence of groundwater in carbonate rocks, p. 73–91, Copyright 1964, with permission from Elsevier Science.

available to recharge the system. Next, fractures need to be present to transmit water. Finally, ground water must be able to drain out of the system (Stringfield and LeGrand, 1966). Commonly, these three conditions develop in unconfined carbonate aquifers.

The conduits are created as water begins to enlarge the discontinuities close to the recharge area of the flow system (Figure 4.16). Eventually, a few pathways develop, and aggressive ground water flows rapidly through the rock to the discharge point. At the time that this flow regime develops, the conduits are only about 1 cm wide (Palmer, 1990). When the flow paths become kilometers in length, it is possible for selected conduits to grow to the size of caves (conduits accessible to humans, thus generally larger than 1 m in diameter). There is an extensive literature related to the human exploration of caves.

Apart from this network of conduits and caves, other secondary porosity can be present but not enlarged. Thus, significant portions of the aquifer might have hydraulic conductivities and porosities comparable to fractured rocks.

Not all carbonate rocks will develop this kind of secondary porosity and enhanced permeability. Stringfield and LeGrand (1966) found that secondary porosity will not develop (1) if the deposits overlying a carbonate terrain are of low permeability and are thick, or (2) if the carbonate rock was never elevated into a ground-water circulation system. Features of the carbonate rock are also important. For example, the required circulation will not develop in a carbonate rock that is essentially unfractured. Similarly, a carbonate rock like chalk can be so porous that flow is diffuse and not concentrated along preferential paths (Brahana et al., 1988). In these cases, pores grow equally. Figure 4.17 summarizes this kind of information in terms of grain size, porosity, and hydraulic conductivity for several types of carbonate rocks. The shaded area indentifies conditions indicated by Ford (1980) as being conducive to the formation of conduits and caves.

Karst Landscapes

Such extensive modifications to carbonate or other rocks have a significant influence on the landscape. The term *karst* describes landscapes shaped by dissolution processes to provide spectacular landforms and underground drainage. Sinkholes are probably the most well-known landform associated with karst. White (1990) describes other landforms.

Sinkholes are bowl-shaped depressions on the land surface. They include: (1) the classical solution sink, which involves a closed depression in the bedrock with or without a soil cover, (2) a cavern collapse sink that develops when the roof falls in on a near-surface conduit, (3) subsidence sinks caused by upward stoping of a collapsed solution cavity through a substantial thickness of bedrock, and (4) soil piping sinkholes, which represent the abrupt collapse of a soil arch on mechanically sound bedrock through soil

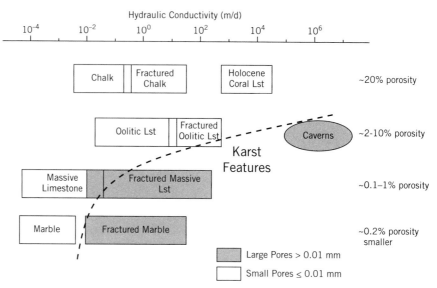

Figure 4.17 Primary and secondary porosity, pore size, and theoretical hydraulic conductivities of selected carbonate rocks, karst features, and caverns (information from Brahana et al., 1988).

loss into solution cavities (White and White, 1987). Sinkholes are a serious geologic hazard. Considerable damage has resulted from their formation beneath highways, railroads, dams, reservoirs, pipelines, and vehicles.

Ground-Water Resources of Carbonate Terranes

Brahana and others (1988) provided a comparison of properties of 27 carbonate-rock units from a variety of different terranes in North America, including the Caribbean. They point out the following important facts:

1. Carbonate rocks serve as significant aquifers throughout North America. They are not limited by location or the age of the formation.

2. Carbonate rocks show a total range of hydraulic conductivity of ten orders of magnitude, from the tightest confining beds to the most prolific aquifers.

3. The hydrogeologic response of carbonates at any time is related to the rock permeability, which is affected most by the dynamic process of dynamic freshwater circulation and solution of the rock.

4. The significant control on the dynamic freshwater circulation is the integrity of the hydraulic circuit, which requires that recharge, flow-through and discharge must be maintained. Without each of these, the system is essentially static and does not act as a conduit. Its evolution rate approaches that of noncarbonate rocks. Secondary controls on dynamic circulation include lithologic, structural, geomorphic, hydrologic and chronologic aspects.

5. The significant controls on the process of solution are (a) rock solubility and (b) the chemical character of the ground water.

A good example of carbonate aquifers is provided by the Winnipeg Carbonate aquifer, Winnipeg, Canada (Case Study 4-7).

WINNIPEG CARBONATE AQUIFER

Carbonate rocks often turn out to be important and prolific aquifers. Here we examine a simple system, the Winnipeg Carbonate aquifer. It has an areal extent of 3400 km² and is located in the vicinity of Winnipeg, Canada. The aquifer is formed in limestones and dolostones of the Red River Formation, which is Ordovician in age. Solutional weathering has created a zone of enhanced hydraulic conductivity along the top of the bedrock surface (Figure 4.18; Render, 1970). The so-called Upper Carbonate aquifer is covered by glacial till, as well as silt and clay, which act as confining beds. The aquifer is best developed in the uppermost 7.5 m due to solutional enhancement along joints, fractures, and bedding planes. The hydraulic conductivity

Render, 1970

gradually diminishes with depth, and the aquifer dies out at depths of 15 to 30 m. Transmissivities of the aquifer range from 25 to 2500 m²/d. Storativities range from 1×10^{-3} to 1×10^{-6}. The aquifer historically has been heavily exploited.

Ford (1983) describes the system as a mature, preglacial pavement karst. The system is presently being flushed by ground water. However, given the thick surface cover, ground water is not aggressively continuing to enlarge the system (Ford, 1983). Unlike many karsts in Canada, it survived glaciation. What makes this aquifer productive in a ground-water sense is the addition of confining beds, which enabled the solutionally enhanced zone to become fully saturated and pressurized.

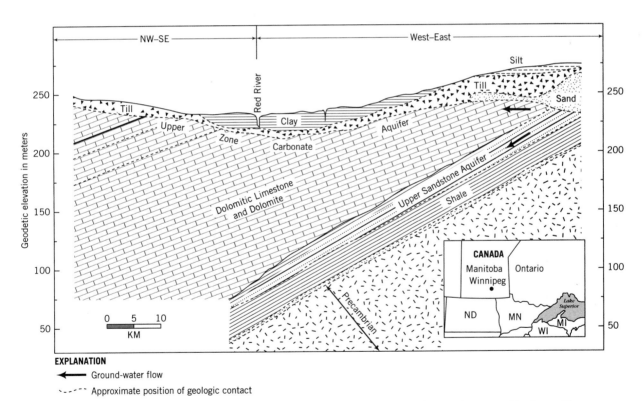

Figure 4.18 Hydrogeological cross section of the Winnipeg carbonate aquifer, located in Manitoba, Canada. The best well yields are developed along and close to the bedrock surface (modified from Render, 1970).

Regional Variability in Carbonate Units

Clearly, the extent to which porosity is enhanced depends on features like the topographic setting, the fracture development, and the nature of confining beds that control

FLORIDAN AQUIFER SYSTEM

The Floridan aquifer system is a major carbonate aquifer that covers much of Florida as well as parts of Alabama, Georgia, and other states in the southeastern United States. It consists of permeable zones comprised mainly of fractured and solutionally enhanced carbonate rocks with various lower permeability confining beds (Figure 4.19). The geology of this system is complex given that it is hundreds of meters thick and involves many separate stratigraphic units. Interested readers can pursue these details in a summary paper by Miller (1990). The karst is best developed in zones where the overlying confining beds are thin or have been stripped away (Johnston and Miller, 1988). For

example, transmissivities in central and northern Florida are often greater than 93,000 m²/d. Here, the carbonate rocks are characterized by caves, sinkholes, and other solutional features. During glacial times, the much lower sea levels enhanced the development of these karst systems.

Enhancement of the hydraulic conductivity by solutional processes is much less marked where thick, confining beds cover the carbonates. For example, in panhandle Florida and southernmost Florida, transmissivities are generally less than 5600 m²/d because the aquifer is confined by thick beds of clay and contains low-permeability limestone (Johnston and Miller, 1988).

Miller, 1990

---

the circulation of recharge through the system. Regional variability in these factors can result in obvious differences in the extent to which permeability and porosity are enhanced. A good example is the Floridan aquifer system (Case Study 4-8).

Sandstone and Carbonate-Rock Aquifers

In a few places in the United States, carbonate rocks are interbedded with almost an equal thickness of permeable sandstone. Where these two rock types are interbedded, usually carbonate rocks yield much more water than the sandstone. Most carbonate rocks originate as sedimentary deposits in marine environments. Compaction, cementation, and dolomitization processes act on the deposits as they lithify and greatly change their porosity and permeability. However, the principal postdepositional change in carbonate rocks is the enhancement of secondary porosity by circulating ground water.

Examples of sandstone and carbonate-rock aquifers include Edwards-Trinity aquifer system (Texas and Oklahoma), Valley and Ridge aquifers (eastern United States), Mississippian aquifers (central and eastern United States), and Paleozoic aquifers (northern Great Plains). Let's examine the Edwards-Trinity aquifer system with Case Study 4-9.

Basaltic and Other Volcanic-Rock Aquifers

A typical Pliocene and younger-age basaltic lava flow of the northwestern United States contains layers of varying permeability. Permeability is higher near the top and bottom of the flow and lower in the dense, center part of the flow. Volcanic rocks exhibit a wide range of chemical, mineralogic, structural, and hydraulic properties, owing mostly to variations in rock type and the way the rock was ejected and deposited. Unaltered pyroclastic rocks, for example, could have porosity and permeability values similar to poorly sorted sediments. Hot pyroclastic material, however, might become welded as it settles, providing an extremely low hydraulic conductivity. Silicic lavas tend to be extruded as thick, dense flows, and they have low permeability except where they are fractured. Basaltic lavas tend to be fluid and occur as thin flows that have considerable pore space at the tops and bottoms of the individual flows. Basalt flows are commonly stacked with permeable soil zones or alluvial material in between flows. Columnar joints that develop in the central

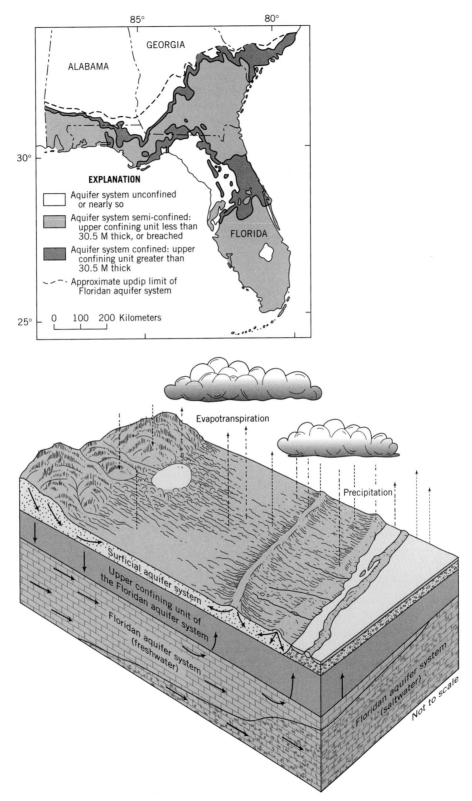

Figure 4.19 Floridan aquifer system (from Kuniansky, 1990).

EDWARDS-TRINITY AQUIFER SYSTEM

The Edwards-Trinity aquifer system covers a 77,000 mi^2 area that extends from southern Oklahoma to western Texas. Carbonate rocks, sandstones, and sands are the predominant water-yielding units. The aquifer system consists of Edwards-Trinity aquifer, Edwards aquifer, and Trinity aquifer (Figure 4.20). The Edwards-Trinity aquifer and the Trinity aquifer are stratigraphically equivalent in part and are hydraulically connected in some places. In the Balcones Fault Zone, the Edwards aquifer overlies the Trinity aquifer with a confining unit between them. The Edwards-Trinity aquifer is limestone in the upper part, and sand and sandstone in the lower part. The Edwards aquifer consists of highly faulted and fractured carbonate rocks. The Trinity aquifer consists of sandstone, sand, silt, clay, conglomerate, shale, limestone, dolomite, and marl. The rocks dip to the southeast, with thickness increasing from a few tens of feet (10 m) to more than 1000 feet (300 m).

Producing wells in the Edwards-Trinity aquifer typically are from 150 to 300 feet deep. Well yields commonly range from 50 to 200 gallons per minute, with yields of up to 3000 gallons per minute in the jointed and cavernous limestone. The Edwards aquifer is the most transmissive of all the aquifers in Texas and Oklahoma, with values ranging between 200,000 and 2,000,000 ft^2/day in the San Antonio area. Some wells in the aquifer yield more than 16,000 gallons per minute. In the Austin area, the transmissivity is less—2 to 40,000 ft^2/day. The average specific yield in the unconfined zones in the San Antonio area is between 3 and 4 percent, while storativity in the confined zones is about 10^{-4} and 10^{-5}. The Trinity aquifer has a transmissivity of about 80 to 5700 ft^2/day, a hydraulic conductivity of 1 to 31 ft/day, a storativity of about 2×10^{-5} to 0.026, and well yields between 50 and 500 gallons per minute. Withdrawals of freshwater from the Edwards-Trinity aquifer system for public water supplies and agricultural totaled about 794 million gallons per day during 1985. The aquifer system provides water for many cities in Texas and Oklahoma.

Ryder, 1996

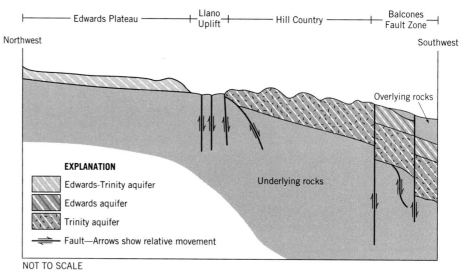

Figure 4.20 Edwards-Trinity aquifer system (from Crain et al., 1975).

parts of basalt flows create passages that allow water to move vertically through the basalt. Basaltic rocks are the most productive aquifers in volcanic rocks. Under favorable conditions, basaltic and other volcanic rock provide water supply for local communities. These aquifers in the United States include southern Nevada volcanic-rock aquifers, volcanic- and sedimentary-rock aquifers (northwestern United States), Snake River Plain aquifer system

VOLCANIC AQUIFERS IN HAWAII

Volcanic rocks that range in age from Miocene to Holocene are the most productive aquifers in the State of Hawaii. The rocks vary widely in origin, chemical composition, and texture, as well as in their ability to transmit water. Volcanic-rock aquifers are found on all the islands and in some places are overlain with sedimentary deposits. The permeability of volcanic rocks is variable and depends mainly on their mode of emplacement. Three main groups of volcanic rocks exist: lava flows, dikes, and pyroclastic deposits (Figure 4.21*a* and 4.21*b*). Weathering reduces the permeability in all three types.

Lava flows are mainly pahoehoe, which has a smooth, undulating surface with a ropy appearance, and aa, which has a surface of coarse rubble (clinker) and an interior of massive rock. A typical sequence of lava flows contains both aa and pahoehoe flows. Aa clinker zones are found above and below the massive central core of the aa flow. Pahoehoe flows commonly occur in a sequence of numerous thin flows. Void spaces in a layered sequence of lava flows include vesicles (due to gas bubbles), fractures, interflows (separations between flows), intergranular (fragmental rock), and conduit (lava tubes) porosity. Pahoehoe flows are fluid, flow rapidly, and tend to spread out.

The void spaces in a sequence of pahoehoe flows impart a high permeability. Pahoehoe lava commonly grades into aa lava with increasing distance from the eruptive vent. The layers of clinker at the top and bottom of aa flows commonly form productive aquifers, with permeability similar to that of coarse-grained gravel. However, the lava in the core of an aa flow typically cools as a massive body of rock with much lower permeability. The most productive and most widespread aquifers consist of thick sequences of numerous thin lava flows. In volcanic-rock aquifers composed mainly of flat-lying lava flows, the hydraulic conductivity is greatest parallel to the direction of the flows and is least perpendicular to the layering. Dikes are thin, near-vertical sheets of massive, low-permeability rock that intrude existing rocks, commonly the permeable lava flows (Figures 4.21*b* and 4.22).

The hydraulic conductivity of clinker zones ranges from several hundred to several thousand feet per day, which is similar to that of coarse, well-sorted gravel. The dikes lower overall rock porosity and permeability because the hydraulic conductivity of a dike complex can be as low as 0.01 ft per day. Pyroclastic rocks include ash, cinder, spatter, and larger blocks. These deposits have hydraulic conductivity between 1 and 1000 feet per day.

Oki et al., 1999

(Idaho), Columbia Plateau aquifer system, Miocene basaltic-rock aquifers, Pliocene and younger basaltic-rock aquifers, northern California volcanic-rock aquifers, and volcanic-rock aquifers overlain by sedimentary deposits (Hawaii). The volcanic rocks in Hawaii are examples of these aquifers (Case Study 4-10).

▶ 4.4 HYDRAULIC PROPERTIES OF GRANULAR AND CRYSTALLINE MEDIA

Historically, most textbooks try to develop relationships between geology and hydraulic properties in terms of deposit or aquifer types. For example, knowing the hydraulic conductivity of shale at one site is developed as a basis for understanding similar deposits at other places. In this respect, we have introduced ranges for hydraulic properties in the previous sections. However, the major problem with this approach is the difficulty in ultimately unraveling the influence of geological processes that affect the porosity and hydraulic conductivity. Simply knowing the type of unit or the geological setting doesn't usually provide enough information to estimate hydraulic parameters. Various geologic processes (chemical weathering, cementation) act to create tremendous variability with each type of deposit or aquifer, making generalizations difficult.

Part of the difficulty in creating a broad understanding of Earth materials is that few retain their primary porosity and permeability. As an example, let us consider sediments like clays, carbonate muds, or fluvial sands. When first deposited, they have porosities greater than 30%. In most depositional environments, sediments often become buried as more and more sediments are added on top. Loading causes the porous medium to consolidate,

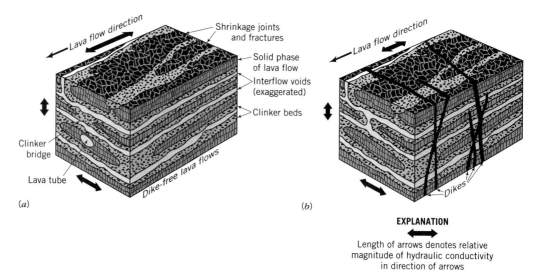

EXPLANATION

Length of arrows denotes relative
magnitude of hydraulic conductivity
in direction of arrows

Figure 4.21 Structural features associated with lava flows: (*a*) hydraulic conductivity is highest in the lava-flow direction, and (*b*) hydraulic conductivity is significantly lowered by dikes intruded at right angles to the lava-flow direction (from Takasaki and Valenciano, 1969).

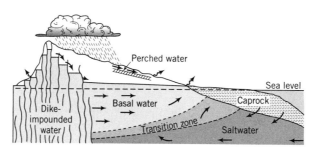

Figure 4.22 Impound effect of dikes on the ground-water flow. (From *Hydrogeology*, Hunt et al., 1988). Reproduced with permission of the publisher, the Geological Society of America, Boulder, Colorado, USA. Copyright ©1988 by the Geological Society of America, Inc.

leading to a loss of porosity and permeability. The greatest loss in porosity as a function of depth occurs in materials containing abundant clay minerals. As loading occurs, the pore structure will change irreversibly as the "house-of-cards" structure of the clay platelets is collapsed. With quartz sands, loading compresses the grains rather than disrupting the entire pore structure. Thus, the effects of loading are less marked in sandstones than in shales. The Frio Sandstone of Texas loses about 1.2% of its porosity for every 1000 ft of burial (Loucks et al., 1984). In the case of shale, porosity decreases from 0.3 to 0.5 near the ground surface, and to 0.05 to 0.15 at depths of 5000 m or more (Davis, 1988).

It is often difficult to separate porosity loss due to compaction alone from the combination of compaction and mineralogical alterations. For example, as temperatures increase, the reactive components in sandstones (e.g., feldspars) can change to clay minerals, which contribute to a more marked loss of porosity with depth. Chemically more mature sandstones (composed of quartz) are less susceptible to porosity losses with depth.

Porosity and permeability are also reduced in sandstone and carbonate rocks through cementation, wherein new minerals form in the pore spaces. Typical cements found in sedimentary rocks include quartz, carbonates, clay minerals, various oxides, sulfides, and sulfates (Schwartz and Longstaffe, 1988). For example, lime sands quickly lose their porosity and permeability due to compaction and cementation.

Our comments here have just scratched the surface of how processes, collectively referred to as diagenesis, can affect the porosity and permeability of sediments and sedimentary rocks. The topic is of considerable importance in the field of Petroleum Geology, where the formation, migration, and trapping of oil and gas depend on porosity and permeability. The key point we would make here is that information on rock types can't replace actual porosity and permeability measurements.

Pore Structure and Permeability Development

One way to understand the development of permeability is to look at the pore structure rather than the type of rock. The major resistances to flow in a porous network are related to the size distributions of the entrances and exits to a pore and the length of the pore wall. The exact quantitative relationship has remained elusive for more than 60 years. Equations like the well-known Kozeny relationship (see Table 3.5), however, provide a basis for presenting key relationships in a simplified manner. This equation is given as:

$$k = \frac{c}{M_s^2} \frac{n^3}{(1-n)^2} \tag{4.16}$$

where k is the permeability, c is the Kozeny constant relating shape factor and tortuosity factor to flow path, M_s is the specific area of solids, defined as the total surface area of solids per unit volume of solid, and n is porosity. The specific surface is essentially a reflection of the particle surface area found in a given volume of material. As particle sizes get smaller in a medium, the surface areas become larger.

We can use Eq. (4.16) in a qualitative manner to help explain the hydraulic behavior of some geologic materials. Since the values of c and $(1-n)^2$ span a relatively limited range, their particular value won't influence k very much. Values of n^3 and M_s, however, can range over orders of magnitude and are the key variables controlling permeability. The equation shows that those small porosity values and/or large surface areas promote low permeabilities.

For the purposes of this discussion, a few common geologic materials are categorized into one of three types (Table 4.7). Media-type I includes salt and granite that has almost no pore space present in the rock. There are a few micron-sized spaces present in the rock between the crystals. The very low porosities and modestly large specific surfaces of these media make these rocks essentially impermeable. Media-type II includes fine-grained sediments that contain abundant silt and clay-size particles. Typical examples would be shale, glacial till, or silty sediments. These media are porous, but the tiny grains give them a huge specific surface and thus a relatively low permeability. Media-type III includes media with both relatively high porosities and relatively small specific surfaces. These characteristics promote the high permeabilities that we associate with sands and gravels. For these media, the empirical equations presented in Chapter 3 provide a useful way to estimate k given estimates of particle diameters.

This classification scheme runs into problems because of the tremendous variety of geologic media. The main point was to illustrate how features of the pore system can help to understand permeability development in common media.

▶ 4.5 HYDRAULIC PROPERTIES OF FRACTURED MEDIA

In previous sections, we saw how many geologic media were characterized by relatively low permeabilities. To some extent, these observations contradict real-world experience that rocks and sediments are generally permeable—much more so than the numbers in

Table 4.7 would suggest. In many parts of the world, glacial-till aquifers are developed with dug wells to supply modest domestic needs. In the eastern part of the United States and elsewhere, productive wells are developed in metamorphic rocks, and recrystallized limestone. Pervasive fracturing of rocks and sediments has the potential to increase their hydraulic conductivities significantly with a small increase in porosity.

A *fracture* is a planar discontinuity in a rock or cohesive sediment. Commonly, it provides a pathway for flow through a rock. A *joint* is a type of fracture that forms near the ground surface. Joints are described by Trainer (1988) as macrofractures (sometimes you can stick your hand into a joint) along which dilation has occurred without movement parallel to the fracture surface. Figure 4.23 is sketch of a jointed and fractured rock.

A single fracture can be described in terms of its attitude (orientation) in space, its size (that is, dimensions), and its aperture. A collection of fractures having roughly the same

TABLE 4.7 Hydraulic properties of some porous and crystalline materials

Porosity and Permeability Type	Material Type	Porosity	Permeability (m^2)	Hydraulic Conductivity (m/s)
I. Small Porosity	salt	0.03–0.001	10^{-19}–10^{-24}	10^{-12}–10^{-17}
Small Per-	granite	0.003	10^{-20}–10^{-21}	10^{-13}–10^{-14}
meability	granite	—	3.5×10^{-19}	3.5×10^{-12}
II. Large Porosity	shale[a]	—	10^{-19}–10^{-21}	10^{-12}–10^{-14}
Small Per	tuff[b]	0.37	5×10^{-14}	5×10^{-7}
meability	clay till[c]	0.33	1×10^{-18}	1×10^{-11}
	clayey silt[d]	0.31–0.36	4–8×10^{-17}	4–8×10^{-10}
III. Large Porosity	glass beads[e]	0.38	1.2×10^{-11}	1.2×10^{-4}
Large Per-	glass beads[f]	0.38	3×10^{-10}	3×10^{-3}
meability	sand[g]	0.31–0.36	1×10^{-10}	1×10^{-3}

[a] Pierre Shale.
[b] Paintbrush (nonwelded) tuff.
[c] 14% sand, 54% silt, 32% clay.
[d] 6% sand, 76% silt, 18% clay.
[e] Mean particle size 0.15 mm.
[f] Mean particle size 0.65 mm.
[g] Mean particle size 0.60 mm.

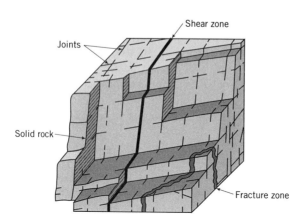

Figure 4.23 Shear zones, fracture zones, joints in solid rocks from *Recent trends in hydrogeology*, Gale, 1982. Reproduced with permission of the publisher, the Geological Society of America, Boulder, Colorado, USA. Copyright ©1982 by the Geological Society of America, Inc.

attitude in space is referred to as a fracture *set*. The *aperture* (*b*) is a measure of the width of the fracture opening (Figure 4.24). As the figure shows, the apertures of real fractures are quite variable because of the roughness of the fracture walls. Often, fractures contain secondary minerals (for example, calcite and clay minerals) that formed as water continued to flow through the medium. These fracture-filling materials can be important because they can reduce the capability of fractures to transmit water, or even plug up the fracture. Thus, the presence of a fracture does not necessarily mean that a permeable pathway for flow exists.

Because fractures have a finite size, flow through a rock requires that a connected network of fractures exist. Figure 4.25 is an example of an idealized network created by the intersection of two fracture sets. The distance between the two adjacent fractures is referred to as *fracture spacing*. The relative abundance of fractures can also be measured as *fracture density*—the number of fractures per volume of material, or as *fracture frequency*—the number of fractures intersecting a unit length of borehole. As a general rule, the greater the density of open fractures, the higher the permeability of the rock.

We reserve the detailed quantitative treatment of flow through fractured media until Chapter 6. However, with the help of the following simple equations, we will illustrate how fracture aperture (*b*) and fracture spacing (*s*) controls the permeability and porosity of a fracture network. The ratio of the fracture aperture to the fracture spacing is defined as *fracture porosity*. Consider the fracture network shown in Figure 4.26 with three sets of uniform fractures oriented at right angles to each other. This simple model is also called the sugar cube model—similar to the network that would be constructed by piling up sugar cubes into a large block.

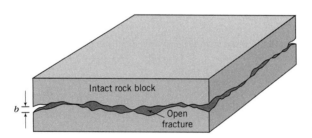

Figure 4.24 Sketch showing a rough-walled fracture with an aperture in a rock.

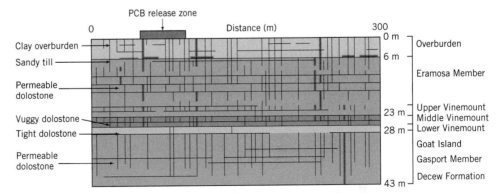

Figure 4.25 Idealization of a two-dimensional fracture network developed in carbonate rocks. The intersection of a vertical and horizontal set provides a permeable network (unpublished figure provided by E. A. Sudicky).

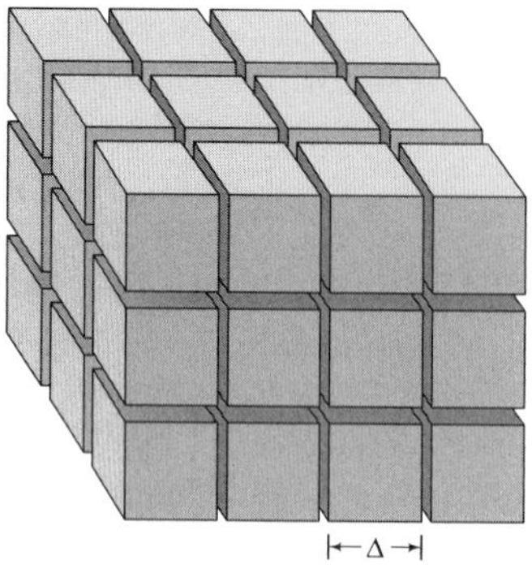

Figure 4.26 "Sugar-cube" model for fractured rocks. Three sets of fractures are assumed to intersect at right angles forming cubic rock blocks.

According to Snow (1968), the permeability (k) and porosity (ϕ) for the fracture set is given as:

$$k = \frac{b^3}{12s} \tag{4.17}$$

$$\phi = \frac{b}{s} \tag{4.18}$$

where b is the fracture aperture and s is the distance between fractures. Equation (4.17) shows clearly how permeability is proportional to the cube of the aperture and inversely proportional to the spacing between fractures. Porosity is proportional to aperture and inversely proportional to fracture spacing.

▶ **EXAMPLE 4.4**

Assume that unfractured granite has a permeability of 1×10^{-17} m^2. Calculate what the permeability and porosity would be for cubic networks of fractures having the following characteristics: (a) $b = 100 \, \mu$m, $s = 1.0$ m; (b) $b = 25 \, \mu$m, $s = 1$ m.

SOLUTION

(a) $b = 100 \, \mu$m, $\quad s = 0.1$ m.

$$k = \frac{(100 \times 10^{-6} \, \text{m})^3}{12(0.1 \, \text{m})} = 8.3 \times 10^{-13} \, \text{m}^2$$

$$\phi = \frac{(100 \times 10^{-6} \, \text{m})}{(0.1 \, \text{m})} = 10^{-3}$$

(b) $b = 25 \, \mu$m, $\quad s = 1$ m.

$$k = \frac{(25 \times 10^{-6} \, \text{m})^3}{12(1 \, \text{m})} = 1.3 \times 10^{-15} \, \text{m}^2$$

$$\phi = \frac{(25 \times 10^{-6} \, \text{m})}{(1 \, \text{m})} = 2.5 \times 10^{-5}$$

In both cases, fracturing has substantially increased the permeability of the granite.

Factors Controlling Fracture Development

This section examines some of the key factors influencing the distribution of fractures. We begin with depth and rock lithology. In many environments, there is a noticeable tendency for the permeability of fractured rocks to decrease with depth. Most explanations point to a reduction in the frequency of fracturing coupled to a decrease in the aperture (*b*) caused by loading.

Francis (1981) presented data for rocks from Prince Edward Island, Canada, that illustrate (1) how fracturing enhances the permeability of a low-permeability sandstone and (2) how the frequency of fractures decreases with depth. The unfractured matrix has a hydraulic conductivity of about 1×10^{-7} m/s. In the most fractured zones, the hydraulic conductivity is some five orders of magnitude higher. Generally, the largest hydraulic conductivities are measured in zones with the largest fracture frequencies (Figure 4.27). With depth, the bulk hydraulic conductivity decreases as the fracture frequency declines. In this and other systems, near-surface weathering and the removal of overburden stresses contributed to the formation of the shallow fracture network.

A variety of data compilations illustrate the general decline in hydraulic conductivity in fractured igneous and metamorphic rocks as a function of depth (for example, see Figure 4.28). What is lost in the averaging is detail that shows how fractures tend to occur in zones with less fractured rock intervening. This feature is evident from the Francis data presented in Figure 4.27, where several highly fractured zones are evident.

Fractures are often highly deformable. When subject to stresses, the apertures decrease, making the fractures and the network less permeable. In granitic rocks, burial to 1000 feet may reduce apertures to cause a two-order-of-magnitude reduction in the overall hydraulic conductivity. Thus, both the frequency of fractures and the fracture apertures decline with increasing depth.

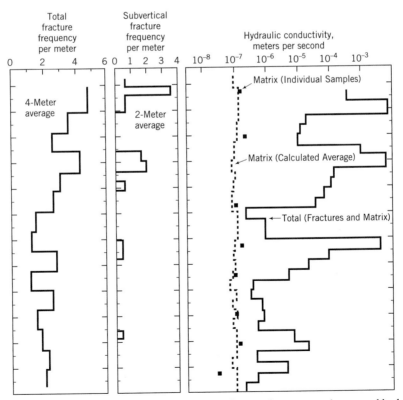

Figure 4.27   Comparison of two measures of fracture frequency and measured hydraulic conductivities as a function of depth (from Francis, 1981).

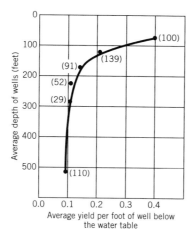

Figure 4.28 Decrease in well yields (gpm/ft of well below the water table) with depth in crystalline rocks of the Statesville area, North Carolina. Numbers near points indicate the number of wells used to obtain the average values that define the curve (from LeGrand, 1954).

TABLE 4.8 Features of the geologic environment and the scale at which they are manifested

Scale	Feature
μm	fracture wall roughness
	aperture variations in fractures
mm	soil grains
	laminations in soils
	fracture spacing
m	layering of sedimentary deposits
	clay lenses
	faults
km	lithologic units
	depositional systems
100s km	sedimentary basins
	structural features related to plates

Source: Modified from unpublished presentation of Karsten Pruess, 1998.

It is difficult to generalize how fractures can develop in a given geologic setting. In clay-rich deposits, deep erosion of the overburden creates fracturing due to differential stresses. With igneous rocks, shrinkage cracks that form as hot rocks cool. Other fracturing can be related to rock deformation associated with folding and faulting. With tectonic fractures, fracture density is related to the brittleness of the rocks. Typically, rocks like quartzite and dolostone will contain more fractures than limestone and shale. Going beyond these simple generalizations requires intensive, site-specific investigations that are capable of characterizing fracture properties.

Variability in hydraulic properties can develop on a variety of scales (Table 4.8). Variability at the μm and mm scales has begun to be explored with the help of new tools like the micropermeameter. Moving up to the cm to m scale, we find other features like changes in fracture spacing and patterns of layering giving rise to variability. At larger scales, variability due to fault development and changing lithologies of geologic units becomes evident (Table 4.8). At scales of 10 s to 100 s of km, depositional systems and plate tectonics contribute to variability in hydraulic parameters.

There has always been the hope that some understanding of the geological setting will help bound the expected variability in hydraulic properties. In the past, detailed studies of geological examples have mostly underscored the complexity of natural systems and the myriad natural processes at work. However, knowledge of geological setting occasionally can provide a useful experience base for dealing with some deposits. The question of variability boils down to organizing or classifying rocks and sediments into units having a consistent set of hydraulic properties. Such units can be characterized in terms of their thickness, lateral continuity, and internal variability in textural or fracture styles.

Variability in the geometry of fractured-rock units can be developed at a succession of scales. As suggested in Table 4.8, fractures at a small scale and faults at a larger scale are features that can exert significant control on hydraulic properties. The interaction of fractures and faulting on a large scale can produce complexity in the geometry of aquifers and confining beds.

▶ **EXERCISES**

4.1. The volume of a wet specimen is 2000 cm^3. After the specimen is drained by gravity, the volume of the specimen is 1650 cm^3. After the specimen is oven-dried, the volume becomes 1500 cm^3. Calculate the specific retention, specific yield, and porosity of the specimen as a percentage.

4.2. The transmissivity calculated from the analysis of aquifer test is 0.0134 m^2/s for a confined aquifer of 13 m thick. Calculate hydraulic conductivity in gpd/ft^2 and intrinsic permeability in darcys.

4.3. In a ground-water basin of 12 square miles, there are two aquifers: an upper unconfined aquifer 500 ft in thickness and a lower confined aquifer with an available hydraulic head drop of 150 ft. Hydraulic tests have determined that the specific yield of the upper unit is 0.12 and the storativity of the lower unit is 4×10^{-4}. What is the amount of recoverable ground water in the basin?

4.4. Calculate the permeability and porosity for a cubic network of fractures having a fracture aperture of 50 μm and a block size of 0.25 m.

▶ **CHAPTER HIGHLIGHTS**

1. An aquifer is a lithologic unit or combination of lithologic units capable of yielding water to pumped wells or springs. In a confined aquifer, confining beds form the lower and upper boundaries. Confining beds are units of low permeability that can transmit water but are not sufficiently permeable to supply water to a well. In a water table or an unconfined aquifer, the water table forms the upper boundary.

2. There are two main aquifer parameters of importance: transmissivity and storativity. Transmissivity is the rate at which water of prevailing kinematic viscosity is transmitted through a unit width of the aquifer under a unit hydraulic gradient and is the product of hydraulic conductivity and aquifer thickness. Storativity of an aquifer is the volume of water that an aquifer releases from or takes into storage per unit surface area of the aquifer per unit change in head.

3. Water is released from storage in different ways in an unconfined versus a confined aquifer. An unconfined aquifer yields water through drainage of pores. A confined aquifer yields its water through the drainage of saturated pores. An unconfined aquifer provides its water due to expansion of the water and compression of the granular matrix as the hydraulic head is reduced. For a unit decline in hydraulic head, much more water is released from storage in an unconfined aquifer than in a confined aquifer.

4. Aquifers may be classified into one of five types: unconsolidated sediments, semi-unconsolidated sediments, carbonate rocks, sandstone rocks, and volcanic and other crystalline rocks. Aquifers formed from unconsolidated sediments and some limestones are the most productive aquifers.

5. Simply knowing what type of sediment or rock comprises an aquifer does not usually provide sufficient information on which to base an estimate of hydraulic conductivity. Geologic processes modify the pore structure, create fractures, or enlarge existing fractures through dissolution.

Permeabilities of granular and crystalline media are best determined from empirical relationships like the Kozeny equation or from careful field and laboratory testing. Permeabilities of fractured media may be estimated using information on fracture aperture and spacing between fractures.

6. Variability in hydraulic properties can develop on a variety of scales. In some cases, basic geological knowledge is helpful in understanding how this variability is manifested.

5

THEORY OF GROUND-WATER FLOW

In this chapter, we describe the theory of flow in saturated, ground-water systems and develop basic equations of ground-water flow. These equations are fundamental to the quantitative treatment of flow and provide the basis for calculating hydraulic heads, given an idealization of some hydrologic system, boundary, and initial conditions. This chapter also provides simple approaches for solving such equations. For example, flownet theory provides a straightforward graphical way to determine a hydraulic-head distribution and the resulting pattern of flow, especially for problems that have a simple geometry. This chapter also shows how analytical solutions are developed and used for solving problems of steady-state flow.

▶ 5.1 DIFFERENTIAL EQUATIONS OF GROUND-WATER FLOW IN SATURATED ZONES

What sets hydrogeology apart from many of the other geosciences is an emphasis on treating problems mathematically. For example, one might be interested in calculating how much water levels will fall in the vicinity of a well after 10 years of pumping, or how contaminant concentrations change after five years of aquifer remediation. These mathematical approaches also help us interpret measurements made in the field (for example, aquifer tests and slug tests).

Basically, the mathematical approach involves representing a ground-water process by an equation and solving that equation. Let's illustrate this idea with the simple ground-water flow problem shown in Figure 5.1a. For this two-dimensional section, assume we know the pattern of layering, the hydraulic conductivity of the various units, and the configuration of the water table. Given this information, can one calculate what the pattern of flow would look like? Developing this problem from a mathematical viewpoint requires (1) finding and using the appropriate equation to describe the flow of ground water, (2) establishing a domain or region where the equation is to be solved, and (3) defining flow conditions along

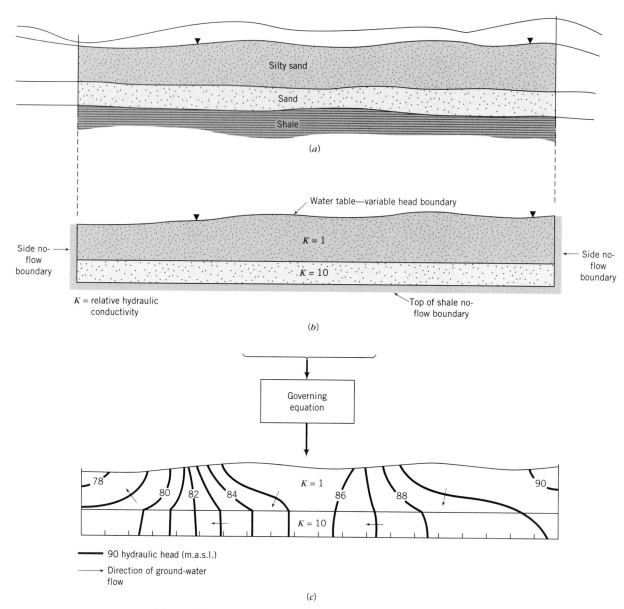

Figure 5.1 A geological problem (*a*) is conceptualized as a formal mathematical problem (*b*), which provides the basis for calculating the hydraulic head distribution (*c*) (from Domenico and Schwartz, 1998. *Physical and chemical hydrogeology*). Copyright ©1998 by John Wiley & Sons, Inc. Reprinted by permission of John Wiley & Sons, Inc.

the boundaries (the so-called boundary conditions)(Figure 5.1*b*). With this information we can calculate the hydraulic head at a large number of specified locations (*x* and *z*) within the domain. In principle, this step is like taking readings from a large number of hypothetical piezometers. Contouring the hydraulic-head distribution provides the equipotential distribution, from which we can deduce the patterns of flow (Figure 5.1*c*). This simple example helps to highlight some of the new knowledge that is required for the quantitative treatment of ground-water flow.

Useful Knowledge about Differential Equations

Many students of ground-water hydrology have difficulty in dealing with the quantitative aspects of this subject. At first glance, a differential equation describing ground-water flow immediately implies a need to understand advanced mathematical concepts:

$$\frac{\partial}{\partial x}\left(T_x \frac{\partial h}{\partial x}\right) + \frac{\partial}{\partial y}\left(T_y \frac{\partial h}{\partial y}\right) + \frac{\partial}{\partial z}\left(T_z \frac{\partial h}{\partial z}\right) = S \frac{\partial h}{\partial t} \tag{5.1}$$

Fortunately, at this introductory level, one doesn't need to do much with this equation except to recognize it. The basics of hydrogeology are set up to provide simplified approaches for dealing with these equations.

Let's consider this idea that these equations contain recognizable features. Even as toddlers we all could recognize things. Shown a picture of a rhinoceros, we could say—yes, that is a rhino because of the four stubby legs and two horns on an ugly-looking head. In looking at this picture, one wouldn't become consumed with trying to figure out why the legs were stubby and why there were two horns. Let's consider Eq. (5.1). What are its distinguishing features? Well, first, like all equations, it has some unknown that we are trying to evaluate. The unknown is hidden in the derivative terms, for example:

$$\frac{\partial h}{\partial x} \text{ or } \frac{\partial h}{\partial y} \text{ or } \frac{\partial h}{\partial z} \tag{5.2}$$

where h is the unknown. In other words, a solution to Eq. (5.1) will be of the form

$$h = \dots \text{stuff} \tag{5.3}$$

where the "stuff" on the right-hand side are terms that are simple functions of the time and space variables (t, x, y, z) and various parameters like T and S. For a solution to exist, all the terms on the right-hand side need to be known. Fortunately, for most of our applications, the "stuff" is algebraic in form and easy to evaluate.

Here are some simple steps in examining a differential equation. First look at the equation and determine the unknown. This step is straightforward—an equation containing

h : (hydraulic head) makes the equation a ground-water flow equation,

C : (concentration) makes the equation a mass transport equation, and

T : (temperature) makes the equation an energy-transport equation.

Thus, Eq. (5.1) with h as the unknown is a ground-water-flow equation. We know that it is used to apply to aquifers because it also contains the expected hydraulic parameters (T and S). Similarly, a mass transport equation will contain parameters (for example, D, a dispersion coefficient) related to processes involved with mass transport.

Equation (5.1) has other distinguishing features as well. For example, you can look at the space variables to determine the dimensionality of the problem. The dimensionality of a problem describes in how many directions the unknown (hydraulic head) is changing. For example, in Eq. (5.1), there are three space variables, x, y, and z. Three space variables make this equation a three-dimensional equation. Later in this chapter, you'll encounter one-dimensional flow equations that imply that values of hydraulic head change in one direction but not the other two.

Next, you need to decide whether hydraulic head changes with time. To figure this out, look and see whether there is a time variable (t) in the equation. In Eq. (5.1), t is there. With hydraulic head changing with time, the equation of flow is transient. If there is no t term, the equation describes a steady-state problem, where hydraulic head doesn't change with time. A partial differential equation is a concise way to represent hydrogeological processes. By looking at the unknown parameters, dimensionality, and transient nature, you will under-stand something about the problem that the equation is trying to portray. Such equations are still difficult to solve, but in looking at a partial differential equation, you'll discover plenty of useful information. The following example illustrates how to use these ideas.

▶ **EXAMPLE 5.1**

Shown here are two different differential equations that can be applied to ground-water problems. Look at each equation and determine what kind of problem it applies to, what dimensionality is involved, and whether the equation is a transient or steady-state form.

$$\frac{\partial^2 h}{\partial x^2} = 0$$

The unknown is h; therefore, it is a ground-water-flow equation. Only one space dimension (x) is included; therefore, it is one-dimensional. There is no time term; therefore, the equation is a steady-state form. There are no parameters (like K); therefore, you can conclude that the hydraulic-head distribution in this case doesn't depend on the parameter values.

$$D\frac{\partial^2 C}{\partial x^2} - v_x \frac{\partial C}{\partial x} = \frac{\partial C}{\partial t}$$

The unknown in this equation is C; therefore, it is a mass transport equation. With x as the only space dimension and t present, it is a one-dimensional transient equation.

More about Dimensionality

This chapter has made the point that the solution to ground-water-flow equations describes how hydraulic head varies within some domain or region of interest. Yet, it is not exactly clear how the dimensionality of the equation matches the dimensionality of the flow region. For example, is it necessary to use a three-dimensional form of a flow equation to calculate hydraulic head values in a three-dimensional domain?

In general, there is no requirement that the dimensionality of the equation be the same as the dimensionality of the region. A one-dimensional equation could be applied to a three-dimensional domain. The number of directions in which the ground water actually moves determines the dimensionality of the equation. For any problem, it is the combination of the region shape, along with the boundary conditions and/or heterogeneity, which determines how the ground water is likely to move and the dimensionality of the equation.

Let's consider some simple examples. Figure 5.2a sets up a problem that we will be coming back to later in this chapter. Two parallel rivers fully penetrate a confined aquifer that receives no recharge. The river stages are assumed to be constant but different from each other. This difference in stage sets up a flow through the aquifer from one river to the other. As the arrows in Figure 5.2b imply, flow through the ground-water system is in only one direction. Thus, this problem can be represented mathematically using a one-dimensional flow equation, even though the aquifer itself has three dimensions.

With a few changes, it is not hard to make the flow more complicated and to require that the dimensionality of the governing equation be increased. For example, if the river didn't fully penetrate the aquifer, flows out of and into the rivers would have components in both the x- and z-directions requiring a two-dimensional flow equation. Adding a high-permeability lens in the middle of the aquifer would produce flow in all three coordinate directions, requiring a three-dimensional flow equation.

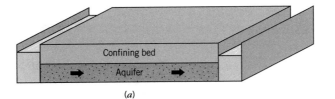

(a)

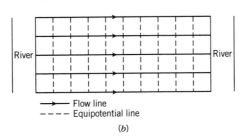

River

River

→ Flow line
- - - - Equipotential line

(b)

Figure 5.2 Panel (a) depicts ground-water flow in an aquifer due to inflow and outflow from rivers. Panel (b) suggests that although the system is three dimensional, hydraulic head varies in only one direction. Thus, flow is described by a one-dimensional flow equation.

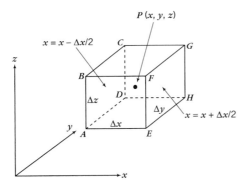

Figure 5.3 Conservation principles applied in relation to this representative elementary volume provide the basis for the development of ground-water flow equations.

Deriving Ground-Water-Flow Equations

The differential equations for ground-water flow are developed from principles of mass conservation in a representative elementary volume (Figure 5.3). In words, such a conservation statement can be written as

$$\text{mass inflow rate} - \text{mass outflow rate} = \text{change of mass storage with time} \qquad (5.4)$$

A representative elementary volume (REV) is defined as a volume exhibiting the average properties of the porous media around a point $P(x, y, z)$, which is the center of the volume. For reference purposes, let's label the six faces of the REV as x_1, x_2, y_1, y_2, z_1, and z_2. Now, assume that the water mass fluxes through the six faces are M_{x1}, M_{x2}, M_{y1}, M_{y2}, M_{z1}, and M_{z2}, respectively. Furthermore, ρ_w is the fluid density [ML^{-3}] and n is the porosity in the REV. Equation (5.4) may be rewritten in mathematical terms as

$$M_{x1} - M_{x2} + M_{y1} - M_{y2} + M_{z1} - M_{z2} = \frac{\partial}{\partial t}(n\rho_w \Delta x \Delta y \Delta z) \qquad (5.5)$$

The mass flux of water in the direction i is expressed as

$$M_i = \rho_w q_i \Delta S_i$$

where q_i is the i ($= x$, y, and z) component of the Darcy velocity vector, and ΔS_i is the area perpendicular to the flow direction i. To obtain the mass inflow and outflow rates in the x-direction for the REV, we derive an expression for $(\rho_w q_x)$ across the faces x_1 and x_2. A Taylor's expansion series of $(\rho_w q_x)$ at the outflow face x_2 is given by Thomas (1972) as

$$(\rho_w q_x)|_{x=x+\Delta x/2} = (\rho_w q)|_{x=x} + \frac{\partial(\rho_w q_x)}{\partial x}\bigg|_{x=x} \left(\frac{\Delta x}{2}\right)$$
$$+ \frac{1}{2}\frac{\partial^2(\rho_w q_x)}{\partial x^2}\bigg|_{x=x} \left(\frac{\Delta x}{2}\right)^2 + \cdots \tag{5.6}$$

Similarly, the Taylor's expansion series at the inflow face x_1 is given by

$$(\rho_w q_x)|_{x=x-\Delta x/2} = (\rho_w q)|_{x=x} - \frac{\partial(\rho_w q_x)}{\partial x}\bigg|_{x=x} \left(\frac{\Delta x}{2}\right)$$
$$+ \frac{1}{2}\frac{\partial^2(\rho_w q_x)}{\partial x^2}\bigg|_{x=x} \left(\frac{\Delta x}{2}\right)^2 + \cdots \tag{5.7}$$

By neglecting the terms of higher orders, the mass flow rate out of the REV in the x-direction is expressed as

$$M_{x2} = \left[\rho_w q_x + \frac{1}{2}\frac{\partial(\rho_w q_x)\Delta x}{\partial x}\right]\Delta y \Delta z \tag{5.8}$$

and the mass inflow rate is rewritten as

$$M_{x1} = \left[\rho_w q_x - \frac{1}{2}\frac{\partial(\rho_w q_x)\Delta x}{\partial x}\right]\Delta y \Delta z \tag{5.9}$$

The net inflow rate into the REV in the x-direction is the difference between the inflow and the outflow rates.

$$M_{x1} - M_{x2} = -\frac{\partial(\rho_w q_x)\Delta x \Delta y \Delta z}{\partial x} \tag{5.10}$$

Similarly, the net flow rates into the REV in the y- and z-directions are

$$M_{y1} - M_{y2} = -\frac{\partial(\rho_w q_y)\Delta x \Delta y \Delta z}{\partial y} \tag{5.11}$$

$$M_{z1} - M_{z2} = -\frac{\partial(\rho_w q_z)\Delta x \Delta y \Delta z}{\partial z} \tag{5.12}$$

The sum of water inflow rate minus the sum of water outflow rate for the REV is

$$M_{x1} - M_{x2} + M_{y1} - M_{y2} + M_{z1} - M_{z2} = -\left[\frac{\partial(\rho_w q_x)}{\partial x} + \frac{\partial(\rho_w q_y)}{\partial y} + \frac{\partial(\rho_w q_z)}{\partial z}\right]\Delta x \Delta y \Delta z \tag{5.13}$$

The change in ground-water storage within the REV is

$$\text{change of water storage per unit time} = \frac{\partial(\rho_w n)}{\partial t}\Delta x \Delta y \Delta z \tag{5.14}$$

According to Eq. (5.4), the net rate of water inflow is equal to the change in storage. Collecting Eqs. (5.13) and (5.14) and dividing both sides by $\Delta x \Delta y \Delta z$ gives

$$-\left[\frac{\partial(\rho_w q_x)}{\partial x} + \frac{\partial(\rho_w q_y)}{\partial y} + \frac{\partial(\rho_w q_z)}{\partial z}\right] = \frac{\partial(\rho_w n)}{\partial t} \tag{5.15}$$

By making a further assumption that the density of the fluid does not vary spatially, the density term on the left-hand side can be taken out as a constant so that Eq. (5.15) becomes

$$-\left[\frac{\partial q_x}{\partial x} + \frac{\partial q_y}{\partial y} + \frac{\partial q_z}{\partial z}\right] = \frac{1}{\rho_w}\frac{\partial(\rho_w n)}{\partial t} \tag{5.16}$$

The right side of Eq. (5.16) is related to the specific storage of an aquifer by

$$\frac{1}{\rho_w}\frac{\partial(\rho_w n)}{\partial t} = S_s\frac{\partial h}{\partial t} \tag{5.17}$$

where S_s is the specific storage and h is the hydraulic head.

Let's recall Eqs. (3.33) through (3.35), which relate the Darcy velocity to the hydraulic head. Making use of these relationships, Eq. (5.16) becomes

$$\frac{\partial}{\partial x}\left(K_x\frac{\partial h}{\partial x}\right) + \frac{\partial}{\partial y}\left(K_y\frac{\partial h}{\partial y}\right) + \frac{\partial}{\partial z}\left(K_z\frac{\partial h}{\partial z}\right) = S_s\frac{\partial h}{\partial t} \tag{5.18}$$

Equation (5.18) is the main equation of ground-water flow in saturated media. It can be written in many forms that apply to a variety of different conditions. Here are some of these alternative equations and the conditions under which they apply.

1. Under steady-state flow conditions $\left(\frac{\partial h}{\partial t} = 0\right)$ Eq. (5.18) simplifies to

$$\frac{\partial}{\partial x}\left(K_x\frac{\partial h}{\partial x}\right) + \frac{\partial}{\partial y}\left(K_y\frac{\partial h}{\partial y}\right) + \frac{\partial}{\partial z}\left(K_z\frac{\partial h}{\partial z}\right) = 0 \tag{5.19}$$

   If the porous medium is isotropic (K_x, K_y, K_z) and homogeneous $(K_{x,y,z} = \text{constant})$, Eq. (5.19) simplifies to the well-known Laplace equation

$$\frac{\partial^2 h}{\partial x^2} + \frac{\partial^2 h}{\partial y^2} + \frac{\partial^2 h}{\partial z^2} = 0 \tag{5.20}$$

2. With the same assumptions about isotropicity and homogeneity, Eq. (5.18) can be rewritten as

$$\frac{\partial^2 h}{\partial x^2} + \frac{\partial^2 h}{\partial y^2} + \frac{\partial^2 h}{\partial z^2} = \frac{S}{K_s}\frac{\partial h}{\partial t} \tag{5.21}$$

3. By dividing both sides of Eq. (5.18) by S_s, the equation is transformed into

$$\frac{\partial}{\partial x}\left(\frac{K_x}{S_s}\frac{\partial h}{\partial x}\right) + \frac{\partial}{\partial y}\left(\frac{K_y}{S_s}\frac{\partial h}{\partial y}\right) + \frac{\partial}{\partial z}\left(\frac{K_z}{S_s}\frac{\partial h}{\partial z}\right) = \frac{\partial h}{\partial t} \tag{5.22}$$

   where K_x/S_s, K_y/S_s, and K_z/S_s are called hydraulic diffusivities in the x-, y-, and z-directions, respectively. A constant specific storage is assumed in Eq. (5.22). Writing the equation in this form shows that the ground-water-flow equation is a form of diffusion equation in which the hydraulic diffusivities and hydraulic gradients in the x-, y-, and z-direction are the determining factors.

4. Multiplying both sides of Eq. (5.18) by the aquifer thickness (b) gives

$$\frac{\partial}{\partial x}\left(T_x\frac{\partial h}{\partial x}\right) + \frac{\partial}{\partial y}\left(T_y\frac{\partial h}{\partial y}\right) + \frac{\partial}{\partial z}\left(T_z\frac{\partial h}{\partial z}\right) = S\frac{\partial h}{\partial t} \tag{5.23}$$

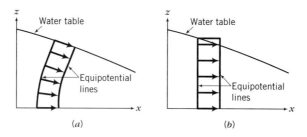

Figure 5.4 This figure illustrates how Dupuit's assumption simplifies flow in an unconfined aquifer. Panel (*a*) depicts the actual flow direction. Panel (*b*) depicts the flow with Dupuit's assumption.

where T is transmissivity and S is storativity.

$$T_x = K_x b, \quad T_y = K_y b, \quad T_z = K_z b, \quad S = S_s b \tag{5.24}$$

This form of the ground-water-flow equation is solved in numerical models like MODFLOW to predict hydraulic-head changes due to pumping in complex systems of aquifers and confining beds.

5. If there is no change of hydraulic head in the vertical direction, Eq. (5.23) is simplified to a two-dimensional ground-water-flow equation of the following form:

$$\frac{\partial}{\partial x}\left(T_x \frac{\partial h}{\partial x}\right) + \frac{\partial}{\partial y}\left(T_y \frac{\partial h}{\partial y}\right) = S\frac{\partial h}{\partial t} \tag{5.25}$$

6. The next equation relies on simplifications stemming from *Dupuit's assumption* where the direction of ground-water flow is assumed to be horizontal because the vertical hydraulic gradient is small and negligible (Figure 5.4). The differential equation for ground-water flow in an unconfined aquifer is

$$\frac{\partial}{\partial x}\left(K(x,y)h\frac{\partial h}{\partial x}\right) + \frac{\partial}{\partial y}\left(K(x,y)h\frac{\partial h}{\partial y}\right) = S_y\frac{\partial h}{\partial t} \tag{5.26}$$

where h is the hydraulic head, $K(x,y)$ is the average hydraulic conductivity, and S_y is the specific yield of the water-table aquifer. Equation (5.26) is also known as Boussinesq's equation. The equation only applies to ground-water regions where the vertical hydraulic gradient is very small. For general problems of flow in an unconfined aquifer, Eq. (5.18) should be used.

7. If a ground-water source exists, Eq. (5.18) is written as

$$\frac{\partial}{\partial x}\left(K_x \frac{\partial h}{\partial x}\right) + \frac{\partial}{\partial y}\left(K_y \frac{\partial h}{\partial y}\right) + \frac{\partial}{\partial z}\left(K_z \frac{\partial h}{\partial z}\right) + Q(x,y,z,t) = S_s\frac{\partial h}{\partial t} \tag{5.27}$$

where Q is the volumetric source rate per unit volume $[L^3 T^{-1}\ L^{-3}]$. Equation (5.27) may be expressed in words as

Inflow rate − outflow rate + source rate = change of storage $\quad$ (5.28)

▶ 5.2 BOUNDARY CONDITIONS

In order to solve a ground-water-flow equation, it is necessary to specify boundary conditions. When you define a simulation domain (for example, Figure 5.1*b*), you are selecting a small piece of some larger hydrologic system for detailed analysis. Unfortunately, the conditions outside of the domain can influence what is going on inside the domain. The job of the *boundary conditions* is to carry information as to how the simulation domain is impacted by flow conditions outside of the simulation domain. In other words, boundary conditions are the price you pay for attempting to analyze just a piece of a large, continuous system.

Boundary conditions for ground-water-flow problems are one of three types:

1. First-type boundary condition (Dirichlet boundary condition).

    The *Dirichlet boundary condition* provides a value of hydraulic head at a boundary. Mathematically, this boundary condition can be expressed as

    $$h(x, y, z)|_\Gamma = h_1(x, y, z, t), \quad (x, y, z) \in \Gamma \tag{5.29}$$

    where $h_1(x, y, z)$ is the specified value of hydraulic head at the boundary Γ. For example, a constant head of 120 m is specified at the face *EFGH* (Figure 5.5*a*). Physically, a river, a lake, or other location where the hydraulic head in the system is known can provide a constant-head boundary.

2. Second-type boundary condition (Neumann boundary condition):

    The *Neumann boundary condition* gives the water flux at a boundary. This boundary condition is written as

    $$q_n|_\Gamma = q_1(x, y, z, t), \quad (x, y, z) \in \Gamma \tag{5.30}$$

    or

    $$Q_n|_\Gamma = Q_1(x, y, z, t), \quad (x, y, z) \in \Gamma \tag{5.31}$$

    where n is an outward direction normal to the boundary, q_n is the outflow Darcy velocity [L/T], Q_n is the volumetric outflow rate [L³/T], $q_1(x, y, z)$ is the specified outflow Darcy velocity [L/T], and Q_1 is the specified outflow volumetric flux rate [L³/T]. An injection well or withdrawal well can also be considered as an inner boundary condition of second type. In this case, the boundary is the wall of the well. An example of a second-type boundary is found at the top of an aquifer where there is recharge or discharge. A *no-flow boundary* ($q_n = 0$) is a special case of a second-type boundary. Such a condition could occur at the boundary between

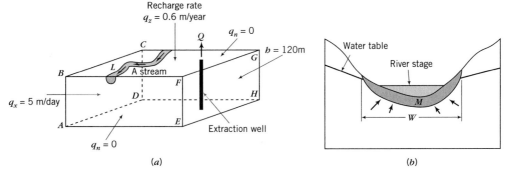

(a)   (b)

Figure 5.5 This figure illustrates how boundary conditions are applied at the sides and internal to the flow region. Panel (*a*) illustrates Dirichlet and Neumann conditions. Panel (*b*) illustrates a Cauchy boundary condition at the interface between ground-water and surface-water systems.

an aquifer and a low-permeability unit, or at ground-water divides and boundaries along flow lines where there is no flow perpendicular to the flow line.

3. Third-type boundary condition (Cauchy boundary condition):

   The *Cauchy boundary condition* relates hydraulic head to the water flux and is expressed as

   $$q_n = \frac{K_m}{M}[h(x, y, z, t) - h_m(x, y, z, t)], \quad (x, y, z) \in \Gamma \tag{5.32}$$

   or

   $$Q_n = C_m[h(x, y, z, t) - h_m(x, y, z, t)], \quad (x, y, z) \in \Gamma \tag{5.33}$$

   where q_n is the outflow Darcy velocity [L/T], Q_n is the volumetric outflow rate [L^3/T], K_m is the hydraulic conductivity of the boundary, M is the thickness of the boundary, C_m is the conductance of the boundary, h_m is the hydraulic head outside the boundary, and h is the hydraulic head inside the boundary.

   One example of where a Cauchy boundary condition applies is in the stream-aquifer interaction (Figure 5.5b). The Darcy velocity from an aquifer to a stream is

   $$q_n = \frac{K_m}{M}(h - h_m) \tag{5.34}$$

   The volumetric discharge rate from the aquifer to the stream is

   $$Q_n = \frac{LW}{M}K_m(h - h_m) \tag{5.35}$$

   where K_m is the hydraulic conductivity of the streambed, L is the length of the streambed, W is the width of the streambed, M is the thickness of the streambed, h is the hydraulic head in the aquifer, and h_m is the river stage.

▶ **EXAMPLE 5.2**

Determine the type of boundary conditions in Figures 5.5a and 5.5b. Assuming that $L = 500$ m, $W = 20$ m, $M = 0.5$ m, $K_m = 0.005$ m/day, $h = 20$ m, and $h_m = 18$ m, what are the Darcy velocity and the volumetric discharge rate from the aquifer to the stream?

SOLUTION
Dirichlet boundary conditions:

   The constant head on the right side of the box: $= 120$ m

Neumann boundary conditions:

   No flow boundaries on the front and back sides: $q_n = 0$
   Inflow velocity on the left side: $q_x = 5$ m/day
   Recharge rate on the top: $q_z = 0.6$ m/year
   Volumetric flow rate in the extraction well in the box: $Q = 20$ gpm

The Darcy velocity and the volumetric discharge from the aquifer to the stream are

$$q_n = \frac{K_m}{M}(h - h_m) = \frac{(0.005 \text{ m/day})}{(0.5 \text{ m})}[(20 \text{ m}) - (18 \text{ m})] = 0.02 \text{ m/day}$$

and

$$Q_n = \frac{LW}{M}K_m(h - h_m) = \frac{(500 \text{ m})(20 \text{ m})}{(0.5 \text{ m})}(0.005 \text{ m/day })[(20 \text{ m}) - (18 \text{ m})] = 200 \text{ m}^3/\text{day}$$

▶ 5.3 INITIAL CONDITIONS FOR GROUND-WATER PROBLEMS

For solving steady-state equations of ground-water flow, only boundary conditions need to be specified. For transient equations, initial conditions also have to be specified. The *initial condition* provides the hydraulic head everywhere within the domain of interest before the simulation begins (that is, $t = 0$). The initial condition is written as

$$h(x, y, z, 0) = h_0(x, y, z) \qquad (5.36)$$

where h_0 is the initial hydraulic head in the domain considered. Equation (5.36) states that the hydraulic head at any point (x, y, z) at time zero should be set equal to h_0. In modeling the response of an aquifer due to pumping, the initial condition would be the original head distribution in the aquifer before pumping began.

▶ 5.4 FLOWNET ANALYSIS

Having specified a flow equation and boundary/initial conditions, the next logical step is to solve the equation. One of the simplest procedures is a graphical approach that lets you sketch the unique set of streamlines and equipotential lines that describe flow within a domain. The *streamlines* (or flow lines) indicate the path followed by a particle of water as it moves through the aquifer in the direction of decreasing head. The *equipotential lines* representing the contours of equal head in the aquifer intersect the streamlines.

Flownets in Isotropic and Homogeneous Media

A flownet for an isotropic and homogeneous system provides a graphical solution to the Laplace equation. Recall that the Laplace equation describes steady-state flow in isotropic and homogeneous media and is written for two dimensions as

$$\frac{\partial^2 h}{\partial x^2} + \frac{\partial^2 h}{\partial y^2} = 0 \qquad (5.37)$$

In order to construct a flownet, you need to understand the general features of flow in a two-dimensional domain. These principles form the basis for the set of "rules," which provide guidance in sketching the streamlines and equipotential lines. The distribution of equipotential lines describes the hydraulic head in the domain, which is the unknown in Eq. (5.37).

1. Streamlines are perpendicular to the equipotential lines. If the hydraulic-head drops between the equipotential lines are the same, the streamlines and equipotential lines form curvilinear squares. A *curvilinear square* has curved sides that are tangent to an inscribed circle (Figure 5.6).

2. The same quantity of ground water flows between adjacent pairs of flow lines, provided no flow enters or leaves the region in the internal part of the net. It follows then that the number of flow channels (also known as stream tubes) must remain constant throughout the net.

3. The hydraulic-head drop between two adjacent equipotential lines is the same.

Figure 5.6 illustrates these principles with flow in an *x-z* plane through a pervious rock unit beneath a dam. Flow occurs because the hydraulic head of water in the pool above the dam is higher than in the pool below the dam. The bottom surface of the reservoir can be taken as an equipotential line, that is, a constant head boundary across which the flow is directed downward. The bottom of the flownet is the boundary between the impermeable

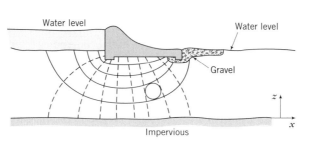

Figure 5.6 Here is an example of a flownet depicting seepage under a dam. It illustrates how flow lines and equipotential lines meet at right angles and form curvilinear squares (from Domenico and Schwartz, 1998. *Physical and chemical hydrogeology*). Copyright ©1998 by John Wiley & Sons, Inc. Reprinted by permission of John Wiley & Sons, Inc.

and permeable rock unit. This boundary is an example of a no-flow boundary. Because flow right next to a no-flow boundary must be parallel to the boundary, a no-flow boundary is a flow line. Similarly, the base of the dam is also a flow line. The pool below the dam receives discharge and provides another constant-head boundary or another equipotential line. Because the system is isotropic and homogeneous, flow lines and equipotential lines intersect to form curvilinear squares. Given all this information, the flownet is a theoretical representation of flow beneath the dam.

Bennett (1962) provides strategies for learning to sketch a flownet, which were developed from a paper by Casagrande (1937).

1. Study the appearance of well-constructed flownets and try to duplicate them by independently reanalyzing the problems that they represent.

2. In a first attempt at sketching, use only four or five flow channels.

3. Observe the appearance of the entire flownet; do not try to adjust details until the entire net is approximately correct.

4. Be aware that frequently parts of a flownet consist of straight and parallel lines, which result in uniformly sized true squares. By starting the sketching in such areas, the solution can be obtained more readily.

5. In a flow system that has symmetry, only a section of the net needs to be constructed because the other parts are images of that section.

6. During the sketching of the net, keep in mind that the size of the rectangle changes gradually; all transitions are smooth, and where the paths are curved they are of elliptical or parabolic shape.

It is very useful to keep several simple flow cases in mind before you begin studying complex flow systems. Here are a few rules that should help you prepare a flownet.

1. A no-flow boundary is a streamline.

2. The water table is a streamline when there is no flow across the water table, that is, no recharge or evapotranspiration. When there is recharge, the water table is neither a flow line nor an equipotential line.

3. Streamlines end at extraction wells, drains, and gaining streams, and they start from injection wells and losing streams.

4. Lines dividing a flow system into two symmetric parts are streamlines.

5. In natural ground-water systems, streamlines often begin and end at the water table in areas of ground-water recharge and and discharge, respectively.

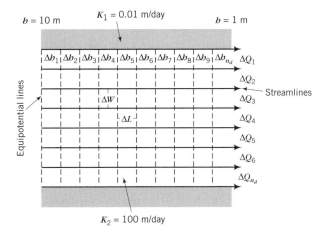

Figure 5.7 Streamlines and equipotential lines under homogeneous, and isotropic conditions. The flow is due to the constant heads applied at the left and right sides of the domain.

Flownets usually are drawn only for a two-dimensional flow field. Applying Darcy's equation to one flow channel in the two-dimensional space (Figure 5.7), we get

$$\Delta Q = T \Delta h \frac{\Delta W}{\Delta L} \tag{5.38}$$

where ΔQ is the flow rate between the equipotential lines in the flow channel, T is the transmissivity of the aquifer, Δh is the hydraulic-head drop between two equipotential lines, ΔW is the width of the flow channel, and ΔL is the distance between two equipotential lines. If the equipotential lines and streamlines result in "squares" in a flownet, $\Delta W/\Delta L$ is equal to 1. If there are a total number of n_f flow channels and n_d hydraulic-head drop, Darcy's equation is written as

$$Q = \frac{n_f}{n_d} T \Delta H \tag{5.39}$$

where Q is the total flow rate and is equal to $n_f \Delta Q$, and ΔH is the total hydraulic head drop and is equal to $n_d \Delta h$.

▶ **EXAMPLE 5.3**   Bennett and Meyer (1952) found that the average discharge due to ground-water withdrawals from the Patuxent Formation in the Sparrows Point District in 1945 was 1 million ft³/day (Figure 5.8). Calculate the transmissivity of the formation.

SOLUTION   The number of flow tubes surrounding this area of pumping is 15. The number of hydraulic-head drops is 3 between the contour lines 30 and 60 ft for a contour interval of 10 ft. From Eq. (5.39), we get

$$T = \frac{n_d Q}{n_f \Delta H} = \frac{(3)(10^6 \,\text{ft}^3/\text{day})}{(15)(30 \,\text{ft})} = 6670 \,\text{ft}^2/\text{day}$$

Flownets in Heterogeneous Media

Equipotential lines and flow lines don't necessarily intersect to form "squares" in heterogeneous media. When flow traverses two adjacent media with differing hydraulic conductivities (Figure 5.9), a condition of constant discharge in a stream tube provides

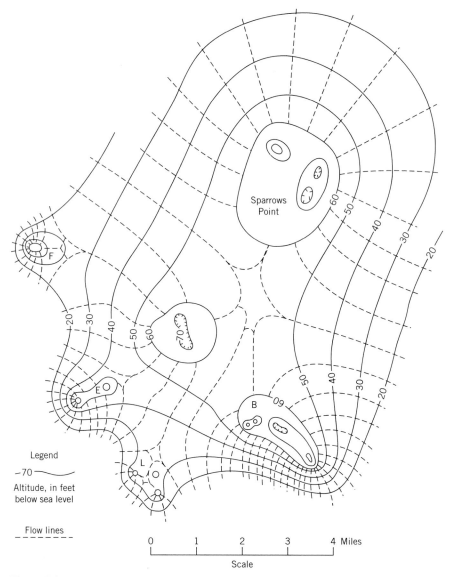

Figure 5.8 Flownet for the Patuxent Formation (from Bennett and Meyer, 1952).

$$\Delta Q = T_1 \Delta h_1 \frac{\Delta W_1}{\Delta L_1} = T_2 \Delta h_2 \frac{\Delta W_2}{\Delta L_2} \tag{5.40}$$

For the same hydraulic-head drop ($\Delta h_1 = \Delta h_2$), Eq. (5.40) is written as

$$\frac{T_1}{T_2} = \frac{\Delta L_1 \Delta W_2}{\Delta L_2 \Delta W_1} \tag{5.41}$$

If the width of the stream tube remains constant, the segment length of the stream tube in the higher transmissivity zone is longer than that in the lower transmissivity zone.

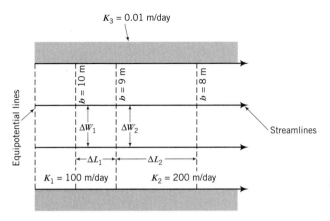

Figure 5.9 An example of flownet in heterogeneous media with a change in hydraulic conductivity in the direction of flow. Note how the distance between the equipotential lines changes across this boundary.

► **EXAMPLE 5.4**   With reference to Figure 5.9, calculate the flow rate and the length ΔL_2. Assume that $\Delta W_1 = \Delta W_2 = \Delta L_1 = 10$ m and that the thickness of the aquifer is 50 m.

SOLUTION   The flow rate is

$$\Delta Q = T \Delta h \frac{\Delta W}{\Delta L} = (100 \text{ m/day})(50 \text{ m})(1 \text{ m})\frac{(10 \text{ m})}{(10 \text{ m})} = 5000 \text{ m}^3/\text{day}$$

The segment length (ΔL_2) is

$$\Delta L_2 = \frac{T_2}{T_1} \Delta L_1 = \frac{(200 \text{ m/day})(50 \text{ m})}{(100 \text{ m})(50 \text{ m})}(10 \text{ m}) = (20 \text{ m})$$

When flow crosses interface between media with differing hydraulic conductivities, it is refracted according to the tangent law (see Eq. 3.47), or

$$\frac{K_1}{K_2} = \frac{\tan (\alpha_1)}{\tan (\alpha_2)} \tag{5.42}$$

An example of flow-line refraction is provided by Heath (1983) for flow across aquifers and confining bed, which have quite different hydraulic-conductivity values (Figure 5.10). Notice how the flow lines are nearly vertical in the less permeable layer.

For a geological medium consisting of a number of subareas, each of which is homogeneous and isotropic, the flownet consists of squares in one subarea and rectangles in other subareas. The streamlines change directions according to the refraction equation (5.42).

Flownets in Anisotropic Media

In anisotropic media, streamlines do not intercept equipotential lines at right angles except when flow is aligned with one of the principal directions of hydraulic conductivity or transmissivity. Assuming that the principal directions of transmissivity are aligned in the x-, and y-directions, the flow equation in homogeneous, anisotropic media is written as

$$\left(\frac{K_x}{K_y}\right) \frac{\partial^2 h}{\partial x^2} + \frac{\partial^2 h}{\partial y^2} = 0 \tag{5.43}$$

By transforming the horizontal coordinate using $X = (K_y/K_x)^{1/2} x$, Eq. (5.43) can be rewritten as

$$\frac{\partial^2 h}{\partial X^2} + \frac{\partial^2 h}{\partial y^2} = 0 \tag{5.44}$$

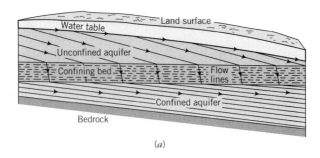

(a)

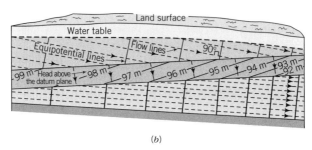

(b)

Figure 5.10 Flownets illustrating how flow lines refract in moving across hydraulic conductivity boundaries in systems of aquifers and confining beds (from Heath, 1983).

The result of this mathematical manipulation implies that a flownet in an anisotropic medium may be constructed by transforming the flow field. Details of this procedure are summarized as follows.

1. Determine the directions of the maximum and minimum hydraulic conductivity and designate the direction of maximum transmissivity as the x-direction and the direction of minimum transmissivity as the y-direction.

2. Multiply the dimension in the x-direction by a factor of $(K_y/K_x)^{1/2}$. Sketch the flownet in the transformed flow domain.

3. Project the flownet back to the original dimension by dividing the x-coordinates of the flownet by a factor of $(K_y/K_x)^{1/2}$.

Figures 5.11a and b are the flownets in original and transformed coordinate systems for flow parallel to the principal direction of hydraulic conductivity, respectively. In the transformed system, the flownet consists of squares (Figure 5.11b). In the original system, the flownet consists of rectangles (Figure 5.11a). If the flow direction is not parallel to one of the principal directions, the streamlines are no longer perpendicular to the equipotential lines (Figure 5.11c), although the flownet consists of squares in the transformed coordinate system (Figure 5.11d).

▷ 5.5 MATHEMATICAL ANALYSIS OF SOME SIMPLE FLOW PROBLEMS

The first half of this book emphasizes the use of flow equations to address problems of flow and the interpretation data from various field tests. Forms of Eq. (5.18) can be solved using both analytical and numerical approaches. Analytical methods are based on classical methods, which have been around for more than 100 years for solving differential equations. They have been used in ground-water applications since the 1930s. For a problem to be amenable for an analytical solution, it needs to be simple. Thus, analytical approaches are usually applied to problems with a regular geometry, homogeneous aquifer properties, and

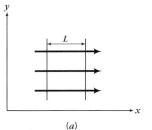 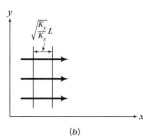 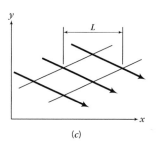 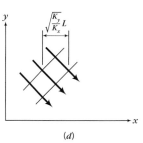

(a)   (b)   (c)   (d)

Figure 5.11 Flownet in anisotropic media. In Panel (a), the direction of flow is parallel to the principal direction (x) of hydraulic conductivity: the horizontal dimension of the porous medium cross section is changed by a ratio of $\sqrt{K_y/K_x}$. In Panel (b), a transformed cross section with square nets is formed when the same ratio is applied to horizontal dimension. In Panel (c), the direction of flow is at some angle to the principal direction (x) of hydraulic conductivity: the angles formed by the streamlines and the equipotential lines are no longer 90 degrees. In Panel (d), a transformed cross section with square nets can be formed by applying the ratio of $\sqrt{K_y/K_x}$ to the horizontal dimension of the flownet. Note: arrows point to the direction of flow.

simple initial and boundary conditions. Computers are now used to help evaluate analytical solutions. However, they still can be treated with a calculator and various tables of functions.

Numerical approaches developed in concert with modern digital computers require the computational capabilities of these machines. These techniques are tremendously powerful and can be applied to evaluate the most complicated, real-world problems. They can handle variability in hydraulic properties, large numbers of wells, and complicated boundary conditions, which might include variable recharge/evaporation and ground-water/surface-water interactions.

In this section, we will introduce some simple applications of analytic approaches for solving steady-state flow problems, which are applied to flow in aquifers and flow to wells. Chapters 9 to 11 will address the application of more complicated transient analytic solutions to problems of well hydraulics. Numerical approaches to solving problems involving complex aquifer systems will be treated in Chapter 15.

Ground-Water Flow in a Confined Aquifer

We illustrate the mathematical analysis with a simple problem of steady-state ground-water flow in a confined aquifer. The flow is one dimensional, produced by imposing different constant heads on the opposite sides of a rectangular flow domain (Figure 5.12a). Assuming that the aquifer is homogeneous and that the system is at steady state, one can describe flow by

$$\frac{\partial^2 h}{\partial x^2} = 0 \tag{5.45}$$

with these boundary conditions:

$$h|_{x=0} = h_0 \tag{5.46}$$

and

$$h|_{x=l} = h_L \tag{5.47}$$

The resulting analytical solution provides the hydraulic head as a function of the x-position in the aquifer

$$h = h_0 + (h_L - h_0)\frac{x}{L} \tag{5.48}$$

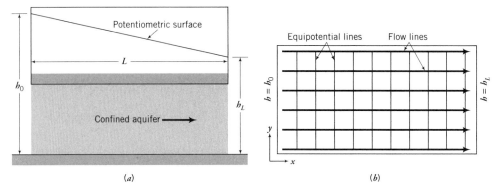

Figure 5.12 Unidirectional flow in a confined aquifer. Panel (*a*) is a cross section showing the region of interest, the aquifer, and the potentiometric surface. Panel (*b*) is a plan view showing the flownet.

Because the flow is one-dimensional and the medium is homogeneous, calculation of the Darcy flow velocity is straightforward:

$$q_x = K \frac{h_0 - h_L}{L} \tag{5.49}$$

Given the solution in Eq. (5.48), one can construct equipotential lines and sketch flow lines, as shown in Figure 5.12*b*.

▶ **EXAMPLE 5.5**

Two rivers 1000 m apart penetrate a confined aquifer 20 m thick. The hydraulic conductivity is 20 m/day. The stages of the two rivers are 500 m and 495 m above sea level, respectively. What is the Darcy flow velocity for ground water in the aquifer? If the reaches of the rivers are 600 m long and there are no pumped wells between them, what is the volume of ground-water outflow/inflow per year from the rivers? If a well were installed at a point located exactly between the rivers, what would be the hydraulic head before any pumping?

SOLUTION The calculation of inflow/outflow from the rivers starts with the Darcy velocity

$$q = (20 \text{ m/day}) \frac{(500 \text{ m}) - (495 \text{ m})}{1000 \text{ m}} = 0.1 \text{ m/day}$$

Given q, the inflow/outflow from the rivers is determined by multiplying q by the area of inflow/outflow areas, or

$$Q = (600 \text{ m})(20 \text{ m})(0.1 \text{ m/day}) \left(365 \frac{\text{days}}{\text{year}} \right) = 4.38 \times 10^5 \text{ m}^3/\text{year}$$

The hydraulic head in the aquifer midway between the two rivers is given from the analytic solution (Eq. 5.48):

$$h = 500 \text{ m} - \frac{(500 \text{ m}) - (495 \text{ m})}{(1000 \text{ m})} (500 \text{ m}) = 497.5 \text{ m}$$

Ground-Water Flow in an Unconfined Aquifer

This example is similar to the previous one except that now the aquifer is unconfined. The equation for steady-state, ground-water flow in a homogeneous, unconfined aquifer (Figure 5.13*a*) is

$$\frac{\partial}{\partial x} \left(h \frac{\partial h}{\partial x} \right) = 0 \tag{5.50}$$

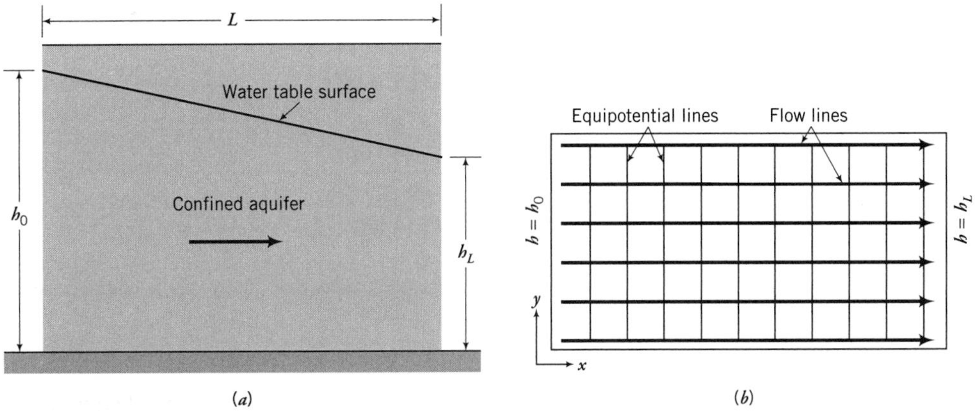

Figure 5.13   Unidirectional flow in an unconfined aquifer. Panel (*a*) is a cross section showing the region of interest and the water-table surface. Panel (*b*) is a plan view showing the flownet.

The boundary conditions are the same as before with

$$h|_{x=0} = h_0 \tag{5.51}$$

and

$$h|_{x=l} = h_L \tag{5.52}$$

The solution with Dupuit's assumption and these boundary conditions is

$$h = \sqrt{h_0^2 + (h_L^2 - h_0^2)\frac{x}{L}} \tag{5.53}$$

The Darcy flow velocity is

$$q = K\frac{h_0^2 - h_L^2}{2Lh} \tag{5.54}$$

The Darcy velocity will be different for different x locations because in a confined aquifer the thickness of the aquifer changes as a function of the hydraulic head. However, the flow rate through a unit width of the aquifer will be the same.

$$Q_{\text{unit width}} = K\frac{h_0^2 - h_L^2}{2L} \tag{5.55}$$

The solution provides the flow lines and equipotential lines shown in Figure 5.13*b*.

► **EXAMPLE 5.6**   Return to Example 5.5, but now assume that the aquifer is unconfined. Calculate the hydraulic head in the aquifer at the midpoint, the Darcy flow velocity at $x = 500$ m, and the inflow/outflow from the rivers.

SOLUTION   The hydraulic head at the midpoint (x=500 m) is calculated by appropriate substitution into the analytic solution to give

$$h = \sqrt{(500\,\text{m})^2 + [(495\,\text{m})^2 - (500\,\text{m})^2]\frac{(500\,\text{m})}{(1000\,\text{m})}} = 497.5\,\text{m}$$

The Darcy flow velocity at x=500 m is given as

$$q = (20 \text{ m/day}) \frac{(500 \text{ m})^2 - (495 \text{ m})^2}{2(1000 \text{ m})(497.5 \text{ m})} = 0.1 \text{ m/day}$$

Discharge to and from the river is given as

$$Q = (600 \text{ m})(20 \text{ m})(0.1 \text{ m/day}) \left(365 \frac{\text{days}}{\text{year}}\right) = 4.38 \text{ x } 10^5 \text{ m}^3/\text{year}$$

Comparing Examples 5.4 and 5.5 show that the hydraulic characteristics are essentially the same.

Ground-Water Flow in an Unconfined Aquifer with Recharge

Here, we examine a problem of flow in an unconfined aquifer with recharge (Figure 5.14). The differential equation for this example is

$$\frac{\partial}{\partial x}\left(K_x h \frac{\partial h}{\partial x}\right) + q_R = 0 \tag{5.56}$$

with boundary conditions:

$$h|_{x=0} = h_0 \tag{5.57}$$

and

$$h|_{x=1} = h_L \tag{5.58}$$

The solution with Dupuit's assumption is:

$$h = \sqrt{h_0^2 + (h_L^2 - h_0^2)\frac{x}{L} + \frac{q_R}{K_x}(L-x)x} \tag{5.59}$$

The solution now has become more complicated, with the hydraulic head varying in a nonlinear manner in the x-direction. The Darcy velocity is

$$q_x = K\frac{h_0^2 - h_L^2}{2Lh} - \frac{1}{2}q_R\frac{L-2x}{h} \tag{5.60}$$

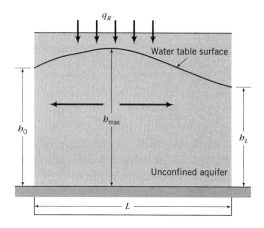

Figure 5.14 This cross section shows ground-water flow in an unconfined aquifer with recharge. Recharge leads to the formation of a drainage divide causing ground water to flow to each of the rivers.

The Darcy velocity in this case is dependent not only on hydraulic head but also on the x-position. The flux through a unit width of the aquifer and its full thickness is

$$Q_{\text{unit width}} = K\frac{h_0^2 - h_L^2}{2L} - \frac{1}{2}q_R(L - 2x) \tag{5.61}$$

Comparing Eqs. (5.60) and (5.61) shows that the water flux in a unit width of the aquifer is not constant but changes with the x-position. For ground-water flow in the unconfined aquifer with recharge, a ground-water divide exists in the recharge area. The location of the ground-water divide (x_d) is determined by

$$x_d = \frac{L}{2} - \frac{K}{q_R}\frac{h_0^2 - h_L^2}{2L} \tag{5.62}$$

▶ **EXAMPLE 5.7**

To avoid the salinization of soil in a farming area, parallel drainage canals are to be installed to maintain the water table at a depth of at least 4 m below the ground surface. It is thought that no evapotranspiration would occur below that depth. This water-table aquifer has hydraulic conductivity of 1 m/day. The canals fully penetrate the aquifer and have a stage of 3 m. The design depth from the ground surface to the bottom of the canals is 10 m. The daily recharge from precipitation and irrigation to the aquifer is 0.015 m. What is the optimal distance between the two drainage canals to achieve the desired water-table control? What is the discharge rate per unit width of aquifer to the canals?

SOLUTION Assume that the bottom of the canal is the datum and has an elevation of zero feet. The desired hydraulic head at the ground-water divide is

$$h = 10 - 4 = 6\,\text{m}$$

The hydraulic head is equal in the two drainage canals. The location of hydraulic divide is

$$x_d = \frac{L}{2}$$

Considering the hydraulic heads (h=6 m and $h_L = h_0 = 3$ m), Eq. (5.59) at $x = x_d$ may be written as

$$6\,\text{m} = \sqrt{(3\,\text{m})^2 + \frac{0.015\,\text{m/day}}{1\,\text{m/day}}\left(L - \frac{L}{2}\right)\frac{L}{2}}$$

The distance between the canals is

$$L = \frac{2}{0.015}\sqrt{(6\,\text{m})^2 - (3\,\text{m})^2} = 693\,\text{m}$$

The discharge rate in the unit width of a canal ($x = 0$) is

$$Q_{\text{unit width}} = -\frac{1}{2}q_R L = \frac{(0.015\,\text{m/day})(693\,\text{m})}{2} = 5.2\,\text{m}^3/\text{day/m}$$

The aquifer discharges 10.4 m³/day of water per unit length along the two canals.

▶ **EXERCISES**

5.1. For the following situations, determine whether a Dirichlet, Neumann, or Cauchy boundary condition is evident.

(a) Surfaces marked by the label "water level" above and below the dam in Figure 5.6.

(b) The boundary between two media with different hydraulic conductivities through which ground water is flowing (for example, Figure 5.10).

(c) The water-table surface in Figure 5.14.

(d) The ground-water divide in Figure 5.14.

5.2. Figure 5.15 shows a dike beside a flow channel. The line *IJ* is the bottom of the channel. The dike *ABDEF* and the sheet *BC* are very-low-permeable material. The line *EK* is the water level in the aquifer, and the line *GH* is the water level in the channel. Draw a flownet that depicts the pattern of flow under the dike.

5.3. Figure 5.16 is a map view of an irregularly shaped aquifer bounded by two streams and two no-flow boundaries. The principal directions of hydraulic conductivity are *x* and *y*.

(a) Assume that $K_y/K_x = 1$ and transmissivity of the aquifer is 3000 m²/day; draw a flownet for the aquifer and calculate the total rate of flow from stream *AB* to stream *CD*.

(b) Assume that $K_y/K_x = 1/9$, and draw a flownet for the aquifer.

5.4. Please judge whether or not the following statements are correct.

(a) In a heterogeneous hydraulic-conductivity field, equipotential lines are closer together in areas where the hydraulic conductivity is lower.

(b) When a streamline crosses into a very-low-permeability layer bounded by very-high-permeability formations above and below, the flow in the lower permeability unit tends to be perpendicular to the boundary.

(c) A no-flow boundary is not a streamline.

(d) A stream is not a constant-head boundary.

5.5. The hydraulic head at a point in a confined aquifer bounded by horizontal formations is 80 m, and the horizontal hydraulic gradient in the aquifer is 0.03. What is the hydraulic head at 300 m downgradient from the point? If the hydraulic conductivity is 20 m/day, what is the Darcy velocity?

5.6. In a water-table aquifer bounded on the ends by rivers, the hydraulic conductivity is 50 m/day. The hydraulic heads at $x = 0$ and $x = 1000$ m are 100 m and 90 m, respectively. Calculate the Darcy velocity at $x = 0, 100, 500,$ and 1000 m, and the flow rate in a unit width of the aquifer.

5.7. Repeat the calculation in Exercise 5.6 with a recharge rate of 0.02 m/day. Given that the streams are located at $x = 0$ and 1000 m, determine where the ground-water divide is located.

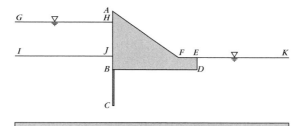

Figure 5.15 Dike along a flow channel.

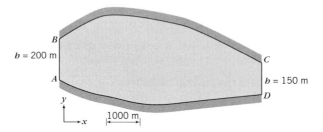

Figure 5.16 Map view of a buried valley aquifer.

► CHAPTER
HIGHLIGHTS

1. Ground-water hydrologists rely on quantitative mathematical approaches in analyzing test data and in making predictions about how systems are likely to behave in the future. The mathematical approach involves representing the flow process by an equation and solving it. The solution within some domain or region of interest defines how the hydraulic head varies as a function of space and time.

2. The flow equations are complicated partial differential equations. Fortunately, at this introductory level, all one really needs to do is to identify the equation and extract a few details. In most applications, the solutions are available in simplified forms. To find the unknown in an equation, simply find the variable residing in the derivative term. For example, in a flow equation, we find terms like $\partial h/\partial x$, making h or hydraulic head the unknown. The dimensionality of a problem can be found simply by counting the space variables, x, y, z. The presence or absence of time terms determine whether the equation is a transient or steady-state form, respectively.

3. Equations of ground-water flow can be developed, starting with an appropriate conservation statement of this form

$$\text{mass inflow rate} \; - \; \text{mass outflow rate} \; = \; \text{change of mass storage with time}$$

The general approach is to apply this equation to a block of porous medium called a representative elementary volume. It is possible to replace the words in this equation by mathematical expressions, transforming it to a form that can be developed to the main equation of ground-water flow in a porous medium

$$\frac{\partial}{\partial x}\left(K_x \frac{\partial h}{\partial x}\right) + \frac{\partial}{\partial y}\left(K_y \frac{\partial h}{\partial y}\right) + \frac{\partial}{\partial z}\left(K_z \frac{\partial h}{\partial z}\right) = S_s \frac{\partial h}{\partial t}$$

where $K_{x,y,z}$ are hydraulic conductivities in the x-, y-, z-coordinate directions, S_s is the specific storage, h is hydraulic head, and t is time. Other flow equations can be derived from this basic form.

4. The solution of differential equations requires boundary conditions. Boundary conditions define the relationship between the simulation domain and the rest of the world outside of the simulation domain. In effect, the boundary conditions stand in for the conditions outside of the simulation domain and effectively let one concentrate the modeling on the simulation domain.

5. There are three types of common boundary conditions. The first type or Dirichlet condition involves providing known values of hydraulic head along the boundary. The second type or Neumann condition requires specification of water fluxes along a boundary. A no-flow boundary (water flux zero) is the most well-known second-type boundary condition. The third-type or Cauchy boundary condition relates hydraulic head to water flux. This boundary condition is commonly used to represent ground-water/surface-water interactions.

6. For transient equations, in which the hydraulic head can change as a function of time, it is necessary to define the initial condition. The initial condition provides the hydraulic head everywhere within the domain of interest before the simulation begins (that is, at time zero).

7. A variety of mathematical and graphical approaches are available to solve ground-water-flow equations. One approach that is emphasized in this chapter is called the flownet analysis. For relatively simple, two-dimensional, steady-state flow problems, you can determine the distribution of equipotential lines graphically. Starting with an outline of the simulation domain, one adds streamlines and equipotential lines following a set of rules. For example, streamlines and equipotential lines must intersect at right angles to form a set of curvilinear squares. If you are careful, you can develop the unique pattern (and reproducible pattern) that describes flow in the domain.

8. This chapter demonstrates how simple analytical solutions can be used to describe some simple steady-state problems of flow. We will return to analytical solutions again in Chapters 8–11 on well hydraulics and regional ground-water flow.

6

THEORY OF GROUND-WATER FLOW IN UNSATURATED ZONES AND FRACTURED MEDIA

This chapter continues the exploration of basic principles of ground-water flow. We begin by looking at the vadose or unsaturated zone, which occurs between the water table and the ground surface (Figure 1.8). The vadose zone is important for the growth of land plants and is the interface between the ground-water and surface hydrologic systems. As will become evident, water flow in the vadose zone is complicated because pores contain both soil gas and water. Parameters like hydraulic conductivity and storativity, which are hydraulic "constants" in saturated systems, become hydraulic functions. In essence, the transmissive and storage properties of an unsaturated medium change dramatically depending on the relative proportions of soil gas and water. Not surprisingly, then, the governing equations for flow become more complicated to solve because parameters turn into nonlinear functions.

This chapter also explores issues of fracture flow that we introduced earlier in the book. It is important to know something about fractured media because they figure so prominently in studies of ground-water resources and contaminant hydrology. This topic is vast, and, unfortunately, within the scope of this introductory book we can only begin to scratch the surface.

▶ 6.1 BASIC CONCEPTS OF FLOW IN UNSATURATED ZONES

With the exception of parts of the capillary fringe, pores in the unsaturated zone contain both water and soil gases. The quantity of water in a partially saturated medium can be represented in terms of the *volumetric water content* (θ), which is defined as

$$\theta = \frac{V_w}{V_T} \tag{6.1}$$

where V_T is the unit volume of soil or rock and V_w is the volume of water.

The volumetric water content is a key property of unsaturated media. Wierenga et al. (1993) describe a measurement technique in the laboratory as follows: (1) Place the soil in a can, capping it tightly to prevent evaporation, and weight the soil; (2) take the lid off and dry the soil in a forced-draft oven for 10 hours or in a convection oven for 24 hours; (3) remove the soil from the oven, replace the lid, and put the soil into a desiccating jar, with desiccant, until the soil cools; and (4) weigh the sample again with the lid. The water content θ_w, expressed as a weight fraction, is

$$\theta_w = \frac{W_{\text{wet-soil}}}{W_{\text{dry-soil}}} - 1 \tag{6.2}$$

where $W_{\text{wet-soil}}$ and $W_{\text{dry-soil}}$ are the weights of wet and dry soils, respectively. The volumetric water content is related to θ_w as

$$\theta = \theta_w \frac{\rho_b}{\rho_w} \tag{6.3}$$

where ρ_b is the bulk density of the soil and ρ_w is the density of water. This method for determining the volumetric water content is a direct method. The volumetric water content can be determined indirectly by measuring soil properties related to water content. The possible methods include electrical conductivity, time-domain reflectometry, and gamma ray attenuation (Gardner, 1986).

If the volume of the void space in the sample volume V_T is V_{void}, the *water saturation* (*s*) is defined as

$$s = \frac{V_w}{V_{\text{void}}} \tag{6.4}$$

In the unsaturated zone, the volume of water present in the sample is usually less than the volume of void space. Thus, the volumetric water content is usually less than the porosity $(0 < \theta < n)$, and the water saturation is less than 1 $(0 < s < 1)$.

Water in the unsaturated zone can be characterized in terms of a hydraulic head that determines the direction of fluid flow. Like its saturated counterpart, a hydraulic head also involves two components.

$$h = z + \Psi \tag{6.5}$$

where z is the elevation head and Ψ is the pressure head. The pressure head is negative in the unsaturated zone $(\Psi < 0)$, whereas the pressure head in the saturated zone is positive $(\Psi > 0)$ (Figure 6.1). Right at the water table, the pressure head is zero $(\Psi = 0)$. Water pressures in the unsaturated zone are less than the atmospheric pressure. For this reason, the pressure head in the unsaturated zone is also called the tension head or suction head, acknowledging the capillary forces that bind water to solids. It is this "negative" pressure head in the unsaturated zone that explains why water present in partially saturated soils cannot flow into a borehole.

A *tensiometer* is a device used to measure the pressure head in the unsaturated zone (ASTM, 1992; Stannard, 1986). It consists of a porous ceramic cup connected by a water column to a manometer, or a vacuum gage, or a pressure transducer (Figure 6.2). The very

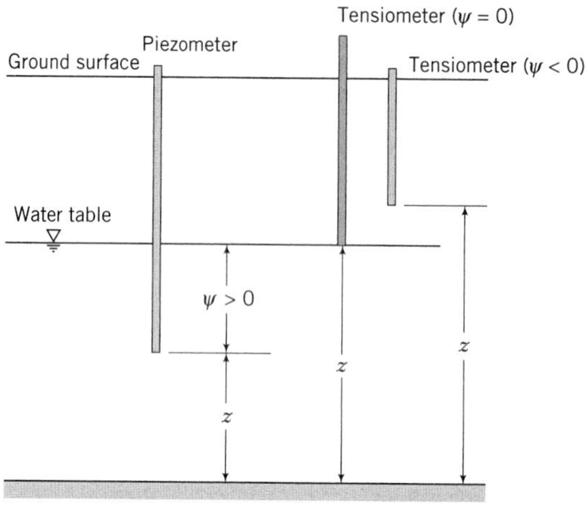

Figure 6.1 Pressure heads are negative in the unsaturated zone, zero along the water table, and positive in the saturated zone. As shown, the total hydraulic head is the algebraic sum of the elevation head and the pressure head (from Domenico and Schwartz, 1998. *Physical and chemical hydrogeology).* Copyright ©1998 by John Wiley & Sons, Inc. Reprinted by permission of John Wiley & Sons, Inc.

fine pores of the ceramic cup fill with water, which provides a hydraulic connection between the soil water and the water column. As the pressure head changes in the soil, water flows into or out of the tensiometer to maintain hydraulic equilibrium.

Figure 6.3 illustrates what the pressure head and hydraulic head would look like with a constant and continuous rainfall on the soil surface (Figure 6.3*a*). The pressure head distribution is shown in Figure 6.3*b*. Near the bottom of this field is the *water table*, which is defined by the $\Psi = 0$ contour. The pressure heads above the water table are negative, and those below the water table are positive. The total head distribution is shown in Figure 6.3*c*. As expected, downward flow through the unsaturated zone reflects total head values that decrease with depth (Figure 6.3*c*).

▶ **EXAMPLE 6.1**

A soil sample was taken from an unsaturated zone. The wet and dry weights of the sample are 105 g and 100 g, respectively. The bulk density of the sample is 1.65 g/cm^3, and the density of water is 1 g/cm^3. What is the volumetric water content?

SOLUTION   The water content expressed as a weight percentage is

$$\theta_w = \frac{(105\,\text{g})}{(100\,\text{g})} - 1 = 0.05 = 5\%$$

The volumetric water content is

$$\theta = (0.05)\frac{(1.65\,\text{g/cm}^3)}{(1.0\,\text{g/cm}^3)} = 0.083$$

Changes in Moisture Content during Infiltration

The water content of an unsaturated medium changes as a function of space and time. We illustrate this idea using an example showing how infiltration from rainfall is redistributed. Water entering the vadose zone increases the water content at early time (Figure 6.4*a*). Capillary forces dominate this initial wetting. With time the infiltration stops, and that water moves downward (Figure 6.4*b*). Both capillary and gravitational forces are in action at this stage. Notice that a distinct wetting front is evident as the pulse of infiltrated water moves downward. The *wetting front* is defined as the narrow zone that marks the beginning

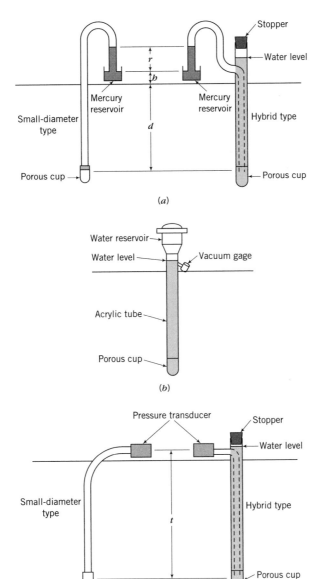

Figure 6.2 Three common types of tensiometers: (*a*) manometer; (*b*) vacuum gage; (*c*) pressure transducer (From ASTM, 1992). Copyright ASTM. Reprinted with permission.

of elevated water content due to the infiltration event. Once infiltration ceases, evapotranspiration can take place near the ground surface. The final stage of water movement in the unsaturated zone occurs as soil water enters saturated zones (Figure 6.4*c*).

► 6.2 CHARACTERISTIC CURVES

Flow in the unsaturated zone is complicated because there are generally two fluid phases (air and water) present together. Both the volumetric moisture content (θ) and the unsaturated hydraulic conductivity (K) depend on the pressure head, or the capillary pressure. The pressure head/moisture content relationship describes how a sample behaves as water

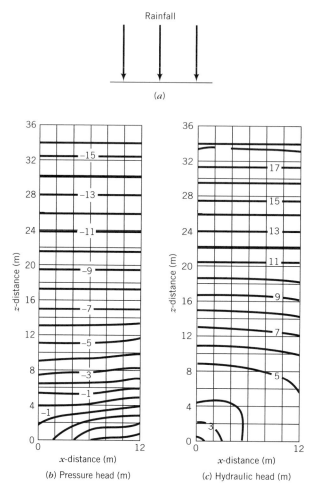

Figure 6.3 Head distribution in the unsaturated zone: (*a*) rainfall; (*b*) pressure head; (*c*) hydraulic head.

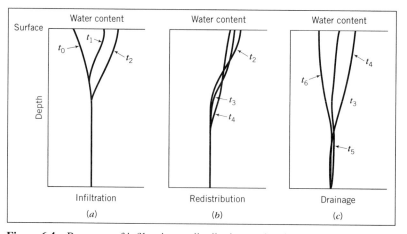

Figure 6.4 Processes of infiltration, redistribution, and recharge (Ravi and Williams, 1998).

is added or removed. With hydraulic conductivity, a relationship between hydraulic conductivity and pressure head would not be surprising. As a soil dries out, an increasingly large negative pressure head means that the mostly air-filled system has a large resistance to flow (or small K).

Water-Retention or $\theta(\psi)$ Curves

Generally, as the water content of a soil decreases, the pressure head becomes more negative, or alternatively, the capillary pressure increases. This response is due to the tendency for the water to find itself located in smaller and smaller voids. The relationship between negative pressure head and volumetric water content for a sample is called a *water-retention curve*. An example of one of these curves is shown in Figure 6.5. In practice, a volumetric water content is plotted on an arithmetic scale, while the negative pressure head is plotted on either an arithmetic scale or a logarithmic scale, as shown. The curves are typically nonlinear regardless of how they are plotted. At both large and small water content, small changes in water content are accompanied by extremely large changes in pressure head. The behavior at low water content reflects the fact that soils never lose all of their water. This lower limit in water content is termed the *residual volumetric water content* (θ_r).

The shape of the water-retention curve changes depending on whether the soil is drying or wetting. The term *hysteretic* is used to describe this effect. *Drying* involves air entering the soil to replace water that is draining. *Wetting* involves the entry of water and the displacement of the air.

The actual shape of the water-retention curve depends on several factors, with pore-size distribution being the most important. Figure 6.6 illustrates curves for sand, fine sand, and silt loam. The sand has the most uniform distribution of large pores, whereas the silt loam with a broad, grain-size distribution contains small pores, which will lead to small negative pressure heads as the soil dries.

In modeling applications, it is common to represent $\theta(\psi)$ curves using various types of mathematical relationships. Here, we illustrate two of the commonly used relationships, the Brooks–Corey (Brooks and Corey, 1964) and van Genuchten (1980) equations. These equations are written in terms of a dimensionless soil moisture content termed the *effective saturation* (s_e). From the mathematical definition that follows, note how the effective saturation varies between zero and one for every sample

$$s_e = \frac{\theta - \theta_r}{n - \theta_r} \tag{6.6}$$

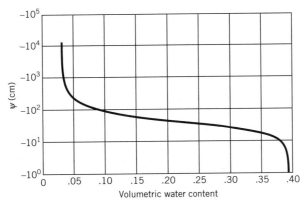

Figure 6.5 Water-retention curve for the Berino fine sandy loam (modified from Wierenga et al., 1986).

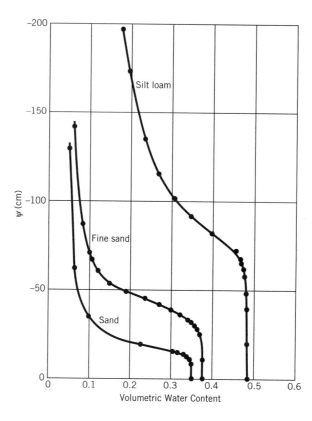

Figure 6.6 Water-retention curves for sand, fine sand, and silt loam. (from Brooks and Corey, 1966, *Properties of porous media affecting fluid flow.*) Copyright by ASCE and reproduced by permission of the publisher, ASCE.

where s_e is the effective saturation, θ_r is the residual volumetric water content, and n is the porosity.

The Brooks–Corey equation is given by

$$s_e = \begin{cases} \left(\frac{\psi_b}{\psi}\right)^\lambda, & \psi < \psi_b \\ 1, & \psi \ge \psi_b \end{cases} \tag{6.7}$$

where ψ_b is the bubbling or air-entry pressure head [L] and equal to the pressure head to desaturate the largest pores in the medium, and λ is a pore-size distribution index.

The van Genuchten equation is expressed as

$$s_e = \frac{1}{[1 + (\alpha|\psi|)^\beta]^\gamma} \tag{6.8}$$

Where α is coefficient [1/L], β is the exponent, and $\gamma = 1 - 1/\beta$.

The coefficients ψ_b and λ in the Brooks–Corey equation, and α and β in the van Genuchten equation, can be determined by fitting the measured water-retention curve with the calculated water-retention curves using Eqs. (6.16) through (6.18).

▶ **EXAMPLE 6.2**   Water-retention curves (Figure 6.7a and 6.7b) for sand and Yolo light clay were measured. Determine the parameters for Brooks–Corey and van Genuchten models. This example is adapted from Lappala et al. (1987).

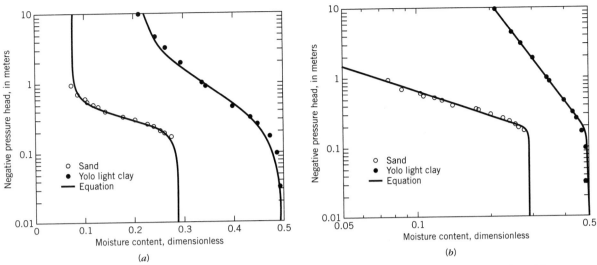

Figure 6.7 Determination of hydraulic properties from water-retention curves (adapted from Lappala et al., 1987). (*a*) Water-retention curves of sand and Yolo light clay fitted by van Genuchten equation; (*b*) water-retention curves of sand and Yolo light clay fitted by Brooks–Corey equation.

SOLUTION The calculated and measured water-retention curves of sand and Yolo light clay for the van Genuchten model are shown in Figure 6.7*a*. The fitted parameters include: $K_s = 8.2$ m/day, $n = 0.435$, $\theta_r = 0.069$, $\alpha = 3.07$ (1/m), and $\beta = 3.9$.

The calculated and measured water-retention curves of sand and Yolo light clay for the Brooks–Corey model are shown in Figure 6.7*b*. The fitted parameters include: $\theta_r = 0$, $\psi_b = -0.196$ m, and $\lambda = 0.84$.

$K(\psi)$ Curves

For unsaturated media, hydraulic conductivity is not a constant but is strongly dependent on the degree of saturation. When a medium is near saturation with a pressure head close to zero, the hydraulic conductivity takes on its maximum value. As the water content declines and pores become filled with air, the pressure head becomes more and more negative and the hydraulic conductivity decreases. As the volumetric water constant approaches residual, the water phase may not even be continuous through the sample, providing a hydraulic conductivity that is near zero.

In problems with more than one flowing phase (for example, air and water), the concept of *relative permeability* or *relative hydraulic conductivity* has proven useful in capturing the relationship existing between hydraulic conductivity (K) and negative pressure head ($-\psi$). The relative hydraulic conductivity of an unsaturated medium is defined as

$$K_r(\psi) = \frac{K(\psi)}{K_s} \tag{6.9}$$

where K_r is the relative hydraulic conductivity, which varies between 0 and 1, K_s is the hydraulic conductivity when the medium is saturated, and K is the unsaturated hydraulic conductivity. The product $K_r K_s$ points out how the unsaturated hydraulic conductivity is really some fraction of the saturated hydraulic conductivity. The specific fraction is determined by $-\psi$.

Not unexpectedly, relative hydraulic conductivity can be written as a function of the pressure head. For the Brooks–Corey model, it is represented by

$$K_r(\psi) = \begin{cases} (\frac{\psi_b}{\psi})^{2+3\lambda}, & \psi < \psi_b \\ 1, & \psi \geq \psi_b \end{cases} \tag{6.10}$$

For the van Genuchten model, relative hydraulic conductivity is written as

$$K_r(\psi) = \frac{\{1 - (\alpha|\psi|)^{\beta-1}[1 + (\alpha|\psi|)^\beta]^{-\gamma}\}^2}{[1 + (\alpha|\psi|)^\beta]^{\gamma/2}} \tag{6.11}$$

The relative hydraulic conductivity may also be expressed as a function of saturation. For example, the relative hydraulic conductivity for the van Genuchten model is

$$K_r(s_e) = s_e^l[1 - (1 - s_e^{1/\beta})^\beta]^2 \tag{6.12}$$

where l is pore connectivity and equal to about 0.5 for many soils (Mualem, 1976).

Moisture Capacity or C(ψ) Curves

The storage properties of an unsaturated soil are represented by a parameter called the *specific moisture capacity, c_m*, which is defined as the change in moisture content divided by the change in pressure head, or

$$c_m = \frac{d\theta}{d\psi} \tag{6.13}$$

Mathematically, c_m is the slope of the $\theta(\psi)$ characteristic discussed previously. Qualitatively, an increase in pressure head, for example, from -100 cm to -50 cm with the fine sand (Figure 6.6) is accompanied by an increase in the volumetric water content from 0.08 to 0.18. Notice how the increase in storage changes as a function of the material type. In the case of sand, a change in pressure head of this magnitude is accompanied by a minimal change in water storage (Figure 6.6).

As was the case with the other characteristics, specific moisture capacity can be defined as a function of various parameters. For the Brooks-Corey model, specific moisture capacity is given by

$$c_m(\psi) = \begin{cases} -(n - \theta_r)\dfrac{\lambda}{\psi_b}\left(\dfrac{\psi}{\psi_b}\right)^{-(\lambda+1)}, & \psi \leq \psi_b \\ 0, & \psi > \psi_b \end{cases} \tag{6.14}$$

For the van Genuchten model, it is expressed as

$$c_m(\psi) = \begin{cases} \dfrac{\alpha\gamma\beta(n - \theta_r)(\alpha|\psi|)^{\beta-1}}{[1 + (\alpha|\psi|)^\beta]^{\gamma+1}}, & \psi < 0 \\ 0, & \psi > 0 \end{cases} \tag{6.15}$$

► **EXAMPLE 6.3**   The relative hydraulic-conductivity curves in Figures 6.8*a* and 6.8*b* were measured for sand and Yolo light clay. Determine the parameters for Brooks-Corey and van Genuchten models and calculate the specific moisture capacity curves. This example is adapted from Lappala et al. (1987).

SOLUTION   The calculated and measured relative hydraulic conductivity curves for the Brooks–Corey and van Genuchten models are shown in Figures 6.8*a* and 6.8*b*. Also shown in Figure 6.8 are the fitting curves of the Haverkamp model (Haverkamp et al., 1977), which is a simplified form of the van Genuchten model. The fitted parameters for the Brooks–Corey model are: $K_s = 8.2$ m/day, $n = 0.435$, $\theta_r = 0$, $\psi_b = -0.196$ m, and $\lambda = 0.84$. The fitted parameters for the van Genuchten model are: $\theta_r = 0.069$, $\alpha = 3.07$ (1/m), and $\beta = 3.9$. The calculated specific moisture capacity curves using the two models for sand and Yolo light clay are shown in Figures 6.9*a* and 6.9*b*.

Table 6.1 shows the hydraulic properties of 11 soils for Brooks–Corey and van Genuchten models compiled by Lappala et al. (1983).

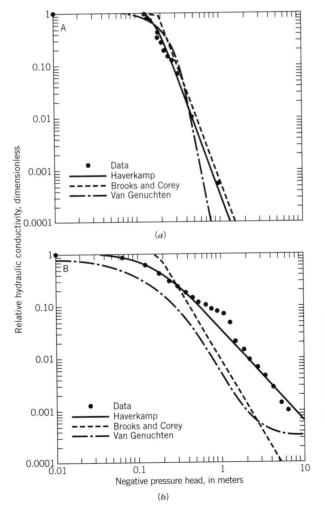

(*a*)

(*b*)

Figure 6.8 Determination of hydraulic properties from relative hydraulic-conductivity curves (adapted from Lappala et al., 1987). (*a*) Relative hydraulic-conductivity curve of sand fitted by Havekamp, Brooks–Corey, and van Genuchten equations; (*b*) relative hydraulic conductivity curve of the Yolo light clay fitted by Havekamp, Brooks–Corey, and van Genuchten equations.

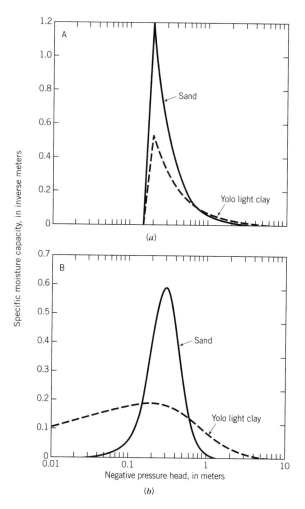

Figure 6.9 Calculated specific moisture capacity as a function of pressure head in sand and yolo light clay (adapted from Lappala et al., 1987). (*a*) using Brooks–Corey equation; (*b*) using van Genuchten equation.

TABLE 6.1   Values of hydraulic properties of 11 soils that best fit Brooks–Corey and van Genuchten models

Soil or Rock	K_s (m/day)	n	Brooks and Corey (1964)			van Genuchten (1980)		
			θ_r	$-h_b$(m)	λ	θ_r	α(1/m)	β
Del Monte Sand (20 mesh)	7000	0.36	0.011	0.112	2.5	0.036	7.04	6.3
Fresno medium sand	400	0.375	0	0.149	0.84	0.02	4.31	3.1
Unconsolidated sand	8.5	0.424	0.09	0.114	4.4	0.051	7.46	9
Sand	8.2	0.435	0	0.196	0.84	0.069	3.07	3.9
Fine sand	2.1	0.377	0.063	0.82	3.7	0.072	1.04	6.9
Columbia sandy loam	0.7	0.496	0.11	0.85	1.6	0.15	0.85	4.8
Touchet silt loam	0.22	0.43	0.095	1.45	1.7	0.17	0.51	7
Hygiene sand stone	0.15	0.25	0.13	1.06	2.9	0.15	0.79	10.6
Adelanto loam	0.039	0.42	0.13	1.41	0.51	0.16	0.36	2.06
Limon silt (imbibition data)	0.013	0.449	0	0.338	0.22	0.001	1.54	1.3
Yolo light clay	0.011	0.495	0.055	0.181	0.25	0.175	2.49	1.6

Source: From Lappala et al. (1983).

▶ 6.3 FLOW EQUATION IN THE UNSATURATED ZONE

The derivation of the partial differential equation for flow in the unsaturated zone is similar to that in the saturated zone. In the unsaturated zone, the Darcy equation is written as

$$q = -K(\psi)\nabla h \tag{6.16}$$

where q is the Darcy velocity vector [L/T], $K(\psi)$ is the hydraulic conductivity tensor [L/T], and h is the hydraulic head [L]. The hydraulic-conductivity tensor in the unsaturated zone is expressed as

$$K(\psi) = K_r(\psi)K_s \tag{6.17}$$

where $K_r(\psi)$ is the *relative hydraulic conductivity* [dimensionless] $(0 < K_r < 1)$, and K_s is the saturated hydraulic-conductivity tensor as given in (3.27).

$$K_s = \begin{bmatrix} K_{xx} & K_{xy} & K_{xz} \\ K_{yx} & K_{yy} & K_{yz} \\ K_{zx} & K_{zy} & K_{zz} \end{bmatrix}$$

By replacing the porosity n in Eq. (5.16) with soil moisture θ, we obtain the water conservation equation in the unsaturated zone.

$$-\left[\frac{\partial q_x}{\partial x} + \frac{\partial q_y}{\partial y} + \frac{\partial q_z}{\partial z}\right] = \frac{1}{\rho_w}\frac{\partial(\rho_w\theta)}{\partial t} \tag{6.18}$$

Substitution of Eqs. (6.16) and (6.17) into Eq. (6.18) provides the equation for flow in the unsaturated zone.

$$\frac{\partial}{\partial x_i}\left[K_r(\psi)\left(K_{ij}\frac{\partial h}{\partial x} + K_{iz}\right)\right] = \frac{\partial \theta}{\partial t} \tag{6.19}$$

where x_1, x_2, and x_3 are x, y, and z coordinates, $K_{ij}(i, j = x, y, \text{and } z)$ are components of the hydraulic-conductivity tensor, and K_{iz} $(i = x, y, \text{and } z)$ are hydraulic components in the z-direction. In words, the above equation may be expressed as

$$\text{Inflow rate} - \text{Outflow rate} = \text{Moisture storage change rate}$$

Equation (6.19) is a modified form of the Richards equation. In Eq. (6.19), if the principal directions of hydraulic conductivity are $x-$, $y-$, and $z-$ directions, the off-diagonal terms in Eq. (6.19) are zero. Compared with the flow equation in the saturated zone, the hydraulic conductivity in the unsaturated zone is a function of pressure head, and the storage change is the change of volumetric water content. The right side of the equation can be rewritten as

$$\frac{\partial \theta}{\partial t} = [c_m + s\,S_s]\frac{\partial h}{\partial t} \tag{6.20}$$

(see Lappala et al., 1983) where c_m is the specific moisture capacity, s is the saturation, and S_s is specific storage given by Eq. (5.17).

Combining Eqs. (6.19) and (6.20), we get

$$\frac{\partial}{\partial x}\left(K_r(\psi)K_x\frac{\partial h}{\partial x}\right) + \frac{\partial}{\partial y}\left(K_r(\psi)K_y\frac{\partial h}{\partial y}\right) + \frac{\partial}{\partial z}\left(K_r(\psi)K_z\frac{\partial h}{\partial z}\right) = [c_m + sS_s]\frac{\partial h}{\partial t} \tag{6.21}$$

To solve Eq. (6.21), volumetric water content, hydraulic conductivity, and specific capacity as functions of pressure head need to be known. These functional relations have been discussed in previous sections.

For one-dimensional vertical flow, Eq. (6.19) can be simplified and written as

$$\frac{\partial \theta}{\partial t} = \frac{\partial}{\partial z}\left(K_r(\psi)K_z\left(\frac{\partial \psi}{\partial z}+1\right)\right) \tag{6.22}$$

where z is positive downward.

The linear ground-water velocity in the unsaturated zone is related to the Darcy velocity by

$$v = \frac{q}{\theta} = \frac{q}{sn} \tag{6.23}$$

where v is the linear ground-water velocity or ground-water pore velocity vector, s is the saturation, and n is the porosity.

▶ **EXAMPLE 6.4**   The Darcy velocity in the unsaturated zone is 3 cm/hour. The porosity of the medium is 0.36. Under steady-state infiltration, the average saturation of soil is 0.8. For an unsaturated zone of 20 m, calculate the time required for a drop of water at the ground surface to travel to the water table.

SOLUTION   The ground-water velocity is the Darcy velocity divided by the volumetric water content.

$$v = \frac{(3\,\text{cm/h})}{(0.35)(0.8)} = 10.7\,\text{cm/h}$$

The time required is

$$t = \frac{(20\,\text{m})}{(10.7\,\text{cm/h})} = 18.7\,\text{h}$$

▶ 6.4 INFILTRATION AND EVAPOTRANSPIRATION

Infiltration and evapotranspiration are important processes providing flow in and out of unsaturated zones through the ground surface. To describe these processes, numerous mathematical equations have been developed. The Green–Ampt model (Green and Ampt, 1911) was the first physically based equation to describe the infiltration of water into a soil. The infiltration rate in the Green–Ampt model can be expressed explicitly as (Salvucci and Entekhabi, 1994)

$$q = \left(\frac{\sqrt{2}}{2}\tau^{-1/2} + \frac{2}{3} - \frac{\sqrt{2}}{6}\tau^{1/2} + \frac{1-\sqrt{2}}{3}\tau\right)K_s \tag{6.24}$$

where q is infiltration rate (cm/h), K_s is saturated hydraulic conductivity (cm/h), and τ is related to time as,

$$\tau = \frac{t}{t+\chi} \tag{6.25}$$

with

$$\chi = \frac{(h_s - h_f)(\theta_s - \theta_0)}{K_s} \tag{6.26}$$

where t is time, h_s is ponding depth, or capillary pressure head at the ground surface (cm), h_f is capillary pressure head at the wetting front (cm), θ_s is saturated volumetric water content, and θ_0 is initial volumetric water content.

The capillary pressure head (h_f) at the wetting front can be estimated from the air-entry head or the bubbling pressure head.

$$h_f = \frac{2 + 3\lambda}{1 + 3\lambda} \frac{\psi_b}{2} \tag{6.27}$$

where λ is the exponent in the Brooks–Corey water-retention model and ψ_b is the bubbling pressure head. The Green–Ampt model parameters and corresponding water-content profile are illustrated in Figures 6.10a and 6.10b (Ravi and Williams, 1998). A sharp wetting front is assumed in the model. As time t becomes large, the τ in Eq. (6.25) is 1 and the infiltration rate is equal to the saturated hydraulic conductivity. This relationship is expressed as

$$q = K_s, \quad \text{for} \quad t \to \infty \tag{6.28}$$

There are several other approaches for modeling infiltration or evapotranspiration rates. For example, the U.S. Soil Conservation Service (SCS) model relates the infiltration rate to precipitation rate, based on site-specific field data (USDA, 1972). Infiltration/exfiltration (evapotranspiration) models (Eagleson, 1978; Philip, 1957) can be used to estimate the water infiltration during the wetting season and evapotranspiration during the drying season.

▶ **EXAMPLE 6.5** Calculate infiltration rates for the duration of infiltration of 24 hours (adapted from Williams et al., 1998). Assume that $\psi_b = -13.8$ cm, $\lambda = 1.68$, $\theta_s = 0.43$, $\theta_0 = 0.05$, $K_s = 21$ cm/h, and $h_s = 1$ cm.

SOLUTION The capillary pressure head at the wetting front is

$$h_f = \frac{2 + 3(1.68)}{1 + 3(1.68)} \frac{(-13.8 \text{ cm})}{2} = -8.04 \text{ cm}$$

The χ is calculated as

$$\chi = \frac{[(1 \text{ cm}) - (-8.04 \text{ cm})](0.43 - 0.05)}{(21 \text{ cm/h})} = 0.1636 \text{ h}$$

The infiltration rates are in Table 6.2 for time = 1, ..., 24 hours.

▶ 6.5 EXAMPLES OF UNSATURATED FLOW

In practice, solutions to Eqs. 6.19 and 6.21 require the development of powerful finite-element, finite-difference, or other numerical approaches. A variety of codes are available for such applications, including HYDRUS (a one-dimensional finite-element code—Vogel et al., 1996), VS2D (a two-dimensional finite-difference code—Lappala et al., 1987, and

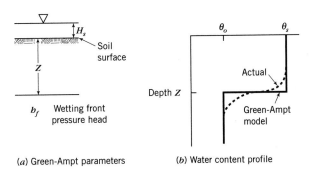

(a) Green-Ampt parameters

(b) Water content profile

Figure 6.10 Green–Ampt model. (*a*) Green–Ampt model parameters; (*b*) Conceptualized volumetric water content profile in the Green–Ampt model (from Vogel et al., 1996).

TABLE 6.2   Calculation of infiltration rates using explicit Green–Ampt model

Time(h)	τ	q(cm/h)	Time(h)	τ	q(cm/h)
1	0.8594	22.937	13	0.9876	21.160
2	0.9244	22.005	14	0.9885	21.149
3	0.9483	21.679	15	0.9892	21.139
4	0.9607	21.513	16	0.9899	21.130
5	0.9683	21.412	17	0.9905	21.123
6	0.9735	21.344	18	0.9910	21.116
7	0.9772	21.296	19	0.9915	21.110
8	0.9800	21.259	20	0.9919	21.104
9	0.9821	21.230	21	0.9923	21.099
10	0.9839	21.208	22	0.9926	21.095
11	0.9853	21.189	23	0.9929	21.091
12	0.9866	21.173	24	0.9932	21.087

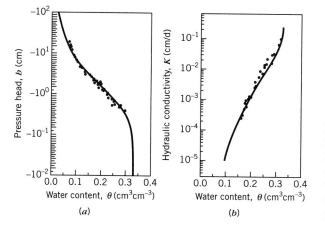

Figure 6.11 Determination of hydraulic properties from the hydraulic testing results for Crushed Bandelier Tuff (from Vogel et al., 1996). (*a*) Soil water retention curve; (*b*) hydraulic-conductivity curve.

Healy, 1990), SWMS-3D (a three-dimensional variable-saturated, finite-element code—Simunek et al, 1995), TOUGH2 (Pruess, 1987), and FEHM (Zyvoloski et al., 1997). This section will illustrate examples of several of these codes that are used to describe unsaturated flow.

Infiltration and Drainage in a Large Caisson

An experiment was conducted in a caisson 6 m in depth and 3 m in diameter at Los Alamos National Laboratory to study the hydraulic properties of the Bandelier Tuff (Abeele et al., 1984). The experimental data were modeled using HYDRUS (Vogel et al., 1996). Observed $\psi - \theta$ and $K - \theta$ relationships (Figure 6.11) were matched using the simulation model to determine the hydraulic properties in the van Genuchten model. The resultant soil hydraulic parameters are: $K_s = 25.0$ cm/day, $n = 0.3308$, $\theta_r = 0.0$, $\alpha = 0.01433$ (1/cm), and $\beta = 1.506$. Figure 6.12 shows the downward progression of a wetting front during transient infiltration in the Bandelier Tuff. Note how the pressure head is close to zero behind the wetting front and strongly negative ahead of the front, indicating dry soil. As drainage occurs, the water content declines everywhere over the next 100 days (Figure 6.13).

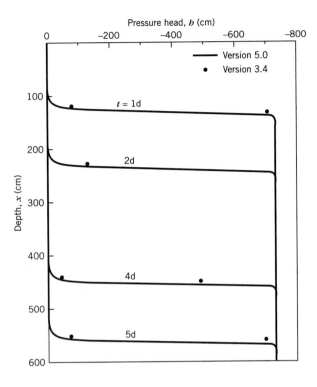

Figure 6.12 Predicted water-content profiles during transient infiltration in Bandelier Tuff (from Vogel et al., 1996).

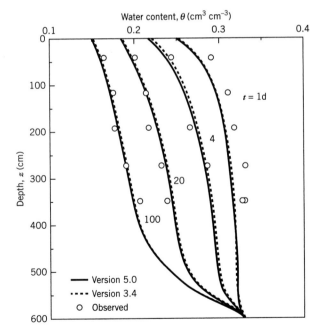

Figure 6.13 Predicted and observed water-content profiles during drainage of Bandelier Tuff (from Vogel et al., 1996).

Unsaturated Leakage from a Ditch

This example looks at the leakage of water from a ditch 4 ft wide and 10 ft deep. Water infiltrated into the unsaturated zone at a rate of 5 ft/year. The background soil moisture content is 0.118 produced by a steady annual infiltration rate of 0.011 ft/year from the ground surface. The space-time variation in pressure head and volumetric water content was simulated using the code VS2DT (Figure 6.14). The pressure head distributions at times of 25, 225, and 525 years are shown in Figures 6.14a through 6.14c, respectively. Figures 6.14d through 6.14f depict the calculated distributions in volumetric water content. The hydraulic properties are characteristic of sand: $K_s = 16$ ft/day, $n = 0.33$, $\theta_r = 0.072$, $\alpha = 72(1/\text{ft})$, and $\beta = 1.7$ for the van Genuchten model. The ratio of vertical hydraulic conductivity to the horizontal hydraulic conductivity is 0.1. The results illustrate: (1) how high water contents correspond with low negative pressure heads; (2) that flow vertically is larger than flow horizontally; and (3) as time increases, water flows down to the water table.

▶ 6.6 GROUND-WATER FLOW IN FRACTURED MEDIA

In fractured rocks, the interconnected network of fractures provides the main pathway for fluid flow. The solid rock blocks in most cases are much less permeable than the network.

Cubic Law

A single fracture is commonly represented using the idealized parallel plate model. The plates represent the rock matrix, with the fracture defined by the open space between the two plates (Figure 6.15). The thickness of the fracture is described in terms of an aperture b. The volumetric flow in a fracture is a function of the aperture cubed—known as the *cubic law* (Romm, 1966):

$$Q = -\frac{\rho_w g b^2}{12\mu}(bw)\frac{\partial h}{\partial L} \tag{6.29}$$

where Q is volumetric flow rate, ρ_w is density of water, g is gravitational acceleration, μ is viscosity, b is aperture opening, w is fracture width perpendicular to the flow direction, and $\partial h/\partial L$ is head gradient along the fracture. This equation is of the form $Q = KiA$, where i is the gradient $(-\partial h/\partial L)$ and the area A is (bw). The hydraulic conductivity for this single fracture is

$$K = \frac{\rho_w g b^2}{12\mu} \tag{6.30}$$

Equation (6.30) is confirmed by experiments for smooth optical glass. Real fractures, however, have a variable aperture due to the roughness of the walls of the fracture. In some places, the two walls of the fracture actually touch. In other places, in the same fracture, the aperture can be large. Flow in a fracture network is also commonly reduced by the presence of secondary minerals that form in the fracture and plug it. The hydraulic conductivity for a rough aperture may be written as

$$K = \frac{\rho_w g b^2}{12\mu[1 + C(x)^n]} \tag{6.31}$$

where C is some constant larger than one, x is a group of variables that describe the roughness, and n is some power greater than one. Hence, roughness causes a decrease in hydraulic conductivity.

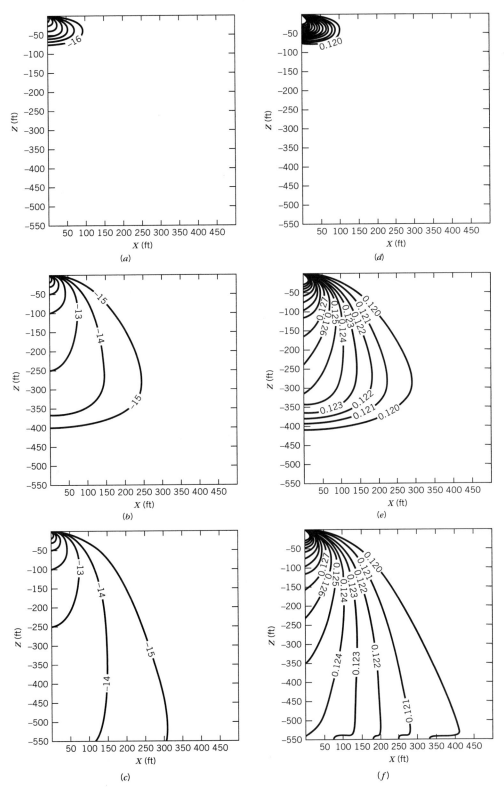

Figure 6.14 Water leakage into the unsaturated zone through a ditch (simulated using VS2DT).
(*a*) Pressure head at 25 years; (*b*) pressure head at 225 years; (*c*) pressure head at 525 years;
(*d*) volumetric water content at 25 years; (*e*) volumetric water content at 225 years; (*f*) volumetric
water content at 525 years.

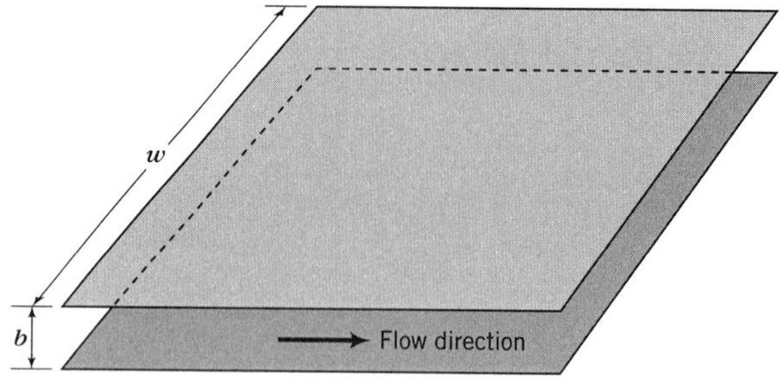

Figure 6.15 A parallel-plate model representing a single fracture.

▶ **EXAMPLE 6.6**

The aperture of a fracture is 0.001 m. The fracture itself is 0.1 m wide. If the hydraulic gradient in the fracture is 0.001, what is the volumetric flow rate in the fracture?

SOLUTION The hydraulic conductivity in the fracture is

$$K = \frac{\rho_w g b^2}{12\mu} = \frac{(998.2 \text{ kg/m}^3)(9.8 \text{ m/sec}^2)(0.001 \text{ m})^2}{12(1.002 \times 10^{-3} \text{ kg/m/sec})} = 0.81 \text{ m/sec}$$

The volumetric flow rate in the fracture is

$$Q = -K(bw)\frac{\partial h}{\partial L} = (0.81 \text{ m/sec})(0.001 \times 0.1 \text{ m}^2)(0.001) = 8.1 \times 10^{-8} \text{ m}^3/\text{sec}$$

Flow in a Set of Parallel Fractures

Calculating the quantity of flow in a complex network of fractures is difficult and usually requires a sophisticated computer model. For scoping calculations, however, some simple fracture systems are amenable for analysis. One example is a network of equally spaced parallel fractures with no ground-water flow in the matrix (Figure 6.16) (Reeves et al., 1986). The density of fractures is defined as the *fracture frequency* (N), which is the number of fractures per unit length. A term called *fracture porosity* is the ratio of the fracture volume over the matrix volume and is written as

$$\phi_f = \frac{b}{s} \tag{6.32}$$

where ϕ_f is fracture porosity, b is fracture aperture, and s is termed fracture spacing, which is the distance between fractures. Snow (1968) derived an expression relating the permeability of a fracture network to fracture porosity and fracture aperture.

$$k_f = \frac{b^3}{12}N \quad or \quad K_f = \frac{\rho_w g b^3}{12\mu}N \tag{6.33}$$

The fracture spacing and fracture frequency can be converted by

$$N = \frac{\cos(\theta_f)}{s} \tag{6.34}$$

where θ_f is the fracture orientation, which is the angle between the fractures and flow directions, or the principal directions of hydraulic conductivity. Assuming θ_f is zero, the permeability can be expressed in terms of fracture porosity as

$$k_f = \frac{b^2}{12}\phi_f \quad or \quad K_f = \frac{\rho_w g b^2}{12\mu}\phi_f \tag{6.35}$$

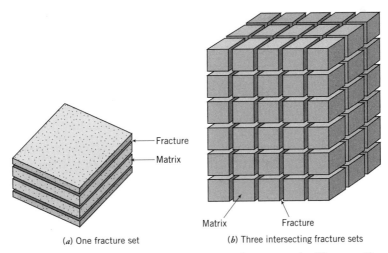

Figure 6.16 A double-porosity model representing a network of fractures (from Reeves et al., 1986).

Equation (6.35) indicates that the permeability of a network of equally spaced fractures is the product of permeability in a single fracture and fracture porosity.

▶ **EXAMPLE 6.7**   Consider a network of parallel fractures with a fracture spacing of 10 m and individual fracture apertures of 0.001 m. The fracture network is oriented parallel to the direction of flow, and the geological medium is homogeneous. Calculate the permeability, hydraulic conductivity, and fracture porosity of the medium.

SOLUTION   Assuming that the fractures are parallel to the flow direction, the fracture frequency of the medium is

$$N = \frac{1}{s} = \frac{1}{(10\,\text{m})} = 0.1(1/\text{m})$$

The permeability of the medium is

$$k_f = \frac{b^3}{12}N = \frac{(0.001\,\text{m})^3}{12}(0.1\,\text{m}^{-1}) = 8.3 \times 10^{-12}\,\text{m}^2$$

The hydraulic conductivity is

$$K_f = \frac{\rho_w g k_f}{\mu} = \frac{(1000\,\text{kg/m}^3)(9.8\,\text{m/s}^2)(8.3 \times 10^{-12}\,\text{m}^2)}{(1.002 \times 10^{-3}\,\text{kg/m/s})} = 8.1 \times 10^{-5}\,\text{m/s}$$

The fracture porosity is

$$\phi_f = \frac{b}{s} = \frac{(0.001\,\text{m})}{(10\,\text{m})} = 1.0 \times 10^{-4}$$

Equivalent-Continuum Approach

So far, we have considered fracture flow in terms of single fractures or a network of parallel fractures. For most problems, it is not practical to worry about flow in terms of the characteristics of individual fractures or simple sets. More commonly, a fractured medium is represented as an equivalent porous medium. In other words, if the medium is sufficiently fractured and one "stands back" far enough, the medium looks like a porous medium with large grains. The equivalent permeability of the continuum in each principal direction can be obtained from the permeability of the fracture and matrix network.

$$
k = \frac{k_m + \dfrac{A_f}{A_m} k_f}{1 + \dfrac{A_f}{A_m}}
\tag{6.36}
$$

where k_m is the permeability of the matrix, k_f is the permeability of the fracture network, and A_m and A_f are the cross-sectional contact areas of matrix and fractures in the principal directions, respectively. For a fractured rock with fracture components in three coordinate directions (Figure 6.16b), Eq. (6.36) applies to each direction, respectively.

▶ **EXAMPLE 6.8**

In a network of fractures, the permeability of the fractures is 1.0×10^{-8} m^2, and the permeability of the matrix is 1.0×10^{-14} m^2. The ratio of the cross-sectional contact areas between a fracture and matrix block is 1.0×10^{-6}. What is the equivalent permeability of the medium? If the ratio is 10^{-4}, what is the equivalent permeability?

SOLUTION   The equivalent permeability for $A_f/A_m = 1.0 \times 10^{-6}$:

$$
k = \frac{k_m + \dfrac{A_f}{A_m} k_f}{1 + \dfrac{A_f}{A_m}} = \frac{(1.0 \times 10^{-14}\, \text{m}^2) + (1.0 \times 10^{-6})(1.0 \times 10^{-8}\, \text{m}^2)}{1 + 1.0 \times 10^{-6}} = 2.0 \times 10^{-14}\, \text{m}^2
$$

The equivalent permeability of the effective continuum for $A_f/A_m = 1.0 \times 10^{-4}$:

$$
k = \frac{k_m + \dfrac{A_f}{A_m} k_f}{1 + \dfrac{A_f}{A_m}} = \frac{(1.0 \times 10^{-14}\, \text{m}^2) + (10^{-4})(1.0 \times 10^{-8}\, \text{m}^2)}{1 + 10^{-4}} = 10^{-12}\, \text{m}^2
$$

▶ **EXERCISES**

6.1. The density of water is 1.0 g/cm^3, the bulk density of soil is 1.6 g/cm^3, and the grain density is 2.67 g/cm^3. Calculate the porosity of the soil. If the weight of dry soil is 200 g and the weight of the wet soil is 230 g, calculate the volumetric water content and water saturation of the sample.

6.2. The porosity, volumetric water content, and residual water content of a sample are 0.45, 0.21, and 0.05, respectively. Calculate the water saturation and the effective water saturation.

6.3. Infiltration is very important for water flow in the unsaturated zone. For a two-hour precipitation event, the precipitation rate is 5 in./hour. The infiltration rate attains steady state after 10 minutes. The saturated hydraulic conductivity is 5 ft/day. Draw a plot of infiltration rate versus time in the unsaturated zone. Assume the infiltration rate is zero before the storm.

6.4. Compare the ground-water flow equation (5.18) in the saturated zone with the flow equation (6.21) in the unsaturated zone. Write down and explain the similarity and difference between the two equations.

6.5. Calculate and plot the water retention curves, relative hydraulic conductivity curves, and specific moisture capacity curves for Del Monte sand, Columbia sandy loam, and Limon silt in Table 6.1 using Brooks–Corey and van Genuchten models. Discuss the characteristics of the curves for different geological materials and different models.

6.6. Numerical models are available for the simulation of water flow in the unsaturated zone. An infiltration experiment was carried out in the farming field to study the potential pesticide contamination of ground water. The infiltration rate is 10 in./hour in a control area. The thickness of the unsaturated zone is 20 ft. Assuming that a one-dimensional model will be used for this project, what boundary and initial conditions would you like to specify? What hydraulic parameters do you need in the model? What do saturation and pressure head profiles look like in the unsaturated zone?

6.7. If the model is a two-dimensional vertical section, answer the same questions in Exercise 6.6.

6.8. For an unconsolidated sand (Table 6.1), calculate the infiltration rate using the Green-Ampt model for a ponding depth of 2 cm. Assume that a precipitation event lasts two hours. If the precipitation rate is 40 cm/hour, plot the precipitation rate, infiltration rate, and runoff rate on the same graph paper.

6.9. Briefly describe the discrete fracture, the double-porosity, and the effective continuum approaches for dealing with water flow in fractured geological media.

6.10. If the aperture of a fracture is 0.001 m, what are the permeability and hydraulic conductivity in the fracture? For a network of equally spaced fractures with fracture porosity of 1.0×10^{-4}, calculate the permeability and hydraulic conductivity of the fractured media.

6.11. If the permeabilities of fracture and matrix blocks are 10^{-4} and 10^{-8} m^2, respectively, what is the effective permeability if an effective continuum model is used to describe the fractured medium? Assume that the ratio of cross-sectional contact areas between the fracture and the matrix blocks is 10^{-3}.

► CHAPTER HIGHLIGHTS

1. The volumetric water content is defined as the ratio of the volume of water to the total volume of a medium. The water saturation is the ratio of the volume of water to the volume of voids. Thus, the water content is equal to or less than the porosity, and the water saturation falls in a range between zero and one.

2. The hydraulic head is the sum of elevation and pressure heads. The pressure head is positive in the saturated zone, negative in the unsaturated zone, and zero at the water table. The negative pressure head in the unsaturated zone explains why water present in partially saturated soils cannot flow into a borehole.

3. A tensiometer is a device that measures pressure head in the unsaturated zone. It consists of a porous ceramic cup connected by a water column to a manometer or vacuum gage.

4. There are three stages of infiltration. At early time, infiltration of precipitation increases the water content. Capillary forces that pull the water into the dry soil cause initial downward spreading. Once infiltration stops, water continues to move downward due to both capillary and gravitational forces. Eventually, the water moves into the saturated zone.

5. Flow in the unsaturated zone is complicated by the fact that there are two fluid phases (air and water) present together. Both the volumetric moisture content (θ) and unsaturated conductivity are functions of pressure head. When a medium is near saturation with a pressure head close to zero, the hydraulic conductivity takes on its maximum value. As the water content is lowered and pores become filled with air, the pressure head becomes more and more negative and the hydraulic conductivity decreases. Brooks–Corey and van Genuchten equations describe the empirical relations between these variables.

6. One can derive an equation to describe patterns of flow in a partially saturated medium. The unsaturated flow equation is similar to the saturated flow equation, with soil moisture replacing porosity and unsaturated hydraulic conductivity now a function of pressure head. Because the parameters of the equation vary as a function of the pressure head, it is difficult to solve unsaturated flow equations. Fortunately, several useful mathematical models are available.

7. Flow in a single fracture is described by the so-called cubic law that states that the volumetric flow rate is proportional to the aperture cubed. Thus, small changes in fracture aperture can produce large changes in flow rates.

8. There are two approaches for dealing with fractured media. The discrete fracture approach conceptualizes the network as a collection of individual fractures having some length, width, and aperture. The continuum approach considers the network to be equivalent to a porous medium with a set of

porous-medium type parameters. This latter approach is used in most practical problems involving fractured media.

9. The hydraulic conductivity of a single fracture is given by

$$K = \frac{\rho_w g b^2}{12\mu}$$

where K is hydraulic conductivity, ρ_w is density of water, g is gravitational accelaration, μ is viscosity, and b is fracture aperture.

7

BASIC GEOLOGIC AND HYDROGEOLOGIC INVESTIGATIONS

Ground-water studies depend on an ability to investigate the subsurface and make key measurements. This chapter examines basic investigative and measurement techniques. We begin by looking at methods for describing the geologic setting. The geologic framework provides the foundation on which a study is created. The chapter continues with a discussion of methods for making hydraulic-head measurements in the field. We will reserve our discussion of hydraulic-conductivity measurements for a series of chapters beginning with Chapter 9. The present chapter concludes with an overview of key geophysical approaches and a discussion of the overall site-investigation process.

▶ 7.1 KEY DRILLING AND PUSH TECHNOLOGIES

Field investigations depend on drilling holes in the ground or pushing sampling/measurement probes into the ground without a hole. Various types of drilling rigs and newly emerging "push" technologies accomplish these tasks.

Many different types of drilling methods may be used. Here, we focus on auger drilling and mud/air rotary drilling, which are the most common techniques. Readers interested in other methods can refer to a detailed summary in Barcelona and others (1985) or Campbell and Lehr (1973).

> Key drilling and push
> technologies are examined in
> more detail in the PowerPoint
> show (show BasicDrillPush.ppt)
> provided on www.wiley.com/college/schwartz

Auger Drilling

Augers can be used to drill holes in unlithified sediments or, in other words, materials other than rock. The rig rotates an auger with a drill bit attached into the ground (Figure 7.1). The augers carry material cut by the drill bit up the hole to the ground surface. The hole is deepened by adding augers to the string of augers already in the ground. This process is shown schematically and illustrated by photographs on the PowerPoint show. There are two main types of augers: solid-stem and hollow-stem augers. *Solid-stem augers* look like large versions of the drill bits that many of us have used to drill holes in wood. They work best in cohesive deposits (like clays) to provide holes that will stay open once the augers are removed. This open hole provides access for sampling or for installing a piezometer. Some saturated deposits like sand are noncohesive. When solid-stem augers are withdrawn from such deposits, the hole caves in, blocking further access.

Hollow-stem augers provide a nifty solution to this caving problem. The drill bit is constructed in two parts, with the central part held in place by a set of small-diameter drilling rods that are inserted through the hollow core of the augers. When it is time to sample or to install a piezometer, the bit is disassembled to provide an opening from the surface to the porous medium in front of the bit. The augers then act as a temporary casing that holds the hole open. Drilling begins again, once the bit is reassembled.

This feature is ideal for collecting periodic soil samples as drilling continues. A coring device, like a split-spoon sampler or Shelby tube (Figure 7.2a, b), is added to the small-diameter drilling rods, run down the hole, and hammered into the porous medium just below the tip of the auger.

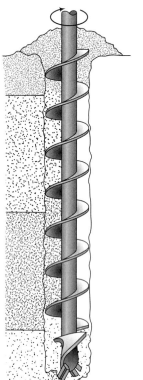

Figure 7.1 This figure shows a solid-stem auger drilling a hole. The bit chops up the porous medium, which is brought up the auger to the ground surface (from Scalf et al., 1981).

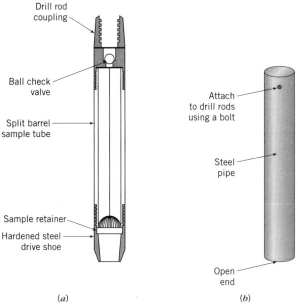

Drill rod
coupling

Ball check
valve

Split barrel
sample tube

Sample retainer

Hardened steel
drive shoe

(a)

Attach
to drill rods
using a bolt

Steel
pipe

Open
end

(b)

Figure 7.2 Examples of equipment used to collect solid samples of unlithified media. The barrel split-spoon sampler (Panel *a*) can be broken in half along its length to obtain the sample. With the Shelby tube (Panel *b*), the sample has to be extruded from the steel cylinder. Panel *a* is from Rehm et al., 1985 ©Electric Power Research Institute, EPRI EA-4301, reprinted with permission.

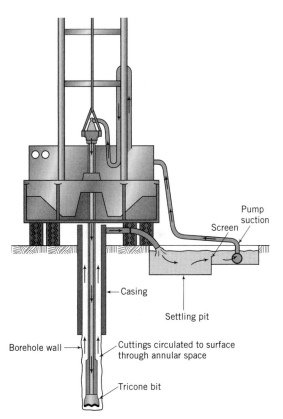

Pump
suction

Screen

Casing

Settling pit

Borehole wall

Cuttings circulated to surface
through annular space

Tricone bit

Figure 7.3 Shown here is an example of a rotary drill rig. The arrows illustrate how the drilling mud circulates from the mud pit, down the casing, and out the bit. The fluid returning up the borehole carries the cuttings up and out of the hole (from Aller et al., 1991).

Mud/Air Rotary Drilling

Mud/air rotary drilling involves turning a string of drill casing with a bit attached (Figure 7.3). The bit can come in several different styles depending on the medium being drilled.

The tricone bit is a familiar style with three "gear-like" rollers attached. Pieces of the material that the bit chops or grinds up have to be removed, or drilling cannot continue. A drilling fluid, air or water, is circulated under pressure down the drill casing and through the drill bit. The drill cuttings are brought back up the hole with the return flow of drilling water or air. Schematics of the drilling process along with examples of the drilling rigs are shown in show/BasicDrillPush.ppt.

With mud rotary rigs, circulation of the drilling fluid is occasionally "lost" when water or air goes into a permeable zone rather than back up the hole. With mud rotary rigs, it is necessary to plug off this permeable zone using drilling fluid thickened by additives (e.g., bentonite). This *drilling mud* must eventually be removed if the zone is subsequently developed to provide ground water. With air rotary rigs, circulation is maintained by driving a casing as the drilling proceeds.

Rotary drill rigs are capable of drilling through both drift and bedrock. This capability makes it the rig of choice for installing wells in rock to virtually any depth of interest. With a mud rotary rig, the drilling mud usually keeps the hole open. When air rotary is used for "soft" formations, a casing needs to be driven to keep the hole open. Drill cuttings returned with the mud can be examined to provide a geologic log. It is also possible to collect split-spoon samples with a rotary rig. Commonly, borehole geophysical logs are run in these holes to provide a more complete picture of the stratigraphy. One drawback of air rotary rigs is the possibility that small quantities of oil from the compressor will be introduced to the subsurface. Filters can be attached to avoid this problem.

Push Rigs

Push technologies provide useful alternatives in hydrogeological investigations. They involve pushing or hammering a string of casing into the ground. The casing can carry various types of sampling tools, sensors, or permanent monitoring devices. This approach is faster than more conventional drilling, provides access without drill cuttings (an advantage at contaminated sites), and in some cases, provides near real-time electronic data. The main disadvantage of these rigs is their inability to drill rock and some types of sediments.

Push rigs are available in all shapes and sizes. Handheld electric hammers are capable of emplacing small-diameter water-sampling wells in shallow settings. *Soil-probing machines* (Figure 7.4) (show/BasicDrillPush.ppt) provide capabilities for pushing or hammering a casing string to about 20 m under good conditions. These rigs are relatively inexpensive, are easy to operate, and provide access to congested sites. They are the tool of choice for rapid sampling of solids at contaminated sites, but they have capabilities for collecting water and soil gas samples. Tools can be added to provide various types of geophysical measurements while pushing.

Cone penetrometer rigs are powerful, sophisticated tools for push-type measurements. In its most basic form, the cone-penetration test uses a hydraulic ram to push a casing string into the soil (Figure 7.5). Electronic sensors on the penetrometer measure various shear resistances (Chiang et al., 1989). Typically, this information is used for geotechnical design purposes and for developing information on the stratigraphy of a site. Refinements of this technology provide water-pressure measurements (yielding hydraulic heads) as a function of depth and hydraulic conductivities at specific depths, using the rate of pore-pressure dissipation (Chiang et al., 1989).

There are two electronic sensors on a cone penetrometer (Strutynsky and Sainey, 1990). One measures the shear resistance to soil penetration on the conical tip (also called cone-end bearing resistance), and the other measures the shear resistance along the side of the

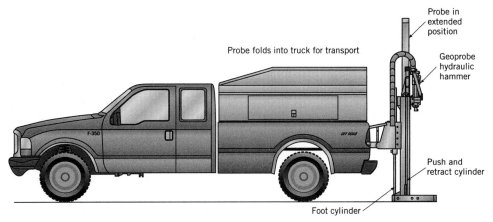

Probe in
extended
position

Probe folds into truck for transport

Geoprobe
hydraulic
hammer

Push and
retract cylinder

Foot cylinder

Figure 7.4 A soil-probing machine mounted in a van can be easily transported and set up at a site (by permission of Geoprobe Systems ®/Kejr, Inc).

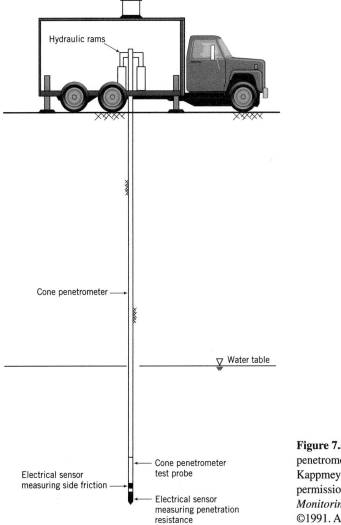

Hydraulic rams

Cone penetrometer →

▽ Water table

Cone penetrometer
test probe

Electrical sensor
measuring side friction →

Electrical sensor
measuring penetration
resistance

Figure 7.5 Schematic of a cone penetrometer (from Smolley and Kappmeyer, 1991). Reprinted by permission of *Ground Water Monitoring Review*. Copyright ©1991. All rights reserved.

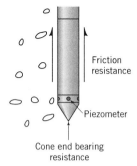

Figure 7.6 A cone penetrometer measures frictional resistance, cone end-bearing resistance, and pressure (modified from Strutynsky and Sainey, 1990). Reprinted by permission of Conference on Petroleum Hydrocarbons and Organic Chemicals in Ground Water—Prevention, Detection and Restoration. Copyright ©1990. All rights reserved.

penetrometer (Figure 7.6). These measurements are made continuously and recorded in digital form. The cone end-bearing resistance can respond to layers just a few inches thick. The frictional resistance has a resolution of about 6 in. (Strutynsky and Sainey, 1990). Some tools carry an electrical transducer to provide pressure measurements and ultimately estimates of hydraulic heads.

The cone penetrometer measurements provide detailed stratigraphic information about a site. The cone end-bearing resistance increases exponentially with grain size. Strutynsky and Sainey (1990) indicate that the cone end-bearing resistance is 7 to 15 tons per square foot (TSF) for stiff clay and 150 to 300 TSF for dense sand.

The ratio of the frictional resistance to the cone end-bearing resistance is the *friction ratio*. The friction ratio is related to the fines content of the medium—low in sands (e.g., 0.02) and high in clays (e.g., 0.05). Figure 7.7 illustrates the typical results available from a cone penetrometer carrying the two shear sensors and a transducer that is capable of measuring pore pressures. Note how it is able to resolve small changes in lithology moving down the section.

The key measurements from a cone penetrometer, the cone end-bearing resistance, and the friction ratio can also be interpreted in terms of the Unified Soil Classification System (USCS) that is commonly used in civil engineering works. Douglas and Olsen (1981) developed correlations between measured soil properties from sampled boreholes and cone penetrometer data. The correlation chart in Figure 7.8 provides the basis for converting cone penetrometer measurements to USCS soil types.

▶ 7.2 PIEZOMETERS AND WATER-TABLE OBSERVATION WELLS

One of the main reasons for drilling at a site is to install piezometers or water-table observation wells. As we saw in Chapter 3, a *piezometer* in its simplest form is a borehole or standpipe that is installed to some depth below the water table (Figure 7.9). Water entering the piezometer rises up the casing and reaches some stable elevation. The elevation of the water in the casing is the hydraulic head at the point of measurement—the midpoint of the screen at the bottom of the standpipe (Figure 7.9).

A *water-table observation well* has a slightly different design and provides somewhat different information. The standpipe is screened across the water table (Figure 7.9). The water-level elevation in this well provides the elevation of the water table and the hydraulic head at the top of the ground-water system. The point of measurement is the water table, which is the top of the saturated ground-water system. Thus, as the water table rises or falls, the point of measurement also moves.

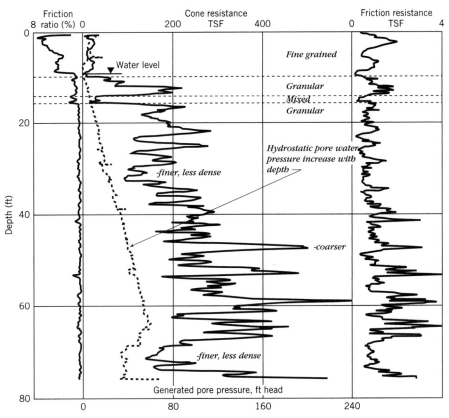

Figure 7.7 Data from a cone penetrometer are plotted in relation to lithologic data. Note particularly how the cone resistance increases with increasing grain size (from Strutynsky and Sainey, 1990). Reprinted with permission of Conference on Petroleum Hydrocarbons and Organic Chemicals in Ground Water—Prevention, Detection and Restoration. Copyright ©1990. All rights reserved.

Basic Designs for Piezometers and Water-Table Observation Wells

Real piezometers don't look much like the idealization in Figure 7.9a. Actual designs (Figure 7.10) provide

1. a screen to create a relatively large surface area for water to enter the standpipe
2. a sand pack around the screen to increase the effective size of the screen and to support material placed above
3. a seal above the screen to prevent water from leaking along the casing
4. screen and casing materials that won't react with the ground water or contaminants carried in the ground water; and
5. a casing protector to finish the top of the piezometer and to prevent unauthorized access.

Let's describe a piezometer that is typically installed in the field. It is constructed with a screen at the end of the casing. A screen is a piece of the casing with holes or slots cut to let water flow into the casing. Manufactured screens are designed both to be strong and to maximize the open area. The screen is surrounded by a *sand pack* that supports the

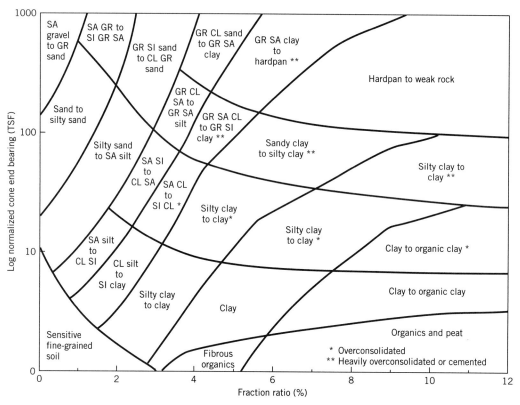

Figure 7.8 Cone end-bearing resistance can be used together with the friction ratio to classify soils according to the Unified Soil Classification System (from Strutynsky and Sainey, 1990). Reprinted with permission of Conference on Petroleum Hydrocarbons and Organic Chemicals in Ground Water—Prevention, Detection and Restoration. Copyright ©1990. All rights reserved.

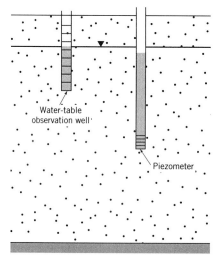

Figure 7.9 Schematic of a piezometer and a water-table observation well.

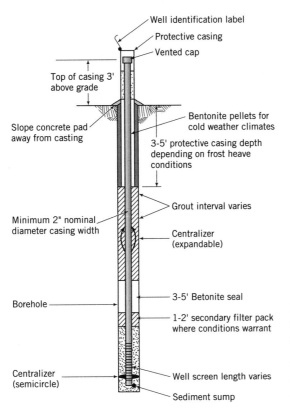

Figure 7.10 This figure shows the construction details for a basic standpipe piezometer (from Nielsen, 1996). Reprinted by permission of Tenth National Outdoor Action Conference and Exposition. Copyright ©1996. All rights reserved.

screen structurally and provides a foundation for all the other materials that are added to the borehole above.

For water levels to change in a piezometer, a relatively large volume of water has to flow into or out of the standpipe. The larger the surface area of the intake, the faster these inflows or outflows are accomplished. There is a big difference in surface area provided by a 5.1 cm (2 in.) diameter open end in a piece of pipe as compared to a 1.5 m long screen packed with sand in a 15.2 cm (6 in.) borehole (Figure 7.11). In effect, the sand pack makes the effective screen's diameter the same as the borehole diameter.

A screen that is 1 m long is adequate for most applications. In practice, however, screens are often constructed 3 to 5 m long, with piezometers bordering on small wells. These longer screens are not inherently bad except that a hydraulic-head measurement is supposed to be a "point" measurement rather than a formation average. With wells used primarily for water-quality monitoring, the long screens give one an ability to "detect" contaminants but not to estimate "point" concentration values.

A piezometer can work properly only if the water moving into or out of the standpipe comes from the zone adjacent to the intake. If water leaks down the borehole from above the intake, then hydraulic-head and water-quality measurements can be erroneous. Piezometers usually incorporate a seal and back fill above the sand pack, which prevent borehole leakage. The *seal* is usually constructed from *granular bentonite*, a clay mineral that expands and has a very low hydraulic conductivity, once it absorbs water. Above the seal, the hole can be backfilled with bentonite grout or appropriate natural materials that support the casing and plug the borehole.

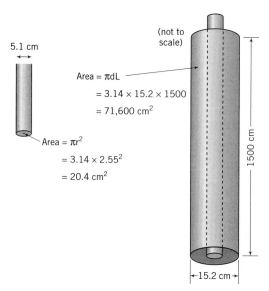

Figure 7.11 The surface area afforded by the open end of 5.1 cm (2 in.) pipe compared to the open area of a 1 m long, 15.2 cm (6 in.) diameter sand-packed hole.

The upper part of the borehole is completed using concrete and a metal casing protector. The concrete seal prevents surface drainage from moving down the old borehole and anchors the casing protector. The casing protector can be locked to prevent unauthorized access to the piezometer and sometimes can prevent damage from collisions with vehicles. In high-traffic areas, piezometers are completed below grade level to prevent damage.

Piezometers are usually constructed with a nominal casing diameter of 5.1 cm (2 in.). This diameter is large enough to facilitate development following installation, measurement of hydraulic head with an electric tape, and sampling with small pumps or bailers. Common casing materials include polyvinyl chloride (PVC), stainless steel, or Teflon®. PVC pipes with threaded joints sealed by o-rings are used most commonly. Applications involved with water-quality monitoring might necessitate these other materials. On occasion, piezometers can be constructed with up to 15.2 cm (6 in.) casings. However, as we will see, these piezometers can be insensitive in low-permeability units.

The water-table monitoring well (Figure 7.12) has many of the same characteristics as the piezometer. The main difference, however, is in the length of the screen. The water-table observation well has a long screen that extends above and below the water table. Thus, if the water-table level rises or falls, it will remain within the screened section. Providing a seal with this well is less critical than the piezometer because the zone above the screen is unsaturated. Good practice would provide a low-permeability backfill to prevent surface drainage from moving down the old borehole.

▶ 7.3 INSTALLING PIEZOMETERS AND WATER-TABLE WELLS

The description of the basic design for wells and piezometers doesn't really explain how they are installed. We will deal with the installation process using a combination of written summary here and a PowerPoint show on the website. In addition, the point needs to be made that many different variations on these basic designs are implemented to respond to specific issues related to hole depth, geological setting, and requirements of the study. For example, it is difficult to provide good seals in formations that cave (that is, saturated sands and silts), or at depths greater than about 50 m.

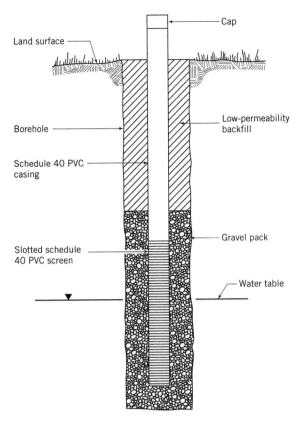

Figure 7.12 Construction details for a basic water-table observation well (from Nielsen, 1996). Reprinted by permission of Tenth National Outdoor Action Conference and Exposition. Copyright ©1996. All rights reserved.

A PowerPoint show is included on the website as show/BasicPiezInstall.ppt to illustrate how the following types of piezometers are installed: (1) shallow piezometer in noncaving materials, (2) shallow piezometer in caving materials, and (3) deep piezometer in rock.

Shallow Piezometer in Noncaving Materials

The basic designs for the piezometer or water-table observation well (Figures 7.10 and 7.12) are the types typically used for noncaving materials. In rock or clayey sediments (for example, lacustrine clay or glacial till), a borehole will remain open once the drill stem or augers are removed. The screen and casing can be lowered down the hole, and proper amounts of filter sand, bentonite, and backfill are added in order. In a shallow borehole, these materials can be poured down the hole. A tape with a heavy weight attached is used to measure where these materials end up in the borehole.

Once holes become deeper than about 30 m, it is difficult to get the materials all the way to the bottom of the hole. For these deeper holes, alternative approaches are necessary.

Shallow Piezometer in Caving Materials

Caving materials, like water-saturated sands or silts, pose problems, especially with auger drilling techniques. Mud rotary holes often will stay open in these materials, but the mud

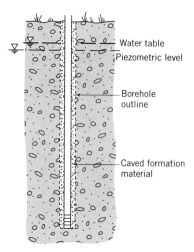

Figure 7.13 A shallow piezometer constructed in caving sediments. Note that materials filling the borehole are assumed to prevent downward leakage around the casing (from Cherry, 1983).

present in the hole can contaminate samples collected for chemical analysis. Solid-stem augers are also problematical. Once the augers are removed, the borehole will collapse, making it virtually impossible to install the casing. One approach with caving materials is to drill a hole to the desired depth with a hollow-stem auger, install the casing down the hollow center of the auger, and back out the auger. The hole will collapse around the casing to provide a natural seal (Figure 7.13). A variation on this approach involves hammering a small-diameter steel screen and casing to the desired depth.

These types of piezometers are commonly used in nonregulatory applications or research investigations, where cost is a factor. Most regulatory agencies find these wells unsatisfactory because the possibility of leakage along the casing cannot be ruled out.

Techniques are available with a hollow-stem auger or soil probing machines to install both a sand pack and a seal. Although it is difficult to send sand down to the screen, down the middle of a hollow-stem auger, it is possible if one is patient and careful. The sealing material can be sent down as pumped slurry or as poured solid granules. The website shows the steps involved with a hollow-stem auger. Figure 17.14 shows how sand packs and seals are installed with wells emplaced with a *Geoprobe® brand* soil probing machine.

In very complex settings, it is often necessary to case off caving intervals and to provide access for deeper holes. For example, at some sites, auger holes are cased to bedrock to keep the hole open for core drilling into bedrock. Temporary casing used for this purpose can be jacked out of the ground once the piezometer is completed. Cable-tool drilling is ideally suited for problems where a temporary casing needs to be installed while drilling. The advantage of a cable tool rig in this application as compared to a hollow-stem auger is (1) a larger diameter hole to work in and (2) the ability to drill through rock. Cable-tool rigs, however, are often slow in drilling.

Deep Piezometers

Installing piezometers to depths greater than 30 m in rock is difficult. There is less of a problem with caving because the drilling technique usually can be counted on to keep the hole open. The greater problem is in placing the sand pack and grout seal. The website shows an approach with no sand pack, where the seal and backfill are kept away from the screen by a metal or plastic basket attached to the casing above the screen. The seal is emplaced as pumped grout from above. These types of piezometers can be installed to about 100 m.

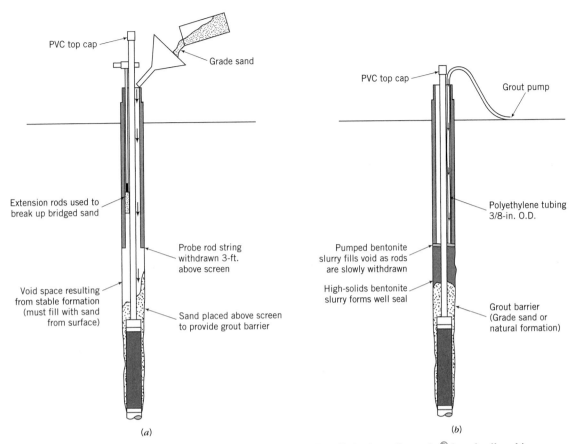

Figure 7.14 Permanent piezometers can be installed using a Geoprobe® brand soil probing machine. If materials around the screen are not caving, sand can be added (Panel *a*). A seal can be installed by pumping grout down through a hose as rods are withdrawn (Panel *b*) (by permission of Geoprobe Systems ®/Kejr, Inc).

Some applications, for example, the design of a nuclear waste repository, could require piezometers up to 1000 m deep. Westbay-type systems have been used for this purpose. They are described in Chapter 16, where additional requirements in well design for chemical sampling is discussed.

▶ 7.4 MAKING WATER-LEVEL MEASUREMENTS

Several techniques are available to measure the elevations of water levels in observation wells or piezometers. Commonly, an *electric tape* is used to measure the distance from a fixed measurement point on the well casing to the water. An electric tape is a plastic tape (feet or meters), which has electrical wires running inside and a weighted electrode at the end. It is lowered down the well. When it hits the water surface, an electrical circuit is completed, turning on a buzzer or light (Figure 7.15). The electrode is constructed so that in mineralized water current is conducted to complete the circuit. In air, the circuit remains open and the buzzer remains quiet.

The actual elevation of the water surface is determined by subtracting the measurement on the tape from elevation of the fixed measurement point at the top of the casing.

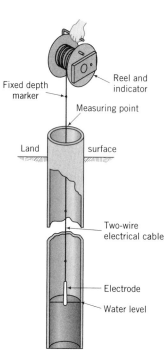

Fixed depth marker

Reel and indicator

Measuring point

Land   surface

Two-wire electrical cable

Electrode

Water level

Figure 7.15 Measuring the depth to water using an electric tape. This depth is subtracted from the elevation of the measuring point at the top of the hole to provide the water-level elevation in the well (from Heath, 1983).

Usually, a surveying crew accurately measures the well-top elevation in relation to nearby benchmarks.

This technique works well in most applications. Care must be taken to completely clean the tape between measurements to avoid cross-contaminating piezometers. One disadvantage is the time required to lower the tape and to make an actual measurement. During the initial stages of an aquifer test with several wells, it may be difficult for a single observer to measure all the wells frequently. One way to deal with this problem is to provide essentially continuous water-level measurements in all of the wells at the same time. The most modern systems use a pressure transducer placed in the well to measure the pressure of a column of water above it (Figure 7.16). Recall that knowledge of the pressure (converted to pressure head) and the elevation head yield the hydraulic head (see example calculations in Figure 7.16).

Output from the transducer is monitored at specified time intervals, that may be as small as a few seconds. This information can be stored in specialized data-acquisition systems or actually transmitted directly to the office. The Hermit 3000, manufactured by In-Situ Inc., is commonly used in aquifer tests for monitoring water-level variations in a series of wells. Data collected in digital form is amenable for the plotting of water-level hydrographs or other types of analyses.

▶ 7.5 GEOPHYSICS APPLIED TO SITE INVESTIGATIONS

A variety of geophysical techniques are used in site investigations. Geophysical methods are based on the physical properties of materials below Earth surface. Surface geophysical methods are commonly used to map features of the geological setting and the location of abandoned hazardous waste disposal sites. Borehole geophysics provide useful stratigraphic and hydrogeologic data.

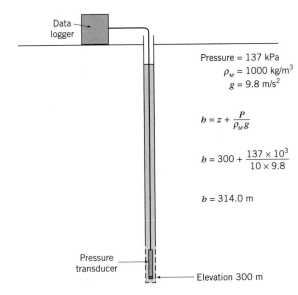

Pressure = 137 kPa
$\rho_w = 1000$ kg/m^3
$g = 9.8$ m/s^2

$$b = z + \frac{P}{\rho_w g}$$

$$b = 300 + \frac{137 \times 10^3}{10 \times 9.8}$$

$$b = 314.0 \text{ m}$$

Figure 7.16 Hydraulic heads can also be determined in a well or piezometer using a pressure transducer and a data logger. The equation illustrates how the elevation of the pressure transducer and the measured pressure are used to calculate hydraulic head.

Surface Geophysical Techniques

Electrical resistivity, electromagnetics, ground penetration radar, seismic reflection, seismic refraction, and magnetic techniques are commonly applied surface geophysical methods that are used in ground-water investigations. In the following sections, we'll introduce the basic theory of the various methods and explain how they are used in ground-water studies.

Electric Resistivity Method

Electrical methods are often applied to describe the geologic setting and patterns of ground-water contamination. The various electrical measurements usually measure *electrical conductivity*—the ability of a material to conduct electricity, or *resistivity*—the reciprocal of electrical conductivity. Rocks and sediments conduct electricity as a consequence of ions in solution in the pore fluid, the charged layer present on clay minerals, and rarely, electrical conduction through metallic minerals. Because metallic minerals typically do not provide long, continuous circuit paths for conduction in the host rock, the conductance of electricity is controlled by the content of dissolved mass in ground water (total dissolved solids) and the relative abundance of clay minerals. In quartz sand, increasing the total dissolved solids content of the pore fluid reduces the resistivity.Increasing the clay content of sediment with the same pore-water chemistry causes a reduction in resistivity.

The field-survey approach to determining the electric resistivity method involves measuring the electric potential difference between two electrodes in an electrical field, as induced by two current electrodes (Figure 7.17). *Apparent resistivity* is then calculated from the measured potential difference and other known parameters as

$$\rho_a = K\frac{\Delta V}{I} \tag{7.1}$$

where ρ_a is the apparent resistivity, ΔV is the voltage difference, I is the electric current induced into the ground, and K is a geometric factor that depends on the pattern of electrode spacing.

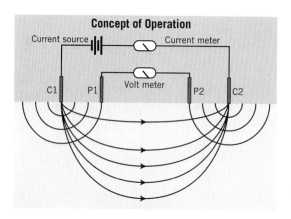

Figure 7.17 An illustration of electric current field induced by electrodes (http://water.usgs.gov/ogw/bgas/, 2001).

There are two modes of field operations: profiling and sounding. The *profiling method* keeps the electrode spacing constant, providing the resistivity change within a fixed volume of earth material. Thus, only one measurement is made at each station. A survey is conducted by moving from station to station along a cross section or a two-dimensional surface grid. Once all the data are plotted, it is possible to interpret differences in the apparent resistivity as a function of geology or human impacts. The sounding method involves making a series of measurements at a single station with increasing electrode spacing. As the spacing between the current electrodes increases, the electrical field extends deeper into the earth to provide resistivity as a function of depth. This information can be interpreted to provide an interpretation of layering at the site.

Several commonly used arrays of electrodes include Schlumberger, Wenner, and Dipole-Dipole. Figure 7.17 illustrates the Schlumberger Array, where two potential electrodes measure the voltage difference at the center of the two current electrodes. The geometric factor (K) in Eq. (7.1) varies depending on the arrangement of electrodes.

Early applications in contamination studies (for example, Carwright and McComas, 1968) involved electrical resistivity surveys to detect changes caused by the presence of contaminant plumes distinguished by elevated salinity. In recent years, these types of surveys have fallen out of favor because terrain conductivity methods (that is, electromagnetic methods) are much more rapid and offer possibilities for higher resolution.

Electromagnetic Methods

Electromagnetic methods induce a current in the ground with an alternating current transmitting coil. The magnetic field around the coil induces an electrical field in the earth to depths that are largely controlled by the background properties of the medium, the moisture content, and the relative difference in the conducting properties of the medium and the target. In the Geonics EM31, two coils are mounted 3.7 m apart on a rigid boom. The equipment is highly portable, and an operator can collect a large number of closely spaced electrical conductivity measurements in a short time. The depth of signal penetration is about 0.75 to 1.5 times the coil separations (Zalasiewicz et al., 1985). Other instruments provide larger coil separations and greater signal penetration.

Terrain conductivity methods have been frequently applied to subsurface mapping, especially to the investigation of contaminated sites. Stewart and Gay (1982) have evaluated electromagnetic soundings for deep detection of conducting fluids, and Greenhouse and Slaine (1983) have applied the technique to mapping contaminants.

The most important application of electromagnetic methods is to detect buried objects or other features at waste disposal sites. Examples of buried objects are filled waste trenches or lagoons, buried steel drums, and lost underground piping or storage tanks. A case study in the application of very-high-resolution electromagnetic surveying is provided by Jordan and others (1991). The method was used to examine a vacant lot that was previously used for industrial hazardous waste processing. During its operation, several unlined lagoons may have existed. The survey was designed to identify the filled lagoons and possibly buried metals. Glacial till covered the site.

The conductivity survey collected 4823 electromagnetic (EM) data points along lines 3.8 m apart in areas suspected of contamination. Stations were about 0.6 to 0.9 m apart. The resulting conductivity map (Figure 7.18) indicates a broad range of conductivity. Suspected

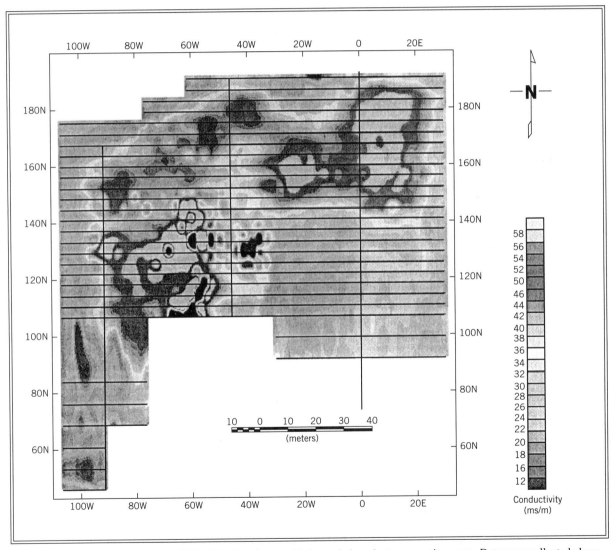

Figure 7.18 Results of a very-high-resolution electromagnetic survey. Data were collected along lines approximately 3.8 m apart (from Jordan et al., 1991). Reprinted by permission of Fifth National Outdoor Action Conference on Aquifer Restoration, Ground Water Monitoring and Geophysical Methods. Copyright ©1991. All rights reserved. Electronic data file kindly provided by Geosoft Inc., Toronto.

lagoon locations are indicated by conductivity values greater than 50 milliSiemens per meter (mS/m), while near-zero conductivities are thought to represent areas of buried metallic debris. The locations of the lagoons correspond with locations determined from old aerial photographs.

Ground Penetrating Radar (GPR)

Ground penetrating radar (GPR) is a rapidly evolving technology for subsurface investigations (Daniels, 1989). The method is used to delineate features of the geologic setting, to map the distribution of buried objects (drums, pipelines etc.), to define the configuration of the water table and stratigraphic boundaries, and to establish the distribution of liquids like gasoline. Thus, GPR is well suited for surveying abandoned waste disposal sites.

Ground penetrating radar is based on the reflection of radio waves from discontinuities under the Earth's surface. A transmitting antenna at the surface radiates short pulses of radio waves into the ground. An antenna that is moved along the surface recovers the reflected energy from the subsurface (Figure 7.19). In principle then, GPR works like a reflection seismic method except that electromagnetic reflections, rather than acoustic returns, are recorded at the surface.

Radar energy is reflected due to changes in dielectric constants and electrical conductivity. Such changes are usually related to variation in properties of the minerals present, degree of saturation, and material density. By moving the GPR along at a slow speed, it is possible to obtain an almost continuous profile of the subsurface. The radar energy reflected from various surfaces (for example, geologic boundaries, water table) is recorded as a wiggle trace or other type of display (Figure 7.20a). Successive traces can be stacked side-by-side (Figure 17.20b) or can be processed to give a picture like a fish finder. In many respects, the interpretation methods used for GPR data are similar to the method for seismic data.

Under ideal conditions (that is, dry sandy units or bedrock), GPR provides a highly resolved picture of subsurface site conditions. The investigative depth of the method is

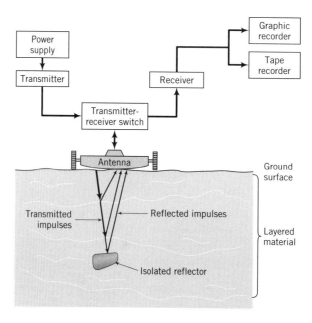

Figure 7.19 With ground-penetrating radar, the transmitted pulse is reflected by geologic boundaries and certain types of buried waste (from Daniels, 1989).

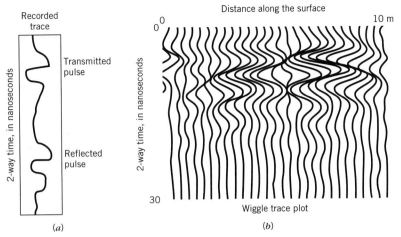

Figure 7.20 The intensity of the reflected impulses returning to the surface is recorded as a function of time (Panel *a*). As the radar unit is moved along the ground surface, a succession of these wave forms are arranged to form a "wiggle-trace" plot (Panel *b*). This plot provides a two-dimensional picture of radar reflectors in the subsurface, which can be interpreted in terms of the geologic setting and underground features (from Daniels, 1989).

determined mainly by the electrical conductivity of the earth materials and can be roughly estimated by

$$Z = \frac{35}{\sigma} \text{(meter)} \tag{7.2}$$

where σ is the electric conductivity in mS/m. Experience shows that the depth of investigation of GPR ranges from a few meters to about 100 m. In most cases, it is less than 30 m. Because clayey sediments are highly conductive, GPR doesn't work well at sites underlain by moist lake clay or glacial till.

A study incorporating numerical modeling, physical modeling, and field surveys at the U.S. Geological Survey Fractured Rock Research Site at Mirror Lake, Grafton County, New Hampshire, was conducted to test the use of GPR surveying to delineate fractures in heterogeneous bedrock (Buursink and Lane, Jr., 1999). Results of 1- and 2.5-dimensional numerical modeling correlate with results of laboratory-scale physical modeling and establish different GPR reflection characteristics for saturated and unsaturated (dry) fractures. Saturated fractures generate higher amplitude reflections than unsaturated fractures and have an opposite phase. GPR reflection data collected over a highway bedrock outcrop near Mirror Lake were processed to reduce noise and clutter, to correct geometric and topographic distortions, and to enhance weak reflections from structures more than 15 m deep. GPR reflection surveys were conducted at the I-93 highway outcrop near Mirror Lake. The outcrop, along the center median of the highway, is more than 200 m long and 20 m wide. The average height of the outcrop is about 6 m above land surface, with the lowest point at land surface and the highest at 12 m. The GPR surveys were conducted along the centerline of the median outcrop, about 10 m from the outcrop wall. GPR data were collected with a portable battery-operated, commercial radar field-instrument using 200 MHz center-frequency antennas.

For this study, two types of GPR surveys were conducted on the bedrock outcrop: (1) continuous profiling common-offset (CO) reflection surveys (Figure 7.21*a*) and (2) common-midpoint (CMP) reflection surveys (Figure 7.21*b*). The continuous profiling

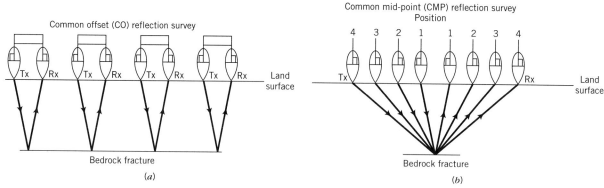

Figure 7.21 Diagram showing (*a*) the antenna geometry for a common-offset (CO) ground-penetrating radar reflection survey and (*b*) the antenna geometry for a common-midpoint (CMP) survey used to derive the average subsurface radar propagation velocity. In both, T_x and R_x indicate the locations of the transmitting and receiving antennas, respectively (from Buursink and Lane, 1999).

surveys were conducted to image the subsurface; the CMP surveys were conducted to establish the radar propagation velocity in bedrock, a value required for data processing.

The data processing methods used in this study included (1) filtering to reduce random noise, instrument noise, and clutter, (2) application of a time-range gain to enhance deep reflections, (3) depth conversion and topographic correction to compensate for changes in surface elevation, and (4) migration to focus reflection energy and to correct for geometric distortion.

Figure 7.22 shows a 40 m section of the field data from the outcrop before (Figure 7.22*a*) and after (Figure 7.22*b*) data processing. Processing improved the field data in several ways. First, true dipping reflectors were differentiated from background clutter caused by small-scale bedrock heterogeneities. Second, deep reflectors were enhanced, increasing the apparent depth of penetration of the GPR survey to at least 17 m below the top of the outcrop. Third, reflectors were correctly positioned and oriented.

The state-of-practice in radar technology is moving toward highly resolved three-dimensional studies involving a large number of parallel survey lines on the ground surface. With the help of appropriate visualization software, one can interpret and display the three-dimensional relationships among various radar reflectors on a site (Thompson et al., 1995).

Reflection Seismic Method

Seismic methods utilize both reflected and refracted energy waves to measure the travel time of seismic waves through different types of lithologic units. Of the two approaches, the reflection seismic is most useful in environmental applications. However, research has been required to refine reflection schemes for studies of relatively shallow depths (Hunter et al., 1984, 1988; Kaida et al., 1995). Reflection seismic methods are often most useful in reconnaissance-type studies for determining the top of the bedrock surface, the topology of structural features, and the pattern of stratigraphic layering. Computer workstations facilitate these interpretations.

Seismic methods are all based on measuring how fast and what paths seismic waves follow in the subsurface. Energy to create various seismic waves is added to the ground by an explosion, vibrating a metal plate on the ground (Vibroseis), or rifle bullet. The amplitudes and travel times of the resulting waves are measured with an array of receivers

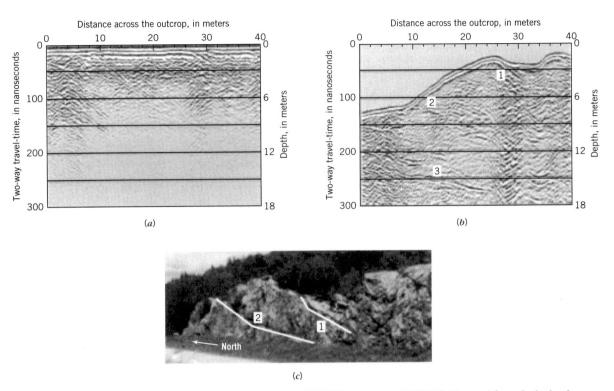

Figure 7.22 Diagram showing (*a*) 200 MHz unprocessed GPR field record from the bedrock outcrop. (*b*) Data shown in (*a*) after processing and topographic correction. Interpreted reflections are labeled 1 through 3. (*c*) Photograph of the section of outcrop surveyed with GPR. Fractures 1 and 2 are annotated in the photograph. The depth scale was calculated assuming a radar propagation velocity of 117 m/μs. The vertical exaggeration is 1.7 times (from Buursink and Lane, 1999).

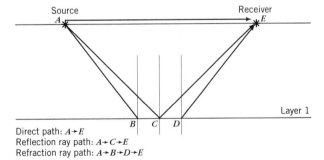

Direct path: $A \rightarrow E$
Reflection ray path: $A \rightarrow C \rightarrow E$
Refraction ray path: $A \rightarrow B \rightarrow D \rightarrow E$

Figure 7.23 Illustration of seismic waves.

and are then recorded. The density and seismic velocity of earth materials are important in determining the behavior of seismic waves. When a seismic wave is generated on the ground, there are several paths along which the seismic waves travel from a source to a receiver (Figure 7.23). The wave transmitted from the source to the receiver along the ground surface is called the *surface wave* (path $A \rightarrow E$). The wave following a pathway down to the lower boundary, along the boundary for some distance, and back to the ground receiver is the *refraction wave* ($A \rightarrow B \rightarrow D \rightarrow E$). The wave, reaching the lower boundary and reflected directly back to the ground surface, is called a *reflection wave* ($A \rightarrow C \rightarrow E$).

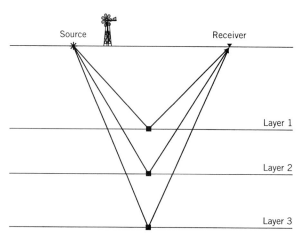

Figure 7.24 Reflection from multiple layers (from Miller and Xia, 1997).

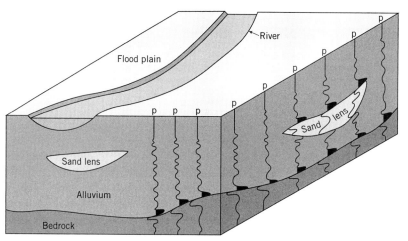

Figure 7.25 A reflection seismic profile may be used to construct geological cross section (from Miller and Xia, 1997).

Each boundary with different seismic velocity can reflect seismic waves (Figure 7.24). With many receivers equally spaced along a profile, a geological cross section can be constructed based on a reflection profile (Figure 7.25).

Seismic reflection surveys routinely involve three basic parts: data acquisition, processing, and interpretation. The basic instrument for seismic studies is a *seismograph*, which records and enhances high-frequency sound waves to detect geologic features. Selection of the frequencies to be enhanced and the amplifier gain necessary to maximize the recorded relevant geologic information depends on the depth and size of the underground geologic features of interest and the acoustic properties of the near-surface material. Receivers for detecting reflected acoustic signals in the ground are called *geophones*, which are very specialized microphones similar in principle to those used in voice recording. The operation of a geophone is based on the voltage induced in a coil of wire when it moves through a magnetic field. The most site-dependent part of a seismic survey is the character of the acoustic energy source. A wide variety of sources have been developed and are used routinely on shallow seismic reflection projects. Shallow seismic surveys mostly employ weight drop

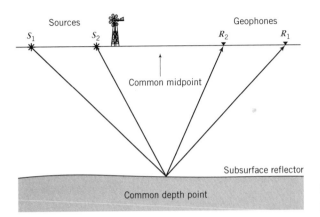

Figure 7.26 Common depth point (from Miller and Xia, 1997).

(accelerated) or explosives as the energy source. In more conventional oil-field applications, vibratory sources are used.

With shallow reflection surveys, the source and receivers are placed so that the path S1–R1 reflects from the same location in the subsurface as path S2–R2 (Figure 7.26). The subsurface point that is common for both source and receiver pairs is called a *common-reflection point* (CRP), *a common-depth point* (CDR), or a *common-midpoint* (CMP), depending on the preference of the author. The purpose of acquiring and processing seismic reflection data in a CDP format is to enhance reflections. More specifically, the CDP approach provides reflections from the same subsurface points with different combinations of surface source and receiver positions. These signals can be combined or stacked. The number of seismic traces produced for each shot is equal to the total number of seismograph channels. Thus, a seismic instrument with more channels provides better flexibility for signal enhancement in processing.

A geophone provides information on the amplitude of waves as a function of time. The deeper a geologic layer causing a reflection, the longer it takes for a wave return to be recorded. For a sequence of reflecting layers, a sequence of arrivals will be recorded. By plotting the data from a geophone with time on the y-axis (zero at top) and amplitude on the x-axis, one can produce what are known as wiggle traces. These days, a wide variety of filtering, display, and static correction techniques are employed to improve the quality of the raw seismic data. A number of processing techniques are routinely applied to produce a seismic section corresponding to a geological section.

The interpretation of seismic reflection data is based on the processed seismic sections. In some cases, seismic data are reprocessed based on new information coming out of the initial interpretation. The most important step in seismic interpretation is to convert the distance-time seismic reflection cross section into a distance-depth geological cross section. Accurate seismic velocities are needed for this conversion. Most processing methods rely on other sources to provide this needed information on seismic velocities. Acoustic logging, vertical seismic profiling, and even seismic refraction surveys are useful in this respect.

Let's examine the reflection seismic survey performed by the Kansas Geological Survey at the Northern Ordnance Plant (1997), currently the Naval Industrial Reserve Ordnance Plant (NIROP) in Fridley, Minnesota. The seismic survey includes a walkaway Vertical Seismic Profiling (VSP) (seismic field operations with seismic sources at ground surface and geophones in boreholes), three walkaway noise tests (VSP survey with different test configurations to optimize signal-to-noise ratio), and 1441 shotpoints of 24-fold CDP data on three different lines. The objective of the survey was to delineate discrete layering within

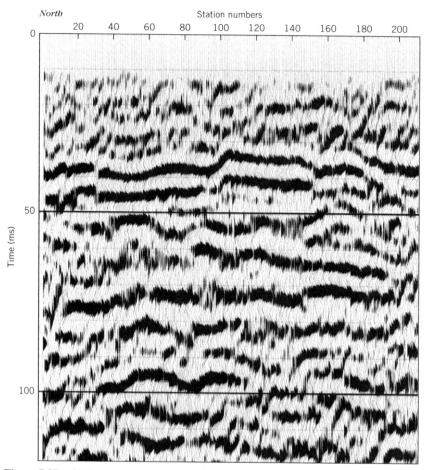

Figure 7.27 CDP stacked section of line 1 along East River Road. Display scale designed to be consistent with display parameters of other engineering reports from this site (from Miller and Xia, 1997).

the approximately 120-ft sequence of glacial drift, which overlies the St. Peters Sandstone and/or the Prairie du Chien dolomites. The seismograph was a 96-channel Geometrics Strata View. CDP data were recorded with a sampling interval of 1/4 ms and a 250 ms record length. The layout of the surveys was designed to maximize productivity without adversely affecting the resolution potential or velocity control of the CDP stacked data. A down hole rifle provided the energy source. The geophones were closely spaced, given the shallow nature of the investigation, with the source 4 ft from the nearest geophone and 100 ft from the furthest geophone.

Seismic data were processed following common petroleum exploration techniques to enhance the signal-to-noise ratio. A stacked seismic reflection section is shown in Figure 7.27. Seismic velocities from VSP were used to convert seismic reflection time to geological depth. A geological section constructed from the seismic section is shown in Figure 7.28.

Borehole Geophysical Methods

Borehole geophysical techniques have been developed to help solve exploration and production problems in the petroleum industry and have become virtually indispensable

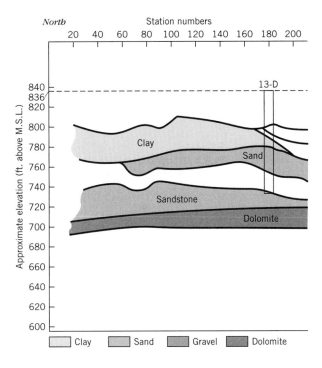

Figure 7.28 Interpreted CDP stacked section of line 1 along East River Road incorporating borehole information. Geologic interpretations were made only if extrapolation from well identifications was possible (from Miller and Xia, 1997).

tools. Many of these same advantages are available in ground-water applications. Unfortunately, the economics of the ground-water industry are such that the most sophisticated tools are unaffordable in most studies. The equipment that is used in ground-water application pales by comparison to that available to the petroleum industry.

There are probably 10 to 15 types of logs that could be used to provide useful information about different aspects of the subsurface. Keys (1990) discusses these techniques in detail. In our brief survey, we will consider four different logging techniques: caliper logs, resistivity/spontaneous potential (SP) logs, and the natural gamma log.

The *caliper log* is a tool designed to record the variation in the diameter of the borehole with depth. According to Dyck et al. (1972), caliper logs are used

- to interpret other logging methods that are affected by hole size,
- to provide information on fracture distributions and lithologies, and
- to estimate quantities of cement or gravel required to complete a water well.

A typical probe has three arms (120° apart) that maintain contact with the sides of the borehole, as the probe is raised up the hole.

Resistivity/SP logs are usually run together. There are a variety of different types of resistivity tools, the simplest of which is the so-called single-point resistance device. This tool measures the change in resistance between a lead electrode in the borehole and a fixed electrode at the surface. The log provides information on how the resistivity (measured in ohms) changes along the mud-filled borehole. Resistivity is the inverse of electrical conductivity.

This log provides information useful in distinguishing different types of lithologies (for example, sand versus lacustrine clay) and helps in correlating units between boreholes. Typically, sand and gravel units have high resistivities and produce a log deflection to the right. Fine-grained deposits containing clay minerals (for example, glacial till or shale) are

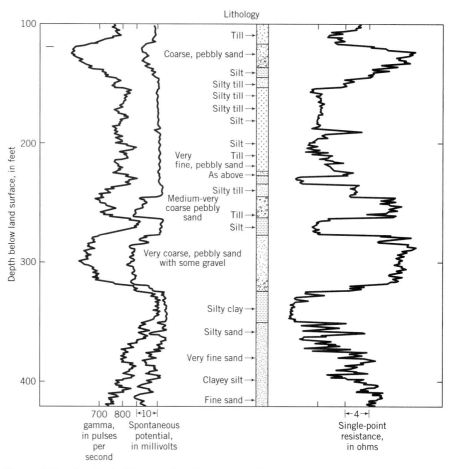

Figure 7.29 Geophysical logs run in a deep, uncased borehole in Saskatchewan, Canada. The detailed lithologic log can be used to understand the lithologic controls on the log response (modified from Dyck et al., 1972).

much less resistive. The most important drawback of this log is that it does not provide quantitative measurements of resistivity. The petroleum industry uses more sophisticated and quantitative resistivity.

Figure 7.29 shows logs from several tools that are run down a 600-ft-deep uncased borehole in Saskatchewan (Dyck et al., 1972). The sequence of glacial drift contains till, sand, and clayey silt. A comparison of the log trace with the detailed lithologic log illustrates how well the single-point resistance log is able to differentiate between adjacent units. Note how the sand units are indicated by deflections of the trace to the right. Till units with a greater content of clay minerals are less resistive (Figure 7.29).

The *spontaneous potential* or SP log measures natural electrical potentials or voltages that develop in boreholes at the contacts between clay (also shale) beds and sands (also sandstones) (Keys, 1990). These currents develop because of differences in lithologies and differences in chemistry between the drilling water and the formation water. With drilling water that is less saline than the formation water, the presence of a sand unit causes a deflection in the SP log to the left. This characteristic response is evident in Figure 7.29. Note how the sand unit at 300 ft is indicated by a deflection of the SP log to the left.

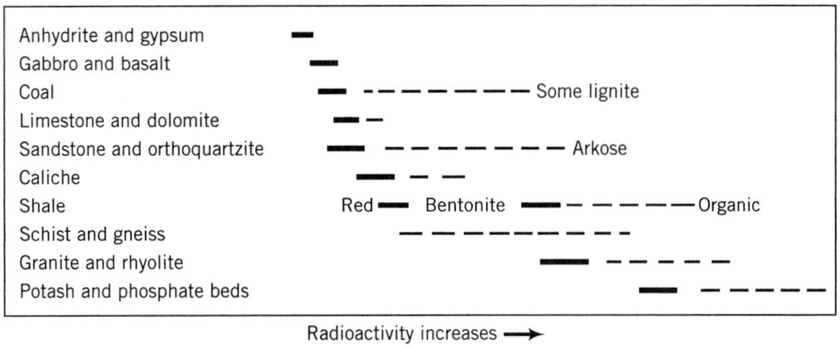

Figure 7.30 Relative radioactivity of some common rocks (from Keys, 1990)

The SP log is helpful in identifying geologic units and in correlating units between boreholes. This approach also can provide an estimate of the resistivity of the formation water, when the resistivity and temperature of the drilling mud are known. Thus, it has been used in water-quality investigations, particularly in basins where oil exploration wells have been drilled and logged.

The *gamma log* measures the total intensity of natural gamma radiation present in rocks or sediments (Dyck et al. 1972). The most important radionuclides found in rocks are potassium-40 and daughter products of the thorium and uranium decay series (Keys, 1990). The relative abundance of these radionuclides in various rock types is summarized in Figure 7.30. For these logs, the American Petroleum Institute (API) gamma-ray unit is preferred, although there are many "older" units used like counts/second. Unlike the other logs discussed here, gamma logs can be run in wells cased with plastic.

Looking at the example log in Figure 7.29, note the higher count rate in the finer grained units (that is, till and clay), as compared to the coarser units (that is, sand). Clay minerals in the fine-grained deposits typically contain relatively higher concentrations of radionuclides. Gamma logs are useful for determining the clay content of units and the boundaries between units. They are also useful for regional correlation between boreholes.

Figure 7.31 illustrates how Norris and Fidler (1973) used gamma logs to help correlate bedrock units in western Ohio. In general, shale in the section is more radioactive than either limestone or dolomite. For example, note how the log deflection to the right (increasing radioactivity) differentiates the Osgood shale from overlying and underlying carbonate units.

► 7.6 GROUND-WATER INVESTIGATIONS

Ground-water investigations can be carried out at a variety of scales. A report by the U.S. Environmental Protection Agency (EPA) (U.S. EPA, 1990) distinguishes the following kinds of investigations:

regional investigation—study may encompass hundreds or thousands of square miles and typically provides an overall evaluation of ground-water conditions.

local investigation—the area of interest typically involves a few tens or hundreds of square miles, and the study provides a more detailed look at the geology, hydrogeology, and water quality.

site investigation—a detailed look at a specific site of concern such as a well field, a leaking refinery, or an abandoned industrial site. Usually, site investigations are

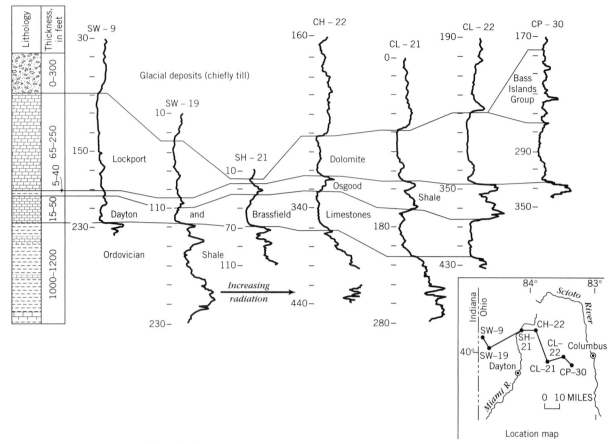

Figure 7.31 Gamma logs for a series of wells in western Ohio help with the correlation of shale and carbonate units. The carbonate rocks comprise the most important bedrock aquifer in this part of Ohio (from Norris and Fidler, 1973).

conducted in conjunction with other investigations, such as risk assessments, air-quality monitoring, and patterns of usage of hazardous chemicals.

The type of investigative tools used in a study changes as a function of scale. The regional scale is studied using reconnaissance-type approaches that usually emphasize the synthesis of existing information concerning the climate, geology, and hydrogeology, with selected field operations to fill in data gaps or to obtain more specialized information (for example, age dates from water samples). The site scale is studied through field studies, such as those presented in this chapter.

Operationally, all investigations should involve the same six steps (U.S. EPA, 1990): (1) establishing objectives, (2) preparing a workplan, (3) collecting data, (4) interpreting data, (5) developing conclusions, and (6) presenting results. The first two steps in the process are designed to make clear exactly what the goals and objectives of the study are and exactly how they will be achieved. *A workplan* is commonly employed to make all those involved aware of how the study will be conducted and what approaches will be followed. Workplans lay out in detail the standard operating procedures (SOPs) to be followed, for example, in installing wells, measuring water levels, collecting water samples, and maintaining chain-of-custody of samples. Issues of worker health and safety and emergency procedures are also addressed in these plans. Step 3 encompasses the process of data collection. The next section in this chapter will summarize the different approaches that one can follow in assembling relevant data for an investigation. The last three steps involve assembling the data and presenting a logical set of conclusions.

Investigative Methods

Some of the approaches for investigating ground-water problems at a variety of scales have been discussed in this chapter. Others come from previous course experience. Many investigations, in particular those on a regional scale, depend on information compiled in previous geologic or hydrogeologic surveys, as well as data existing in the form of well drillers' reports and water-quality reports for individual domestic or industrial wells. Table 7.1 provides examples of information sources that might be useful in hydrogeologic investigations.

Not included in Table 7.1 are the chemical and isotope tools that provide very useful information about ground-water systems. These will be discussed later in the book.

► EXERCISES

7.1. Given the following data on resistances from a cone penetrometer test, calculate the USCS soil type.

Cone Resistance (TSF)	*Friction Resistance (TSF)*
200	1.25
15	0.75

7.2. An electric tape was utilized to measure water level for determining the hydraulic head in a piezometer. The elevation of the measurement point is 735 m a.s.l., and the casing length to the midpoint of the screen is 20 m. Given the depth to water in the well is 12.2 m, calculate (a) the hydraulic head for the piezometer, and (b) the pressure at the midpoint of the screen.

► CHAPTER
 HIGHLIGHTS

1. A variety of different drilling methods can be used to sample the subsurface and to provide boreholes for the installation of wells or piezometers. Hollow-stem and solid-stem augers are well suited for drilling in unlithified sediments. The augers carry material chopped up by the drill bit out of the hole.

TABLE 7.1   Sources of information in ground-water investigations

Existing Information	Geologic Test Data	Hydrogeologic Test Data
Topographic maps	Outcrop mapping	Hydraulic head
Soil maps	Borehole logs	Hydraulic conductivity
Geologic maps	Core samples	slug tests
Aerial photographs	Geophysical logs	aquifer tests
Satellite images	Textural analyses	permeameter
Well/water-quality data		Tracer tests
Climate data		
Stream-discharge data		
Reports of previous studies		
Government reports		
Agencies		
Personal interviews		

Source: Based on U.S. EPA (1990).

Mud or air rotary drilling methods work in sediments and rock. They involve turning a string of drill rods with a bit attached. Drill cuttings are brought up the hole with the return flow of drilling water or air.

2. Push technologies involve pushing or hammering a casing string into the ground, carrying sensors, sampling tools, or permanent monitoring devices. This equipment can range from handheld electric hammers to soil probing machines to dedicated cone penetrometer rigs. A cone penetrometer system can provide real-time information about the site stratigraphy hydraulic-conductivity distributions and water chemistry.

3. A piezometer is a standpipe that is installed to some depth below the water table. Water entering the piezometer rises up the casing and reaches a stable elevation, which is a measure of the hydraulic head at the open part of the casing at the bottom. A water-table observation well is a standpipe with a large screened section spanning the water table. The water-level elevation provides the elevation of the water table at that location.

4. In practice, piezometers are designed to have a screen, attached to the bottom of the casing, a sand-pack to support the screen, and a seal to prevent leakage of water down the borehole. The seal is usually constructed with granular bentonite, a low-permeability material that expands when it becomes wet. The piezometer is finished at the top with a metal casing protector to prevent unauthorized entry.

5. Water-level elevations in wells or piezometers are determined using an electric tape or a transducer system that provides a continuous measurement of pressure with time. An electric tape is a tape measure that buzzes when the electrode on the end touches the water surface. The elevation of the water surface is determined by subtracting the depth to water from the elevation of the top of the casing.

6. A variety of geophysical techniques are used in hydrogeological investigations. The most useful surface surveying methods include electrical resistivity, electromagnetics, ground penetrating radar, and seismic. These methods provide information about site geology or the presence of buried features like tanks, sludge ponds, or former landfills.

7. Geophysical logs are also run in boreholes. Key logs used in ground-water applications include single-point resistance, spontaneous potential (SP), and gamma. They are useful in distinguishing among different geological materials and in correlating units between boreholes.

8. Ground-water investigations are carried out at regional, local, and site scales. The type of investigative approach changes as a function of scales. Regional-scale studies emphasize reconnaissance-type approaches that emphasize the synthesis of existing information. Sites are studied using many of the investigative approaches presented in this chapter. Regardless of the scale of study, detailed planning is required to make sure that the investigative approaches are appropriate to meet the study objectives and that the field program follows standard operating procedures in measurement and data analysis.

CHAPTER

8

REGIONAL GROUND-WATER FLOW

Ground-water hydrologists have generally studied flow at two different scales. Historically, a major emphasis has been on small-scale problems, concerned with the transient response of an aquifer to pumping from one or several closely spaced wells. A less intensive but nevertheless important effort has been made to examine the manifestations of natural ground-water flow over large regions. This topic is referred to as regional ground-water flow. Here, we begin with Hubbert's classical study (1940) and touch on the important theoretically based studies of Tóth (1962, 1963) and Freeze and Witherspoon (1966, 1967). This body of science forms the foundation for the modern study of natural ground-water flow at a regional scale.

This chapter also will take up issues of how ground-water systems interact with other components of the hydrologic cycle. We explain the processes of recharge and discharge, which operate to add and remove water from active flow systems, and we also examine features of how ground-water and surface-water systems interact. Finally, we describe the special case in which seaward-moving ground water meets an ocean and the unique flow physics that are involved.

▶ 8.1 GROUND-WATER BASINS

In the study of regional ground-water flow, the ground-water basin provides a convenient unit for analysis. The *ground-water basin* was defined by Freeze (1969a) as a three-dimensional closed system, which contains the entire flow paths followed by all the water recharging the basin. Let's examine this concept in more detail by looking at M. King Hubbert's famous figure (Figure 8.1) showing the regional steady-state flow of ground water from an upland to nearby streams. This figure also displays one of the two complete ground-water basins that are present. This basin is *closed* in the sense that no flow lines pass through the left- and right-side boundaries or the bottom (Figure 8.1). Water enters

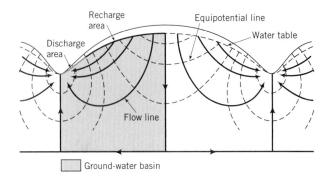

Figure 8.1 Topographically controlled flow pattern. The shaded area defines a ground-water basin (from Hubbert, 1940, "The theory of ground water motion," v. 48, no. 8, pt. 1, p. 785–944). Reprinted from *The Journal of Geology*, by permission of the University of Chicago Press. Copyright ©1940.

the basin at the water table, as recharge in the upland area, and leaves as discharge in topographically low areas. Although this figure is only two dimensional, this same idea applies to a three-dimensional system.

An interesting feature of Figure 8.1 is that the water table is a subdued replica of the ground surface. In other words, the water table is higher under topographically high areas and lower under topographically low areas. However, the relief on the water table is somewhat less marked than the relief on the ground surface. This tendency for the water table to follow the topography of the ground surface but with smaller ups and downs is what is meant by "subdued replica." Because of the similarity between the water table and the ground surface, investigators often use the ground surface as a proxy to the water-table configuration. This approximation works best in areas with plenty of recharge. In arid areas, the water table often does not resemble the ground surface.

The land-surface topography can help define ground-water basins, when the water table is a replica of the ground surface. As an initial assumption, major topographic highs are taken as ground-water divides, as are major topographic lows. In effect, the surface-water divides and ground-water divides are the same. For example, in Figure 8.1, if information is only available on the topography of the ground surface, you could have done a good job in defining the ground-water basins. We make use of topographic data in many studies to imply what the ground-water flow conditions are likely to look like when ground-water data are sparse.

The term *recharge* refers to water, which percolates down through the unsaturated zone and enters the dynamic ground-water flow system (Freeze, 1969a). A *recharge area* is an area where the flow of water is directed downward away from the water table. The recharge area in Figure 8.1 encompasses most of the upland area of the system. The term *discharge* refers to water, which is lost from the dynamic ground-water system by means of stream baseflow, springs, seepage areas, or evapotranspiration. A *discharge area* is an area where the flow of water is directed upward with respect to the water table (see Figure 8.1).

▶ 8.2 MATHEMATICAL ANALYSIS OF REGIONAL FLOW

Mathematical models have proved to be invaluable in developing an understanding of regional ground-water flow and a capability for analyzing real problems quantitatively. A quantitative approach applied to regional ground-water flow requires a ground-water flow equation, a simulation domain, and boundary conditions. The ground-water system is usually assumed to be at steady state so that no initial conditions are required. The simulation domain is a ground-water basin. Conveniently, the vertical side boundaries are no-flow boundaries because by definition a ground-water basin has no flow through the sides. In reality, the side boundaries are flow lines, which in many model studies are adopted as no-flow boundaries. The bottom boundary is also a no-flow boundary. This boundary is more easily understood in the sense that if you go deep enough into the earth, you will eventually find a unit with such a low permeability that essentially no flow is moving into it. The top boundary of the system is a specified head boundary. It is assumed that hydraulic heads are fixed values all along this boundary. Figure 8.2 shows how a simulation domain with boundary conditions can be extracted from Hubbert's figure (8.1).

Water-Table Controls on Regional Ground-Water Flow

One of the important contributions, illustrating the power of mathematical tools, was Tóth's (1962, 1963) work on the influence of water-table configuration on flow in ground-water basins. His analytical approach let him test different water-table configurations and determine what types of ground-water flow patterns emerged.

The first case that Tóth considered was flow in a basin with a linear, sloping water table. This kind of water-table configuration would develop in a ground-water basin with a gently sloping ground surface. Tóth (1962) formulated this problem for a two dimensional cross section oriented parallel to the likely direction of regional flow in a ground-water basin (Figure 8.3). To simplify the mathematics, he assumed that the simulation domain was rectangular. However, as can be seen in Figure 8.3, the hydraulic head is varying along the top boundary. Furthermore, Tóth assumed that the porous medium was homogeneous and isotropic.

Hydraulic head is calculated for this problem as a solution of a two-dimensional form of Laplace's equation:

$$\frac{\partial^2 h}{\partial x^2} + \frac{\partial^2 h}{\partial z^2} = 0 \tag{8.1}$$

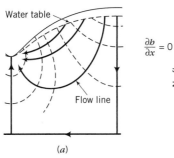

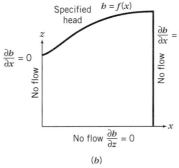

Figure 8.2 Mathematical representation of a ground-water basin. The ground-water basin provides the simulation domain (Panel *a*). The boundary conditions for the analysis are shown in Panel *b*.

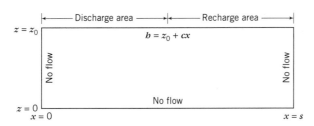

Figure 8.3 Two-dimensional region with boundary conditions for regional flow.

The boundary conditions have no flow boundary conditions at left, right, and bottom boundaries:

$$\left.\frac{\partial h}{\partial x}\right|_{x=0} = 0, \quad \left.\frac{\partial h}{\partial x}\right|_{x=s} = 0, \quad \left.\frac{\partial h}{\partial z}\right|_{z=0} = 0 \tag{8.2}$$

The sloping linear water table along the top boundary is represented by a specified head boundary with

$$h = z_0 + cx \tag{8.3}$$

where z_0 is the hydraulic head at $x = 0$ and c is the slope of the water table.

An analytical solution to this problem was given by Tóth (1962) and provides the steady-state distribution of the hydraulic head in the two-dimensional domain

$$h = z_0 + \frac{cs}{2} - \frac{4cs}{\pi^2} \sum_{m=0}^{\infty} \frac{\cos\left[(2m+1)\,\pi x/s\right] \cosh\left[(2m+1)\,\pi z/s\right]}{(2m+1)^2 \cosh\left[(2m+1)\,\pi z_0/s\right]} \tag{8.4}$$

Figures 8.4a,b are examples of the contoured hydraulic-head field that comes from evaluating the solution to Eq. (8.4). With this water-table configuration, only one large flow system develops in the domain. For a positive value of c, the right side of the flow domain is the recharge area and the left side is the discharge area. The *hinge line*, a line separating

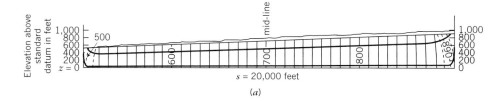

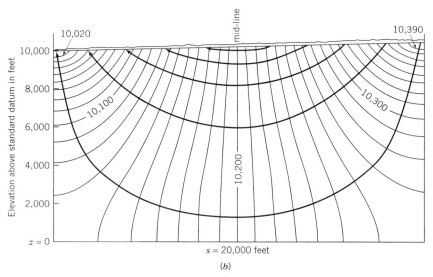

Figure 8.4 Examples of topographically controlled flow pattern derived as a solution to Eq. (8.4). The two panels illustrate how the regional flow changes as a function of region shape (from Tóth, *J. Geophys. Res.*, v. 67, p. 4375–4387, 1962). Copyright by American Geophysical Union.

the recharge and discharge areas, is located at $x = s/2$. The hydraulic head is between z_0 and $z_0 + cs$.

▶ **EXAMPLE 8.1**

A regional flow system is characterized by a sloping linear water table. The hydraulic heads at the left and right sides of the flow domain are 20 m and 40 m, respectively. The total length of the domain is 2000 m. Find all of the parameters in Eq. (8.4) so that the hydraulic head in the domain can be calculated.

SOLUTION It is known that $s = 2000$ m and $z_0 = 20$. The parameter c may be calculated by solving the following equation.

$$40\,\text{m} = 20\,\text{m} + c \times 2000\,\text{m}$$

The result is $c = 0.01$.

In a subsequent study, Tóth (1963) examined how topography on the water table influenced patterns of flow. Superimposing a sinusoidal fluctuation on a regional slope (Figure 8.5) provided this idealized water table. Hydraulic head along the water table is given as

$$h(x, z_0) = z_0 + x \tan \alpha + a \frac{\sin (bx/\cos \alpha)}{\cos \alpha} \qquad (8.5)$$

where z_0 is the depth of the flow domain, x is the horizontal coordinate, α is the average slope of the water table, a is the amplitude of the sine wave, $b = 2\lambda/\pi$ is the frequency, and λ is the period of the sine wave. The second term on the right-hand side corresponds to the regional slope, and the third term is the local relief superimposed on the regional slope. This third term, s/λ, defines the number of sine waves for a flow domain of length s. A simplified form of Eq. (8.5) is written as

$$h(x, z_0) = z_0 + c'x + a' \sin (b'x) \qquad (8.6)$$

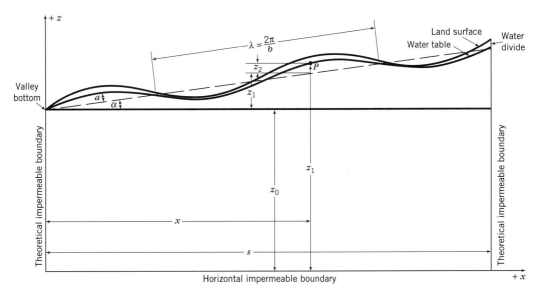

Figure 8.5 Two-dimensional simulation domain for analyzing how a sinusoidal water table with a regional slope affected ground-water flow (from Tóth, *J. Geophys. Res.*, v. 68, p. 4795–4812, 1963). Copyright by American Geophysical Union.

where $c' = \tan \alpha$, $a' = a/\cos \alpha$, and $b' = b/\cos \alpha$. With combinations of z_0, c', a', and b', a variety of different flow fields could be generated.

Tóth (1963) summarized results for a number of water-table configurations. He found that a hierarchical pattern of flow systems developed, which he termed local, intermediate, and regional (Figure 8.6). A *local flow system* has its recharge area at a topographic high and its discharge area in the adjacent topographical low. An *intermediate flow system* has one or more topographic lows intervening between recharge and discharge areas. A *regional flow system* has its recharge area at the highest part of the basin and its discharge area in lowest part of the basin.

Tóth (1963) recognized the presence of stagnation points at the juncture of flow systems (Figure 8.6). A *stagnation point* is a point where ground water flows into two flow systems with opposite flow directions and equal flow magnitudes. Therefore, the ground-water velocity is zero at the stagnation point. Based on his set of simulation experiments, Tóth provided a general set of conclusions with respect to regional flow

1. For an extended flat area, ground-water flow is very slow. The discharge of water would occur mainly by means of evapotranspiration. Water in these areas will have a high concentration of dissolved mineral matter.

2. When local relief is negligible ($a = 0$), the general slope will create a regional flow system by itself. Tóth (1963) cited the case of ground-water flow in southeastern Nassau County, Long Island, New York, as an example (Figure 8.7).

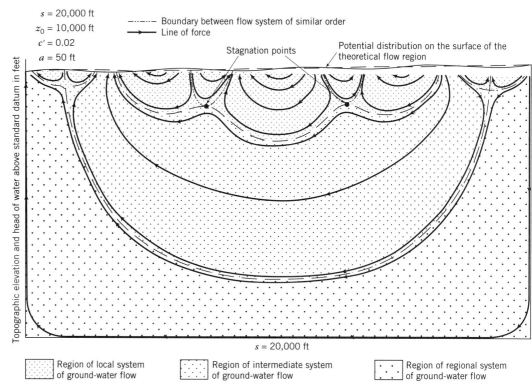

Figure 8.6 Two-dimensional isotropic flow model showing the distribution of local, intermediate, and regional ground-water flow systems (from Tóth, *J. Geophys. Res.*, v. 68, p. 4795–4812, 1963). Copyright by American Geophysical Union.

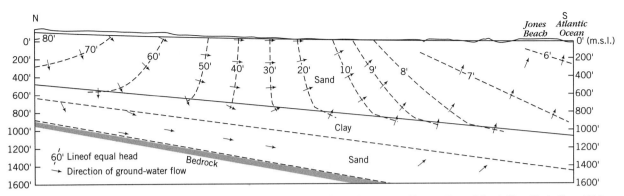

Figure 8.7 Ground-water flow in southeastern Nassau County, Long Island, New York (from Tóth, *J. Geophys. Res.*, v. 68, p. 4795–4812, 1963). Copyright by American Geophysical Union.

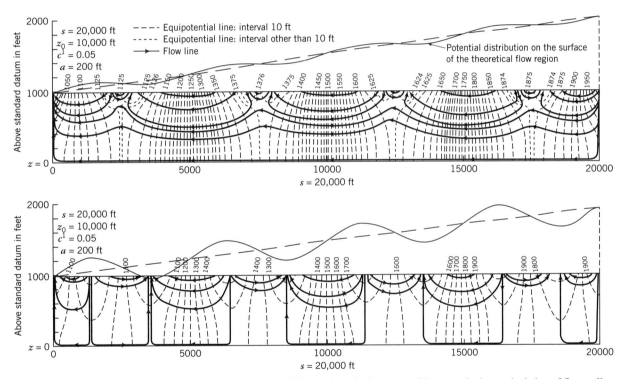

Figure 8.8 The amplitude of fluctuations in the water table controls the vertical size of flow cells that develop (from Tóth, *J. Geophys. Res.*, v. 68, p. 4795–4812, 1963). Copyright by American Geophysical Union.

3. Local flow systems develop when the local relief is well defined. The higher the relief, the deeper the local flow system (Figure 8.8).

4. The flow velocity in a local flow system is often much higher than in a regional system. In addition, major streams in a basin receive most of their ground-water inflows from local flow systems, as compared to regional flow systems. Ground water at shallow depths will likely be influenced by seasonal recharge and discharge, whereas water at depth is relatively stagnant.

Effects of Basin Geology on Ground-Water Flow

In the mid-1960s, Freeze and Witherspoon (1966, 1967) extended Tóth's model-based studies of regional flow. They examined how complexities in the hydraulic-conductivity distribution together with water-table configuration influenced regional flow. These complexities were produced by changing patterns of hydraulic-conductivity layering and the anisotropic character of some units (that is, $K_h \neq K_v$).

Their first set of simulation trials looked at how differing water-table configurations in layered systems impacted patterns of flow. Figure 8.9a shows the pattern of flow that developed with a simple segmented water table. The absence of local topography on the water table promotes the development of a regional flow system. Near the discharge area, the steeper water table focused discharge at the valley bottom. Figure 8.9b shows the complexity that develops in the head field with the addition of local topography on the water table and a flattened discharge area. Local, intermediate, and regional systems now develop. The final panel (Figure 8.9c) shows how local topography causes the development of a set of flow cells, similar to what Tóth found.

The next set of simulations examined the effects of hydraulic-conductivity contrasts among layers on the pattern of flow (Figure 8.10). In the first group of trials (Figures 8.10a, b, and c), a low K layer was assumed to overlie a high K layer. The contrast in hydraulic conductivity between the layers was adjusted by increasing the hydraulic conductivity of the lower, more permeable unit. Increasing the contrast in hydraulic conductivity with

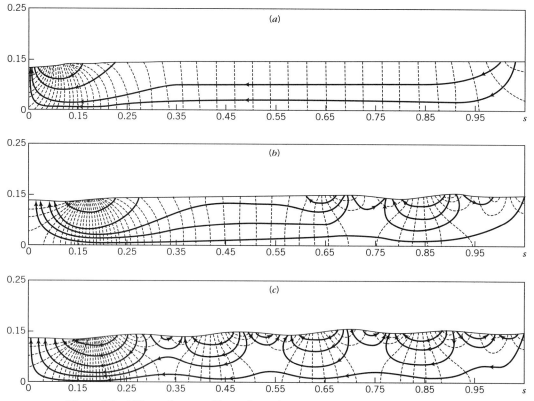

Figure 8.9 Effects of water-table configuration on regional ground water through homogeneous isotropic media (from Freeze and Witherspoon, *Water Resour. Res.*, 3, p. 623–634, 1967). Copyright by American Geophysical Union.

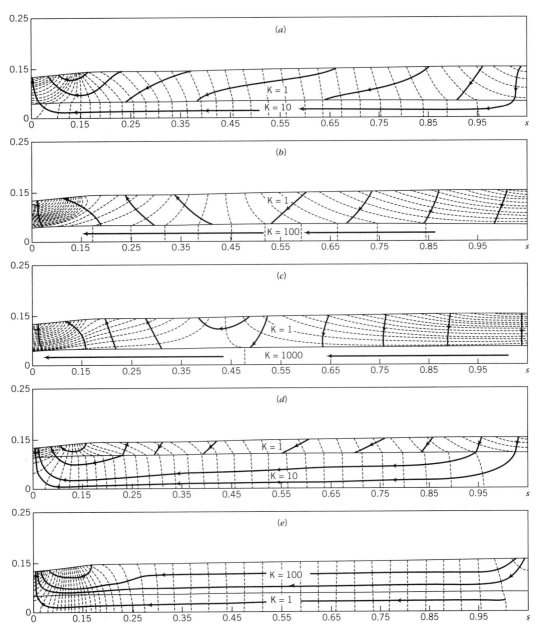

Figure 8.10 Effects of permeability contrast on regional ground water flow in horizontally layered media (from Freeze and Witherspoon, *Water Resour. Res.*, 3, p. 623–634, 1967). Copyright by American Geophysical Union.

this pattern of layering results in an increase in the vertical hydraulic-head gradient in the upper layer and a decrease in the horizontal hydraulic-head gradient in the more permeable layer (Figures 8.10a through 8.10c). Increasing the thickness of the lower layer does not change the pattern of flow appreciably (compare Figures 8.10a and 8.10d). There is mainly downward flow in the low K units and lateral flow in the high K units. When the pattern of layering has a high K unit overlying a low K unit, the flow patterns are not sensitive to layering. Mainly lateral flow develops in both layers, as is illustrated in Figure 8.10e.

Their third set of simulations shows the effect of lenticular bodies of high permeability on flow (Figure 8.11). In general, the lenses tend to gain flow on their upgradient segments and to lose flow on their downgradient segments. Often, discharge areas form in upland settings when lenticular bodies terminate there (for example, Figures 8.11*a* and *c*).

By looking at the simulation results of Freeze and Witherspoon (1966, 1967), one can gradually develop a conceptual understanding of how the pattern of layering influences flow. Flow will proceed from recharge areas to discharge areas by following the most permeable pathways. Thus, flow is mainly lateral in high hydraulic-conductivity layers and locally vertical (either up or down) as flow moves across low hydraulic-conductivity layers.

▶ 8.3 INTRODUCTION TO THE COMPUTER PROGRAM FLOWNETz

Working with models is extremely helpful in developing a better understanding of regional ground-water flow. To facilitate the use of these mathematical approaches, we have developed the computer program, **FLOWNETz**, to calculate the hydraulic-head and stream

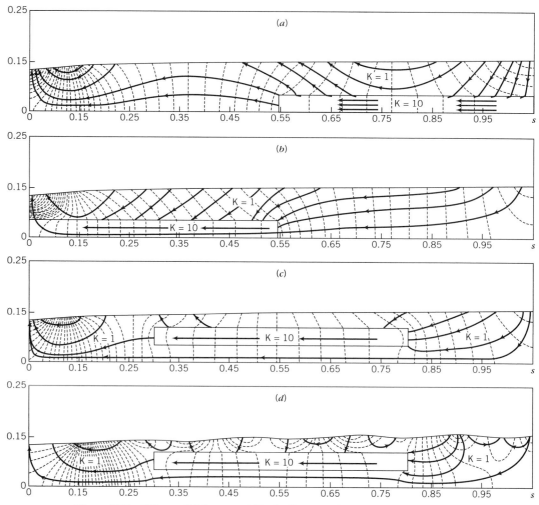

Figure 8.11   Effects of high-permeability bodies on regional ground-water flow (from Freeze and Witherspoon, *Water Resour. Res.*, 3, p. 623–634, 1967). Copyright by American Geophysical Union.

REGIONAL FLOW IN MINNESOTA

This case study is designed to highlight many of the features of regional flow that we have examined in this section. Figure 8.12 shows the equipotential distributions and patterns of groundwater flow in a Cambro-Ordovician system in Minnesota. The sequence of aquifers and confining beds shown in the figure creates patterns of layering somewhat similar but obviously more complicated than that treated by Freeze and Witherspoon. The hummocky topography/water table promotes the development of local and intermediate-flow systems in the upper part of the stratigraphic section. There is a regional system developed from the topographic high in the basin to the valley of the Mississippi River. Flow is mainly vertical through confining units like the St. Lawrence-Franconia or the Eau Claire. Lateral flow is evident in the most permeable units, like the St. Peter-Prairie du Chen-Jordan aquifer and the Mount Simon aquifer (Figure 8.12).

functions for regional flow in two-dimensional vertical cross sections. The program is located in directory PROGRAMS/FLOWNETz on the website. The installation instructions are also located in the directory. Several analytical solutions have been implemented in **FLOWNETz**. Of particular interest here are solutions for cases of (i) a linear water table with a homogeneous vertical cross section (Tóth, 1962), (ii) a linear water table with local fluctuations with a vertical cross section (Tóth, 1963), and (iii) a complex water-table configuration with layered vertical cross section (Freeze and Witherspoon, 1966, 1967). Here are the steps in running **FLOWNETz**.

1. Click the **FLOWNETz** from the **Start/Programs** menu to launch **FLOWNETz**.

2. The main menu for **FLOWNETz** is **Flownet** (Figure 8.13). There are four commands under the **Flownet** menu: **Options, Input, Calculate**, and **Report**. The **Options** command launches a dialog box for users to specify the analytical solutions for the calculation (Figure 8.14).

3. The **Input** command will launch different dialog boxes for different analytical solution options. For case (i), (Tóth, 1962), the slope of the water table and the domain size need to be specified (Figure 8.15). For case (ii), (Tóth, 1963), the head at the right side of the flow domain, the amplitude of the local head, the number of sine waves as defined by Eq. (8.5), and the domain size need to be assigned (Figure 8.16). For case (iii), (Freeze and Witherspoon, 1966, 1967), the hydraulic conductivity, the thickness of each geological layer, and the domain size need to be specified (Figure 8.17).

4. The results of the calculation will be displayed on the screen once the **Calculate** command on the **Flownet** menu is clicked. The flownets are shown in Figures 8.18, and 8.19, respectively, for the input parameters defined in Figures 8.15 through 8.17.

▶ 8.4 RECHARGE

Recharge is the process by which water moves downward from the ground surface through the vadose zone and joins an active flow system. Recharge is promoted by water standing on the ground surface, a relatively shallow water table, and relatively permeable materials in the vadose zone. The following examples will illustrate how these factors work together to control recharge. We will begin with a discussion of recharge in relatively arid settings and progress to moist settings.

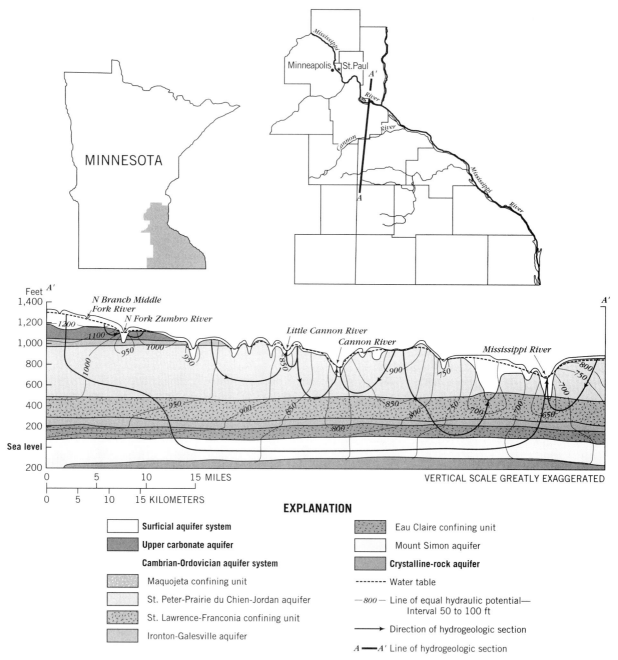

Figure 8.12 Hydraulic-head distribution and patterns of regional ground-water water flow in the Hollandale embayment, southeastern Minnesota (from Delin and Woodward, 1984; Olcott, 1992).

Desert Environments

In desert settings, rates of potential evaporation often greatly exceed precipitation. What is surprising is that any recharge at all reaches the ground water. Usually, two factors work together in promoting recharge—significant but infrequent storms that bring water to very dry areas and mechanisms that locally accumulate this water.

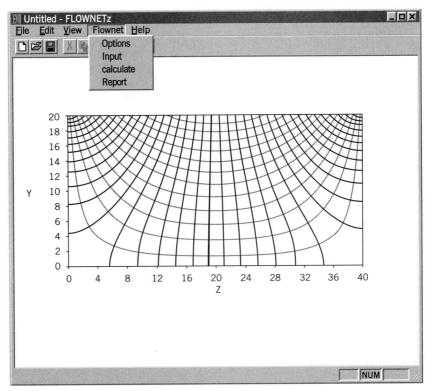

Figure 8.13 FLOWNETz menu showing the operations available on the Flownet dropdown menu and simulation results for two-dimensional regional flow with a linear water table.

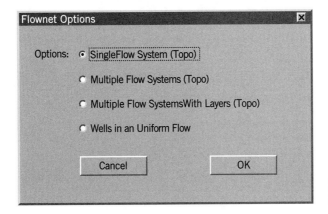

Figure 8.14 Options dialog box in FLOWNETz

The Yucca Mountain area in Nevada is a case in point. The mean annual precipitation is approximately 160 mm (about 6 in.) per year. With so little precipitation and high rates of potential evapotranspiration, there is no recharge in most years. However, on occasion, every few years, a single storm could yield several inches of precipitation. With this precipitation, normally dry riverbeds carry water for a week or so, and limited water becomes available for recharge from leakage from the stream. Figure 8.20 develops the conceptual model of recharge from ephemeral streams in these settings.

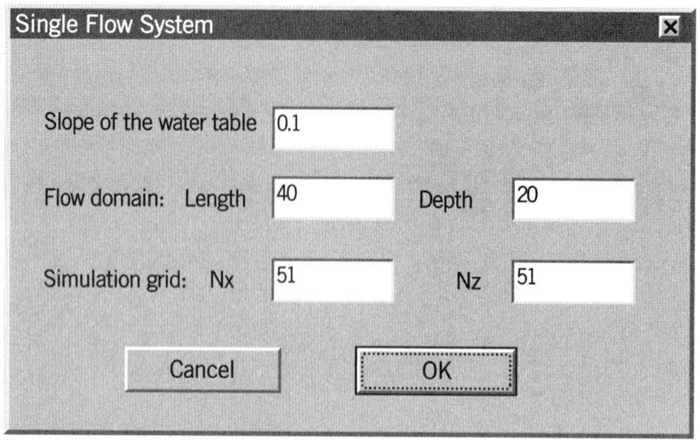

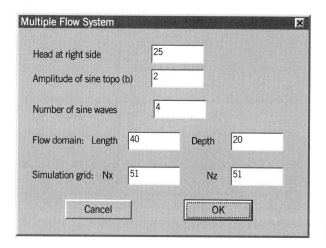

Figure 8.15 Input dialog box for problems with a linear water table and homogeneous, isotropic medium.

Figure 8.16 Input dialog box for problems with a sinusoidally fluctuating water table.

Clearly, not every dry area is this arid. Precipitation can come and go seasonally, with streamflow persisting over periods of months. With water present on the landscape for longer times, there is potential for more significant quantities of recharge.

Recharge in dry settings can also be promoted by the presence of mountains. Typically, precipitation is greater at higher elevations, and the relief promotes rapid and efficient surface runoff. Water running off the mountains effectively infiltrates alluvial fans bordering the mountain ranges (Figure 8.20). Alluvial-fan sediments are highly permeable, and infiltrating water is able to move downward rapidly, away from the reach of surface evaporation.

Semiarid Climate and Hummocky Terrain

Another type of recharge occurs across the semiarid Great Plains of the United States and Canada. Although precipitation amounts are relatively low and for most of the year less than potential evapotranspiration, there is often a window of opportunity of recharge in late spring. The melting of accumulated snow during a time of low temperatures and no plant growth can in some years provide the excess moisture necessary for recharge. Recharge is also promoted by poorly drained glacial and dune landscapes, formed from closely spaced hills and depressions. Because of the poorly organized surface drainage,

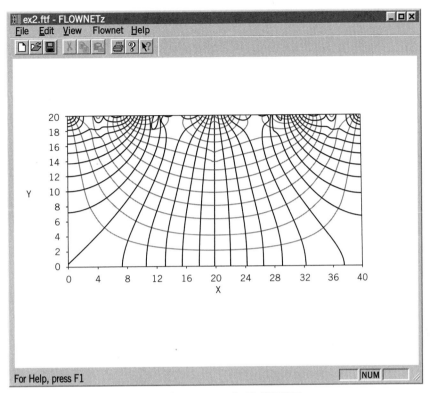

Figure 8.17 Display of multiple flow systems in FLOWNETz.

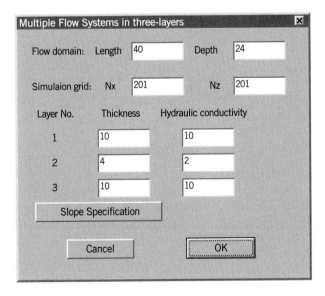

Figure 8.18 Input dialog box for a regional-flow problem involving a three-layer medium with a complex water table.

runoff accumulates and is retained in small pothole lakes and small wetlands. When these lakes and wetlands are located in topographically high settings (Figure 8.21), they provide a source of recharge through the hot summer months, until they eventually dry up. Water bodies at lower elevations last longer as ground-water discharge contributes inflow.

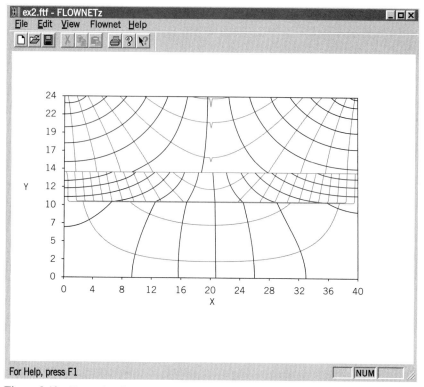

Figure 8.19 Example of a simulation result showing the hydraulic-head distribution and patterns of flow calculated for a three-layer problem.

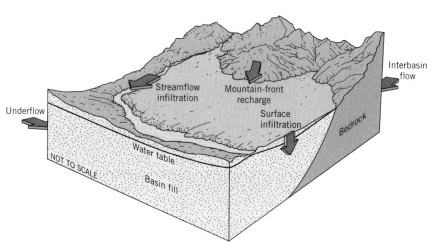

Figure 8.20 In desert settings, recharge is often localized along ephemeral streams and the upstream end of aluvial fans at the base of mountains. Surface infiltration can occur infrequently (from Robson and Banta, 1995).

In dune settings, the relatively high hydraulic conductivity of these deposits promotes relatively active flow or ground water between depressions (Figure 8.21).

These lakes and wetlands are extremely sensitive to both long- and short-term climatic fluctuations because the water available to these water bodies can be markedly different in

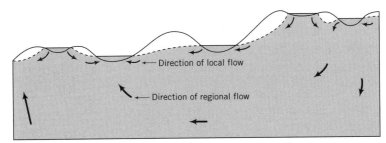

Figure 8.21 Lakes and wetlands in topographically high positions are often a significant source of recharge to ground water. These conditions are fostered in North America by glacial and dune terrain (from Winter et al., 1998).

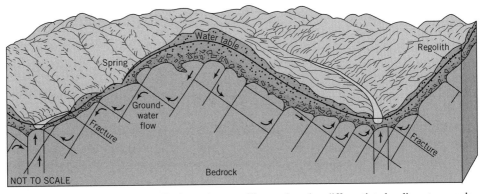

Figure 8.22 Recharge and discharge in crystalline-rock and undifferentiated sedimentary-rock aquifers (Trapp and Horn, 1997).

wet versus dry years. As a consequence, stages fluctuate markedly. They are important in sustaining migratory waterfowl in the central region of the North American continent.

Recharge in Structurally Controlled Settings

In crystalline rocks, the structural setting exerts a major influence on ground-water flow at both local and regional scales (Whitehead, 1996). In particular, fault zones can conduct and store ground water. Where permeable fault zones extend to great depths, ground water can circulate downward and become heated. Ground water in carbonate rocks of the Valley and Ridge aquifers flows along fractures and bedding planes (Figure 8.22) (Trapp and Horn, 1997). Precipitation recharges carbonate aquifers by infiltrating through alluvium and regolith. Commonly, ground water dissolves carbonate rocks and creates a network of large and interconnected openings in carbonate aquifers. Precipitation that falls on the valley recharges the carbonate rocks as well as runoff from adjacent ridges.

When underlain by karst, recharge can be tremendously effective. Fractures and sinkholes create a direct connection between the land surface and deeper aquifers. Streams are also an important source of recharge for carbonate aquifers.

Distributed Recharge in Moist Climates

In many parts of the world, there is abundant precipitation. With the opportunity for water often present across the landscape, recharge occurs in a distributed fashion. Recharge, then, is much more aerially extensive without a need for special situations to concentrate surface water from larger areas. In many more northerly settings, spring snowmelt and accompanying rainstorms provide more water than most systems can actually accommodate. Thus, in many eastern states and provinces in the United States and Canada, potential recharge may be rejected because the water table is so close to the ground surface in spring months. Through the warmer summer months, there is much less recharge due to the increasing utilization of water by growing plants.

Approaches for Estimating Recharge

Water resources investigations often require an estimate of the quantity of recharge added to flow systems. A variety of empirical and measurement-based approaches are available to help answer this question. One crude empirical approach is to estimate recharge based on precipitation. The rule-of-thumb is that about 5% of the total precipitation could be available as recharge. This kind of estimate is essentially an educated guess.

Other more rigorous approaches are available. As noted in Chapter 2, ground-water recharge can be estimated from streamflow hydrographs. In addition, water balance calculations provide another way to estimate recharge rates. Simple water balance calculations for soils, based on simple climatic information, let one determine rates of precipitation and potential/actual evapotranspiration. With knowledge of the soil moisture storage capacity of the soil, calculated soil moisture surpluses represent water that is available for recharge (Freeze, 1969b).

Freeze (1969a) described how quantitative flownets for regional ground-water systems can be used to estimate recharge. As we discussed previously, given a flownet with curvilinear squares, one can calculate the discharge in a collection stream tube using the Darcy equation

$$Q = \frac{n_f}{n_d} K \cdot \Delta H \tag{8.7}$$

where Q is the discharge from a collection of flow tubes, K is hydraulic conductivity, and ΔH is the number of head drops. In the third dimension, which is into and out of the cross section, the section is assumed to have unit thickness. For example, measuring in meters, the flow system would have a thickness of 1 m. We illustrate this approach using an example from Freeze (1969a).

▶ **EXAMPLE 8.2**

Figure 8.23 shows a flownet for a local flow system. Of interest here is the flow system labeled B. Ignore the stream tube A, which discharges elsewhere. The length of the cross section in Figure 8.23 is about 0.4 s or 7300 m, and each head drop is 6.1 m. Given that the hydraulic conductivity of the two units are 0.25 m/d and 2.5 m/d, calculate the quantity of discharge from flow system B.

SOLUTION For flow system B, there are eight stream tubes carrying water. In the discharge area for B, the flow lines and equipotential lines form curvilinear squares. Substitution of the appropriate numbers into Eq. (8.7) gives

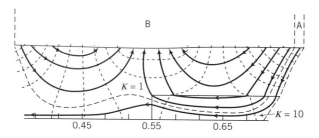

Figure 8.23 Flownet extracted from a larger, regional system (adapted from Freeze, 1969a). Reproduced with the permission of the Minister of Public Works and Government Services Canada, 2001 and courtesy of Environment Canada.

$$Q = \frac{8}{1} \times 0.25 \times 6.1 = 12.2 \, \text{m}^3/\text{d}$$

Given that the flow system is at steady state, the rate of recharge and discharge for flow system B is 12.2 m³/d. Remember: this recharge/discharge applies to a slice 1 m thick into and out of the section.

Chemical approaches have also proven useful for estimating recharge. Readers can refer to Allison and Hughes (1978), Sharma and Hughes (1985), and Phillips et al. (1988) for a discussion of chloride mass balance approaches. Allison et al. (1983) discuss the use of environmental isotopes, deuterium, and oxygen-18. More recently, bomb-pulse tritium and chlorine-36 methods have provided ways to estimate recharge (Zimmerman et al., 1966).

▶ 8.5 DISCHARGE

Discharge is a process by which water leaves a ground-water flow system and returns to the surface hydrologic cycle. The common modes of natural discharge include: (i) outflow to rivers, lakes, and wetlands, (ii) flow as springs and seeps to the ground surface, and (iii) evapotranspiration in settings with a shallow water table.

Inflow to Wetlands, Lakes, and Rivers

In the case of wetlands or lakes, ground-water inflow occurs as subsurface seepage or springs (Figure 8.24). Seepage measurements in lakes show that typically most of the seepage occurs near the shore. In the case of rivers, the discharge situation can be complicated by the

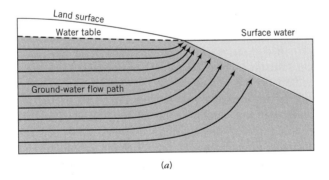

(a)

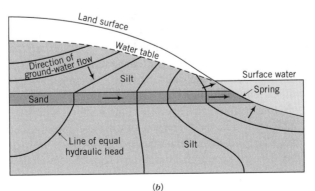

(b)

Figure 8.24 Panel (a) illustrates the pattern of ground-water seepage into surface water. Most of the inflow is concentrated near the shore. Panel (b) illustrates how subaqueous springs can lead to discharge into lakes. In this case, the pattern of geologic layering promoted the concentration of discharge (from Winter et al., 1998).

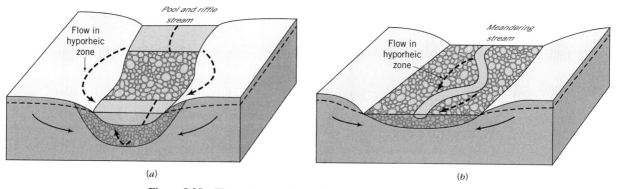

Figure 8.25 The exchange of ground-water and surface water in the hyporheic zone is associated with abrupt changes in streambed slope (*a*) and stream meanders (*b*) (from Winter et al., 1998).

fact that the water in the permeable streambed is moving downstream along with the surface water. Ground water entering this so-called hyporheic zone mixes with the streambed water before actually discharging to the stream. The *hyporheic zone* is the subsurface zone where stream water flows through short segments of its adjacent bed and banks (Figure 8.25) (Winter et al., 1998). Because this water is a mixture of stream water and discharging ground water, it can have a unique chemical and biological character. Given the importance of the interactions between ground water and surface water, we will discuss this topic in more detail in Section 8.6.

Springs and Seeps

A *spring* is a concentrated flow of ground water issuing from the subsurface into a body of surface water or onto the land surface at a rate sufficient to form a current. Springs are classified in terms of temperature, type of aquifer that supplies the spring, and geological structures where the springs are located. For example, springs in the Valley and Ridge Province are commonly categorized as one of three types: contact springs, turbulent springs, and impermeable-rock springs (Figure 8.26) (Brahana et al., 1986; Trapp and Horn, 1997). Flow in large springs in the Valley and Ridge Province is supplied by the carbonate aquifers. *Contact springs* are formed at the contact between units of different hydraulic conductivity; *turbulent springs* are formed from solution channels in the carbonate rocks; and the *impermeable-rock springs* are connected to fractures, joints, and bedding planes in low-permeability layers.

In terms of aquifer types associated with springs, a spring is called an *artesian spring* when the water flow in the spring is supplied by a confined aquifer. With a *gravity spring*, water discharges from an unconfined aquifer. Springs are also classified as cold springs or thermal springs according to their temperatures.

The most important characteristic of a spring is its discharge. Meinzer (1927) devised a classification system based on discharge (Table 8.1). Discharge of a spring depends on many factors, including the aperture of the fractures within the rock, the hydraulic head, the size of the ground-water drainage area which supplies water to the spring, and the quantity of rainfall. Ground water discharges at a spring when the hydraulic head in an aquifer (h_{aq}) is higher than the elevation of the spring (Z_{sp}). The spring flow rate can be approximated by

$$Q_{sp} = C(h_{aq} - Z_{sp}) + D(h_{aq} - Z_{sp})^2 \qquad (8.8)$$

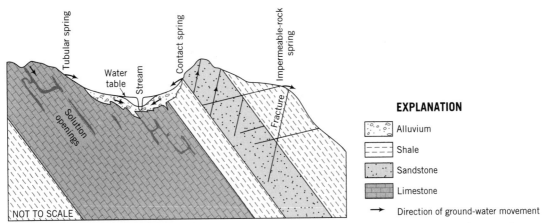

Figure 8.26 The occurrence of springs in the Valley and Ridge Province (from Brahana et al., 1986; Trapp and Horn, 1997).

TABLE 8.1 Classification of springs according to average discharge

Magnitude	Average flow (ft^3/s)	Average flow (m^3/s)
1	100	9.3
2	10–100	0.93–9.3
3	1–10	0.093–0.93
4–8	Less than 1	Less than 0.093

Source: Meinzer (1927).

where Q_{sp} is the volumetric flow rate from the spring [L^3/T], C is the hydraulic conductance between the aquifer and the spring [L^2/T], and D is a proportionality constant. D is usually assumed to be zero for laminar flow.

The time variation in discharge from a spring can provide useful information about changing hydraulic conditions within an aquifer. For example, discharge from a spring will cease when the hydraulic head in the aquifer is lowered by pumping to less than the elevation of the spring. Similarly, a dry year with minimal recharge can lower hydraulic heads, causing a reduction in spring flow. A wet year can lead to an increase in spring flow.

Many springs around the world have stopped flowing as a result of lowered potentiometric surfaces. For example, Jinan City of Shangdong Province, China, has been known as the "City of Springs" for thousands of years. However, many of the springs have stopped flowing in the last few decades because of the overdevelopment of ground water. In the United States, many springs in western Kansas have stopped flowing due to the overproduction of ground water from the Ogallala aquifer.

▶ **EXAMPLE 8.3**

The estimated conductance between an aquifer and a spring is between 10 and 100 m^2/day. The spring has an elevation of 500 m above sea level. The hydraulic head in the aquifer is 505 m above sea level. Calculate the range of the spring flow rates.

SOLUTION The range of spring flow rate is

$$Q_{sp} = (10 \sim 100)(\text{m}^2/\text{day})(505\,\text{m} - 500\,\text{m}) = (50 \sim 500)\,\text{m}^3/\text{day}$$

Evapotranspiration

In many settings, ground water discharges through transpiration and evaporation of soil water. Transpiration involves the utilization of ground water by plants. One particular group of plants called phreatophytes is particularly important in this respect. Meinzer defined *phreatophytes* as plants that habitually obtain their water supply from the zone of saturation, either directly or through the capillary fringe. However, they also can grow under conditions of abundant soil moisture. Phreatophytes are particularly noticeable in dry regions, where locally a shallow water table often provides the stable water supply necessary to sustain the growth of these plants. When a community of phreatophytes is established, it can use tremendous quantities of water and provide an important mechanism for natural discharge.

According to Meyboom (1967), a record of daily water-level fluctuations provides evidence that phreatophytes use water. In general, the water table declines during the day when transpiration by the plants is at a maximum. Water levels bottom out in the early evening and begin to rise as surrounding ground water moves back into the zone of depressed hydraulic head caused by the plants. Phreatophytes depend on this nightly recovery of water levels. If it were not for this cycle of recovery, the water table would fall below the root system and the vegetation would perish (Meyboom, 1967). Phreatophytes are common in ground-water discharge areas because in these areas the water table is often close to the ground surface and able to recover rapidly. Phreatophytes growing along major rivers use not only ground water but also surface water through infiltration of river water.

On the Canadian Prairies, the types of phreatophytes include Manitoba maple, salt grass, baltic rush, alfalfa, poplar, and buffalo berry. In the arid southwestern United States, typical phreatophytes include salt grass, greasewood, and mesquite. Trees that depend on ground water include the willow, cottonwood, and sycamore. With the exception of alfafa, which is used for animal forage, phreatophytes have typically been regarded as nuisance plants that waste water. However, communities of phreatophytes found along perennial streams form unique and important ecological environments. These unique environments found along streams in arid regions are called *riparian zones*.

Any time that mineralized ground water gets within a few meters of the ground surface in an arid climate, the potential exists for water to evaporate. Evaporative rates can be sufficiently high that evaporation is the principal mode of discharge. The processes of evaporation concentrate whatever salts were initially present in the ground water, and eventually salts precipitate to form saline soils. The following example documents these interesting discharge conditions.

CASE STUDY 8-2

EVAPORATIVE DISCHARGE IN SASKATCHEWAN, CANADA

Saline soils that sometimes form in arid areas can be indicators of ground-water discharge. Shown in Figure 8.27 are features of the "Great Salt Plain" that occupies a broad depression between hills on the east and west of an area in east-central Saskatchewan. The shading outlines an area with large patches of salt efflorescence (mainly Na_2SO_4) on which only a few salt-tolerant plants are able to survive. Interestingly, one of the abundant plants is salt grass, one of the phreatophytes we mentioned previously. The area of salt accumulation coincides almost exactly with an area of ground-water discharge. We can see the evidence in Figure 8.27 by observing that salt accumulates in places where the head gradient between the shallow ground water and a deeper artesian aquifer is at a maximum. Thus, water is leaving the flow system across the lowland area.

Meyboom, 1966

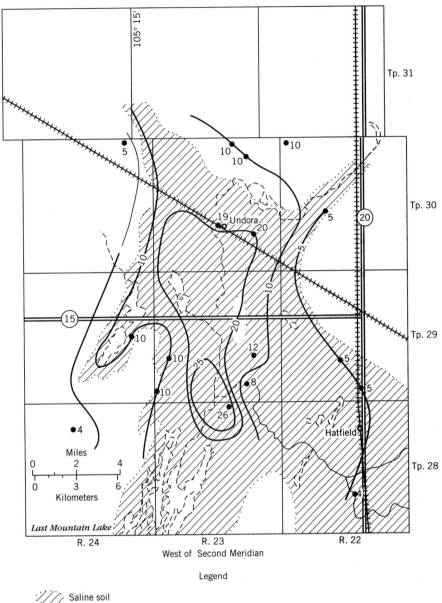

Figure 8.27 Saline soils on the Great Salt Plain, Saskatchewan and their relationship to the hydrogeology (from Meyboom, 1966). Reproduced with the permission of the Minister of Public Works and Government Services Canada, 2001 and courtesy of the Geological Survey of Canada.

▶ 8.6 GROUND-WATER SURFACE-WATER INTERACTIONS

Chapter 1 introduced the concept of gaining and losing streams. A gaining stream receives ground-water discharge; a losing stream provides ground-water recharge. Once hydraulic-head data are available in the vicinity of a stream, one can determine whether the stream is gaining (Figure 8.28*a*) or losing (Figure 8.28*b*). The stage of a gaining stream is lower than that of hydraulic heads in the aquifer immediately adjacent. The stage in a losing stream is higher than adjacent hydraulic heads. The resulting flow patterns show that ground water flows toward a gaining stream and away from a losing stream (compare Figure 8.28*a*, *b*). Numerous studies have shown that the hydraulic conductivity of a streambed, sometimes called a clogging layer, is lower than that in the aquifer immediately below the streambed. Thus, a situation arises where the stream is not all that well connected to the ground-water system and flow has difficulties moving in or out.

Peterson and Wilson (1988) classified streams as connected gaining streams, connected losing streams, disconnected losing streams with a shallow water table, and disconnected

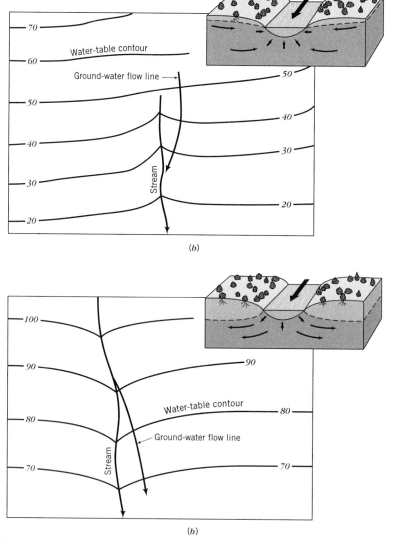

(*b*)

(*b*)

Figure 8.28   Detailed information on the configuration of the water table in the vicinity of a stream enables one to determine whether the stream is gaining (Panel *a*) or losing (Panel *b*) (modified from Winter et al., 1998).

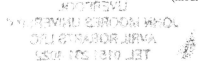

losing streams with a deep water table (Figures 8.29*a* through 8.29*d*). Their studies showed that even if a stream and an aquifer are disconnected, the unsaturated flow under the stream depends on the depth of water table for a disconnected losing stream with shallow water table (Figure 8.28*c*). Overall, unsaturated flow affects the discharge rate from stream to aquifer.

Lakes, like rivers, commonly interact with ground waters and are classified as discharge lakes, recharge lakes, or flow-through lakes depending on their mode of interaction with the ground water. A *discharge lake* receives ground-water discharge throughout its entire bed (Figure 8.30*a*) (Winter et al., 1998). A *recharge lake* loses water as seepage to a ground-water flow system (Figure 8.30*b*). Most lakes are *through-flow* types, receiving inflow over part of their bed and losing water elsewhere (Figure 8.30*c*). For example, the Browns Lake in the southeastern corner of Wisconsin is a through-flow lake (Figure 8.31). The lake receives water from the aquifer at areas of higher elevation and recharges the aquifer at areas of lower elevation.

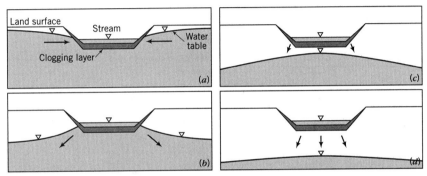

Figure 8.29 This figure shows different types of aquifer–stream interactions, (*a*) a gaining stream; (*b*) a connected losing stream; (*c*) a disconnected losing stream with a shallow water table; and (*d*) a disconnected losing stream with a deep water table (from Peterson and Wilson, 1988).

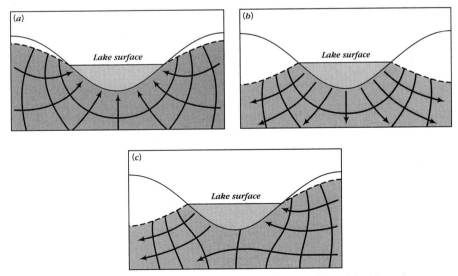

Figure 8.30 Examples of ground-water/lake interactions. A discharge lake (*a*) receives ground-water inflow. A recharge lake (*b*) loses water as seepage to ground water. A through-flow lake (*c*) both gains and loses water (from Winter et al., 1998).

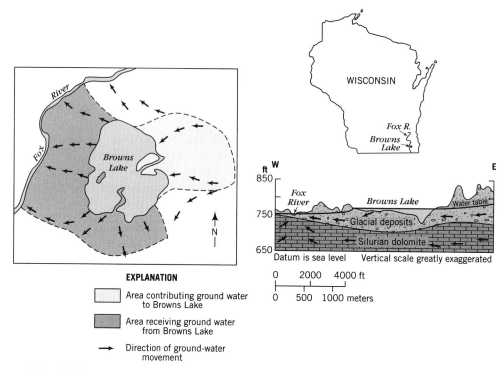

Figure 8.31 Ground-water conditions in the vicinity of Browns Lake in the southeastern corner of Wisconsin (from Cotter et al., 1969; Olcott, 1992).

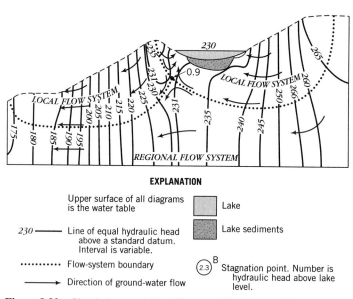

EXPLANATION

Upper surface of all diagrams is the water table

230 ——— Line of equal hydraulic head above a standard datum. Interval is variable.

•••••••• Flow-system boundary

———▶ Direction of ground-water flow

▢ Lake

▢ Lake sediments

(2.3)^B Stagnation point. Number is hydraulic head above lake level.

Figure 8.32 Simulation result that illustrates the pattern of flow in the vicinity of a lake and the relationship to local and regional flow systems. A stagnation point is evident in the vicinity of the lake (from Winter and Pfannkuch, 1984).

    Much has been learned about the interaction between aquifers and lakes using computer models (Winter, 1976, 1978; Winter and Pfannkuch, 1984). Figure 8.32 illustrates how a

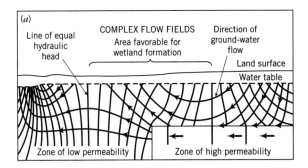
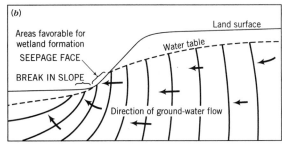
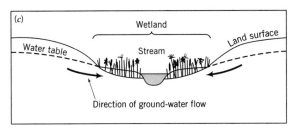
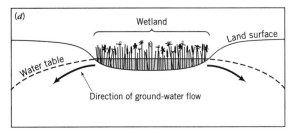

Figure 8.33 Types of wetlands. Panel (*a*) shows ground-water discharge to the ground surface from a complex ground-water flow system. Panel (*b*) illustrates ground-water discharge through seepage face and break in slope. In Panel (*c*), riverine wetlands form along streams. The wetlands in Panel (*d*) are formed from precipitation (from Winter et al., 1998).

discharge lake receives inflow from local flow systems, controlled by the position of the water table in the vicinity of the lake. *A stagnation point* exists for a discharge lake when a local flow system recharges the lake. At the stagnation point, the hydraulic head is higher than the water level in the discharge lake.

 Wetlands are common in countries all around the world. They form when ground water discharges to land surface or when conditions are such that surface water is slow to drain away. The development of wetlands is often intimately related to regional flow conditions. For example, the upland wetland in Figure 8.33a forms in a ground-water discharge area, which itself is a manifestation of the complexities of the regional ground-water flow (Winter et al., 1998). The wetland in Figure 8.33b formed at a break in slope, which fostered ground-water discharge as seepages or springs. Another type of wetland forms along a stream (Figure 8.33c). A riverine wetland depends primarily on the stream and incidentally on ground-water discharge for water (Winter et al., 1998). In some settings, wetlands receive precipitation and surface runoff and function as ground-water recharge areas (Figure 8.33d). Often in these circumstances, water is slow to leak from the wetlands because the organic sediments in the wetlands do not conduct water well (Winter et al., 1998).

► 8.7 FRESHWATER/SALTWATER INTERACTIONS

Freshwater and saltwater interact with each other in a variety of hydrogeological settings. In coastal areas, less dense freshwater tends to override denser saltwater (Figure 8.34). Between the two water masses is a zone of mixing, sometimes called the zone of diffusion, where the fluid has an "in between" chemical composition. Thus, the actual interface between freshwater and saltwater can be more than 1000 ft wide (Figure 8.34), although most mathematical analyses are based on an assumption of a sharp interface. The 1000 mg/L isochlor line commonly delineates the saltwater front (Figure 8.34).

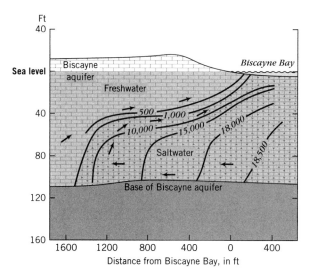

Figure 8.34 Cross section of seawater intrusion into the Biscayne aquifer near Miami, Florida (from Miller, 1992).

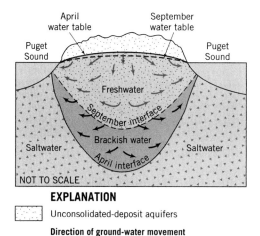

EXPLANATION

☐ Unconsolidated-deposit aquifers

Direction of ground-water movement

→ Freshwater

→ Brackish water

Figure 8.35 The position of the fresh-water interface beneath an island fluctuates seasonally. The interface is deeper during periods with greater-than-normal precipitation (U.S Department of the Interior, 1973, or Miller, 1990).

An island in an ocean provides a slight variation in the distribution of waters. In this case, a lens of freshwater sits on top of the saltwater, which extends beneath the island (Figure 8.35). The position of the mixing zone varies depending on the quantity of freshwater moving through the system. With greater recharge raising the water table, the interface moves deeper (Figure 8.35). With smaller quantities of recharge, the interface retreats. It follows then that in coastal and island settings, diversion of seaward-moving ground water by pumping can cause the interface to retreat and produce a condition known as saltwater intrusion.

On the continents, deep geological basins often contain saltwater in permeable units. When these units crop out and are invaded by freshwater, the condition is set up again where freshwater and saltwater are juxtaposed. Although the setting is different, a zone of mixing develops between the two fluids.

Locating the Interface

Because of saltwater intrusion has the potential to impair the quality of coastal aquifers, studies have continued for more than half a century to understand the behavior of freshwater/saltwater systems and their interfaces. These days, ground-water hydrologists have a powerful array of computational models at their disposal. These models are sufficiently powerful to account for the details of the most complicated settings. Our objective here is to provide an understanding of the processes involved by studying some of the simpler, historically important approaches.

Let's begin by looking at the freshwater/saltwater system shown in Figure 8.36. For this simple analysis, it is assumed that the system is hydrostatic. This assumption means that the weight of a column of freshwater extending from the water table to a point on the interface is the same as a column of saltwater extending from sea level to the same depth (Figure 8.36). This condition can be expressed mathematically as

$$\rho_s g z = \rho_f g (h_f + z) \tag{8.9}$$

where ρ_f is the density of freshwater, ρ_s is the density of saltwater, z is the height of the saltwater column, or the depth below sea level to a point on the interface, h_f is the hydraulic head above sea level, and $h_f + z$ is the height of the freshwater column. Equation (8.9) can be rearranged as

$$z = \frac{\rho_f}{\rho_s - \rho_f} h_f \tag{8.10}$$

If the density of freshwater is taken as 1.0 g/cm^3 and seawater as 1.025 g/cm^3, then

$$z = 40 h_f \tag{8.11}$$

In words, the depth to the interface is approximately 40 times the height of the water table above sea level. Equation (8.11) is referred to as the Ghyben–Herzberg relation because it was first determined by the two scientists independently (Ghyben, 1889; Herzberg, 1901).

This simple theory can be applied to examine some of the features of saltwater/freshwater interactions involving a confined aquifer of thickness b (Figure 8.37). Let's assume we know a little more about the system, like the discharge rate per unit length of coastline (Q'). An improved estimate of the position of the freshwater/saltwater interface is given as

$$x = \frac{1}{2} \frac{(\rho_s - \rho_f) K z^2}{\rho_f Q'} \tag{8.12}$$

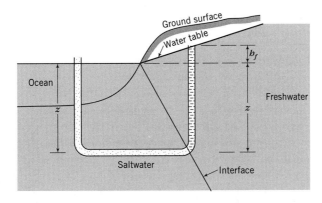

Figure 8.36 Idealization of a coastal saltwater/freshwater system for analysis by the Ghyben–Herzberg relation.

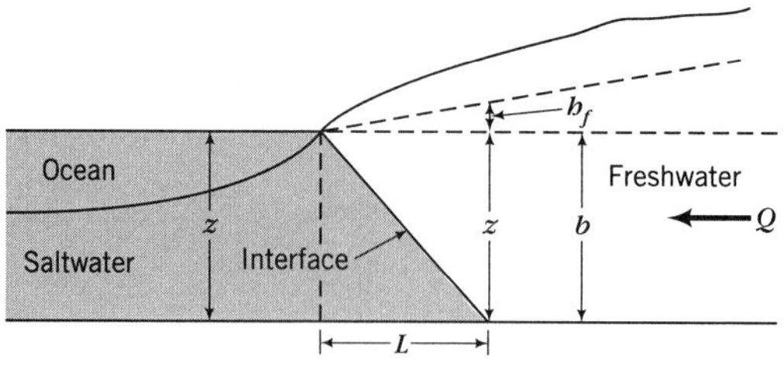

Figure 8.37 Geometry of a coastal aquifer for calculations of the position of the interface and the length of the saltwater wedge for flowing conditions.

The form of Eq. (8.12) indicates that the freshwater–saltwater interface is now no longer a straight line. For an aquifer with thickness of b (Figure 8.37), the length of the saltwater protrusion wedge (L) is expressed as

$$L = \frac{1}{2}\frac{(\rho_s - \rho_f)Kb^2}{\rho_f Q'} \qquad (8.13)$$

where K is the hydraulic conductivity of the aquifer and b is the thickness of the aquifer. Thus, the length of protrusion of saltwater under natural conditions is directly proportional to the hydraulic conductivity and thickness squared and inversely proportional to the flow of freshwater to the sea.

▶ **EXAMPLE 8.4**

The densities of fresh- and saltwater are 1.0 and 1.025 g/cm³, respectively. The water levels in two monitoring wells far from the shoreline are 0.5 and 1.0 m above sea level. The distance between the two wells is 1000 m. The hydraulic conductivity of the coast aquifer is 10 m/day. If the aquifer thickness is 50 m, calculate the length of saltwater wedge and the interface between the freshwater and the saltwater using Eqs. (8.12) and (8.13).

SOLUTION The discharge from the aquifer to the sea per unit length of shoreline ($z = b$) is

$$Q' = Kb\frac{dh}{dx} = (10 \text{ m/day})(50 \text{ m})\frac{(1.0 \text{ m}) - (0.5 \text{ m})}{(1000 \text{ m})} = 0.25 (\text{m}^3/\text{day/m})$$

The equation describing the interface is

$$x = \frac{1}{2}\frac{(\rho_s - \rho_f)Kz^2}{\rho_f Q'} = \frac{1}{2}\frac{(1.025 \text{ g/cm}^3) - (1.0 \text{ g/cm}^3)}{(1.0 \text{ g/cm}^3)}\frac{(10.0 \text{ m/day})}{(0.25 \text{ m}^2/\text{day})}z^2 = 0.5 z^2$$

The length of protrusion of saltwater into the aquifer is

$$L = \frac{1}{2}(0.025)\frac{(10 \text{ m/day})(50)^2}{(0.25 \text{ m}^2/\text{day})} = 1250 \text{ m}$$

Methods Limiting Saltwater Intrusion

In confined and unconfined coastal aquifers, excessive pumping can cause saltwater intrusion. The gradual landward migration of saltwater can cause deterioration in the quality of aquifer water or require mitigation. Four common approaches can be used to limit saltwater intrusion in coastal areas: controls on withdrawals, artificial recharge through ponds, pumping trough barriers, or freshwater injection (Figure 8.38a–d) (Planert and Williams, 1995). Reducing the pumping rates in wells or eliminating some of the wells in the coastal area

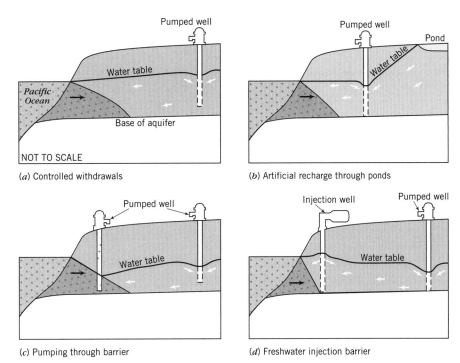

Figure 8.38 Methods to prevent saltwater intrusion (from Planert and Williams, 1995).
Panel (*a*) represents a situation where withdrawals are controlled. In Panel (*b*), fresh water heads are increased by a recharge pond. Panels (*c*) and (*d*) illustrate how barriers can be found by pumping and injection.

increases the freshwater discharge into the ocean. With increased ground-water discharge to the ocean, the extent of saltwater intrusion is reduced (Figure 8.38*a*). Artificial recharge through ponds or water spreading using imported water or reclaimed wastewater replaces some of the ground water lost from storage and partly compensates for the loss of recharge potential that can accompany urbanization (Figure 8.38*b*). A *pumping trough* creates a barrier with a series of pumping wells. They are designed to remove saltwater from the aquifer and form a potentiometric barrier to block the saltwater incursion (Figure 8.38*c*). Injection of freshwater or reclaimed wastewater into the aquifer establishes a seaward freshwater hydraulic gradient (Figure 8.38*d*). In southern California, injection barriers have been successful in halting saltwater intrusion and in reducing the size of the area with ground-water levels below sea level (Figure 8.39*a* and 8.39*b*) (Planert and Williams, 1995).

In Florida, many drainage canals were constructed to remove excess surface and ground water rapidly to prevent flooding in low-lying interior areas. Water is transported to coastal areas from inland to raise ground-water levels and resist saltwater encroachment (Figure 8.40*a*). However, before 1953, the direct connection between the canals and the Biscayne aquifer provided a pathway for saltwater encroachment for appreciable distances inland during the periods of low flow. Salinity-control structures (Figure 8.40*b*) were constructed in coastward reaches of the major canals. These dam-like structures are closed to prevent the inland movement of saltwater up the canals during periods of less than normal precipitation. They worked to halt or reverse saltwater encroachment from the canals. By 1977, additional control structures and effective water management practices had reduced the area of saltwater contamination considerably from its maximum extent in 1953 (Figure 8.41).

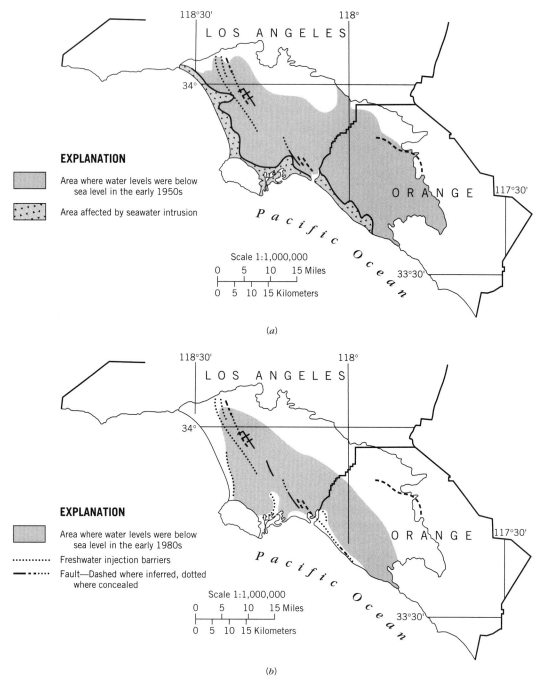

Figure 8.39 In southern California near Los Angeles, seawater intrusion has been controlled using freshwater injection barriers. The extent of areas with water levels below sea level was reduced from the early 1950s (*a*) to the late 1980s (*b*) (from Planert and Williams, 1995).

Upconing of the Interface Caused by Pumping Wells

The Ghyben–Herzberg relationship shows how lowering of water levels causes the interface to rise. In this vicinity of pumping wells where water levels are most affected, the interface

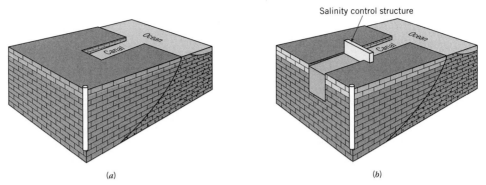

Figure 8.40 Illustration of canals in Florida without a salinity control structure (*a*) and with a salinity control structure (*b*).

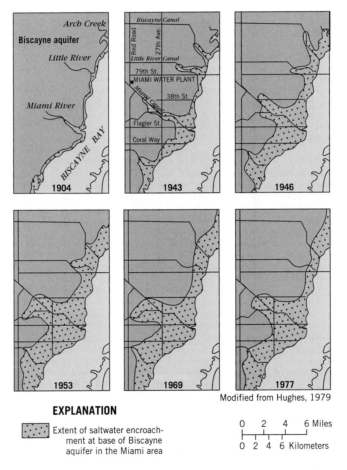

EXPLANATION

Extent of saltwater encroachment at base of Biscayne aquifer in the Miami area

Modified from Hughes, 1979

0 2 4 6 Miles
0 2 4 6 Kilometers

Figure 8.41 Extent of saltwater encroachment at the base of Biscayne aquifer in the Miami area. (*a*) in 1904; (*b*) in 1943; (*c*) in 1946; (*d*) in 1953; (*d*) in 1969; (*e*) in 1977 (from Hughes, 1979).

can rise, a phenomenon referred to as *upconing* (Figure 8.42). The extent of upconing can be calculated with the following equation from Schmorak and Mercado (1969):

$$z = \frac{Q\rho_f}{2\pi dK(\rho_s - \rho_f)} \tag{8.14}$$

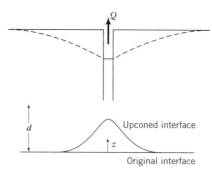

Figure 8.42 Upconing of the saltwater/freshwater interface in response to pumping (from Domenico and Schwartz, 1998. *Physical and chemical hydrogeology*). Copyright ©1998 by John Wiley & Sons, Inc. Reprinted by permission of John Wiley & Sons, Inc.

where z is the vertical change of the freshwater-saltwater interface, Q is the pumping rate, and d is the distance from the base of the well to the original (prepumping) interface (Figure 8.42). Both laboratory and field observations suggest that this relationship holds only for very small rises in the interface, and there exists a critical elevation at which the interface is no longer stable and saltwater flows to the well (Schmorak and Mercado, 1969).

Dagan and Bear (1968) suggest that the interface will be stable for upconed heights that do not exceed one-third of d given in Figure 8.42. Thus, if z is taken as $0.3\, d$, the maximum permitted pumping rate should not exceed

$$Q_{max} \leq 0.6\pi d^2 K \frac{(\rho_s - \rho_f)}{\rho_f} \tag{8.15}$$

▶ **EXAMPLE 8.5**

The distance from the base of a pumping well to the freshwater–saltwater interface is 100 m, the pumping rate is 3000 m³/day, and the hydraulic conductivity is 10 m/day. What will be in the position of the freshwater–saltwater interface? What is the maximum permitted pumping rate?

SOLUTION   The rise in the freshwater–saltwater interface is

$$z = \frac{(3000\,\text{m}^3/\text{day})(1.0\,\text{g/cm}^3)}{(2 \times 3.1416)(100\,\text{m})(10\,\text{m/day})[(1.025\,\text{g/cm}^3) - (1.0\,\text{g/cm}^3)]} = 13.26\,\text{m}$$

The maximum permitted pumping rate is

$$Q_{max} \leq 0.6(3.1416)(100\text{m})^2(10\,\text{m/day})(0.025) = 4.7 \times 10^3\,\text{m}^3/\text{day}$$

▶ **EXERCISES**

8.1. It is known that $A = 10$ m, $B = 50$ m, $z_0 = 50$ m, and $L = 2000$ m. Calculate the flownet using Tóth's equation (8.1). Use FLOWNETz which is provided on the website.

8.2. Assuming that $L = 1000$ m and $z_0 = 100$ m, calculate flownets for the following three cases: (a) $B = 0.2$, $b = 20$ m, and $L/8 = 1$; (b) $B = 0.2$, $b = 20$ m, and $L/8 = 0.25$; (c) $B = 0.2$, $b = 0$ m, and $L/8 = 1$ m using the Tóth's equation (8.2) and FLOWNETz.

8.3. Calculate the flownet for a three-layer regional aquifer system (Freeze and Witherspoon, 1966) using FLOWNETz. With reference to Figure 8.43 assume that $Z_0 = 100$ m, $Z_1 = 75$ m, $Z_2 = 50$ m, $K_1 = 1$ m/day, $K_2 = 10$ m/day, $K_3 = 5$ m/day, $L = 1000$ m, $k = 2$, $X_1 = 500$ m, $C_1 = 0.3$, $X_2 = 1000$ m, and $C_2 = 0.5$.

8.4. The flux rate per unit coastline from an aquifer to the sea is 10 m³/day/m. The hydraulic conductivity and thickness of the aquifer are 10 m/day and 100 m, respectively. Calculate the position of freshwater/saltwater interface.

8.5. A saltwater interface is stable at a position of about 40 m beneath the base of a well in a formation with a hydraulic conductivity of 1×10^{-1} cm/s. Calculate the maximum pumping rate so as not to cause unstable upconing of the interface.

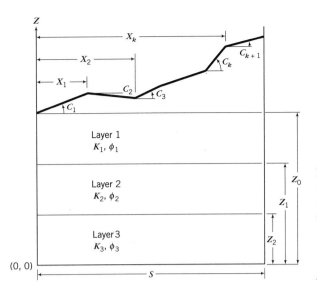

Figure 8.43 Schematic of the simulation domain for an analytic solution involving flow in a regional system with three layers (from Freeze and Witherspoon, *Water Resour. Res.*, 2, p. 641–656, 1966). Copyright by American Geophysical Union.

▶ CHAPTER
HIGHLIGHTS

1. In the study of regional flow, the ground-water basin is a convenient unit for analysis. It is a three-dimensional volume that contains all of the flow paths followed by water recharging the basin. In effect, water enters the subsurface at the water table and discharges at a lower elevation also at the water table. There is no flow through the side and bottom boundaries of the basin.

2. In areas with abundant precipitation, the water table is often a subdued replica of the ground surface. Thus, in the absence of detailed ground-water data, one can infer the location of groundwater divides by finding the major topographic highs and lows.

3. Recharge is water that percolates down through the unsaturated zone and enters the groundwater flow system. In a recharge area, flow is downward, away from the water table. Water leaving the ground-water system is called discharge. Discharge can occur as evapotranspiration, springs, seeps, and baseflow to streams.

4. Important contributions to knowledge of regional ground-water flow have come from the application of numerical models. For example, Tóth (1962, 1963) showed how the water-table configuration controls the pattern of flow. A linear water table leads to the development of a regional flow system that has recharge at the topographically highest portion of the basin and discharge at the topographically lowest portion. When there is local topography on a sloping water table, then local and intermediate-flow systems form. A local flow system has its recharge area at a topographic high and its discharge area at the adjacent topographic low. An intermediate-flow system has one or more topographic lows intervening between recharge and discharge areas.

5. Freeze and Witherspoon (1966, 1967) extended this mathematical approach to examine situations where the pattern of hydraulic-conductivity (K) layering and water-table configuration both change. These simulations show how flow patterns adjust to follow the high-permeability pathways to the extent practicable. High vertical flow gradients develop when low K units overlie high K units. The truncating of permeable layers forms discharge areas in topographically higher parts of the watershed.

6. The pattern of recharge in a basin depends on precipitation, hydrogeologic setting, and topography. In arid climates, recharge can occur due to a combination of a rare precipitation event and a setting that is locally able to concentrate runoff on the landscape. In the Great Plains, recharge is promoted by poorly drained glacial landscapes through small ephemeral lakes and wetlands.

7. The common modes of natural discharge include (i) outflow to wetlands, lakes, and rivers, (ii) discharge from springs and seeps at the surface, and (ii) evapotranspiration. This ground-water

discharge can play a key role in maintaining wetlands and riparian vegetation in arid areas. phreato-phytes are plants that habitually obtain their water from the zone of saturation or zones with continually abundant soil moisture.

8. Along seacoasts, freshwater and saltwater interact. Because freshwater is less dense, it tends to override saltwater. A zone of diffusion that is several thousand feet wide separates these two water types. One can calculate the approximate depth of the interface between fresh- and saltwater using the Ghyben–Herzberg relationship. The depth to the interface is 40 times the height of the water table above sea level.

9. Exploitation of freshwater near coasts can cause saltwater intrusion or the landward migration of seawater toward the wells. Various schemes exist to limit this problem.

9

RESPONSE OF CONFINED AQUIFERS TO PUMPING

One of the important jobs for ground-water hydrologists is to find and to develop water supplies. High-capacity wells installed in productive aquifers are capable of providing thousands of gallons of water per minute. Research in the 1940s and 1950s produced quantitative analytical tools to predict how pumping would impact hydraulic head in the aquifers and to interpret the results of hydraulic tests. From the mid-1960s through the early 1980s, powerful numerical codes like MODFLOW became available to help analyze much more complicated systems.

There are fundamentally two types of problems related to an aquifer's responses to pumping. The so-called forward problem is concerned with predicting what the hydraulic-head distribution will be in an aquifer at times in the future, given boundary conditions, initial conditions, and information about transmissivity, storativity, and pumping rate. As we saw in Chapter 3, one can calculate this hydraulic-head distribution by solving a ground-water flow equation. Forward modeling is essential to designing well systems, analyzing whether drawdowns caused by a well are impacting other wells, and designing dewatering systems.

The inverse problem, as applied to well problems, involves using measurements of hydraulic head in an aquifer as a function of time to calculate values of transmissivity, storativity, specific yield, and so on. In other words, the mathematical theory provides the basis for interpreting the results of an aquifer test. Here and in following chapters, you will learn how to make drawdown predictions for various types of aquifers and how to interpret the results of aquifer tests.

▶ 9.1 AQUIFERS AND AQUIFER TESTS

The study of well hydraulics is complicated. For every situation where the type of aquifer, the pattern of layering of aquifers, or the length of the screen in relation to the aquifer thickness changes, we need a different analytical solution. If our goal was to treat every pumping situation that one might conceivably encounter in the field, we could end up showing 40 or 50 different solutions. Fortunately, we have set a less ambitious goal of looking at only a few of the most common situations. This chapter focuses on a confined aquifer that is homogeneous, isotropic, and infinite in extent.

We have used the term *aquifer test* without discussing it in detail. An *aquifer test* involves pumping a well for the purpose of determining aquifer parameters like T and S. A test typically involves a pumping well and one or more observation wells. A *pumping well* has a relatively large- diameter casing and is screened across all or part of the aquifer (Figure 9.1). A large-diameter casing is necessary because a pump and piping system need to be installed down in the well. *Observation wells* are located at varying distances from the pumping well. They commonly are smaller in diameter and again are screened across all or part of the aquifer. Before an aquifer test is begun, water levels in all of the wells are measured to provide the *prepumping or static water levels* (h_0). In other words, these measurements provide hydraulic heads in the wells at time zero. A test starts with the pumping of water from the well. The *pumping rate* (Q) is the volume of water pumped from a well per unit time [L^3/T].

As the aquifer is pumped, water levels are measured periodically in the pumping and observation wells. Water levels are measured frequently at first because at early times they change rapidly. The term *pumping water level* (h) is used to describe the water level in a well during a test.

By convention, one works with the change in water levels through the test rather than water levels. The term *drawdown* ($s = h_0 - h$) is the difference between the static water level and the pumping water level (Figure 9.1). Once the impact of pumping becomes evident at a well, drawdowns usually increase with time. The zone around the well in which there is a measurable water-level change is called the cone of depression. The *cone of depression* is a water level low in water table or potentiometric surface, which has the shape of an inverted cone, centered on the pumped well. Away from the cone of depression, drawdown caused by pumping is undetectable. The *radius of influence* (R) is the distance from a pumped well to the edge of the cone of depression. Under steady-state conditions, the water discharged by a well is assumed to be coming from sources beyond the radius of influence. Under transient-flow conditions, the water discharged by a well is assumed to be coming

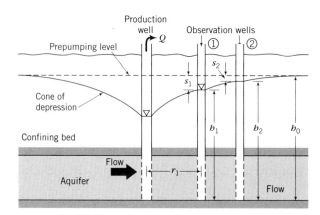

Figure 9.1 A confined aquifer from which ground water is being withdrawn at a constant rate Q. The cone of depression spreads away from the well and produces drawdowns s_1 and s_2.

from the aquifer storage within the radius of influence and sources beyond the radius of influence.

Units

The analytical solutions presented in this and the following chapters are all developed in terms of consistent units. Thus, it does not matter what units of length (for example, feet or meters) or time (seconds, day) you use, as long as all the units are consistent. For example, if meters and days are selected as the consistent units, discharge would have units of m^3/day, distances would be in meters, transmissivities in m^2/day, and so on. Readers will often find it necessary to convert units before using any of the equations in the following chapters. Conversion equations for transmissivity and hydraulic conductivity were introduced in Chapters 3 and 4. The following equations will help provide pumping rates as a consistent unit from the English unit.

$$1 \text{ gpm} = 192.5 \text{ ft}^3/\text{day} = 5.45 \text{ m}^3/\text{day} = 6.3 \times 10^{-5} \text{m}^3/\text{sec} \tag{9.1}$$

$$1 \text{ ft}^3/\text{day} = 5.19 \times 10^{-3} \text{ gpm} = 2.832 \times 10^{-2} \text{m}^3/\text{day} = 3.28 \times 10^{-7} \text{m}^3/\text{sec} \tag{9.2}$$

$$1 \text{ m}^3/\text{day} = 35.31 \text{ft}^3/\text{day} = 0.1835 \text{ gpm} = 1.1574 \times 10^{-5} \text{ m}^3/\text{sec} \tag{9.3}$$

$$1 \text{ m}^3/\text{sec} = 3.051 \times 10^6 \text{ ft}^3/\text{day} = 1.58 \times 10^4 \text{ gpm} = 8.64 \times 10^4 \text{ m}^3/\text{day} \tag{9.4}$$

► 9.2 THIEM'S METHOD FOR STEADY-STATE FLOW IN A CONFINED AQUIFER

Historically, one of the first quantitative approaches for looking at flow in a confined aquifer was that of Thiem (1906). This theory applies to a homogeneous and isotropic aquifer that is infinite in extent. The analysis also assumes that there has been sufficient pumping to allow the ground-water system to achieve steady state. In other words, water levels in the wells do not change with time (Figure 9.2a). The map view in Figure 9.2b shows that the flow in this case is radial, toward the well, with hydraulic heads increasing away from the pumping well.

The hydraulic head in the aquifer can be determined as a solution to a ground-water flow equation, like those we presented in Chapter 3. However, in this case, it is beneficial to use radial coordinates, where distances (r) are measured from the well to some point of interest. A solution to the flow equation with appropriate boundary conditions is

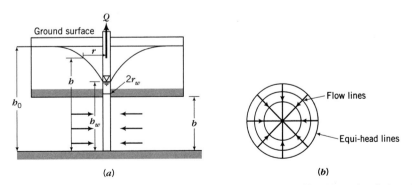

Figure 9.2   Steady-state cone of depression in a confined aquifer, (a) sectional view, (b) map view.

$$h = h_0 + \frac{Q}{2\pi T} \ln \frac{r}{R} \tag{9.5}$$

where h is the hydraulic head at a distance r from a pumped well, h_0 is the prepumping hydraulic head, Q is the pumping rate of the well (positive for withdrawal and negative for injection), T is the transmissivity of the aquifer, and R is the radius of influence of the pumped well. You can determine the hydraulic head in the actual pumping well by assuming that the radius to the observation point is equal to the radius of the pumped well, as

$$h_w = h_0 + \frac{Q}{2\pi T} \ln \frac{r_w}{R} \tag{9.6}$$

where r_w is the radius of the pumped well. In a situation with two observation wells, the hydraulic heads in the wells are related by

$$h_2 = h_1 + \frac{Q}{2\pi T} \ln \frac{r_2}{r_1} \tag{9.7}$$

Remember Eq. (9.7) because it will be used as the basis for a field test to estimate transmissivity. The radius of influence can be estimated from the hydraulic-head measurements.

$$\ln R = \frac{h_0 - h_1}{h_2 - h_1} \ln r_2 - \frac{h_0 - h_2}{h_2 - h_1} \ln r_1 \tag{9.8}$$

▶ **EXAMPLE 9.1**   A well is extracting water from a confined aquifer 50 m thick. At steady state, the hydraulic heads at distances of 100 and 200 m away from the well are 81.7 and 87.2 m, respectively. The hydraulic head before pumping began was 100 m. Calculate the radius of influence.

SOLUTION   The radius of influence is calculated as

$$\ln R = \frac{100 - 81.7}{87.2 - 81.7} \ln 200 - \frac{100 - 87.2}{87.2 - 81.7} \ln 100 = 6.93, \quad R = 1022 \text{ m}$$

Interpreting Aquifer Test Data

The Thiem equation is rarely used for forward calculations because often the requirement that the system be at steady state is not met. However, the theory is used on occasion to calculate aquifer transmissivity from aquifer test data. Because the Thiem equations are developed for steady-state conditions, it is not possible to calculate storativity values. Knowing the pumping rate (Q) and steady-state water levels, h_1 and h_2 for two observation wells located at distances r_1 and r_2, respectively (Figure 9.1), we can calculate transmissivity for a confined aquifer as

$$T = \frac{Q}{2\pi (h_2 - h_1)} \ln \frac{r_2}{r_1} = \frac{2.3Q}{2\pi (h_2 - h_1)} \log \frac{r_2}{r_1} \tag{9.9}$$

where Eq. (9.9) comes by rearranging Eq. (9.7). The drawdowns are related to hydraulic heads by

$$s_1 = h_0 - h_1 \quad \text{and} \quad s_2 = h_0 - h_2 \tag{9.10}$$

where h_0 is the static water level, and s_1 and s_2 are the drawdowns at distances r_1 and r_2, respectively. Thus, Eq. (9.9) can be written in terms of drawdowns as

$$T = \frac{2.3Q}{2\pi (s_1 - s_2)} \log \frac{r_2}{r_1} \tag{9.11}$$

Remember that in applying Eq. (9.11) several assumptions need to be met. First the aquifer should be homogeneous, isotropic, uniform in thickness, and infinite in extent. With experience, you'll find that there is wiggle room in meeting these assumptions because otherwise you would never be able to use Eq. (9.11). Second, the pumping well must be screened across the entire aquifer (that is, fully penetrating) and pumped at a constant discharge rate. Third, sufficient time has elapsed that water levels are no longer changing and the system is at steady state. The following example illustrates how aquifer test data are used to estimate transmissivity.

▶ **EXAMPLE 9.2**

Table 9.1 lists drawdowns measured in five observation wells at steady state in a confined aquifer. The well is being pumped at a constant rate of 100 m^3/day. Use the test data to calculate the transmissivity of the aquifer.

SOLUTION A plot of drawdown versus the logarithm of distance is prepared (Figure 9.3). A best fit line is drawn through the data points. Two points are chosen to calculate the transmissivity.

$$T = \frac{2.3(100\,m^3/\text{day})}{2(2.505 - 0.73)\,m} \log(400/40) = 65\,m^2/\text{day}$$

TABLE 9.1 Drawdowns in a confined aquifer under steady-state flow conditions

r(m)	s(m)
30	2.79
50	2.38
100	1.83
200	1.28
400	0.73

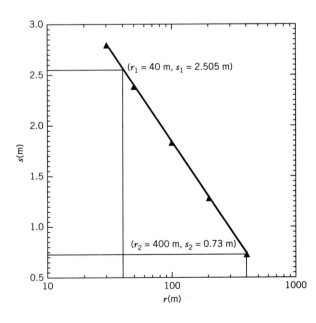

Figure 9.3 Determination of transmissivity using drawdowns measured at multiple observation wells.

▶ 9.3 THEIS SOLUTION FOR TRANSIENT FLOW IN A FULLY PENETRATING, CONFINED AQUIFER

Theis theory lets us evaluate the behavior of a well pumping in a confined aquifer that is under transient (that is, nonsteady-state) conditions. The flow equation describing hydraulic head in a confined aquifer (Figure 9.4) can be written in polar coordinates as

$$\frac{\partial^2 h}{\partial r^2} + \frac{1}{r}\frac{\partial h}{\partial r} = \frac{S}{T}\frac{\partial h}{\partial t} \qquad (9.12)$$

where h is the hydraulic head, r is the radial distance from a pumped well to an observation well, t is time since the pumping started, S is the storativity of the aquifer, and T is the transmissivity of the aquifer. The following initial and two boundary conditions apply to this problem

$$h(r,0) = h_0$$

$$h(\infty, t) = h_0$$

$$\lim_{r \to 0}\left(r\frac{\partial h}{\partial r}\right) = \frac{Q}{2\pi T}$$

In words, the first condition means that at time zero and any distance r from the well, the head is equal to the initial head, h_0. The second condition means that at an infinite radius (one boundary) for all time the hydraulic head is fixed at h_0. The third condition provides a constant withdrawal rate at the pumping well (another boundary).

The solution of Eq. (9.12) was first derived by Theis (1935) and is expressed as

$$h_0 - h = s = \frac{Q}{4\pi T}W(u) \qquad (9.13)$$

where Q is the pumping rate and T is the transmissivity of the aquifer. The well function $W(u)$ and the dimensionless variable u are expressed as:

$$W(u) = \int_u^\infty \frac{e^{-y}}{y}dy = -0.577216 - \ln(u) + u - \frac{u^2}{2!2} + \frac{u^3}{3!3} - \frac{u^4}{4!4} + \cdots \qquad (9.14)$$

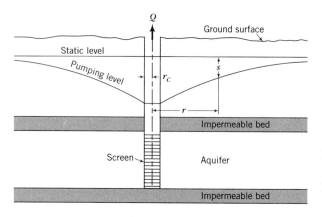

Figure 9.4 Illustration of a nonleaky, confined aquifer being pumped by a fully penetrating well (from Reed, 1980).

and

$$u = \frac{r^2 S}{4 T t} \qquad (9.15)$$

Even though $W(u)$ is a complicated function, it can be evaluated by using well function tables (Table 9.2) or by approximating computer programs. The Theis solution is based on the following assumptions:

1. The pumping well is fully penetrating, with a constant discharge rate, infinitesimal diameter, and negligible storage.

2. The aquifer is confined, infinite in extent, homogeneous, and isotropic.

3. All water pumped by the well comes from the storage and is discharged instantaneously with the decline in head.

▶ 9.4 PREDICTION OF DRAWDOWN AND PUMPING RATE USING THE THEIS SOLUTION

The drawdown in an observation well at some future time can be calculated directly using Eq. (9.13) for known hydraulic parameters. For other problems, we might need to know what pumping rate provides a specified drawdown at a fixed place and time in the future. This calculation requires transforming the Theis equation into the following form.

$$Q = \frac{4\pi T s}{W(u)} \qquad (9.16)$$

▶ **EXAMPLE 9.3**   The transmissivity and storativity of a confined aquifer are 1000 m^2/day and 0.0001, respectively. An observation well is located 500 m away from a pumping well. For a pumping period of 220 min, calculate (a) the drawdown at the observation well if the discharge rate is 1000 m^3/day; (b) the pumping rate required to provide a drawdown of 1 m at that well after 220 minutes.

TABLE 9.2   Values of well function $W(u)$

u	1.0	2.0	3.0	4.0	5.0	6.0	7.0	8.0	9.0
$\times 1$	0.219	0.049	0.013	0.0038	0.0011	0.00036	0.00012	0.000038	0.000012
$\times 10^{-1}$	1.82	1.22	0.91	0.70	0.56	0.45	0.37	0.31	0.26
$\times 10^{-2}$	4.04	3.35	2.96	2.68	2.47	2.30	2.15	2.03	1.92
$\times 10^{-3}$	6.33	5.64	5.23	4.95	4.73	4.54	4.39	4.26	4.14
$\times 10^{-4}$	8.63	7.94	7.53	7.25	7.02	6.84	6.69	6.55	6.44
$\times 10^{-5}$	10.94	10.24	9.84	9.55	9.33	9.14	8.99	8.86	8.74
$\times 10^{-6}$	13.24	12.55	12.14	11.85	11.63	11.45	11.29	11.16	11.04
$\times 10^{-7}$	15.54	14.85	14.44	14.15	13.93	13.75	13.60	13.46	13.34
$\times 10^{-8}$	17.84	17.15	16.74	16.46	16.23	16.05	15.90	15.76	15.65
$\times 10^{-9}$	20.15	19.45	19.05	18.76	18.54	18.35	18.20	18.07	17.95
$\times 10^{-10}$	22.45	21.76	21.35	21.06	20.84	20.66	20.50	20.37	20.25
$\times 10^{-11}$	24.75	24.06	23.65	23.36	23.14	22.96	22.81	22.67	22.55
$\times 10^{-12}$	27.05	26.36	25.96	25.67	25.44	25.26	25.11	24.97	24.86
$\times 10^{-13}$	29.36	28.66	28.26	27.97	27.75	27.56	27.41	27.28	27.16
$\times 10^{-14}$	31.66	30.97	30.56	30.27	30.05	29.87	29.71	29.58	29.46
$\times 10^{-15}$	33.96	33.27	32.86	32.58	32.35	32.17	32.02	31.88	31.76

Source: From Wenzel (1942).

SOLUTION Drawdown can be calculated using the Theis equation. All of the parameters on the RHS of Eq. (9.13) are known except for $W(u)$. To evaluate $W(u)$, we first calculate u as

$$u = \frac{r^2 S}{4Tt} = \frac{500 \times 500 \text{ m}^2 \times 0.0001}{4 \times 1000 \dfrac{\text{m}^2}{\text{day}} \times \dfrac{1}{1440} \dfrac{\text{day}}{\text{min}} \times 220 \text{ min}} = 0.041$$

The well function $W(u)$ at $u = 0.041$ is

$$W(0.041) = 2.66$$

For a pumping rate of 1000 m³/day, the drawdown is calculated as

$$s = \frac{Q}{4\pi T} W(u) = \frac{1000 \text{ m}^3/\text{day}}{4 \times 3.14 \times 1000 \text{ m}^2/\text{day}} (2.66) = 0.21 \text{ m}$$

For a drawdown of 1 m, the pumping rate is calculated as

$$Q = \frac{4\pi T s}{W(u)} = \frac{4 \times 3.14 \times 1000 \text{ m}^2/\text{day} \times 1\text{m}}{(2.66)} = 4.72 \times 10^3 \text{ m}^3/\text{day}$$

▶ 9.5 THEIS TYPE-CURVE METHOD

Another important use of the Theis solution is in determinating the transmissivity and storativity from data collected from an aquifer test. A variety of aquifer testing approaches are available. We begin here with the so-called type-curve matching technique, which is widely used. The test data are a series of drawdown values in an observation well, each matched with a time since pumping began. The approach involves plotting the field data on one graph, which is overlain on a type curve plotted at the same scale. Here are the details:

1. Create the type curve by plotting the well function $W(u)$ versus $1/u$ on log-log graph paper (Figure 9.5). Usually, you can buy a copy of this curve.

2. Define a match point on the type curve. This match point only serves as a reference and can be located anywhere on the graph. However, the math works out best if you choose a match point with "simple" coordinates like $W(u) = 1$ and $1/u = 10^3$, 10^2, or 10.

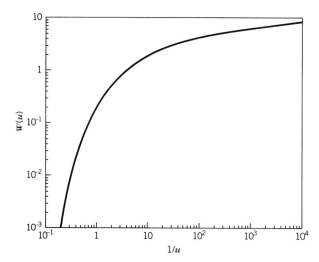

Figure 9.5 A plot of the well function $W(u)$ versus $1/u$ on log-log paper.

3. Prepare a transparent overlay with drawdown (s) plotted versus time (t) on log-log graph paper. This graph paper must be the same as the type curve. This step is where you use the set of field data from the observation well.

4. Superimpose the transparent graph of the field data on the type curve. Adjust the field curve until the collection of field points appears to fall along the type curve underneath. You must keep the axes of the two graphs parallel to each other.

5. Mark the point on the field curve that exactly corresponds with the match point on the type curve underneath. Now you will have points marked on both graphs with coordinates $W(u)$, $1/u$, and $s(t)$, t. These pairs of values will be substituted in step 5.

6. Calculate T and S using the following equations.

$$T = \frac{Q}{4\pi s} W(u) \tag{9.17}$$

and

$$S = \frac{4\,T\,t\,u}{r^2} \tag{9.18}$$

In the case of Eq. (9.17), Q is known from the pump-test data, $W(u)$ is the coordinate value of the match point on the type curve, and s is the coordinate value for the match point on the curve of the field data. In the case of Eq. (9.18), T is known from the previous calculation, r is known from the setup of the aquifer test, and $1/u$ and t are the match-point coordinates obtained via curve matching. Here is an example that illustrates these steps.

▶ **EXAMPLE 9.4**   In a test of a confined aquifer, the pumping rate was 500 m³/day. Drawdown/time data were collected at an observation well 300 m away (Table 9.3). Use the type-curve method to determine the hydraulic conductivity and storativity of the aquifer.

SOLUTION   Figure 9.6 is the plot of drawdown versus time. Superimposing the field curve on to the type curve, as shown in Figure 9.7, gives the match-point coordinates $1/u = 10$, $W(u) = 1.0$, $t = 22$ min, and $s = 0.78$ m. Thus,

$$T = \frac{Q}{4\pi s} W(u) = \frac{(500\,\text{m}^3/\text{day})(1)}{(4\pi)(0.78\,\text{m})} = 51\,\text{m}^2/\text{day}$$

and

$$S = \frac{4Ttu}{r^2} = \frac{4(51\,\text{m}^2/\text{day})\left(22\,\text{min}\dfrac{1\,\text{day}}{1440\,\text{min}}\right)(0.1)}{(300\,\text{m})^2} = 3.46 \times 10^{-6}$$

▶ 9.6 COOPER–JACOB STRAIGHT-LINE METHOD

Using the Theis solution (Eq. 9.13) is complicated by the fact that it contains the function $W(u)$, which is an exponential integral. The value of $W(u)$ is the sum of an infinite series (see Eq. 9.14), which we normally evaluate using the table of well functions. However, when the values of u are small, less than 0.01, the higher order terms of the infinite series are negligible and can be ignored (Cooper and Jacob, 1946; Jacob, 1940). With the Cooper–Jacob's assumption, drawdown is calculated as

$$s(t) = \frac{Q}{4\pi T}(-0.577216 - \ln(u)) = \frac{2.3Q}{4\pi T} \log \frac{2.25T\,t}{r^2 S} \tag{9.19}$$

**TABLE 9.3   Drawdowns measured
at an observation well 300 m away**

Time (min)	S(m)
1.00	0.03
1.27	0.05
1.61	0.09
2.04	0.15
2.59	0.22
3.29	0.31
4.18	0.41
5.30	0.53
6.72	0.66
8.53	0.80
10.83	0.95
13.74	1.11
17.43	1.27
22.12	1.44
28.07	1.61
35.62	1.79
45.20	1.97
57.36	2.15
72.79	2.33
92.37	2.52
117.21	2.70
148.74	2.89
188.74	3.07
239.50	3.26
303.92	3.45
385.66	3.64
489.39	3.83
621.02	4.02
788.05	4.21
1000.0	4.39

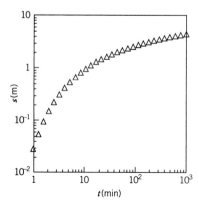

Figure 9.6 A plot of measured drawdown versus time on log-log paper.

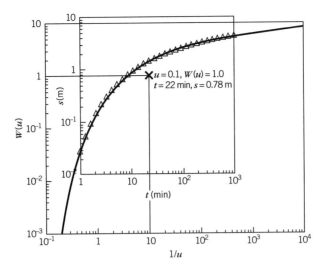

Figure 9.7 Illustration of the method of superposition used to determine transmissivity and storativity.

The result of this simplification is that integral expression in the Theis solution is replaced by a much simpler function. It can be evaluated using a calculator without the need for the table of well functions.

Another benefit of the Cooper and Jacob (1946) modification of the Theis equation is that it leads to a simple, graphical approach to evaluating aquifer test data. If drawdown/time data are plotted as drawdown versus the logarithm of time, the Cooper–Jacob theory predicts that the data will fall along a straight line. By extracting two numbers from the graph, we can solve equations to determine transmissivity and storativity. Here is a summary of the steps involved, followed by an example.

1. Plot drawdown versus time on semi-log graph paper, with time on the x-axis as a logarithmic scale and drawdown on the y-axis as an arithmetic scale. Often, zero drawdown is at the top of the y-axis.

2. Fit a straight line through the data points. If there is difficulty, use the later-time points.

3. Select two points (t_1, s_1 and t_2, s_2) on the line. The equation that is needed can be derived from Eq. (9.19) by writing one equation in terms of s_2 and one equation in terms of s_1 and subtracting them from each other. After some manipulation, the resulting equation is

$$\Delta s = s_2 - s_1 = \frac{2.3Q}{4\pi T} \log \frac{t_2}{t_1} \qquad (9.20)$$

4. Choose t_1 and t_2 one log cycle apart, for example $t_1 = 10$ minutes and $t_2 = 100$ minutes, to give Δs or drawdown per log cycle. This choice simplifies the math. For example, with $t_1 = 10$ minutes and $t_2 = 100$ minutes, $\log(t_2/t_1) = \log(100/10) = 1$. The log term in Eq. (9.20) becomes one, and the equation simplifies to

$$T = \frac{2.3Q}{4\pi \Delta s} \qquad (9.21)$$

where Δs is the drawdown per log cycle. All the terms on the RHS of Eq. (9.21) are known, so it is a simple matter to calculate transmissivity.

5. Find the value of t_0 on the graph by extending the straight line to intersect the line of zero drawdown ($s = 0$). The corresponding time (t_0) is in effect the time that it takes for the cone of depression to reach the observation well. With this value of t_0, all values on the RHS of Eq. (9.21) are known, and storativity can be calculated as

$$S = \frac{2.25T\, t_0}{r^2} \tag{9.22}$$

You can derive Eq. (9.22) by setting the drawdown in Eq. (9.19) to zero and rearranging terms.

6. The final step is a final check to make sure that the Cooper–Jacob simplification applies to this problem or in other words whether $u = r^2 S/(4Tt) < 0.01$. Hopefully, the check will be successful and the problem solved.

▶ **EXAMPLE 9.5**

Determine the transmissivity and storativity of the aquifer in Example 9.4 using the Cooper-Jacob straight-line method.

SOLUTION Figure 9.8 shows a plot of drawdown versus $\log(t)$ for the data set. A line is fitted a late-time section of the curve. On the figure, we find drawdowns corresponding to two times, which are a factor of 10 different from each other. For example, with $t_1 = 100$ min, $s_1 = 2.58$ m, and with $t_2 = 1000$ min, $s_2 = 4.39$ m. Thus, the drawdown per log cycle is $4.39 - 2.58$ or 1.81 m

$$T = \frac{2.3Q}{4\pi\, \Delta s} = \frac{(2.3)(500\,\text{m}^3/\text{day})}{4\pi(1.81\,\text{m})} = 51\,\text{m}^2/\text{day}$$

The next step is to substitute the value of t_0, 3.4 min (determined from the graph) in Eq. (9.22), along with the other known parameters, and to calculate S

$$S = \frac{2.25T\, t_0}{r^2} = \frac{(2.25)(51\,\text{m}^2/\text{day})\left(3.4\,\text{min}\,\dfrac{1\,\text{day}}{1440\,\text{min}}\right)}{(300\,\text{m})^2} = 3.0 \times 10^{-6}$$

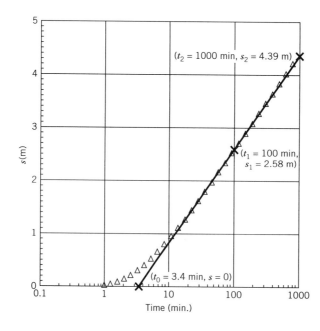

Figure 9.8 Illustration of how the Cooper–Jacob straight-line method is used with observation well data.

The last step is to use the calculated T and S values with other parameters to see whether u is appropriate for the Cooper–Jacob assumption

$$u = \frac{r^2 S}{4Tt} = \frac{(300)^2 \, \text{m}^2 (3.0 \times 10^{-6})}{4(51 \, \text{m}^2/\text{day}) \left(100 \, \text{min} \dfrac{1\,\text{day}}{1440 \, \text{min}} \right)} = 0.02$$

In this example, the maximum u is close to 0.01, and it is acceptable to calculate T and S values using the Cooper–Jacob method.

▶ 9.7 DISTANCE-DRAWDOWN METHOD

The Cooper–Jacob simplification is also useful in determining transmissivity and storativity values in aquifer tests if water levels are measured in two or more observation wells at the same time. For example, assume that two observation wells are located at distances r_1 and r_2 from the pumping well. Knowing the drawdowns at these two wells are s_1 and s_2, respectively, at some time t, we can write this equation based on Eq. (9.19):

$$s_1 = \frac{2.3Q}{4\pi T} \log \frac{2.25T\,t}{r_1^2 S} \tag{9.23}$$

and

$$s_2 = \frac{2.3Q}{4\pi T} \log \frac{2.25T\,t}{r_2^2 S} \tag{9.24}$$

The combination of Eqs. (9.23) and (9.24) yields

$$s_1 - s_2 = \frac{2.3Q}{2\pi T} \log \frac{r_2}{r_1} \tag{9.25}$$

The procedures for determining the transmissivity and storativity for an aquifer using the distance-drawdown method are as follows:

1. For a selected time, plot the drawdown and distance information for the observation wells on semi-log graph paper. Distance is plotted as a logarithmic scale on the x-axis, and drawdown is plotted on a linear scale on the y-axis.
2. Fit a straight line through the data points.
3. Select two points (r_1, r_2) on the line, one log cycle apart, and determine the drawdown Δs. The transmissivity is calculated by

$$T = \frac{2.3Q}{2\pi(\Delta s)} \tag{9.26}$$

4. Extend the straight line to $s = 0$ and determine the distance r_0. The storativity is calculated by

$$S = \frac{2.25T\,t}{r_0^2} \tag{9.27}$$

▶ **EXAMPLE 9.6**   A confined aquifer is pumped at 220 gpm. At time = 220 min, drawdowns were recorded in nine observation wells (Table 9.4). Calculate the transmissivity and storativity of the aquifer.

SOLUTION   Drawdown versus distance is plotted in Figure 9.9. The pumping rate in gpm is first converted to ft³/day using Eq. (9.1).

TABLE 9.4   Values of drawdown versus distance measured at 220 minutes

r(ft)	s(ft)
10	35.20
50	24.35
100	19.68
150	16.96
200	15.03
250	13.54
300	12.32
400	10.42
500	8.97

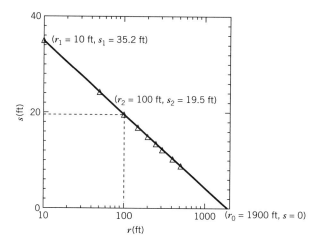

Figure 9.9 Drawdown data for nine observation wells at different radial distances from a pumping well are fit with a straight line to provide Δs and r_0.

$$Q = (220\,\text{gpm})\left(192.5\,\frac{\text{ft}^3/\text{d}}{\text{gpm}}\right) = 42350\,\text{ft}^3/\text{d}$$

Selecting r_1 as 10 ft and $r_2 = 100$ ft, the drawdown per log cycle is $s_1 - s_2$, or $35.2 - 19.5 = 15.7$ ft. The value of $r_0 = 1900$ ft. Thus,

$$T = \frac{2.3Q}{2\pi(\Delta s)} = \frac{2.3(42350\,\text{ft}^3/\text{d})}{2\pi(15.7)\text{ft}} = 987\,\text{ft}^2/\text{d}$$

and

$$S = \frac{2.25(987\,\text{ft}^2/\text{d})(220\,\text{min})\left(\dfrac{1}{1440}\dfrac{\text{d}}{\text{min}}\right)}{(1900\,\text{ft})^2} = 9.4 \times 10^{-5}$$

► 9.8 ESTIMATING T AND S USING RECOVERY DATA (THEIS, 1935)

If pumping of a well is halted, theory predicts that the water level in the aquifer will return to its prepumping level, h_0. Water-level data obtained during this recovery phase of a test provides a basis for determining the transmissivity of the aquifer. The residual drawdown during the recovery period for a confined aquifer is expressed as

$$s' = \frac{Q}{4\pi T} \left[\int_u^\infty \frac{e^{-u}}{u} du - \int_{u'}^\infty \frac{e^{-u'}}{u'} du' \right] \qquad (9.28)$$

where

$$u = \frac{r^2 S}{4 T t} \qquad (9.29)$$

and

$$u' = \frac{r^2 S}{4 T t'} \qquad (9.30)$$

where t is the time since pumping starts, t' is time since pumping stops, and s' is the residual drawdown. Because recovery measurements are made in the pumped well or in a nearby observation well, the radius to the measurement point, r, is typically small. A small r usually leads to a small value u', which enables us to take advantage of the Cooper-Jacob simplification. Under the Cooper–Jacob assumption, Eq. (9.28) reduces to

$$s' = \frac{2.3Q}{4\pi T} \log \left(\frac{t}{t'} \right) \qquad (9.31)$$

and transmissivity can be determined as

$$T = \frac{2.3Q}{4\pi s'} \log \left(\frac{t}{t'} \right) \qquad (9.32)$$

The procedure for determining transmissivity using recovery data is as follows:

1. Plot the residual drawdown (s') on an arithmetic scale versus the time ratio (t/t') on a logarithmic scale.

2. Choose two points on the graph. Again it helps to select the two points one log cycle apart.

The transmissivity is obtained by

$$T = \frac{2.3Q}{4\pi \Delta s'} \qquad (9.33)$$

where $\Delta s'$ is the change in residual drawdown over one log cycle (t/t').

Storativity can be calculated if the recovery data are collected in an observation well rather than the pumping well. The drawdown (s_P) when the pump is turned off at time (t_P) is expressed as

$$s_P = \frac{2.3Q}{4\pi T} \log \frac{2.25 T t_P}{r^2 S} \qquad (9.34)$$

Once T is known, the storativity is obtained by

$$S = \frac{2.25 T t_P}{r^2} 10^{-\frac{4\pi T s_P}{2.3Q}} \qquad (9.35)$$

► **EXAMPLE 9.7**

In an aquifer test reported by USBR in 1995, drawdowns are recorded in the pumped well (Table 9.5) and an observation well (Table 9.6). In both tables, the first column is the time since the pumping started, the second column is the drawdown during the pumping period, the third column is the time since pumping stopped, the fourth column is the time ratio, and the last column is the residual drawdown. A constant pumping rate of 162.9 ft^3/min was maintained during the pumping part of the test.

TABLE 9.5   Aquifer test information from a pumped well

t(min)	s(ft)	t'(min)	t/t'	s'(ft)
3	10.2	3	267.67	−20
8	10.6	8	101.00	−5
13	10.8	13	62.54	−0.5
20	11.3	20	41.00	1.5
80	11.6	80	11.00	1
140	11.8	140	6.71	0.8
195	11.8	195	5.10	0.69
255	11.8	255	4.14	0.59
315	12	315	3.54	0.51
375	12.2	375	3.13	0.49
435	12.2	435	2.84	0.46
495	12.2	495	2.62	0.38
560	12.2	560	2.43	0.34
616	12.3	616	2.30	0.33
668	12.4	668	2.12	0.33
737	12.5	727	2.10	0.22
800	12.5	800	2.00	0.22

Source: Modified from USBR (1995).

TABLE 9.6   Aquifer test information at an observation well

t(min)	s(ft)	t'(min)	t/t'	s'(ft)
5	0.08	5	161.00	1.78
10	0.22	10	81.00	1.64
15	0.33	15	54.33	1.53
20	0.41	20	41.00	1.45
25	0.5	25	33.00	1.37
30	0.55	30	27.67	1.32
40	0.66	40	21.00	1.22
50	0.73	50	17.00	1.15
60	0.8	60	14.33	1.09
70	0.86	70	12.43	1.03
80	0.92	80	11.00	0.97
90	0.96	90	9.89	0.94
100	1	100	9.00	0.9
110	1.04	110	8.27	0.87
120	1.07	120	7.67	0.85
180	1.24	180	5.44	0.7
240	1.35	240	4.33	0.61
300	1.45	300	3.67	0.54
360	1.52	360	3.22	0.49
420	1.59	420	2.90	0.46
480	1.65	480	2.67	0.4
540	1.71	540	2.48	0.36
600	1.73	600	2.33	0.36
660	1.77	660	2.21	0.34
720	1.81	720	2.11	0.31
800	1.86	800	2.00	0.29

Source: Modified from USBR (1995).

The observation well is 100 ft away from the pumped well. Calculate the hydraulic parameters using the recovery data.

SOLUTION Drawdown data for the pumped well and the observation well are plotted in Figures 9.10 and 9.11, respectively. A straight line approximates the drawdown versus time curve in each figure. We have derived $\Delta s' = 0.91$ ft and 0.87 ft for the pumped and observation wells, respectively. Therefore, the transmissivity determined from the recovery data in the pumped well is

$$T = \frac{2.3Q}{4\pi \Delta s'} = \frac{2.3(162.9 \, \text{ft}^3/\text{min})}{4\pi(0.91 \, \text{ft})} = 32.8 \, \text{ft}^2/\text{min} = 4.7 \times 10^4 \, \text{ft}^2/\text{day}$$

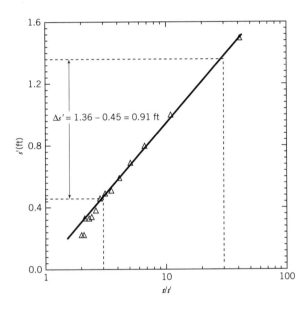

Figure 9.10 This plot illustrates the straight-line methods for determining transmissivity using residual drawdown data from the pumped well.

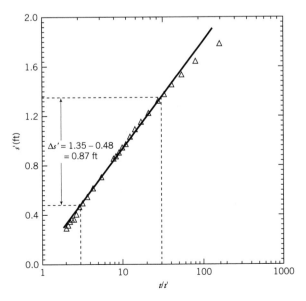

Figure 9.11 This plot illustrates the straight-line methods for determining transmissivity using residual drawdown data from the observation well.

and transmissivity determined from the recovery data in the observation well is

$$T = \frac{2.3Q}{4\pi \Delta s'} = \frac{2.3(162.9\,\text{ft}^3/\text{min})}{4\pi(0.87\,\text{ft})} = 34.3\,\text{ft}^2/\text{min} = 4.9 \times 10^4\,\text{ft}^2/\text{day}$$

Using $t_p = 800$ min, $s_P = 1.86$ ft, $Q = 162.9$ ft^3/min, $r = 100$ ft, and $T = 34.3$ ft^2/min, storativity of the aquifer can be determined from the residual drawdown in the observation well.

$$S = \frac{2.25\,T\,t_P}{r^2}10^{-\frac{4\pi T s_P}{2.3Q}} = \frac{2.25(34.3\,\text{ft}^2/\text{min})(800\,\text{min})}{(100\text{ft})^2}10^{-\frac{4\pi(34.3\,\text{ft}^2/\text{min})(1.86\text{ft})}{2.3(162.9\,\text{ft}^3/\text{min})}} = 0.045$$

It is not possible to determine the storativity of the aquifer using drawdown data from the pumped well.

By mathematically manipulating Eq. (9.28), analytical methods developed for a pumping scheme can also be used to determine the hydraulic parameters from recovery measurements. Equation (9.28) can be rewritten as

$$s^* = \frac{Q}{4\pi T}W(u') = \frac{Q}{4\pi T}W(u) - s' \tag{9.36}$$

where s^* is the recovery drawdown, the first term on the right side of Eq. (9.36) is the drawdown during the pumping period projected to time t', and s' is the residual drawdown. The same techniques for interpreting the drawdown-time curves during a pumping period can be used to interpret the recovery drawdown (s^*) versus recovery time (t') curves. Figure 9.12 shows the relationship between drawdown, residual drawdown, and recovery drawdown.

▶ EXERCISES

9.1. For transmissivity (T) = 2500 m^2/day, storativity (S) = 1.0×10^{-3}, and a pumping rate (Q) = 500 m^3/day, calculate drawdowns in a confined aquifer for r = 10, 50, and 100 m at t = 150 min.

9.2. Use calculations to answer the following questions: (a) How will the cone of depression be changed if the transmissivity is increased while other parameters are kept constant? (b) How will the cone of depression be changed if the storativity is increased while other parameters are kept constant?

9.3. Values of drawdown and time in a pumping test are listed in Table 9.7. Determine the transmissivity and storativity of a confined aquifer using the Theis type-curve technique. Both the pumped and observation wells are fully penetrating. A pumping rate of 10,000 m^3/day is used in the test. The observation well is located 150 m away from the pumped well. You will find two template files for this

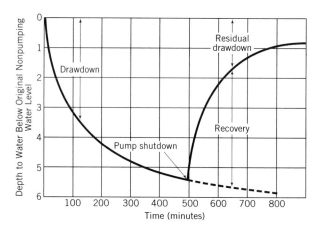

Figure 9.12 Arithmetic plot showing the shape of the drawdown/recovery curve versus time (from Domenico and Schwartz, 1998. *Physical and chemical hydrogeology*). Copyright ©1998 by John Wiley & Sons, Inc. Reprinted by permission of John Wiley & Sons, Inc.

TABLE 9.7 **Values of drawdown versus time in an aquifer test with a confined aquifer**

Time (min)	Drawdown(m)	Time(min)	Drawdown(m)
14.4	0.62	254	1.08
18	0.66	316.7	1.11
22.4	0.69	394.9	1.15
27.9	0.73	492.5	1.18
34.8	0.76	614.1	1.22
43.4	0.8	765.8	1.25
54.2	0.83	955	1.29
67.5	0.87	1190.9	1.33
84.2	0.9	1485.1	1.36
105	0.94	1852	1.4
131	0.97	2309.5	1.43
163.3	1.01	2880	1.47
203.6	1.04		

TABLE 9.8 **Values of drawdown versus distance in an aquifer test with a confined aquifer**

Distance (m)	Drawdown (m)
25	1.51
50	1.29
100	1.07
150	0.94

exercise in the directory Chap9 on the website. The template file 'temp-9a.doc' contains the Theis type curve. The other template file, 'temp-9.doc', is a graphic template in the same log-log scale as the type curves. Print the template files, plot the values of drawdown versus time on the graphic template, and superimpose the two graphic papers to determine hydraulic parameters.

9.4. Use the values of time and drawdown in Table 9.7 and determine the aquifer parameters by the time straight-line technique.

9.5. Values of drawdown and distance in a pumping test in a confined aquifer are listed in Table 9.8. Determine the hydraulic parameters of the aquifer using the distance straight-line technique.

9.6. Table 9.9 lists drawdowns during a pumping period and residual drawdowns during a recovery period in a confined aquifer. The pumping rate is 1.0×10^4 m^3/day. The distance from the observation well to the pumped well is 100 m. Calculate the hydraulic parameters using residual drawdown data.

► **CHAPTER HIGHLIGHTS**

1. Hydrogeologists work on two types of problems related to pumping in aquifers. The forward problem is concerned with predicting what the hydraulic-head distribution will be in an aquifer in the future, given boundary conditions, initial conditions, and aquifer parameters. Forward modeling is helpful in designing well systems or in analyzing whether pumping might impact other wells. The inverse problem, as applied to aquifer-test problems, involves using measurements of hydraulic head in an aquifer as a function of time to calculate values of transmissivity, storativity, specific yield, and so on.

2. An aquifer test involves pumping a well to determine aquifer parameters like T and S. A test involves a pumping well and one or more observation wells. Observation wells are located at varying distances from the pumping well. Before a test begins, water levels in the wells are measured to provide the prepumping (h_0). As the aquifer is pumped, drawdowns are measured as a function of time.

TABLE 9.9   Values of drawdown versus time in an observation well

t(min)	s(ft)	t'(min)	s'(ft)	t(min)	s(ft)	t'(min)	s' (ft)
1.44	0.089	1.44	1.052	53.08	0.595	53.08	0.508
1.64	0.102	1.64	1.032	60.38	0.615	60.38	0.489
1.86	0.115	1.86	1.012	68.68	0.635	68.68	0.470
2.12	0.130	2.12	0.993	78.13	0.656	78.13	0.452
2.41	0.145	2.41	0.973	88.87	0.676	88.87	0.433
2.74	0.160	2.74	0.954	101.09	0.696	101.09	0.415
3.12	0.176	3.12	0.934	114.99	0.717	114.99	0.396
3.55	0.193	3.55	0.915	130.80	0.737	130.80	0.378
4.04	0.210	4.04	0.895	148.78	0.757	148.78	0.360
4.59	0.227	4.59	0.875	169.24	0.778	169.24	0.343
5.22	0.245	5.22	0.856	192.51	0.798	192.51	0.325
5.94	0.263	5.94	0.836	218.98	0.819	218.98	0.308
6.76	0.281	6.76	0.817	249.09	0.839	249.09	0.291
7.69	0.300	7.69	0.797	283.33	0.860	283.33	0.275
8.74	0.318	8.74	0.778	322.29	0.880	322.29	0.259
9.95	0.337	9.95	0.758	366.60	0.900	366.60	0.243
11.31	0.356	11.31	0.739	417.01	0.921	417.01	0.227
12.87	0.376	12.87	0.719	474.35	0.941	474.35	0.212
14.64	0.395	14.64	0.700	539.57	0.962	539.57	0.198
16.65	0.415	16.65	0.681	613.76	0.982	613.76	0.184
18.94	0.435	18.94	0.661	698.15	1.003	698.15	0.170
21.54	0.454	21.54	0.642	794.14	1.023	794.14	0.157
24.51	0.474	24.51	0.623	903.33	1.044	903.33	0.145
27.87	0.494	27.87	0.603	1027.53	1.064	1027.53	0.133
31.71	0.514	31.71	0.584	1168.81	1.085	1168.81	0.122
36.07	0.534	36.07	0.565	1329.52	1.105	1329.52	0.112
41.03	0.554	41.03	0.546	1440.00	1.106	1440.00	0.105
46.67	0.574	46.67	0.527				

3. The cone of depression is a water-level low in water table or potentiometric surface, which has the shape of an inverted cone, centered on the pumped well. The radius of influence (R) is the distance from a pumped well to the edge of the cone of depression.

4. An early approach to analyzing drawdown resulting from pumping in a confined aquifer is that of Thiem (1906). His theory applies to an infinite, homogeneous, and isotropic aquifer. The analysis assumes that drawdown eventually stops increasing and remains constant. Forward calculations provide the steady-state hydraulic head at a distance r from the pumping well. Given the hydraulic heads in two observation wells, one can calculate the aquifer transmissivity.

5. The Theis theory provides a basis for evaluating the transient hydraulic response of an infinite confined aquifer due to pumping. The drawdown s in an observation well at distance r from the pumping well is given as

$$h_0 - h = s = \frac{Q}{4\pi T} W(u)$$

where Q is the pumping rate and T is the transmissivity of the aquifer. The well function $W(u)$ and the dimensionless variable u are expressed as:

$$W(u) = \int_u^\infty \frac{e^{-y}}{y} dy = -0.577216 - \ln(u) + u - \frac{u^2}{2!2} + \frac{u^3}{3!3} - \frac{u^4}{4!4} + \cdots \quad \text{and} \quad u = \frac{r^2 S}{4 T t}$$

where S is storativity and t is time.

6. Several methods for interpreting the results of aquifer tests are based directly on the Theis solution. The Theis-type curve method takes a graph of the log of drawdown (s) versus the log of time (t) and superimposes it on a type curve $W(u)$ versus $1/u$, plotted at the same scale. A curve-matching process yields a set of corresponding values of s, t, $W(u)$ and $1/u$, which are substituted into appropriate equations to yield T and S.

7. The Cooper-Jacob method provides a simple algebraic solution to the Theis equation. Thus, one can calculate drawdown without using well functions. This theory also simplifies the process of interpreting aquifer tests. One can prepare a plot of measured drawdowns as a function of time at a single observation well, or a plot of drawdowns in several wells versus radius at the same time.

CHAPTER

10

LEAKY CONFINED AQUIFERS AND PARTIALLY PENETRATING WELLS

▶ 10.1 TRANSIENT SOLUTION FOR FLOW WITHOUT
STORAGE IN THE CONFINING BED

▶ 10.2 STEADY-STATE SOLUTION

▶ 10.3 TRANSIENT SOLUTIONS FOR FLOW WITH
STORAGE IN CONFINING BEDS

▶ 10.4 EFFECTS OF PARTIALLY PENETRATING WELLS

When a confined aquifer is pumped, *leakage* into the aquifer from adjacent units can sometimes result. This leakage is caused by flow across the confining unit from an adjacent aquifer (Figure 10.1). Besides flow across the confining bed, another component of leakage is the release of water from storage in the confining bed. The net effect of this leakage is to reduce the drawdown in the confined aquifer from that which is expected from a Theis-type response.

This chapter looks at the response of a leaky confined aquifer to pumping. The first case we examine is flow across a confining bed without storage. An idealization of this setting is presented in Figure 10.1. Later, we consider situations where storage in the confining beds is also a source of leakage.

During the initial stage of an aquifer test, the change in water level in an observation well exhibits a Theis-type response. At early times, water flowing to the pumped well comes mainly from storage in the confined aquifer. At late times, water being pumped from the well comes from leakage, causing a deviation from the Theis response.

▶ 10.1 TRANSIENT SOLUTION FOR FLOW WITHOUT STORAGE IN THE CONFINING BED

Hantush and Jacob (1955) derived an analytical solution for flow in a leaky confined aquifer system (Figure 10.1). The governing equation, describing drawdown in the pumped aquifer, is

$$\frac{\partial^2 s}{\partial r^2} + \frac{1}{r}\frac{\partial s}{\partial r} - \frac{K'}{T'b'}s = \frac{S}{T}\frac{\partial s}{\partial t} \tag{10.1}$$

where s is the drawdown, t is time, r is the distance from the pumping well to an observation well, T is the transmissivity of the aquifer, S is the storativity of the aquifer, K' is the

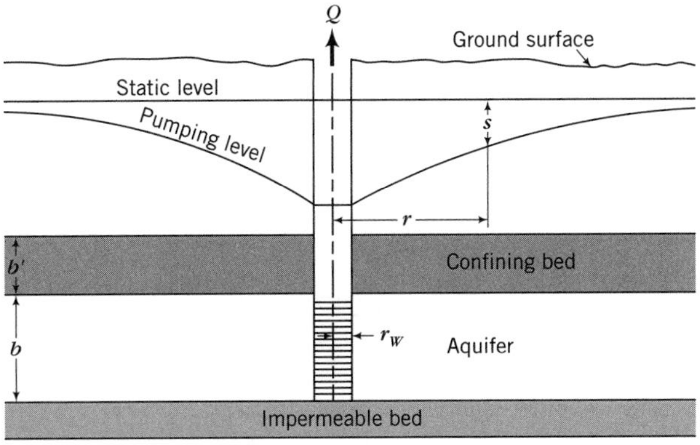

Figure 10.1 Schematic illustration of a leaky, confined aquifer, which is being pumped by a well at a constant rate of discharge (modified from Reed, 1980).

hydraulic conductivity of the confining bed, and b' is the thickness of the confining bed. The extra term in this equation represents leakage into the aquifer through the confining bed from a neighboring aquifer. The solution to Eq. (10.1) provides the drawdown in a leaky confined aquifer due to the pumping of a fully penetrating well (Hantush and Jacob, 1955)

$$s = \frac{Q}{4\pi T} W\left(u, \frac{r}{B}\right) \tag{10.2}$$

where $B = (Tb'/K')^{1/2}$ and $W(u, r/B)$ is expressed as

$$W\left(u, \frac{r}{B}\right) = \int_u^\infty \frac{e^{-z - \frac{r^2}{4B^2 z}}}{z} dz \tag{10.3}$$

Values of $W(u, r/B)$ are tabulated in Table 10.1 (Hantush, 1956) and the function $W(u, r/B)$ is plotted in Figure 10.2. Unlike the last type curve, there is now a family of curves, each labeled with a separate r/B value. The different r/B curves represent differences in the amounts of leakage across the confining bed. For example, the curve for an r/B value of zero applies to the situation where the confining bed is impermeable. This type curve is the same as the $W(u)$ curve presented previously in Chapter 9 and explains the label Theis-type curve in Figure 10.2. Curves with small r/B values (for example, 0.05) exemplify minor leakage and hence minor deviation from the Theis curve. Curves with a large r/B value (for example, 1) exemplify significant leakage with a much greater deviation from the Theis-type curve.

Before proceeding further, let's formally set down the assumption implicit in the solution, Eq. (10.2).

1. The pumping well has a constant rate of discharge and fully penetrates the aquifer.

2. The pumping and observation wells have an infinitesimally small diameter.

3. The confining bed that overlies or underlies the aquifer has a uniform hydraulic conductivity (K') and thickness (b').

4. Storage in the confining bed is negligible.

5. Leakage across the confining bed comes from an aquifer whose head is assumed to remain constant during the test.

6. Flow is vertical in the confining bed and radial in the confined aquifer.

TABLE 10.1   Values of W(u, r/B)

					r/B					
u	0.01	0.015	0.03	0.05	0.075	0.10	0.15	0.2	0.3	0.4
0.000001										
0.000005	9.4413									
0.00001	9.4176	8.6313								
0.00005	8.8827	8.4533	7.2450							
0.0001	8.3983	8.1414	7.2122	6.2282	5.4228					
0.0005	6.9750	6.9152	6.6219	6.0821	5.4062	4.8530				
0.001	6.3069	6.2765	6.1202	5.7965	5.3078	4.8292	4.0595	3.5054		
0.005	4.7212	4.7152	4.6829	4.6084	4.4713	4.2960	3.8821	3.4567	2.7428	2.2290
0.01	4.0356	4.0326	4.0167	3.9795	3.9091	3.8150	3.5725	3.2875	2.7104	2.2253
0.05	2.4675	2.4670	2.4642	2.4576	2.4448	2.4271	2.3776	2.3110	1.9283	1.7075
0.1	1.8227	1.8225	1.8213	1.8184	1.8128	1.8050	1.7829	1.7527	1.6704	1.5644
0.5	0.5598	0.5597	0.5596	0.5594	0.5588	0.5581	0.5561	0.5532	0.5453	0.5344
1.0	0.2194	0.2194	0.2193	0.2193	0.2191	0.2190	0.2186	0.2179	0.2161	0.2135
5.0	0.0011	0.0011	0.0011	0.0011	0.0011	0.0011	0.0011	0.0011	0.0011	0.0011

					r/B				
u	0.5	0.6	0.7	0.8	0.9	1.0	1.5	2.0	2.5
0.000001									
0.000005									
0.00001									
0.00005									
0.0001									
0.0005									
0.001									
0.005									
0.01	1.8486	1.5550	1.3210	0.1307					
0.05	1.4927	1.3955	1.2955	1.1210	0.9700	0.8409			
0.1	1.4422	1.3115	1.1791	1.0505	0.9297	0.8190	0.4271	0.2278	
0.5	0.5206	0.5044	0.4860	0.4658	0.4440	0.4210	0.3007	0.1944	0.1174
1.0	0.2103	0.2065	0.2020	0.1970	0.1914	0.1855	0.1509	0.1139	0.0803
5.0	0.0011	0.0011	0.0011	0.0011	0.0011	0.0011	0.0010	0.0010	0.0009

Source: Hantush, Trans. *Amer. Geophys. Union* v. 37, p. 702–714, 1956. Copyright by American Geophysical Union.

Equation (10.2) can be applied to the prediction of drawdown with time and the interpretation of a aquifer test data in which leaky behavior is evident. The following example illustrates the forward calculation.

▶ **EXAMPLE 10.1**

A leaky confined aquifer has a transmissivity of 2.5×10^{-2} m²/s and a storativity of 2.7×10^{-4}. The vertical hydraulic conductivity (K') of the 3-m-thick confining bed is 1.8×10^{-8} m/s. The well is pumped at a rate of 0.063 m³/s. What is the drawdown in an observation well 150 m away after 1000 minutes?

SOLUTION   Make the units consistent by changing time from minutes to seconds—1000 minutes is 6×10^4 s. Calculate u and r/B.

$$u = \frac{r^2 S}{4Tt} = \frac{(150\,\text{m})^2 \times 2.7 \times 10^{-4}}{4 \times 2.5 \times 10^{-2}\,\text{m}^2/\text{s} \times 6 \times 10^4\,\text{s}} = 0.001$$

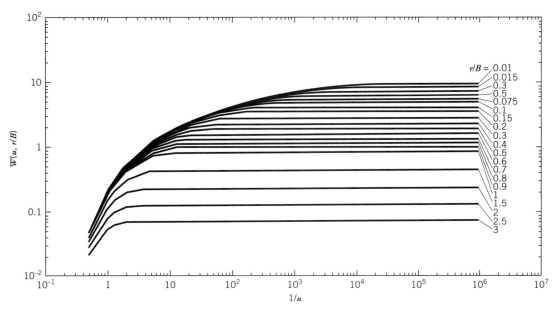

Figure 10.2 Type curve, $W(u, r/B)$, for a leaky, confined aquifer without storage in the confining bed.

$$\frac{r}{B} = r\left(\frac{K'/m'}{T}\right)^{1/2} = 150\,\text{m}\left(\frac{1.8 \times 10^{-8}\,\text{ms}^{-1}/3\text{m}}{2.5 \times 10^{-2}\text{m}^2/\text{s}}\right)^{1/2} = 0.073$$

Determine $W(u, r/B) = W(0.001, 0.075) = 5.30$ from Table 10.1. Calculate the drawdown as

$$s = \frac{Q}{4\pi T}W(u, r/B) = \frac{0.063\,\text{m}^3/\text{s}}{4\pi \times 2.5 \times 10^{-2}\,\text{m}^2/\text{s}} \times 5.30 = 1.06\,\text{m}$$

Interpreting Aquifer-Test Data

Chapter 9 introduced a curve-matching procedure for estimating transmissivity and stora-tivity in a confined aquifer from an aquifer test. The same procedure can be applied with modifications to leaky, confined aquifer systems. One extra benefit is that a test also provides estimates of the vertical hydraulic conductivity of the confining bed K'.
The steps are as follows:

1. Create or find type curves for the well function $W(u, r/B)$ versus $1/u$ on log-log graph paper (Figure 10.2). Mark a match point on this graph with $W(u, r/B) = 1$ and $1/u$ as 10, 100, or 1000.

2. Plot the observed drawdown versus time on a transparent overlay using log-log paper with the same scale.

3. Move the plotted drawdown versus time curve on the type curves and find which of the family of type curves matches best. Record the r/B value for that curve.

4. Plot the match point on the field curve and record the coordinates s and t. T and S are calculated by substituting the appropriate values of $W(u, r/B)$, $1/u$, s, and t as needed in

$$T = \frac{Q}{4\pi s}W(u, r/B) \tag{10.4}$$

and

$$S = \frac{4uTt}{r^2} \tag{10.5}$$

5. Record r/B. The hydraulic conductivity of the confining bed is determined by

$$K' = \frac{Tb'(r/B)^2}{r^2} \tag{10.6}$$

▶ **EXAMPLE 10.2**   A test of a leaky, confined aquifer was conducted at a pumping rate of 7.55 m³/min. Drawdown versus time measurements were collected at an observation well 154 m away (Table 10.2). Calculate the transmissivity and storativity of the aquifer. Assuming the thickness of the semiconfining bed is 20 m, what is the vertical hydraulic conductivity of the confining bed?

SOLUTION   Plot values of drawdown versus time on log-log graph paper at the same scale as the type curves (Figure 10.3). Overlay the plot of the field data on the family of type curves. Find the best match between the field data and one of the type curves (Figure 10.4). The match-point coordinates are $1/u = 100$, $W(u, r/B) = 10$, $t = 6.2$ min, and $s = 6.3$ m. The data best fit the type curve with $r/B = 0.15$. Substitute known values into the appropriate equations

TABLE 10.2   Drawdown versus time in a leaky, confined aquifer test

t(min)	s(m)	t(min)	s(m)
0.10	0.27	11.72	2.54
0.14	0.39	16.10	2.59
0.19	0.53	22.12	2.63
0.26	0.68	30.39	2.64
0.36	0.85	41.75	2.65
0.49	1.03	57.36	2.65
0.67	1.21	78.80	2.65
0.92	1.39	108.26	2.65
1.27	1.58	148.74	2.65
1.74	1.76	204.34	2.65
2.40	1.93	280.72	2.65
3.29	2.09	385.66	2.65
4.52	2.23	529.83	2.65
6.21	2.36	727.90	2.65
8.53	2.46	1000.00	2.65

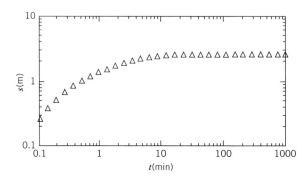

Figure 10.3 Plot of drawdown versus time on a log-log scale from an aquifer test of a leaky, confined aquifer.

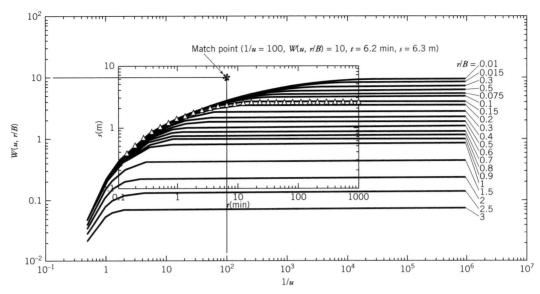

Figure 10.4 Graphical procedure for determining the hydraulic parameters of a leaky, confined aquifer using the type-curve matching technique.

$$T = \frac{Q}{4\pi s}W(u, r/B) = \frac{7.55\,\text{m}^3/\text{min}}{4\pi(6.3\,\text{m})}(10) = 0.95\,\text{m}^2/\text{min} = 1373\,\text{m}^2/\text{day}$$

and

$$S = \frac{4uTt}{r^2} = \frac{(4)(0.01)(0.95\,\text{m}^2/\text{min})(6.2\,\text{min})}{(154\,\text{m})^2} = 1.0 \times 10^{-5}$$

and

$$K' = \frac{Tb'(r/B)^2}{r^2} = \frac{(0.95\,\text{m}^2/\text{min})(20\,\text{m})(0.15)^2}{(154\,\text{m})^2} = 1.8 \times 10^{-5}\,\text{m/min} = 0.026\ \text{m/day}$$

▶ 10.2 STEADY-STATE SOLUTION

At steady state, the drawdown in a leaky, confined aquifer is expressed as

$$s = \frac{Q}{2\pi T}K_0\left(\frac{r}{B}\right) \tag{10.6}$$

(see Jacob, 1946) where $K_0(x)$ is the zero-order modified Bessel function of the second kind. This equation can be used to predict drawdown in a leaky, confined aquifer at steady state. Tabulated values of $K_0(x)$ are available in the literature for this application (Table 10.3).

As before, this solution also provides the basis for evaluating the results of an aquifer test. Steady-state values of drawdown at a set of observation wells can be used to derive a transmissivity value for the aquifer and a vertical hydraulic-conductivity value for the leaky, confining bed. The steps are as follows:

1. Create a type curve by plotting $K_0(x)$ versus x on log-log graph paper (Figure 10.5). Provide a match point on the type curve with simple coordinates (for example, $x = 0.1$; $K_0(x) = 1.0$).

TABLE 10.3   Values of $K_0(x)$ for values of x

x	$K_0(x)$	x	$K_0(x)$	x	$K_0(x)$
0.01	4.7212	0.09	2.531	0.8	0.5653
0.015	4.3159	0.1	2.4271	0.9	0.4867
0.02	4.0285	0.15	2.03	1	0.421
0.03	3.6235	0.2	1.7527	1.5	0.2138
0.04	3.3365	0.3	1.3725	2	0.1139
0.05	3.1142	0.4	1.1145	3	0.0347
0.06	2.9329	0.5	0.9244	4	0.0112
0.07	2.7798	0.6	0.7775	5	0.0037
0.08	2.6475	0.7	0.6605		

Source: From Hantush (1956).

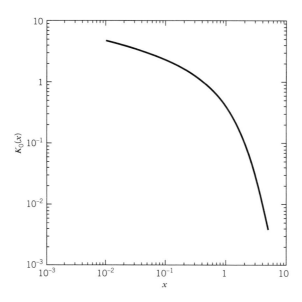

Figure 10.5 Type curve of $K_0(x)$ for steady-state flow in a leaky, confined aquifer.

2. Plot measured values of drawdown versus distance on a transparent overlay at the same log-log scale.

3. Superimpose the field curve on the type curve. Match the two curves, keeping the axes parallel.

4. Plot the match point on the field curve and determine the coordinates r and s. The transmissivity is calculated as

$$T = \frac{Q}{2\pi s} K_0(x) \tag{10.7}$$

and the vertical hydraulic conductivity is

$$K' = \frac{x^2 \, b' \, T}{r^2} \tag{10.8}$$

▶ **EXAMPLE 10.3**   An aquifer test is run with a leaky, confined aquifer at a pumping rate of 220 gpm. At a late time, drawdowns were recorded for a number of wells (Table 10.4). The thickness of the confining bed is 10 ft. Calculate the hydraulic conductivity for the aquifer and K for the confining bed.

SOLUTION   Drawdown versus distance data for the observation wells are plotted on a log-log scale in Figure 10.6. Figure 10.7 shows the field curve superimposed on the type curve. The coordinates of the match points are $x = 0.1$, $K_0(x) = 1.0$, $r = 140$ ft, and $s = 7$ ft. Convert the pumping rate from gpm to ft^3/day.

TABLE 10.4   Values of drawdown versus distance

r(ft)	S(ft)
10	35.2
50	24.352
100	19.683
150	16.957
200	15.027
250	13.535
300	12.321
400	10.42
500	8.965

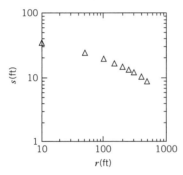

Figure 10.6 Plot of drawdown versus distance for steady-state flow in a leaky, confined aquifer.

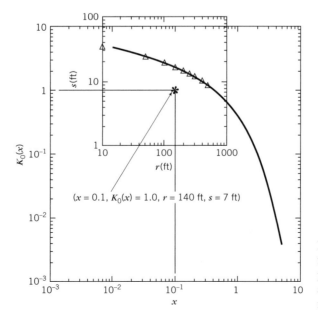

$(x = 0.1, K_0(x) = 1.0, r = 140 \text{ ft}, s = 7 \text{ ft})$

Figure 10.7 Graphical procedure for determining the hydraulic parameters of a leaky, confined aquifer using the type-curve matching technique.

$$Q = (220\text{gpm})(192.5\frac{\text{ft}^3/\text{day}}{\text{gpm}}) = 42350\,\text{ft}^3/\text{day}$$

From Eqs. (10.7) and (10.8),

$$T = \frac{(42350\,\text{ft}^3/\text{day})}{2\pi(7\,\text{ft})}1.0 = 963\,\text{ft}^2/\text{day}$$

and

$$K' = \frac{(0.1)^2(10\,\text{ft})(963\,\text{ft}^2/\text{day})}{140^2} = 0.005\,\text{ft/day}$$

▶ 10.3 TRANSIENT SOLUTIONS FOR FLOW WITH STORAGE IN CONFINING BEDS

Hantush (1960) considered three situations with storage in confining beds overlying or underlying the aquifer (Figure 10.8). Here are the cases: (1) both of the confining beds are bounded by source units having constant heads; (2) both of the confining beds are bounded by impermeable layers; (3) the upper confining bed is bounded by a source unit with a constant head, and the lower confining bed is bounded by an impermeable layer. With assumptions comparable to those for leaky, confined aquifers, Hantush (1960) derived the following analytical solution for early-time drawdown.

$$s = \frac{Q}{4\pi T}H(u,\beta), \quad \text{for} \quad t < \frac{b'S'}{10K'}, t < \frac{b''S''}{10K''} \tag{10.9}$$

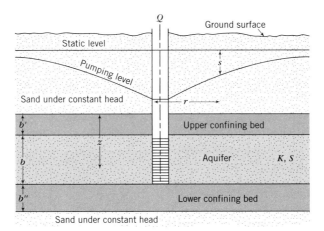

(a)

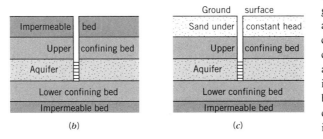

(b)   (c)

Figure 10.8 Three different geometries of a leaky, confined aquifer with storage in the confining beds. (a) Case 1 has constant-head plane sources above and below; (b) Case 2 has impermeable beds above and below; (c) Case 3 has a constant-head plane above and impermeable bed below.

where

$$H(u, \beta) = \int_{u}^{\infty} \frac{e^{-y}}{y} \, erfc \left(\frac{\beta \sqrt{u}}{\sqrt{y(y - u)}} \right) dy \qquad (10.10)$$

$$u = \frac{r^2 S}{4Tt} \qquad (10.11)$$

$$\beta = \frac{r}{4} \left(\sqrt{\frac{K'S'}{b'TS}} + \sqrt{\frac{K''S''}{b''TS}} \right) \qquad (10.12)$$

$$erfc(x) = \frac{2}{\sqrt{\pi}} \int_{x}^{\infty} e^{-y^2} dy \qquad (10.13)$$

where T and S are the transmissivity and storativity of the confined aquifer, respectively, T' and S' are the transmissivity and storativity of the upper confining layer, respectively, and T'' and S'' are the transmissivity and storativity of the lower confining layer, respectively. Tabulated values of well function $H(u, \beta)$ are provided in Table 10.5 (Hantush, 1961b). Figure 10.9 shows the resulting family of type curves, with each curve having its own β value.

This theory can be applied to interpret the results of an aquifer test following the general scheme for superposition developed previously.

1. Create the family of type curves by plotting $H(u, \beta)$ versus $1/u$ on a log-log graph paper (Figure 10.9). Choose a match point that works.

2. Plot the field curve of drawdown versus time on a log-log paper at the same scale as the family of type curves.

3. Superimpose the two graphs and find the best match between the field curve and one of the family of type curves.

4. Record β and $1/u$, $H(u, \beta)$, t, and s for the match point.

5. The transmissivity of the leaky, confined aquifer is determined by

$$T = \frac{Q}{4\pi s} H(u, \beta) \qquad (10.14)$$

6. The storativity of the leaky, confined aquifer is

$$S = \frac{4T \, tu}{r^2} \qquad (10.15)$$

7. If $T'' S'' = 0$, then

$$K'S' = \frac{16 \beta^2 b' TS}{r^2} \qquad (10.16)$$

8. If $T' S' = T'' S''$, then

$$K'S' = \frac{16 \beta^2 TS}{r^2} \frac{b'b''}{b' + b'' + 2\sqrt{b'b''}} \qquad (10.17)$$

TABLE 10.5 Values of $H(u, \beta)$ for selected values of u and β

				β				
u	0.03	0.1	0.3	1	3	10	30	100
1×10^{-9}	12.3088	11.1051	10.0066	8.8030	7.7051	6.5033	5.4101	4.2221
2	11.9622	10.7585	9.6602	8.4566	7.3590	6.1579	5.0666	3.8839
3	11.7593	10.5558	9.4575	8.2540	7.1565	5.9561	4.8661	3.6874
5	11.5038	10.3003	9.2021	7.9987	6.9016	5.7020	4.6142	3.4413
7	11.3354	10.1321	9.0339	7.8306	6.7337	5.5348	4.4487	3.2804
1×10^{-8}	11.1569	9.9538	8.8556	7.6525	6.5558	5.3578	4.2737	3.1110
2	10.8100	9.6071	8.5091	7.3063	6.2104	5.0145	3.9352	2.7858
3	10.6070	9.4044	8.3065	7.1039	6.0085	4.8141	3.7383	2.5985
5	10.3511	9.1489	8.0512	6.8490	5.7544	4.5623	3.4919	2.3662
7	10.1825	8.9806	7.8830	6.6811	5.5872	4.3969	3.3307	2.2159
1×10^{-7}	10.0037	8.8021	7.7048	6.5032	5.4101	4.2221	3.1609	2.0591
2	9.6560	8.4554	7.3585	6.1578	5.0666	3.8839	2.8348	1.7633
3	9.4524	8.2525	7.1560	5.9559	4.8661	3.6874	2.6469	1.5966
5	9.1955	7.9968	6.9009	5.7018	4.6141	3.4413	2.4137	1.3944
7	9.0261	7.8283	6.7329	5.5346	4.4486	3.2804	2.2627	1.2666
1×10^{-6}	8.8463	7.6497	6.5549	5.3575	4.2736	3.1110	2.1051	1.1361
2	8.4960	7.3024	6.2091	5.0141	3.9350	2.7857	1.8074	.8995
3	8.2904	7.0991	6.0069	4.8136	3.7382	2.5984	1.6395	.7725
5	8.0304	6.8427	5.7523	4.5617	3.4917	2.3661	1.4354	.6256
7	7.8584	6.6737	5.5847	4.3962	3.3304	2.2158	1.3061	.5375
1×10^{-5}	7.6754	6.4944	5.4071	4.2212	3.1606	2.0590	1.1741	.4519
2	7.3170	6.1453	5.0624	3.8827	2.8344	1.7632	.9339	.3091
3	7.1051	5.9406	4.8610	3.6858	2.6464	1.5965	.8046	.2402
5	6.8353	5.6821	4.6075	3.4394	2.4131	1.3943	.6546	.1685
7	6.6553	5.5113	4.4408	3.2781	2.2619	1.2664	.5643	.1300
1×10^{-4}	6.4623	5.3297	4.2643	3.1082	2.1042	1.1359	.4763	963(− 4)
2	6.0787	4.9747	3.9220	2.7819	1.8062	.8992	.3287	494(− 4)
3	5.8479	4.7655	3.7222	2.5937	1.6380	.7721	.2570	315(− 4)
5	5.5488	4.4996	3.4711	2.3601	1.4335	.6252	.1818	166(− 4)
7	5.3458	4.3228	3.3062	2.2087	1.3039	.5370	.1412	103(− 4)
1×10^{-3}	5.1247	4.1337	3.1317	2.0506	1.1715	.4513	.1055	390(− 5)
2	4.6753	3.7598	2.7938	1.7516	.9305	.3084	551(− 4)	169(− 5)
3	4.3993	3.5363	2.5969	1.5825	.8006	.2394	355(− 4)	713(− 6)
5	4.0369	3.2483	2.3499	1.3767	.6498	.1677	190(− 4)	205(− 6)
7	3.7893	3.0542	2.1877	1.2460	.5589	.1292	120(− 4)	821(− 7)
1×10^{-2}	3.5195	2.8443	2.0164	1.1122	.4702	955(− 4)	695(− 5)	274(− 7)
2	2.9759	2.4227	1.6853	.8677	.3214	487(− 4)	205(− 5)	226(− 8)
3	2.6487	2.1680	1.4932	.7353	.2491	308(− 4)	888(− 6)	
5	2.2312	1.8401	1.2535	.5812	.1733	160(− 4)	261(− 6)	
7	1.9558	1.6213	1.0979	.4880	.1325	982(− 5)	106(− 6)	
1×10^{-1}	1.6667	1.3893	.9358	.3970	966(− 4)	552(− 5)	365(− 7)	
2	1.1278	.9497	.6352	.2452	468(− 4)	149(− 5)	307(− 8)	
3	.8389	.7103	.4740	.1729	281(− 4)	592(− 6)		
5	.5207	.4436	.2956	.1006	130(− 4)	151(− 6)		
7	.3485	.2980	.1985	646(− 4)	714(− 5)	534(− 7)		
1×1	.2050	.1758	.1172	365(− 4)	337(− 5)	151(− 7)		
2	458(− 4)	395(− 4)	264(− 4)	760(− 5)	487(− 6)			
3	122(− 4)	106(− 4)	707(− 5)	196(− 5)	102(− 6)			
5	108(− 5)	934(− 6)	624(− 6)	167(− 6)	672(− 8)			
7	109(− 6)	941(− 7)	629(− 7)	165(− 7)				
1×10	391(− 8)	339(− 8)	227(− 8)					
2								
3								
5								
7								

Source: From Hantush (1961b).

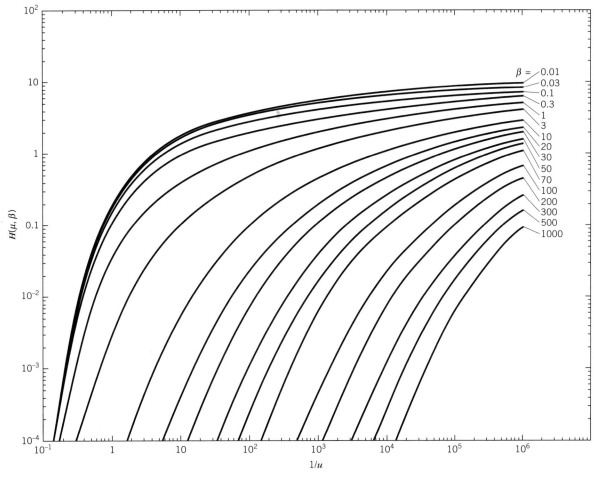

Figure 10.9 Early-time type curves of $H(u, \beta)$ in a leaky, confined aquifer with storage in confining beds (modified from Reed, 1980).

▶ **EXAMPLE 10.4**   A test of a leaky, confined aquifer was run at a pumping rate of 750 gpm (Lohman, 1972). A confining layer 6 ft thick overlies the aquifer. Above the confining layer is an unconfined aquifer 200 ft thick. At an observation well 1400 ft away, values of drawdown versus time were recorded (Table 10.6). Calculate hydraulic parameters using Hantush's method.

SOLUTION   The drawdown versus time data are plotted as shown in Figure 10.10. This field curve (Figure 10.10) is overlain on the type curves in Figure 10.9. The best match comes with the field data and the $\beta = 3$ type curve. With the match point transferred to the field curve (Figure 10.11), the match-point coordinates are $1/u = 10$, $H(u, \beta) = 1.0$, $t = 90$ min, and $s = 6.5$ ft. The pumping rate is converted from gpm to ft^3/day.

$$Q = (750 \, \text{gpm}) \left(192.5 \frac{\text{ft}^3/\text{day}}{\text{gpm}} \right) = 144375 \, \text{ft}^3/\text{day}$$

The transmissivity is

$$T = \frac{Q}{4\pi s} H(u, \beta) = \frac{(144375 \, \text{ft}^3/\text{day})}{4\pi (6.5 \, \text{ft})} 1.0 = 1768 \, \text{ft}^2/\text{day}$$

TABLE 10.6 Values of drawdown versus time in a leaky, confined aquifer

t(min)	s(ft)	t(min)	s(ft)	t(min)	s(ft)
6.37	0.01	41	0.33	315	1.83
8.58	0.02	44	0.36	335	1.87
10.23	0.03	47	0.38	365	1.99
11.9	0.04	50	0.42	390	2.1
12.95	0.05	54	0.46	410	2.13
14.42	0.06	60	0.52	430	2.2
15.1	0.07	65	0.56	450	2.23
16.88	0.08	70	0.6	470	2.29
17.92	0.1	80	0.65	490	2.32
21.35	0.12	90	0.75	510	2.39
21.7	0.13	100	0.82	560	2.48
22.7	0.14	137	1.04	740	2.92
23.58	0.15	150	1.12	810	3.05
24.65	0.17	160	1.17	890	3.19
29	0.21	173	1.24	1255	3.66
30	0.22	184	1.27	1400	3.81
32	0.24	200	1.35	1440	3.86
34	0.26	210	1.4	1485	3.9
36	0.28	278	1.68		
38	0.3	300	1.76		

Source: From Lohman (1972).

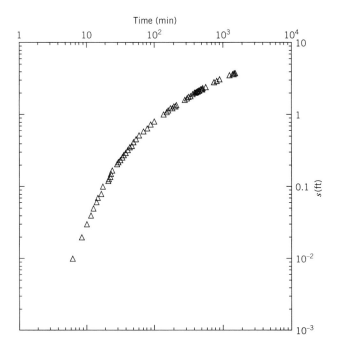

Figure 10.10 Plot of drawdown versus time in a leaky, confined aquifer with storage in the confining beds.

The storativity is

$$S = \frac{4T\,t\,u}{r^2} = \frac{4(1768\,\text{ft}^2/\text{day})(90\,\text{min})\left(\dfrac{1}{1440}\dfrac{\text{day}}{\text{min}}\right)(0.1)}{(1400\,\text{ft})^2} = 2.3 \times 10^{-5}$$

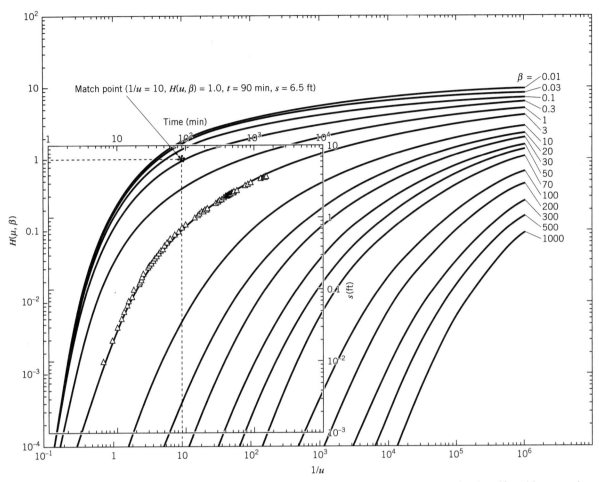

Figure 10.11 Determination of hydraulic parameters in a leaky, confined aquifer with storage in confining beds.

For $T''S'' = 0$ in this example, the hydraulic properties of the upper confining are calculated with Eq. (10.16).

$$K'S' = \frac{16(3)^2(6\,\text{ft})(1768\,\text{ft}^2/\text{day})(2.3 \times 10^{-5})}{(1400\,\text{ft})^2} = 1.8 \times 10^{-5}(\,\text{ft/day}\,)$$

▶ 10.4 EFFECTS OF PARTIALLY PENETRATING WELLS

When the screened or open section of a well casing does not coincide with the full thickness of the aquifer it penetrates, the well is referred to as *partially penetrating*. With many aquifers, this well design is the rule rather than the exception. Under such conditions, the flow toward the pumping well (or observation point) will be three dimensional because of vertical flow components (Figure 10.12). In practice, once you get far enough away from a pumping well, the effects of partial penetration become unimportant. The following equation describes when the effects of partial penetration become negligible

$$r > 1.5\,b\left(\frac{K_r}{K_z}\right)^{\frac{1}{2}} \tag{10.18}$$

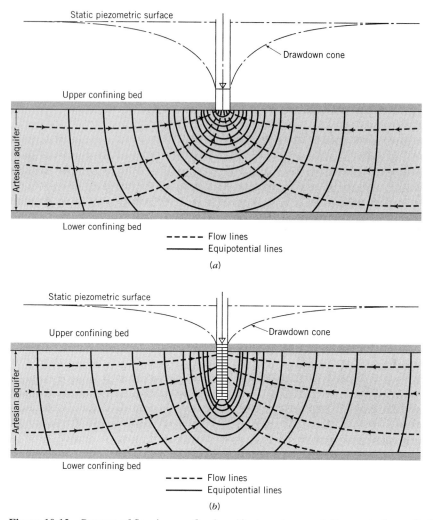

Figure 10.12 Patterns of flow in a confined aquifer to various partially penetrating wells (from U.S. Bureau of Reclamation, 1995). In Panel (*a*), the well penetrates the top of the confined aquifer. In Panel (*b*), the well is screened in the top 50 percent of the aquifer.

where b is the thickness of the aquifer and K_z and K_r are the horizontal and vertical hydraulic conductivities of the aquifer, respectively.

The topic of partial penetration has been the subject of numerous papers (Hantush, 1961c, 1964; Muskat, 1937; Neuman, 1974). Hantush (1964) has conducted most of the work in this area and provides some general guidelines for confined aquifers. Hantush's correction equation of drawdown in a piezometer for a partially penetrating pumping well (Figure 10.13) can be expressed as

$$s_{\text{partially}}(t) = s_{\text{fully}}(t) + \frac{Q}{4\pi T} f\left(u, \frac{ar}{b}, \frac{l}{b}, \frac{d}{b}, \frac{z}{b}\right) \tag{10.19}$$

where t is time, $u = r^2 S/(4Tt)$, $a = sqrt(K_z/K_r)$, r is radial distance from the pumping well to the observation well, b is aquifer thickness, l is depth of the pumped well in the aquifer, and z_1 and z_2 are bottom and top z-coordinates of the screen in the observation

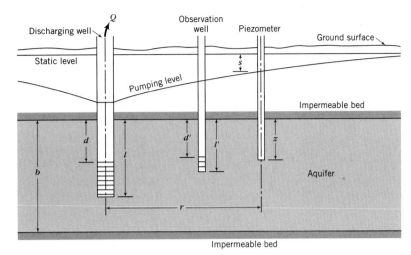

Figure 10.13 Cross section of a confined aquifer with partially penetrating pumping and observation wells.

well, respectively. For a confined aquifer

$$s_{\text{full}} = \frac{Q}{4\pi T} W(u) \tag{10.20}$$

and

$$f\left(u, \frac{ar}{b}, \frac{l}{b}, \frac{d}{b}, \frac{z}{b}\right) = \frac{2}{\pi(l/b - d/b)} \sum_{n=1}^{\infty} \frac{1}{n}\left(\sin\frac{n\pi l}{b} - \sin\frac{n\pi d}{b}\right)\cos\frac{n\pi z}{b}W\left(u, \frac{n\pi a r}{b}\right) \tag{10.21}$$

where $W(u)$ is the Theis well function defined in Section 6.1 and $W(u, \beta)$ is the Hantush and Jacob well function defined in Section 10.2. For a leaky, confined aquifer

$$s_{\text{full}} = \frac{Q}{4\pi T} W(u, r/B) \tag{10.22}$$

and

$$f\left(u, \frac{ar}{b}, \frac{l}{b}, \frac{d}{b}, \frac{z}{b}\right) = \frac{2}{\pi(l/b - d/b)} \sum_{n=1}^{\infty} \frac{1}{n}\left(\sin\frac{n\pi l}{b} - \sin\frac{n\pi d}{b}\right)$$
$$\cos\frac{n\pi z}{b}W\left(u, \sqrt{\beta^2 + (\frac{n\pi a r}{b})^2}\right)$$

For an observation well, the Hantush correction term is the average hydraulic head along a well screen.

$$f\left(u, \frac{ar}{b}, \frac{l}{b}, \frac{d}{b}, \frac{z_1}{b}, \frac{z_2}{b}\right) = \frac{1}{(z_2 - z_1)}\int_{z_1}^{z_2} f\left(u, \frac{ar}{b}, \frac{l}{b}, \frac{d}{b}, \frac{z}{b}\right) dz \tag{10.23}$$

An example of hydraulic testing in a leaky aquifer with partially penetrating wells will be given in Chapter 15 for computer-aided interpretation of aquifer tests.

▶ **EXERCISES**

10.1. An aquifer test was run with a leaky, confined aquifer without storage in the confining bed at a pumping rate of 1440 m³/day. Drawdown time data from an observation well that is located 40 m away from the pumping well are tabulated in Table 10.7.

TABLE 10.7 Values of drawdown versus time obtained from an aquifer test of a leaky, confined aquifer

Time(min)	s(m)	time(min)	s(m)
0.10	0.21	11.25	0.57
0.13	0.23	14.25	0.58
0.16	0.25	18.05	0.60
0.20	0.27	22.85	0.61
0.26	0.29	28.94	0.62
0.33	0.30	36.65	0.63
0.41	0.32	46.42	0.64
0.52	0.34	58.78	0.65
0.66	0.36	74.44	0.65
0.84	0.38	94.27	0.65
1.06	0.40	119.38	0.66
1.34	0.42	151.18	0.66
1.70	0.43	191.45	0.66
2.15	0.45	242.45	0.66
2.73	0.47	307.03	0.66
3.46	0.49	388.82	0.66
4.38	0.50	492.39	0.66
5.54	0.52	623.55	0.66
7.02	0.54	789.65	0.66
8.89	0.55	1000.00	0.66

TABLE 10.8 Values of drawdown versus distance from a an aquifer test of a leaky, confined aquifer at steady state

r(m)	s(m)
20	0.469022
40	0.185689
80	0.038312
100	0.018347
120	0.008944
150	0.003113

Calculate the transmissivity and storativity of the aquifer. If the thickness of the confining bed is 5 m, what is the vertical hydraulic conductivity of the confining bed? You will find a template file for this exercise in the directory Chap 10 on the website. The template file temp-10a.doc contains type curves for a leaky, confined aquifer. The template file temp-10.doc in the directory Chap 10 provides a log-log graph at the same scale as the type curves. Use these templates and a curve-matching procedure to answer this question.

10.2. An aquifer test is conducted with a leaky, confined aquifer by pumping at a constant rate of $2 \text{ m}^3/\text{min}$. The thickness of the upper confining bed is 4 m. Eventually, steady-state conditions are achieved, and drawdowns are measured for a set of observation wells (Table 10.8). Calculate the hydraulic parameters for the aquifer and the confining bed.

10.3. This aquifer test involves a leaky, confined aquifer with storage in the confining beds. The well is pumped at a constant rate of $2880 \text{ m}^3/\text{day}$, and drawdowns are measured in an observation well located 50 m away from the pumping well (Table 10.9). The thickness of the upper and lower confining beds are 10 m. Assuming that $K'S' = K''S''$, calculate the transmissivity, storativity of

**TABLE 10.9 Values of drawdown versus time from an aquifer test
in a leaky, confined aquifer with storage in the confining beds**

Time(min)	s(m)	Time(min)	s(m)
0.10	0.09	14.15	0.40
0.14	0.11	19.69	0.43
0.19	0.12	27.39	0.45
0.27	0.14	38.11	0.47
0.37	0.16	53.02	0.50
0.52	0.18	73.76	0.52
0.73	0.20	102.62	0.55
1.01	0.22	142.76	0.57
1.40	0.24	198.61	0.60
1.95	0.26	276.31	0.63
2.72	0.28	384.41	0.65
3.78	0.31	534.79	0.68
5.26	0.33	744.01	0.70
7.31	0.35	1035.07	0.73
10.17	0.38	1440.00	0.75

the aquifer, and hydraulic parameters of the confining beds. The template file temp-10b.doc contains
type curves for a leaky, confined aquifer with storage in the confining beds.

► **CHAPTER
HIGHLIGHTS**

1. With pumping, a leaky, confined aquifer receives inflow from units above and below. In the simplest
case, a confining bed above the aquifer transmits water from another aquifer higher in the section. In
more complicated cases, water also comes from storage within confining beds above and below the
aquifer being pumped.

2. Leakage reduces drawdowns in the aquifer being pumped, as compared to the Theis-type
behaviour. With significant leakage, the system will reach steady state. With relatively minor leakage,
the drawdown will follow a Theis-type response.

3. The Hantush and Jacob (1955) analytical solution is commonly used to predict drawdowns in a
leaky aquifer or to interpret aquifer tests. A curve-matching procedure with a match point is used to
provide estimates of T and S for the aquifer being pumped and K', the vertical hydraulic conductivity of
the confining bed. The curve-matching process is complicated because now you not only have to match
a curve with the data but also have to select the appropriate type curves from a family of type curves.

4. If the drawdown data from a collection of observation wells indicates that the system has reached
steady state, then one can use steady-state theory developed by Jacob (1946). This theory applies to
forward predictions of drawdown given the aquifer parameters or inverse calculations of aquifer and
confining-bed parameters from drawdown measurements. Again, the interpretation of aquifer-test
data follows a curve-matching procedure.

5. Sometimes, storage within a confining bed(s) can have a significant impact on the drawdown due
to pumping. Hantush (1960) provided analytical solutions to cases where (a) confining beds above
and below the aquifer (having storage) are bounded by source units having constant heads, (b) both
confining beds are bounded by impermeable layers, and (c) the upper confining bed is bounded by a
source unit with a constant head and the lower confining unit is bounded by an impermeable layer.

6. The term *partial penetration* describes a situation when the screened length of a well casing is
less than the saturated thickness of the aquifer. With this well design, water flowing to the well
has both horizontal and vertical flow components. Under this condition, calculations of drawdown
require solutions that can account for the effects of partial penetration. When the distance from the
pumping well to an observation point becomes sufficiently large, $r > 1.5b(\frac{K_r}{K_z})^{1/2}$, the effects of
partial penetration become negligible.

11

RESPONSE OF AN UNCONFINED AQUIFER TO PUMPING

The response of an unconfined aquifer to pumping is complicated. Once a cone of depression forms, it naturally decreases the aquifer thickness and transmissivity because the upper boundary of the aquifer is the water table. Also, the way in which water comes out of storage in the aquifer changes with time. At early time, when the well is first turned on, water is released from storage due to compression of the matrix and expansion of the water. This response is the same Theis behavior that applies to confined aquifers (Chapter 9). Thus, early-time/drawdown data, plotted on log-log graph paper, follow the Theis-type curve (Figure 11.1). Storativity values would be comparable to those for confined aquifers, 10^{-4} or 10^{-5}.

As pumping continues, water comes from the slow gravity drainage of water from pores as the water table falls near the well. The pattern of drawdown depends on the vertical and horizontal hydraulic conductivity and the thickness of the aquifer. Once this delayed drainage begins, drawdown data deviate from the Theis curve. The drawdown is less than expected, resembling the pattern of drawdown in a leaky aquifer (Figure 11.1). Eventually, the contribution of water from delayed drainage ceases. Flow in the aquifer is mainly radial, and drawdown/time data again fall on a Theis-type curve (Figure 11.1). The storativity of the aquifer now is the same as the specific yield (S_y). The *specific yield* is the ratio of the volume of water that drains from a rock or sediment by gravity to the volume of the rock or soil. The curve-fitting approach developed in this chapter for interpreting aquifer test data requires the ability to recognize these differing aquifer responses.

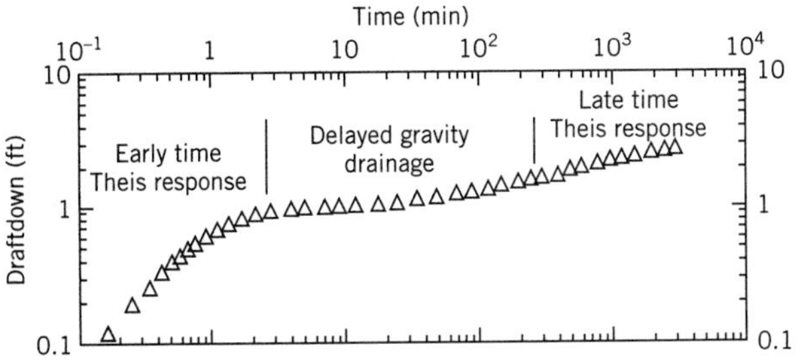

Figure 11.1 An example of the characteristic "S"-shaped drawdown curve obtained from a test with an unconfined aquifer.

► ## 11.1 CALCULATION OF DRAWDOWNS BY CORRECTING ESTIMATES FOR A CONFINED AQUIFER

Because of the intricacies in the response of an unconfined aquifer to pumping, the theory developed for confined aquifers is not generally transferable. However, if the point of interest is far away from the pumped well, one can simply apply confined-aquifer theory without too much error. A slightly more sophisticated approach is to calculate drawdown assuming the aquifer to be confined and correcting the drawdown value with the following equation:

$$s = b - (b^2 - 2s'b)^{1/2} \tag{11.1}$$

where s is the drawdown for the unconfined aquifer, s' is the drawdown for the equivalent confined aquifer, and b is the original thickness of the aquifer. The following example illustrates this calculation.

► **EXAMPLE 11.1**   A well in an unconfined aquifer is pumped at a constant rate of 99 gallons per minute. The aquifer has a hydraulic conductivity of 388 ft/day, an initial saturated thickness of 20 ft, and a specific yield of 0.01. Calculate what the drawdown will be at an observation well 51 ft away after 3180 minutes.

SOLUTION   The first step in solving this problem is to adjust the units to feet and days to make them consistent. Also, T needs to be calculated.

$$Q = 99\,\text{gpm} = 99 \times 192.5\,\text{ft}^3/\text{day} = 19{,}060\,\text{ft}^3/\text{day}$$

$$t = 3180\,\text{minutes} = 3180/1440 = 2.21\,\text{days}$$

$$T = Kb = 388\,\text{ft/day} \times 20\,\text{ft} = 7760\,\text{ft}^2/\text{day}$$

Calculate the drawdown at the observation point, assuming the aquifer to be confined.

$$u = \frac{r^2 S}{4Tt} = \frac{(51\,\text{ft})^2 \times 0.01}{4 \times 7760\,\text{ft}^2/\text{day} \times 2.21\,\text{days}} = 3.79 \times 10^{-4}; \quad W(u) = 7.31$$

$$s' = \frac{Q}{4\pi T}W(u) = \frac{19{,}060\,\text{ft}^3/\text{day}}{4 \times 3.1417 \times 7760\,\text{ft}^2/\text{day}} \times 7.31 = 1.43\,\text{ft}$$

Correct the drawdown with Eq. (11.1) because the aquifer is unconfined.

$$s = b - (b^2 - 2s'b)^{1/2} = 20\,\text{ft} - \{(20\,\text{ft})^2 - 2 \times 1.43\,\text{ft} \times 20\,\text{ft}\}^{1/2} = 1.49\,\text{ft}$$

In this case, the difference between the confined and unconfined estimates is quite small.

These same ideas can also be applied to the interpretation of aquifer tests. One simply converts the observed drawdowns in an unconfined aquifer to drawdowns in an equivalent confined aquifer with this equation

$$s' = s - \frac{s^2}{2b} \tag{11.2}$$

where s is the observed drawdown in an unconfined aquifer and s' is the drawdown in an equivalent confined aquifer. These equivalent values of drawdown can be analyzed by using the techniques presented previously for a confined aquifer. However, this simple approach is typically used with late-time drawdown data, once the contribution from delayed yield ceases (Neuman, 1974). Weeks (1969) suggested the following criteria for estimating the time at which delayed yield becomes negligible:

$$t = b\frac{S_y}{K_z}, \quad \text{if} \quad r < 0.4\,b\left(\frac{K_r}{K_z}\right)^{1/2} \tag{11.3}$$

or

$$t = b\frac{S_y}{K_z}\left[0.5 + 1.25\frac{r}{b}\left(\frac{K_z}{K_r}\right)^{1/2}\right], \quad \text{if} \quad r \geq 0.4\,b\left(\frac{K_r}{K_z}\right)^{1/2} \tag{11.4}$$

▶ 11.2 DETERMINATION OF HYDRAULIC PARAMETERS USING DISTANCE/DRAWDOWN DATA

The distance–drawdown method, introduced in Chapter 9, can also be applied to interpreting test data from unconfined aquifers. This section discusses how values of equivalent drawdown as a function of distance can be used for this purpose. Equations (9.1) and (9.2) can be rewritten in terms of corrected drawdowns as

$$T = \frac{2.3Q}{2\pi(s_1' - s_2')}\log\frac{r_2}{r_1} \tag{11.5}$$

$$S_y = \frac{2.25\,T\,t}{r_0^2} \tag{11.6}$$

where s_1' and s_2' are equivalent or corrected drawdowns at distances r_1 and r_2, respectively. Remember: Eqs. (11.5) and (11.6) only apply for late-time drawdown data.

The procedures for using the corrected drawdown versus distance data are summarized as follows.

1. Calculate corrected drawdowns using Eq. (11.2).
2. Plot corrected drawdown s' against r on semi-log graph paper.
3. Approximate the curve s' versus log (t) with a straight line.
4. Determine hydraulic parameters using Eqs. (11.5) and (11.6).

▶ **EXAMPLE 11.2**

Table 11.1 lists drawdown values measured in wells at various distances from a pumping well in an aquifer test near Grand Island, Nebraska (Wenzel, 1936). Measurements were made after the well had been pumped continuously for 48 hours at 540 gpm. The thickness of the unconfined aquifer is 100 ft. Determine the transmissivity and the specific yield.

SOLUTION Table 11.1 shows the calculations of equivalent drawdowns. These values are plotted in Figure 11.2 as a function of distance on a semi-log scale. The pumping rate needs to be converted from gallons per min to ft³/day.

$$Q = 540 \times 192.5 \, \text{ft}^3/\text{day} = 1.04 \times 10^5 \, \text{ft}^3/\text{day}$$

Two points are selected on the straight line: $r_1 = 30$ ft, $s_1' = 3.66$, $r_2 = 300$ ft, and $s_2' = 0.78$ ft. The transmissivity is

$$T = \frac{2.3(1.04 \times 10^5 \, \text{ft}^3/\text{day})}{2\pi(3.66 - 0.78) \, \text{ft}}(1) = 1.32 \times 10^4 \, \text{ft}^2/\text{day}$$

The specific yield is

$$S_y = \frac{2.25(1.32 \times 10^4 \, \text{ft}^2/\text{day})(2 \, \text{day})}{(550 \, \text{ft})^2} = 0.20$$

TABLE 11.1   Drawdown versus distance data for an aquifer test near Grand Island, Nebraska, after 48 hours of continuous pumping at 540 gpm

r(ft)	s(ft)	$s^2/(2b)$ (ft)	s'(ft)
24.9	4.01	0.08	3.93
59.9	2.79	0.04	2.75
114.4	2.03	0.02	2.01
164.2	1.61	0.01	1.60
229	1.14	0.01	1.13
354	0.65	0.00	0.65
429	0.52	0.00	0.52
479	0.44	0.00	0.44
604	0.26	0.00	0.26
755	0.16	0.00	0.16
904	0.11	0.00	0.11

Source: From Wenzel (1936).

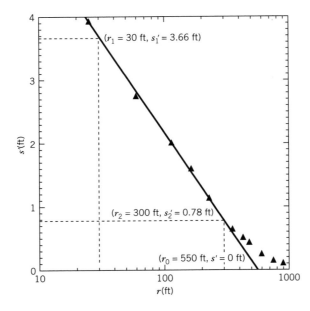

Figure 11.2 Estimation of hydraulic parameters using a distance–drawdown approach. The y-axis is the drawdown of a well in an equivalent confined aquifer.

▶ 11.3 A GENERAL SOLUTION FOR DRAWDOWN

Methods for determining hydraulic parameters in an unconfined aquifer were first introduced by Boulton (1954, 1955, 1963). Since then they have been improved through the effort of numerous investigators (Boulton, 1970; Dagan, 1967; Neuman, 1972, 1973, 1974, and 1975; Prickett, 1965; Streltsova, 1972; Streltsova and Rushton, 1973). In this section, we examine three solutions that differ from each other as a function of where in the aquifer the drawdown is calculated (Figure 11.3). For fully penetrating pumping and observation wells in an unconfined aquifer (Figure 11.3), drawdown is given as

$$s = \frac{Q}{4\pi T} W(u_A, u_B, \beta) \tag{11.7}$$

where $W(u_A, u_B, \beta)$ is the well function for the unconfined aquifer, s is the drawdown, Q is the pumping rate, and T is the transmissivity (Neuman, 1975). The dimensionless parameters (u_A, u_B, and β) are defined as

$$\frac{1}{u_A} = \frac{T t}{S r^2} \quad \text{(for early-time data)} \tag{11.8}$$

$$\frac{1}{u_B} = \frac{T t}{S_y r^2} \quad \text{(for later data)} \tag{11.9}$$

$$\beta = \frac{K_z r^2}{K_r b^2} \tag{11.10}$$

where b is the initial saturated thickness, K_r is the hydraulic conductivity in the radial horizontal direction, and K_z is the vertical hydraulic conductivity. Neuman (1975) provides tables for the function $W(u_A, u_B, \beta)$.

When drawdown is calculated for a piezometer in an unconfined aquifer that is being pumped at a constant rate by a fully penetrating well, the solution depends on where in the aquifer the piezometer is completed (Figure 11.3). Close to the well, head changes vertically because there is a downward flow component. This analytical solution is

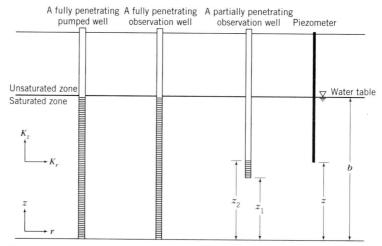

Figure 11.3 This sketch shows a fully penetrating pumping well along with various types of observation wells in an unconfined aquifer.

$$s = \frac{Q}{4\pi T} W(u_A, u_B, \beta, z_D) \qquad (11.11)$$

where $z_D = z/b$, z is the coordinate of the piezometer and b is the thickness of the aquifer. The drawdown in an observation well is the average of the solution expressed by Eq. (11.11) along the screen.

$$W(u_A, u_B, \beta, z_{D_1}, z_{D_2}) = \frac{Q}{4\pi T} \frac{1}{z_{D_2} - z_{D_1}} \int_{z_{D_1}}^{z_{D_2}} W(u_A, u_B, \beta, z_D) dz_D \qquad (11.12)$$

where $z_{D1} = z_1/b$, $z_{D2} = z_2/b$, and z_2 and z_1 are the top and bottom coordinates of the observation well, respectively. Neuman (1972, 1973, 1974, 1975) tabulates these well functions.

Evaluation of Neuman's analytical solutions requires a large amount of computational time as noted by Moench (1993). To improve the efficiency and accuracy of the calculation, Moench (1993, 1995, 1996) computed Neuman's solutions by numerical inversion of the Laplace transform solution (Moench and Ogata, 1984; Stehfest, 1970).

The solutions expressed by Eqs. (11.7), (11.11), and (11.12) are based on the following assumptions (Moench, 1993; Neuman, 1974): (1) the aquifer is homogeneous, infinite in extent, with the principal directions of the hydraulic conductivity tensor oriented parallel to the coordinate directions; (2) the aquifer is bounded by a confining layer at the bottom and a free surface at the top; (3) the diameters of the pumping well, observation wells, or piezometers are infinitesimal; (4) the well is pumped at a constant rate; (5) water is released instantaneously in a vertical direction from a zone above the water table in response to a decline in the elevation of the water table; and (6) the change of the saturated thickness is small compared to the initial saturated thickness.

▶ 11.4 TYPE-CURVE METHOD

There are two common approaches for using aquifer test data to determine the hydraulic parameters of an unconfined aquifer. Both apply to the case of fully penetrating wells and observation wells. The first is a *curve-fitting procedure*, which we discuss in this section. The simpler *straight-line procedure* is explained in Section 11.5.

The curve-fitting approach is quite similar to that used for confined aquifers, though it is more complex because the analysis uses the early-time part of the response (including the delayed drainage) and the later time part separately. The steps in the approach are summarized from Boulton (1963), Prickett (1965), and Neuman (1975).

1. Begin by finding the appropriate type curve. The Neuman (1975) curves $W(u_A, u_B, \beta)$ versus $1/u_A$, and $1/u_B$ work in this case (Figure 11.4). The early/intermediate and late sequence of curves is referred to as the Type-A and the Type-B curves, respectively. The Type-A curves lie to the left of the labels for the β values (Figure 11.4); and the Type-B curves lie to the right. The Type-A and Type-B curves are connected by horizontal asymptotes. The distance between the Type-A and Type-B curves depends on the ratio S/S_y. When this ratio is zero, the two sets of curves are located at the two ends of infinite horizontal asymptotes. For this reason, the two curves are usually plotted on different scales. The top horizontal coordinate axis, $1/u_A$, is the horizontal axis for Type-A curves and the bottom horizontal coordinate axis, $1/u_B$, is the horizontal axis for Type-B curves.

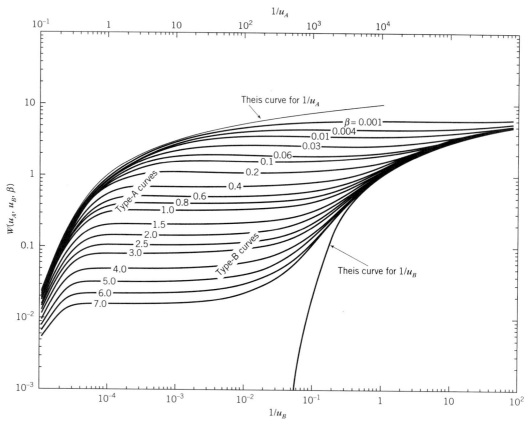

Figure 11.4 Type curves for drawdown in an unconfined aquifer with fully penetrating wells (from Neuman, *Water Resour. Res.*, v. 11, p. 329–342, 1975). Copyright by American Geophysical Union.

2. Plot the field drawdown (s) versus time (t) on (Neuman, 1974) semitransparent log-log graph paper.

3. Select a convenient match point on the type curves and superimpose the s-t curve on the Type-B curves $W(u_B, \beta)$. The match provides values of $W(u_B, \beta)$, $1/u_B$, s, and t.

4. The transmissivity of the aquifer is determined from

$$T = \frac{Q}{4\pi} \frac{W(u_B, \beta)}{s} \tag{11.13}$$

5. From Eq. (11.9), the specific yield is

$$S_y = \frac{T t u_B}{r^2} \tag{11.14}$$

6. Superimpose a field curve of early drawdown versus time curve on the type-A curves.

7. The match provides $W(u_A, \beta)$, $1/u_A$, s, and t values as well as a β value from one of the family of curves. The transmissivity value can be calculated from Eq. (11.13) and should be very close to that calculated from step 4. The storativity is

$$S = \frac{T t u_A}{r^2} \qquad (11.15)$$

8. The horizontal hydraulic conductivity is the ratio of transmissivity to aquifer thickness ($K_r = T/b$). The vertical hydraulic conductivity is given by

$$K_z = \frac{\beta K_r b^2}{r^2} \qquad (11.16)$$

▶ **EXAMPLE 11.3**   A test was conducted with an unconfined aquifer near Fairborn, Ohio (Lohman, 1972). The well was pumped at a constant rate of 1080 gpm. The drawdowns, measured in an observation well 73 ft away, are listed in Table 11.2. The aquifer thickness is 78 ft. Assuming fully penetrating wells, calculate the hydraulic parameters for the aquifer using the type-curve method.

SOLUTION   The drawdown (s) versus time (t) on a log-log scale is shown in Figure 11.5. The curve is first matched with Type B in Figure 11.6. At the match point, $1/u_B = 1$, $W(u_B, \beta) = 1$, $t = 28$ min, and $s = 0.47$ ft. The pumping rate is converted from gpm to ft^3/day using Eq. (9.1).

$$Q = 1080\,\text{gpm} = 1080 \times 192.5\,\text{ft}^3/\text{day} = 2.08 \times 10^5\,\text{ft}^3/\text{day}$$

The transmissivity (T) is obtained by

$$T = \frac{Q}{4\pi} \frac{W(u_B, \beta)}{s} = \frac{(2.08 \times 10^5\,\text{ft}^3/\text{day})}{4\pi} \frac{1}{0.47\,\text{ft}} = 3.5 \times 10^4\,\text{ft}^2/\text{day}$$

TABLE 11.2   Drawdown data obtained from an observation well 73 ft away from the pumping well

Time(min)	s(ft)	Time(min)	s(ft)	Time(min)	s(ft)
0.165	0.12	2.5	0.91	60	1.22
0.25	0.195	2.65	0.92	70	1.25
0.34	0.255	2.8	0.93	80	1.28
0.42	0.33	3	0.94	90	1.29
0.5	0.39	3.5	0.95	100	1.31
0.58	0.43	4	0.97	120	1.36
0.66	0.49	4.5	0.975	150	1.45
0.75	0.53	5	0.98	200	1.52
0.83	0.57	6	0.99	250	1.59
0.92	0.61	7	1.00	300	1.65
1.00	0.64	8	1.01	400	1.70
1.08	0.67	9	1.015	500	1.85
1.16	0.7	10	1.02	600	1.95
1.24	0.72	12	1.03	700	2.01
1.33	0.74	15	1.04	800	2.09
1.42	0.76	18	1.05	900	2.15
1.50	0.78	20	1.06	1000	2.20
1.68	0.82	25	1.08	1200	2.27
1.85	0.84	30	1.13	1500	2.35
2.00	0.86	35	1.15	2000	2.49
2.15	0.87	40	1.17	2500	2.59
2.35	0.9	50	1.19	3000	2.66

Source: From Lohman (1972).

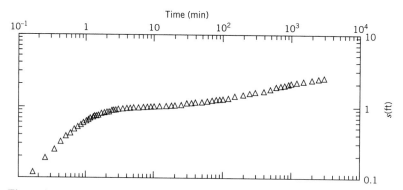

Figure 11.5 Drawdown (s) versus time (t) data plotted on a log-log scale for an aquifer test near Fairborn, Ohio (from Lohman, 1972).

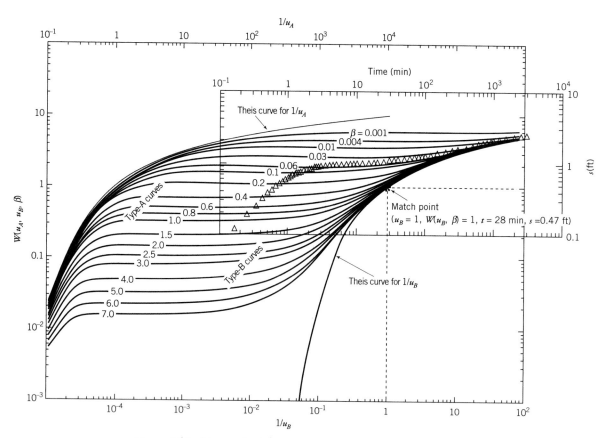

Figure 11.6 The match of the late-time data with the Type-B curve for an aquifer test near Fairborn, Ohio (from Lohman, 1972).

and the specific yield is

$$S_y = \frac{Tt}{r^2} u_B = \frac{(3.5 \times 10^4 \text{ ft}^2/\text{day}) (28 \text{ min}) \left(\dfrac{1}{1440} \dfrac{\text{day}}{\text{min}} \right)}{(73 \text{ft})^2} (1) = 0.13$$

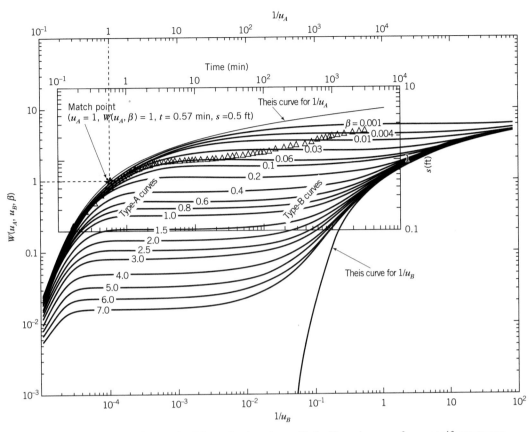

Figure 11.7 The match of the early-time data with the Type-A curves for an aquifer test near Fairborn, Ohio (from Lohman, 1972).

Superimpose the early-time drawdown curve on the Type-A curves and determine the match-point coordinates (Figure 11.7). At the match point, record $1/u_A = 1$, $W(u_A, \beta) = 1$, $t = 0.57$ min, and $s = 0.5$ ft. The field data match the $\beta = 0.06$ curve. From Eq. (11.15),

$$S = \frac{Tt}{r^2} u_A = \frac{3.5 \times 10^4 \text{ ft}^3/\text{day} \, (0.57 \text{min}) \left(\dfrac{1}{1440} \dfrac{\text{day}}{\text{min}} \right)}{(73 \text{ft})^2} (1) = 2.6 \times 10^{-3}$$

and the vertical hydraulic conductivity is

$$K_z = \frac{\beta K_r b^2}{r^2} = \frac{(0.06) \left(\dfrac{3.5 \times 10^4 \text{ ft}}{78} \text{ ft/day} \right) (78 \text{ ft})^2}{(73 \text{ ft})^2} = 31 \text{ ft/day}$$

▶ 11.5 THE STRAIGHT-LINE METHOD

Neuman (1975) illustrated a somewhat simpler method for determining the hydraulic parameters of an unconfined aquifer by straight-line fits to early, intermediate, and late-time drawdown data. The scheme is summarized as follows.

1. Plot values of drawdown versus time on semi-log graph paper.

2. From the late-time segment of the s-log (t) curve, the transmissivity and specific yield of an unconfined aquifer are determined as

$$T = \frac{2.3Q}{4\pi \Delta s'} \tag{11.17}$$

$$S_y = \frac{2.25T \, t_{0y}}{r^2} \tag{11.18}$$

3. From the early-time segment of the s-log (t) curve, the transmissivity and specific yield of an unconfined aquifer are determined by

$$T = \frac{2.3Q}{4\pi \Delta s'} \tag{11.19}$$

$$S = \frac{2.25T \, t_{0S}}{r^2} \tag{11.20}$$

4. fit a straight line through the intermediate part of the s-log (t) curve. The line intercepts the Theis curve at $t = t_\beta$. If the type curves are prepared as a semi-logarithmic plot, the intermediate part of the Type-A and Type-B curves forms a family of straight lines (Figure 11.8). The interception of the straight line and the Theis curve for $1/u_B$ is at $u_{B\beta}$. Neuman (1975) indicated that the $u_{B\beta}$ is related to β by

$$\beta = 0.195 \, u_{B\beta}^{1.1053} = 0.195 \left[\frac{S_y r^2}{T t_\beta} \right]^{1.1053} \quad \text{for} \quad 4 \le u_{B\beta} \le 100 \tag{11.21}$$

Once β is known, the vertical hydraulic conductivity is determined by

$$K_z = \frac{\beta K_r b^2}{r^2} = 0.195 \left[\frac{S_y r^2}{T t_\beta} \right]^{1.1053} \frac{K_r b^2}{r^2} \tag{11.22}$$

However, Eq. (11.7) was derived from the assumption that the decline of the water table is small in comparison to the saturated thickness of an unconfined aquifer. For a large drawdown, Jacob's correction (11.2) should be used with the late drawdown measurements.

▶ **EXAMPLE 11.4**

Let's apply this method to the data from the aquifer test near Fairborn, Ohio (Lohman, 1972). (Information about the test was presented in Example 11.3.) Determine hydraulic parameters for the aquifer using Neuman's straight-line method.

SOLUTION Values of drawdown versus time are plotted on semi-log paper in Figure 11.9. The plot reflects the three types of drawdown response that we described in the introduction. At early time, the aquifer behaves as though it is confined with data falling along a straight line. At some intermediate time, it deviates from this straight line due to delayed gravity drainage. At late time, data again follow a straight line as the aquifer exhibits a Theis-type response with S equivalent to S_y.

The transmissivity and specific yield of the aquifer are determined by fitting a straight line through the late-time data. The early-time response provides another estimate of transmissivity, and a confined storativity value for the aquifer can be obtained. The pumping rate is converted from gpm to ft^3/day using Eq. (9.1).

$$Q = 1080 \, \text{gpm} = 1080 \times 192.5 \, \text{ft}^3/\text{day} = 2.08 \times 10^5 \, \text{ft}^3/\text{day}$$

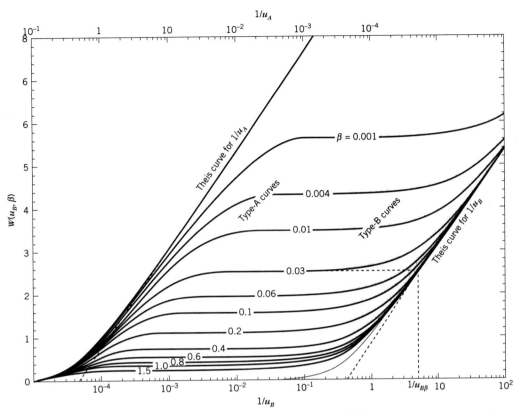

Figure 11.8 Type-A and Type-B curves plotted on a semi-logarithmic scale (from Neuman, *Water Resour. Res.*, v. 11, p. 329–342, 1975). Copyright by American Geophysical Union.

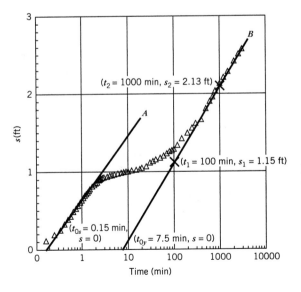

Figure 11.9 Drawdown (*s*) versus time (*t*) data plotted on a semi-log scale for an aquifer test near Fairborn, Ohio (from Lohman, 1972). Straight lines fitted through the data as shown form the basis for the calculation of hydraulic parameters.

From Eq. (9.15), the transmissivity of the aquifer is

$$T = \frac{2.3Q}{4\pi\,\Delta s} = \frac{2.3(2.08 \times 10^5 \text{ ft}^3/\text{day})}{4\pi(2.13 - 1.15)\,\text{ft}} = 3.9 \times 10^4 \text{ ft}^2/\text{day}$$

From Eq. (9.16), the specific yield is

$$S_y = \frac{2.25\, T\, t_{0y}}{r^2} = \frac{2.25(3.9 \times 10^4 \text{ ft}^2/\text{day})\left(7.5 \text{ min}\frac{1\,\text{day}}{1440\,\text{min}}\right)}{(73\,\text{ft})^2} = 0.09$$

From Eq. (9.18), the confined storativity is

$$S = \frac{2.25\, T\, t_{0S}}{r^2} = \frac{2.25(3.9 \times 10^4 \text{ ft}^2/\text{day})\left(0.15 \text{ min}\frac{1\,\text{day}}{1440\,\text{min}}\right)}{(73\,\text{ft})^2} = 1.7 \times 10^{-3}$$

β is calculated by

$$\beta = 0.195 \left[\frac{(0.09)(73\,\text{ft})^2}{(3.9 \times 10^4 \text{ ft}^2/\text{day})(80 \text{ min})\left(\frac{1}{1440}\frac{\text{day}}{\text{min}}\right)}\right]^{1.1053} = 0.037$$

The vertical hydraulic conductivity is

$$K_z = \frac{\beta K_r b^2}{r^2} = \frac{(0.037)(3.9 \times 10^4 \text{ ft}^2/\text{day})(78 \text{ ft})}{(73\,\text{ft})^2} = 21 \text{ ft/day}$$

▶ 11.6 AQUIFER TESTING WITH A PARTIALLY PENETRATING WELL

The well function in Figure 11.4 is valid for fully penetrating discharge and observation wells in an unconfined aquifer. Similar type curves can be calculated for a partially penetrating discharge well (Figure 11.10). The type-curve and straight-line methods will apply to a hydraulic testing with partially penetrating wells when type curves A and B are recalculated to account for the effects of partial penetration. An example of hydraulic testing in an unconfined aquifer with partially penetrating wells will be given in Chapter 15, which deals with computer-aided interpretation of aquifer tests.

▶ EXERCISES

11.1. Drawdown versus distance data are provided in Table 11.3 for an aquifer test near Grand Island, Nebraska. The data were collected after 48 hours of continuous pumping at 540 gpm (Wenzel, 1936). Calculate hydraulic parameters using Jacob's correction equation (11.2). The saturated thickness of the aquifer is 100 ft.

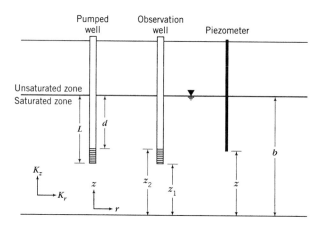

Figure 11.10 Sketch showing partially penetrating pumping and observation wells in an unconfined aquifer.

TABLE 11.3 Values of drawdown versus distance for an aquifer test near Grand Island, Nebraska after 48 hours of continuous pumping at 540 gpm

r(ft)	s(ft)
40.1	3.15
95.1	2.24
144.7	1.71
214	1.24
324	0.77
423	0.51
448	0.46
573	0.28
723	0.15
872	0.10
1073	0.06
1197	0.05

Source: From Wenzel (1936).

TABLE 11.4 Drawdown at an observation of 50 m away from a well pumped at 2 m³/min

Time(min)	s(m)	Time(min)	s(m)
0.10	0.00	19.36	1.06
0.13	0.00	25.19	1.06
0.17	0.01	32.78	1.06
0.22	0.02	42.65	1.06
0.29	0.06	55.49	1.07
0.37	0.11	72.21	1.07
0.49	0.18	93.96	1.08
0.63	0.28	122.26	1.08
0.82	0.39	159.08	1.09
1.07	0.51	206.99	1.10
1.39	0.64	269.34	1.12
1.81	0.75	350.47	1.14
2.36	0.85	456.03	1.16
3.07	0.93	593.38	1.19
3.99	0.99	772.10	1.24
5.19	1.04	1004.66	1.29
6.75	1.05	1307.26	1.36
8.79	1.05	1701.01	1.44
11.43	1.06	2213.34	1.55
14.88	1.06	2880.00	1.68

11.2. Table 11.4 lists drawdown versus time data collected at an observation well 50 m away from a well being pumped at 2 m³/min. The pumped well and observation wells are fully penetrating. The original saturated thickness of the aquifer is 50 m. Calculate the hydraulic parameters using the type-curve method. You will find a template file for this exercise in the directory Chap 11 on the website. The template file temp-11a.doc contains type curves for fully penetrating wells in an unconfined aquifer. The template file temp-11.doc in the directory Chap 11 provides log-log paper at the same scale as the type curves. With the help of these template files, determine hydraulic parameters for the aquifer.

11.3. Recalculate the hydraulic parameters for the test data in Table 11.4 using the straight-line method.

▶ CHAPTER HIGHLIGHTS

1. When an unconfined aquifer is pumped, the water table, which is the upper boundary of the aquifer, falls. This lowering of the water table causes the transmissivity to change with time, especially in the vicinity of the well.

2. The hydraulic response of an unconfined aquifer is also complicated by changes in the way that water is released from storage as a function of time. At early time, water is released due to compression of the matrix and expansion of the water, as is the case with a confined aquifer. Storativity values would be comparable to those for confined aquifers, 10^{-4} or 10^{-5}. As pumping continues, water comes from the slow gravity drainage of water from pores as the water table falls near the well. Once this delayed drainage begins, drawdown is less than expected, resembling the pattern of drawdown in a leaky aquifer. Eventually, this delayed drainage ceases. Flow in the aquifer is mainly radial, and drawdown/time data again fall on a Theis-type curve with a storativity value equivalent to the specific yield (S_y).

3. In predicting the drawdown in an unconfined aquifer, one can calculate the drawdown, assuming the aquifer to be confined and correct values to account for the complexities due to changes in aquifer thickness. This same approach can be used in the interpretation of aquifer tests. One simply converts the observed drawdowns in an unconfined aquifer to drawdowns in an equivalent confined aquifer with the simple correction Eq. (11.2).

4. The classical approach to interpreting results from a test with an unconfined aquifer follows a curve-matching procedure, similar to that for confined aquifers. Field data are plotted as the log of drawdown versus the log of time. In this case, the process is more complicated because there are two sets of type curves to be matched—the early response, including gravity drainage, and the late response once gravity drainage has ceased. This analysis provides transmissivity, specific yield, and vertical hydraulic conductivity of the aquifer.

5. Neuman (1975) provides a simpler, straight-line procedure for estimating the hydraulic parameters, T, S_y, and K_z. It involves plotting corrected drawdown values on semi-log paper, with time as the log scale. The early- and late-time data, which exhibit a Theis-type response, will fall along a straight line. Using information on drawdown per log cycle and values of t_0, one calculates the appropriate aquifer parameters.

12

SLUG, STEP, AND INTERMITTENT TESTS

The "slug test" is a method for determining the hydraulic conductivity of a geological unit using an observation well or piezometer. It involves displacing the water level in a well away from some equilibrium position by adding water or withdrawing a *slug* from the well, which causes an immediate decline in water level (Figure 12.1). A slug is a solid cylinder of metal 5 ft or more long that can be hung in a well on a cable for an appropriate time period before the test. Upon removal, the water level is lowered downward instantaneously from its equilibrium level. Monitoring the return of the water level back to the pretest level provides the data for the analysis. Typically, water levels can be measured using an electric tape. In some cases, where manual measurements cannot be made quickly enough, electrical monitoring systems are used. When water-quality sampling is contemplated, it is advisable not to add water to the well in a slug test.

Hvorslev originally developed this method in 1951. Since then, more rigorous variants of the methods have been derived for a wide range of test conditions (Bouwer, 1989; Bouwer and Rice, 1976; Cooper et al., 1967; Papadopulos and Cooper, 1967; Papadopulos et al., 1973) In practice, these tests are used when quick or inexpensive estimates of hydraulic conductivity are required (Thompson, 1987). They are much simpler than a conventional aquifer test and will work with relatively small-diameter wells or piezometers.

Hydraulic conductivity values obtained from this test are considered to be less representative than those for an aquifer test. Much smaller volumes of water are displaced in a slug test as compared to a conventional aquifer test. Thus, the slug test reflects the hydraulic conductivity of a small volume of the medium near the well. Also, most slug tests do not provide estimates of storativity.

▶ 12.1 HVORSLEV SLUG TEST

In the classical Hvorslev approach, the deviations in water level from the pretest water level are measured in terms of a so-called drawdown ratio (H_t). The *drawdown ratio* is the ratio of the drawdown at any time t to the maximum drawdown when the test is begun or $H_t = s_t/s_0$ (Figure 12.1). In effect, using the drawdown ratios normalizes the drawdown

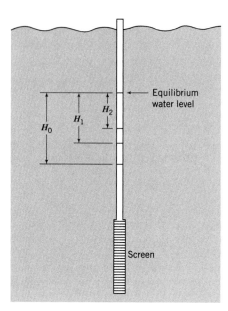

Figure 12.1 In a slug test, the water level in a well or piezometer is displaced away from its equilibrium position. The return of water levels to equilibrium is monitored as a function of time (modified from Thompson, 1989). Reprinted by permission of *Ground Water*. Copyright ©1989. All rights reserved.

between zero and one. Hvorslev found that the return of the water level to equilibrium is exponential—the change per unit time starts out relatively large and slows down. However, the time required to return to equilibrium depends on the hydraulic conductivity. In high K units, water levels return to equilibrium in seconds or minutes; in low K units, the recovery might take months. The recovery rate also depends on how the piezometer is designed. Wells that have a large area for water to enter the casing recover more rapidly than wells with a small open area. Thus, the equation for calculating hydraulic conductivity must account for the construction details of the well in interpreting the recovery rates. Look carefully at the following form of Hvorslev's equation

$$K = \frac{A}{F} \frac{1}{t_2 - t_1} \ln \frac{H_1}{H_2} \tag{12.1}$$

where K is the hydraulic conductivity of an aquifer, A is the cross-sectional area of the well, and F is a shape factor related to the list in Table 12.1 (U.S. Department of Navy, 1992). H_1 and H_2 are the drawdown ratios at time t_1 and t_2, respectively. Information about the well construction is contained in the shape factor F and the area A. The shape factor, as will be evident, can take on different forms as shown in the pictures in Table 12.1. Thus, Eq. (12.1) takes on different forms depending on the shape factor. To find the correct equation, you need to go to Table 12.1 and find the well design that best matches your piezometer or well. The table doesn't contain all the well designs one might encounter. Readers interested in seeing the complete suite of shape factors should refer to Cedergren (1977).

The steps in running the test and determining hydraulic conductivity using the Hvorslev method are summarized as follows:

1. Before displacing the water level from equilibrium, record the initial depth to water. The test begins when the water level is changed by adding water or removing a slug from the well. Measure water level periodically as the water level returns to the pretest level. Keep collecting data until 90% plus of the initial water-level displacement is recovered.

TABLE 12.1   Shape factors, calculational equations, and notes on applicability for various piezometer designs

	CONDITION	DIAGRAM	SHAPE FACTOR, F	PERMEABILITY, K BY VARIABLE HEAD TEST	APPLICABILITY
OBSERVATION WELL OR PIEZOMETER IN SATURATED ISOTROPIC STRATUM OF INFINITE DEPTH	(A) UNCASED HOLE.		$F = 16\pi DSR$	(FOR OBSERVATION WELL OF CONSTANT CROSS SECTION) $$K = \frac{R}{16DS} \times \frac{(H_2 - H_1)}{(t_2 - t_1)}$$ FOR $\dfrac{D}{R} < 50$	SIMPLEST METHOD FOR PERMEABILITY DETERMINATION. NOT APPLICABLE IN STRATIFIED SOILS.
	(B) CASED HOLE, SOIL FLUSH WITH BOTTOM.		$F = \dfrac{11R}{2}$	$$K = \frac{2\pi R}{11(t_2 - t_1)} \ln\frac{(H_1)}{(H_2)}$$ FOR $6'' \le D \le 60''$	USED FOR PERMEABILITY DETERMINATION AT SHALLOW DEPTHS BELOW THE WATER TABLE. MAY YIELD UNRELIABLE RESULTS IN FALLING HEAD TEST WITH SILTING OF BOTTOM OF HOLE.
	(C) CASED HOLE, UNCASED OR PERFORATED EXTENSION OF LENGTH ``L''.		$F = \dfrac{2\pi L}{\ln\left(\dfrac{L}{R}\right)}$	$$K = \frac{R^2}{2L(t_2 - t_1)} \ln\frac{(L)}{(R)} \ln\frac{(H_1)}{(H_2)}$$ FOR $\dfrac{L}{R} > 8$	USED FOR PERMEABILITY DETERMINATION AT GREATER DEPTHS BELOW WATER TABLE.
	(D) CASED HOLE, COLUMN OF SOIL INSIDE CASING TO HEIGHT ``L''		$F = \dfrac{11\pi R^2}{2\pi R + 11L}$	$$K = \frac{2\pi R + 11L}{11(t_2 - t_1)} \ln\frac{(H_1)}{(H_2)}$$	PRINCIPAL USE IS FOR PERMEABILITY IN VERTICAL DIRECTION IN ANISOTROPIC SOILS.
	(E) CASED HOLE, OPENING FLUSH WITH UPPER BOUNDARY OF AQUIFER OF INFINITE DEPTH.		$F = 4R$	(FOR OBSERVATION WELL OF CONSTANT CROSS SECTION) $$K = \frac{\pi R}{4(t_2 - t_1)} \ln\frac{(H_1)}{(H_2)}$$	USED FOR PERMEABILITY DETERMINATION WHEN SURFACE IMPERVIOUS LAYER IS RELATIVELY THIN. MAY YIELD UNRELIABLE RESULTS IN FALLING HEAD TEST WITH SILTING OF BOTTOM OF HOLE.
OBSERVATION WELL OR PIEZOMETER IN AQUIFER WITH IMPERVIOUS UPPER LAYER	(F) CASED HOLE, UNCASED OR PERFORATED EXTENSION INTO AQUIFER OF FINITE THICKNESS: (1) $\dfrac{L_1}{T} \le 0.2$ (2) $0.2 < \dfrac{L_2}{T} < 0.85$ (3) $\dfrac{L_3}{T} = 1.00$ NOTE: R_0 EQUALS EFFECTIVE RADIOS TO SOURCE AT CONSTANT HEAD.		(1) $F = C_s R$	$$K = \frac{\pi R}{C_3(t_2 - t_1)} \ln\frac{(H_1)}{(H_2)}$$	USED FOR PERMEABILITY DETERMINATIONS AT DEPTHS GREATER THAN ABOUT 5 FT.
			(2) $F = \dfrac{2\pi L_2}{\ln(L_2/R)}$	$$K = \frac{R^2 \ln\left(\dfrac{L_2}{R}\right)}{2L_2(t_2 - t_1)} \ln\frac{(H_1)}{(H_2)}$$ FOR $\dfrac{L}{R} \ge 8$	USED FOR PERMEABILITY DETERMINATIONS AT GREATER DEPTHS AND FOR FINE GRAINED SOILS USING POROUS INTAKE POINT OF PIEZOMETER.
			(3) $F = \dfrac{2\pi L_3}{\ln\left(\dfrac{R_0}{R}\right)}$	$$K = \frac{R^2 \ln\left(\dfrac{R_0}{R}\right)}{2L_3(t_2 - t_1)} \ln\left(\frac{H_1}{H_2}\right)$$	ASSUME VALUE OF $R_0/R = 200$ FOR ESTIMATES UNLESS OBSERVATION WELLS ARE MADE TO DETERMINE ACTUAL VALUE OF R_0

2. From the water-level measurements, calculate drawdowns, $s_0, s_1, \ldots \ldots s_n$. Determine the drawdown ratios $H_0, H_1, \ldots \ldots H_n$ by dividing each of the drawdowns by s_0, the drawdown (maximum) at t_1, for example, $H_1 = s_1/s_0$.

3. Plot the head ratios (H_t) on the log scale and time (t) on the linear scale of semi-log graph paper.

4. Fit the best straight line through the set of data points.

5. Choose two points on the straight line and record t_1, H_1, t_2, and H_2.

6. Calculate K using the appropriate form of the equation. In Table 12.1, D is the depth of the well measured from the water table to the bottom of the well, L is the length of well screen or well open, S is the thickness of saturated permeable material above an underlying confining bed, and T is the thickness of the confined aquifer. The C_s can be determined as

$$C_s = \frac{2\pi(L/R)}{\ln(L/R + 1.36)} \tag{12.3}$$

(see Thomson, 1987). This example illustrates how slug-test data are interpreted to provide an estimate of K.

► **EXAMPLE 12.1**   The slug-test data in Table 12.2 were reported by Cooper et al. (1967). The well has a diameter of 7.6 cm with an open hole 98 m in length. The design is similar to F-3 in Table 12.1. Calculate hydraulic conductivity using Hvorslev's method.

TABLE 12.2   Slug-test result

Time(sec)	s(m)	s/s_0
0	0.560	1.000
3	0.457	0.816
6	0.392	0.700
9	0.345	0.616
12	0.308	0.550
15	0.280	0.500
18	0.252	0.450
21	0.224	0.400
24	0.205	0.366
27	0.187	0.334
30	0.168	0.300
33	0.149	0.266
36	0.140	0.250
39	0.131	0.234
42	0.112	0.200
45	0.108	0.193
48	0.093	0.166
51	0.089	0.159
54	0.082	0.146
57	0.075	0.134
60	0.071	0.127
63	0.065	0.116

Source: From Cooper et al. (1967).

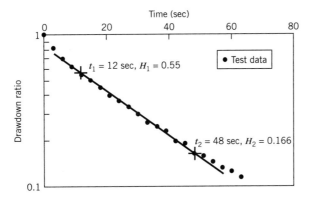

Figure 12.2 A plot of the log of the drawdown ratio versus time provides H_1, t_1 and H_2, t_2 values required to calculate hydraulic conductivity with the Hvorslev theory (from Cooper et al., *Water Resources Res.*, v. 3, p. 263–269, 1967). Copyright by American Geophysical Union.

SOLUTION   The working equation comes from Table 12.1 and is written as

$$K = \frac{R^2 \ln(R_0/R)}{2\,L_3(t_2 - t_1)} \ln \frac{H_1}{H_2}$$

Values for L, 98 m, and r_w, 7.6 cm, are given. The value for $\ln(R_0/R)$ is given as a default parameter on the table with $\ln(200)$. Figure 12.2 shows the semi-log plot of the drawdown ratios versus time with a straight line fitted through the points. We selected two points on the line, $t_1 = 12$ sec, $H_1 = 0.55$ m, $t_2 = 48$ sec, and $H_2 = 0.166$ m. Thus

$$K = \frac{(7.6^2 \text{ cm}^2)(\ln(200))}{2(9800\,\text{cm})(48 - 12)\text{sec}} \ln\left(\frac{0.55}{0.166}\right) = 5.2 \times 10^{-4} \text{ m/sec}$$

The transmissivity is

$$T = Kb = (5.2 \times 10^{-4} \text{ cm/sec})(9800\,\text{cm}) = 5.1 \text{ cm}^2/\text{sec}$$

▶ 12.2 COOPER–BREDEHOEFT–PAPADOPULOS TEST

The Cooper–Bredehoeft–Papadopulos method (Cooper et al., 1967; Papadopulos and Cooper, 1967; Papadopulos et al., 1973) provides a more sophisticated approach to single-well testing. The test setup is shown in Figure 12.3. The following analytical solution provides the drawdown ratio in a confined aquifer for a fully penetrating well:

$$\frac{s}{s_0} = F(\beta, \alpha) = \frac{8\alpha}{\pi^2} \int_0^\infty \frac{e^{-\beta u^2/\alpha}}{u([u\,J_0(u) - 2\alpha\,J_1(u)]^2 + [u Y_0(u) - 2\alpha Y_1(u)]^2)} du \quad (12.4)$$

where s is the head change at time t, and s_0 is the initial head change after the injection or withdrawal. The parameters β and α are expressed as

$$\beta = \frac{Tt}{r_c^2} \quad (12.5)$$

and

$$\alpha = \frac{r_s^2 S}{r_c^2} \quad (12.6)$$

where r_c is the radius of the casing, T is the transmissivity, S is the storativity, and r_s is the effective radius of the well. Tabulated values of $F(\beta, \alpha)$ are available for generating

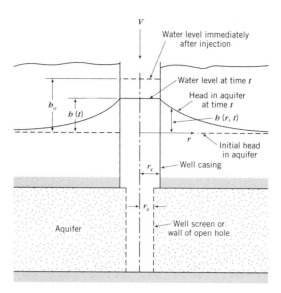

Figure 12.3 Basic geometry for the Cooper–Bredehoeft–Papadopulos slug test (from Cooper et al., *Water Resources Res.*, v. 3, p. 263–269, 1967). Copyright by American Geophysical Union.

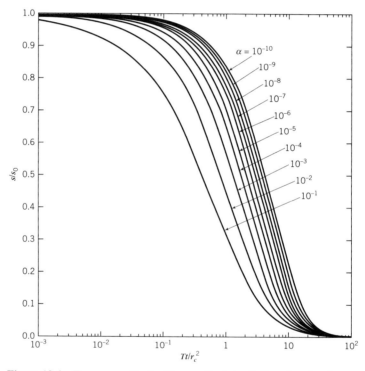

Figure 12.4 Type curves for the Cooper–Bredehoeft–Papadopulos slug test (from Cooper et al., *Water Resources Res.*, v. 3, p. 263–269, 1967). Copyright by American Geophysical Union.

type curves (Figure 12.4). The general approach to applying the Cooper–Bredehoeft–Papadopulos method is similar to the curve-matching schemes of previous chapters. Here is a summary of the steps:

1. Plot the drawdown ratio on a linear scale versus time on a log scale at the same scale as the type curves.

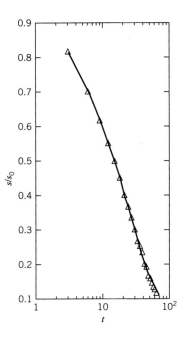

Figure 12.5 A plot of a slug-test result as the log of the drawdown ratio versus the log of time (from Cooper et al., *Water Resources Res.*, v. 3, p. 263–269, 1967). Copyright by American Geophysical Union.

2. Superimpose the plot of field data on the type curves and find the best fit with one of the type curves.

3. Select β (usually 1) and find corresponding t. Transmissivity can be calculated by Eq. (12.4).

$$T = \frac{\beta r_c^2}{t} \qquad (12.7)$$

4. Record α. The storativity can be calculated by Eq. (12.5).

$$S = \frac{r_c^2 \alpha}{r_s^2} \qquad (12.8)$$

▶ **EXAMPLE 12.2** Use the data from Example 12.1 and calculate the transmissivity and the storativity by the Cooper–Bredehoeft–Papadopulos method.

SOLUTION The measured drawdown ratios are plotted versus time in Figure 12.5. The plot is superimposed on the type curve (Figure 12.6). The best match between the field curve and the type curves is with $\alpha = 10^{-3}$. For a β value of 1.0, the corresponding time (t) is 12 sec. It is also known as $r_c = r_s = 7.6$ cm. Thus

$$T = \frac{1.0 r_c^2}{t} = \frac{(1.0)(7.6^2 \, \text{cm}^2)}{(12 \, \text{sec})} = 4.8 \, \text{cm}^2/\text{sec}$$

and

$$S = \frac{r_c^2 \alpha}{r_s^2} = \frac{(7.6 \, \text{cm})^2 10^{-3}}{(7.6 \, \text{cm})^2} = 10^{-3}$$

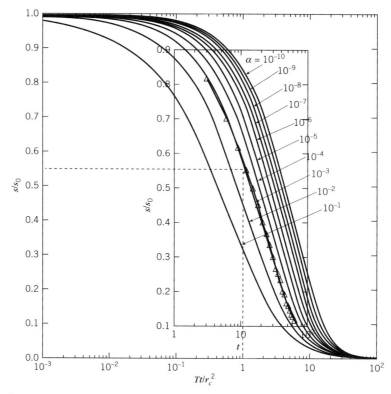

Figure 12.6 Using a type-curve matching procedure to determine α and β values for the Cooper–Bredehoeft–Papadopulos analysis (from Cooper et al., *Water Resources Res.*, v. 3, p. 263–269, 1967). Copyright by American Geophysical Union.

▶ 12.3 BOUWER AND RICE SLUG TEST

Bouwer and Rice (1976) developed a technique for determining the hydraulic conductivity of an unconfined aquifer with a fully or a partially penetrating well. Bouwer (1989) later extended the technique to a confined aquifer. The approach is similar to Hvorslev's method but involves using a set of curves to determine the radius of influence. The rate of water-level change (drawdown or mound) in a slug test (Figure 12.7) is expressed as

$$\frac{ds}{dt} = \frac{Q}{\pi r_c^2} \tag{12.9}$$

where r_c is the casing of the well, Q is the inflow or outflow rate of water into or out of the well after a sudden water-level change, and s is the head change. At steady state, the flow rate is

$$Q = 2\pi K L_e \frac{s}{\ln(R_e/r_w)} \tag{12.10}$$

where K is the hydraulic conductivity of the aquifer, L_e is the screen length, R_e is the radius of influence, and r_w is the radius of the well. By inserting Eq. (12.10) into (12.9) and integrating, a working equation is obtained for calculating the hydraulic conductivity as

$$K = \frac{r_c^2 \ln(R_e/r_w)}{2 L_e} \frac{1}{t} \ln\frac{s_0}{s} \tag{12.11}$$

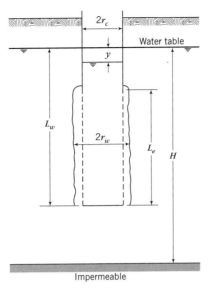

Figure 12.7 Basic geometry for the Bouwer and Rice slug test (from Bouwer, 1989). Reprinted by permission of *Ground Water*. Copyright ©1989. All rights reserved.

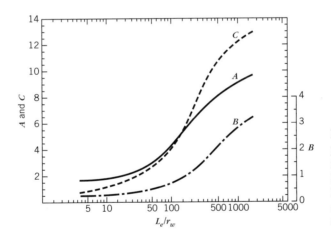

Figure 12.8 Dimensionless parameters A, B, and C as a function of L_e/R_w for the Bouwer and Rice slug test. Reprinted by permission of *Ground Water*. Copyright ©1996. All rights reserved.

where s_0 is the initial water-level change and s is the water-level change at t. An empirical equation relates $\ln(R_e/r_w)$ to the geometry of the system.

$$\ln\frac{R_e}{r_w} = \left[\frac{1.1}{\ln(L_w/r_w)} + \frac{A + B\ln[(H - L_w)/r_w]}{L_e/r_w}\right]^{-1} \qquad (12.12)$$

where L_w is the length of the well in the aquifer, A and B depend on the ratio L_e/r_w (Figure 12.8), and H is the thickness of the saturated material. When $L_w = H$, a simpler form of the equation is written as

$$\ln\frac{R_e}{r_w} = \left[\frac{1.1}{\ln(L_w/r_w)} + \frac{C}{L_e/r_w}\right]^{-1} \qquad (12.13)$$

where C is a function of L_e/r_w (Figure 12.8).

The steps in determining the hydraulic conductivity using the Bouwer and Rice method are as follows:

1. Plot water-level change s on a log scale versus time t on a linear scale using semi-log graph paper.

2. Approximate the straight portion of the plotted curve by a straight line and extend the line to $t = 0$.

3. Calculate $\ln[(H - L_w)/r_w]$ for $L_w \neq H$. If $\ln[(H - L_w)/r_w] > 6$, set $\ln[(H - L_w)/r_w] = 6$.

4. Find A and B for $H \neq L_w$ for C for $H = L_w$ in Figure 12.8.

5. Calculate $\ln(R_e/r_w)$ using Eq. (12.12) or (12.13).

6. Record s_0, and s and t for one other point on the line. Calculate K using Eq. (12.11).

Although this technique is developed for an unconfined aquifer, it can also be used for a confined aquifer that receives water from an overlying confining layer (Bouwer, 1989).

▶ **EXAMPLE 12.3**

Use the data set from the previous problem to calculate hydraulic conductivity with the Bouwer and Rice method.

SOLUTION   Recall that we are given that $r_w = r_c = 7.6$ cm, $L_e = L_w = H = 98$ m. Determine the ratio L_e/r_w, which is equal to 1290. We use this ratio with Figure 12.8 to determine that $C = 12.4$. Thus,

$$\ln\frac{R_e}{r_w} = \left[\frac{1.1}{\ln(9800/7.6)} + \frac{12.4}{9800/7.6} \right]^{-1} = 6.13$$

Plot the drawdown versus time in Figure 12.9 and record $s_0 = 0.49$ m, $t = 33$ sec, and $s = 0.149$ m. These latter coordinates are points on the line. The hydraulic conductivity is

$$K = \frac{7.6^2 \text{ cm}^2 (6.13)}{2(9800 \text{ cm})} \frac{1}{(33 \text{ sec})} \ln\frac{0.49}{0.149} = 6.5 \times 10^{-4} \text{ cm/sec}$$

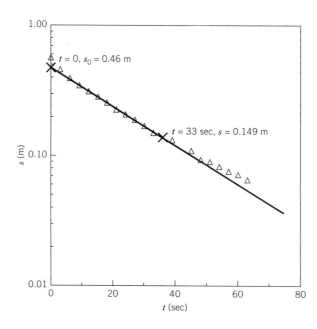

Figure 12.9 Selecting points on the best-fit line for estimating hydraulic conductivity with the Bouwer and Rice slug test (data from Cooper et al., 1967).

With $H = 98$ m, the transmissivity is

$$T = HK = (9800 \text{ cm})(6.5 \times 10^{-4} \text{ cm/sec}) = 6.4 \text{ cm}^2/\text{sec}$$

This value of transmissivity value compares favorably to values of 5.1 and 4.8 cm^2/sec, calculated, respectively, from the Hvorslev and the Cooper–Bredehoeft–Papadopulos methods.

► 12.4 STEP AND INTERMITTENT DRAWDOWN TESTS

Step and intermittent drawdown tests are useful in determining the aquifer parameters and the efficiency of a pumped well (Birsoy and Summers, 1980s; Brown, 1963; Harrill, 1971; Theis, 1963). The intermittent drawdown tests consist of a number of pumping and recovery periods with variable pumping rates (Figure 12.10) whereas step drawdown tests consist of a pumping period with variable pumping rates and a recovery period (Figure 12.11). Birsoy and Summers (1980) derived a convenient equation to express the drawdown for the step and intermittent drawdown tests.

$$\frac{s}{Q_n} = \frac{1}{4\pi T} \log\left[\beta_n(t)\frac{t - \tau_n}{t - \tau_n'}\right] \quad \text{for} \quad t > \tau_n' \text{ during recovery} \tag{12.14}$$

and

$$\frac{s}{Q_n} = \frac{1}{4\pi T} \log\left[\frac{2.25T}{r^2 S}\beta_n(t)(t - \tau_n)\right] \quad \text{for} \quad t > \tau_n \text{ during pumping} \tag{12.15}$$

where n is the pumping rate uncer consideration, Q_i is the ith rate, and τ_i and τ_i' are the times when the ith pumping rate starts and ends, respectively. For $n = 1, \beta_n(t) = 1$. For $n > 1, \beta_n(t)$ is expressed as

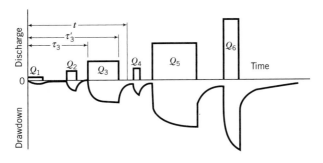

Figure 12.10 Time history of pumping rates for an intermittent drawdown test (from Birsoy and Summers, 1980). Reprinted by permission of *Ground Water.* Copyright ©1980. All rights reserved.

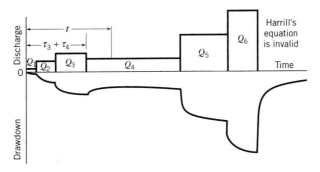

Figure 12.11 Time history of pumping rates for a step drawdown test (from Birsoy and Summers, 1980). Reprinted by permission of *Ground Water.* Copyright ©1980. All rights reserved.

$$\beta_n(t) = \prod_{i=1}^{n-1} \left(\frac{t - \tau_i}{t - \tau_i'}\right)^{\frac{Q_i}{Q_n}} = \left(\frac{t - \tau_1}{t - \tau_1'}\right)^{\frac{Q_1}{Q_n}} \cdots \left(\frac{t - \tau_{n-1}}{t - \tau_{n-1}'}\right)^{\frac{Q_{n-1}}{Q_n}} \quad \text{for} \quad n > 1 \quad (12.16)$$

Determination of Transmissivity and Storativity

The procedure for determining the transmissivity and storativity of an aquifer using the step and intermittent drawdown tests is as follows:

1. Calculate $\beta_n(t)$ at each measurement time for the entire period.
2. Plot s/Q_n in linear scale versus adjusted time $\beta_n(t)(t - \tau_n)$ (or adjusted dimensionless time $\beta_n(t)(t - \tau_n)/(t - \tau_i')$) on the logarithmic scale of semi-log graph paper. The points should fall on a straight line. Selecting two points separated by one log cycle, we calculate the transmissivity as

$$T = \frac{2.3}{4\pi \, \Delta(s/Q)} \quad (12.17)$$

3. To determine storativity you need to find the intercept of the straight line connecting the plotted points and $s/Q_n = 0$ and substituting into the following equation.

$$S = \frac{2.25 \, T \, \beta_n(t_0)(t_0 - \tau_n)}{r^2} \quad (12.18)$$

where $\beta_n(t_0)(t_0 - \tau_n)$ is the intercept.

▶ **EXAMPLE 12.4**

In a step drawdown test reported by Birsoy and Summers (1980), the initial pumping rate of 2.3 gpm was maintained for 30 min. Pumping rates, drawdowns at the pumped well, and starting and ending times for the other four steps are summarized in Table 12.3. The length of well screen is 200 ft. Calculate the transmissivity of the aquifer.

SOLUTION The adjustment time factor $\beta_n(t)$ and adjusted time $\beta_n(t)(t - \tau_n)$ are calculated using Eq. (12.16) and listed in Table 12.3. Because the aquifer is unconfined, Jacob's correction is applied to the measured drawdowns. The corrected drawdown on a linear scale versus adjusted time on a logarithmic scale is shown in Figure 12.12. Theoretically, the data should fall on a straight line. The departure from the straight line may be caused by the assumption that the Jacob modification of the Theis equation is only valid for large time, or the well was still developing at early times. Late-time data are used for the determination of transmissivity. In one log cycle, the drawdown difference is 0.28 ft. The transmissivity is calculated by

$$T = \frac{2.3}{4\pi \, \Delta(s/Q)} = \frac{2.3}{4\pi \, (0.28 \text{ ft/gpm})} = 126 \, \text{ft}^2/\text{day}$$

For the pumping well, Eq. (12.18) cannot be used to obtain storativity because the effective radius of the well is not known.

Estimating Well Efficiency

As a well is pumped, water moves from the aquifer through the gravel pack and screen and finally into the casing. Usually, the large volume of water moving toward the well has difficulties passing through the gravel pack and screen, creating a substantial drop in hydraulic head between the aquifer and the inside of the well. Just how well this water passes through the gravel pack and screen is measured by the *well efficiency*. Well efficiency (E) is defined as the ratio of the drawdown (s_a) in the aquifer at the radius just outside of the pumped well to the drawdown (s_t) inside the well (Health, 1989):

$$E = \frac{s_a}{s_t} \times 100 \quad (12.19)$$

TABLE 12.3 Step pumping test data

Step 2: $\tau_2 = 30$ min, $\tau'_2 = 60$ min, $Q_2 = 14.7$ gpm						Step 4: $\tau_4 = 90$ min, $\tau'_4 = 120$ min, $Q_4 = 38.5$ gpm					
T(min)	s(ft)	$\beta(t)$	$\beta(t-\tau_n)$ (min)	s/Q (ft/gpm)	$[s-s^2/(2b)]/Q$ (ft/gpm)	t(min)	s(ft)	$\beta(t)$	$\beta(t-\tau_n)$ (min)	s/Q (ft/gpm)	$[s-s^2/(2b)]/Q$ (ft/gpm)
30	1.04					90	14.41				
30.5	2.05	1.90	0.95	0.139	0.139	90.5	14.99	41.86	20.93	0.389	0.375
31	2.53	1.71	1.71	0.172	0.171	91	15.36	23.65	23.65	0.399	0.384
31.5	2.85	1.61	2.42	0.194	0.192	91.5	15.61	17.00	25.51	0.405	0.390
32	3.09	1.54	3.09	0.210	0.209	92	15.84	13.49	26.99	0.411	0.395
32.5	3.29	1.49	3.73	0.224	0.222	92.5	16.04	11.30	28.26	0.417	0.400
33	3.45	1.46	4.37	0.235	0.233	93	16.17	9.80	29.39	0.420	0.403
33.5	3.58	1.42	4.98	0.244	0.241	93.5	16.31	8.69	30.42	0.424	0.406
34	3.71	1.40	5.59	0.252	0.250	94	16.45	7.85	31.38	0.427	0.410
34.5	3.81	1.38	6.19	0.259	0.257	94.5	16.57	7.17	32.29	0.430	0.413
35	3.91	1.36	6.78	0.266	0.263	95	16.67	6.63	33.15	0.433	0.415
35.5	4.01	1.34	7.36	0.273	0.270	95.5	16.77	6.18	33.98	0.436	0.417
36	4.10	1.32	7.94	0.279	0.276	96	16.86	5.80	34.77	0.438	0.419
36.5	4.18	1.31	8.51	0.284	0.281	96.5	16.95	5.47	35.54	0.440	0.422
37	4.25	1.30	9.08	0.289	0.286	97	17.03	5.18	36.29	0.442	0.424
37.5	4.32	1.29	9.65	0.294	0.291	97.5	17.12	4.94	37.02	0.445	0.426
38	4.39	1.28	10.21	0.299	0.295	98	17.19	4.72	37.74	0.446	0.427
38.5	4.46	1.27	10.77	0.303	0.300	98.5	17.27	4.52	38.44	0.449	0.429
39	4.51	1.26	11.32	0.307	0.303	99	17.41	4.35	39.13	0.452	0.433
39.5	4.56	1.25	11.87	0.310	0.307	99.5	17.44	4.19	39.81	0.453	0.433
40	4.63	1.24	12.42	0.315	0.311	100	17.51	4.05	40.47	0.455	0.435
41	4.72	1.23	13.51	0.321	0.317	101	17.64	3.80	41.78	0.458	0.438
42	4.82	1.22	14.60	0.328	0.324	102	17.78	3.59	43.06	0.462	0.441
43	4.92	1.21	15.68	0.335	0.331	103	17.90	3.41	44.31	0.465	0.444
44	5.01	1.20	16.75	0.341	0.337	104	18.03	3.25	45.54	0.468	0.447
45	5.11	1.19	17.81	0.348	0.343	105	18.13	3.12	46.75	0.471	0.450
46	5.19	1.18	18.87	0.353	0.348	106	18.24	3.00	47.95	0.474	0.452
47	5.27	1.17	19.93	0.359	0.354	107	18.33	2.89	49.13	0.476	0.454
48	5.35	1.17	20.99	0.364	0.359	108	18.41	2.79	50.30	0.478	0.456
49	5.43	1.16	22.04	0.369	0.364	109	18.56	2.71	51.46	0.482	0.460
50	5.51	1.15	23.08	0.375	0.370	110	18.65	2.63	52.61	0.484	0.462
52	5.64	1.14	25.17	0.384	0.378	112	18.84	2.49	54.89	0.489	0.466
54	5.79	1.14	27.25	0.394	0.388	114	19.03	2.38	57.13	0.494	0.471
56	5.91	1.13	29.32	0.402	0.396	116	19.22	2.28	59.35	0.499	0.475
58	6.04	1.12	31.38	0.411	0.405	118	19.39	2.20	61.55	0.504	0.479
60	6.16	1.11	33.44	0.419	0.413	120	19.56	2.12	63.73	0.508	0.483

Step 3: $\tau_3 = 60$ min, $\tau'_3 = 90$ min, $Q_3 = 32.3$ gpm						Step 5: $\tau_5 = 150$ min, $\tau'_5 = 150$ min, $Q_5 = 46.0$ gpm					
t(min)	s(ft)	$\beta(t)$	$\beta(t-\tau_n)$ (min)	s/Q (ft/gpm)	$[s-s^2/(2b)]/Q$ (ft/gpm)	t(min)	s(ft)	$\beta(t)$	$\beta(t-\tau_n)$ (min)	s/Q (ft/gpm)	$[s-s^2/(2b)]/Q$ (ft/gpm)
60	6.16					120	19.56				
60.5	7.66	6.82	3.41	0.237	0.233	120.5	20.42	58.24	29.12	0.444	0.421
61	8.4	5.01	5.01	0.260	0.255	121	20.86	32.83	32.83	0.453	0.430
61.5	8.93	4.19	6.29	0.276	0.270	121.5	21.16	23.55	35.32	0.460	0.436
62	9.30	3.70	7.40	0.288	0.281	122	21.39	18.64	37.28	0.465	0.440
62.5	9.62	3.37	8.42	0.298	0.291	122.5	21.6	15.57	38.92	0.470	0.444
63	9.90	3.12	9.36	0.307	0.299	123	21.8	13.46	40.37	0.474	0.448

(Continued)

TABLE 12.3 Continued

Step 3: $\tau_3 = 60$ min, $\tau_3' = 90$ min, $Q_3 = 32.3$ gpm						Step 5: $\tau_5 = 150$ min, $\tau_5' = 120$ min, $Q_5 = 46.0$ gpm					
t(min)	s(ft)	$\beta(t)$	$\beta(t-\tau_n)$ (min)	s/Q (ft/gpm)	$[s-s^2/(2b)]/Q$ (ft/gpm)	t(min)	s(ft)	$\beta(t)$	$\beta(t-\tau_n)$ (min)	s/Q (ft/gpm)	$[s-s^2/(2b)]/Q$ (ft/gpm)
63.5	10.16	2.93	10.24	0.315	0.307	123.5	21.96	11.91	41.67	0.477	0.451
64	10.35	2.77	11.08	0.320	0.312	124	22.09	10.72	42.87	0.480	0.454
64.5	10.53	2.64	11.89	0.326	0.317	124.5	22.23	9.78	43.99	0.483	0.456
65	10.70	2.53	12.67	0.331	0.322	125	22.35	9.01	45.05	0.486	0.459
65.5	10.86	2.44	13.42	0.336	0.327	125.5	22.47	8.37	46.05	0.488	0.461
66	10.99	2.36	14.16	0.340	0.331	126	22.49	7.83	47.01	0.489	0.461
66.5	11.11	2.29	14.88	0.344	0.334	126.5	22.68	7.37	47.93	0.493	0.465
67	11.23	2.23	15.58	0.348	0.338	127	22.77	6.97	48.82	0.495	0.467
67.5	11.35	2.17	16.27	0.351	0.341	127.5	22.87	6.62	49.68	0.497	0.469
68	11.47	2.12	16.95	0.355	0.345	128	22.96	6.31	50.52	0.499	0.470
68.5	11.57	2.07	17.61	0.358	0.348	128.5	23.03	6.04	51.33	0.501	0.472
69	11.68	2.03	18.27	0.362	0.351	129	23.12	5.79	52.13	0.503	0.474
69.5	11.78	1.99	18.92	0.365	0.354	129.5	23.19	5.57	52.91	0.504	0.475
70	11.88	1.96	19.56	0.368	0.357	130	23.28	5.37	53.67	0.506	0.477
71	12.07	1.89	20.82	0.374	0.362	131	23.42	5.01	55.16	0.509	0.479
72	12.24	1.84	22.05	0.379	0.367	132	23.58	4.72	56.60	0.513	0.482
73	12.41	1.79	23.27	0.384	0.372	133	23.58	4.46	58.01	0.513	0.482
74	12.55	1.75	24.46	0.389	0.376	134	23.68	4.24	59.38	0.515	0.484
75	13.13	1.71	25.65	0.407	0.393	135	23.78	4.05	60.72	0.517	0.486
76	12.86	1.68	26.81	0.398	0.385	136	23.96	3.88	62.04	0.521	0.490
77	13.00	1.65	27.97	0.402	0.389	137	24.11	3.73	63.34	0.524	0.493
78	13.13	1.62	29.12	0.407	0.393	138	24.23	3.59	64.61	0.527	0.495
79	13.26	1.59	30.25	0.411	0.397	139	24.35	3.47	65.87	0.529	0.497
80	13.39	1.57	31.38	0.415	0.401	140	24.46	3.36	67.12	0.532	0.499
82	13.62	1.53	33.61	0.422	0.407	142	24.69	3.16	69.57	0.537	0.504
84	13.84	1.49	35.82	0.428	0.414	144	24.91	3.00	71.97	0.542	0.508
86	14.02	1.46	38.01	0.434	0.419	146	25.11	2.86	74.34	0.546	0.512
88	14.22	1.43	40.18	0.440	0.425	148	25.32	2.74	76.67	0.550	0.516
90	14.41	1.41	42.33	0.446	0.430	150	25.51	2.63	78.97	0.555	0.519

Source: From Birsoy and Summers, 1980. Reprinted with permission of *Ground Water*. Copyright ©1980. All rights reserved.

The theoretical drawdown (s_a) can be calculated from known aquifer parameters obtained from other sources. Once s_a is known, the well efficiency can be calculated from s_a and drawdown in the pumping well.

Good practice in well construction is to provide the maximum efficiency. A well that is efficient can be pumped at a greater rate than a comparable inefficient well. In effect, the productivity of the well depends on minimizing head losses through the gravel pack and the screen. Using "best" well construction and development techniques, we find that a well efficiency of 80% is the maximum possible, and a well efficiency of 60% may be achievable under unfavorable conditions for most screened wells (Heath, 1983). Use of a proper well screen with a large open area for flow and appropriate slot sizes, as well as complete development of the gravel pack (removal of fines), are key.

Step drawdown tests are used in the field to estimate well efficiency. The theoretical drawdown (s_a) is linearly proportional to the pumping rate. The total drawdown in a pumped

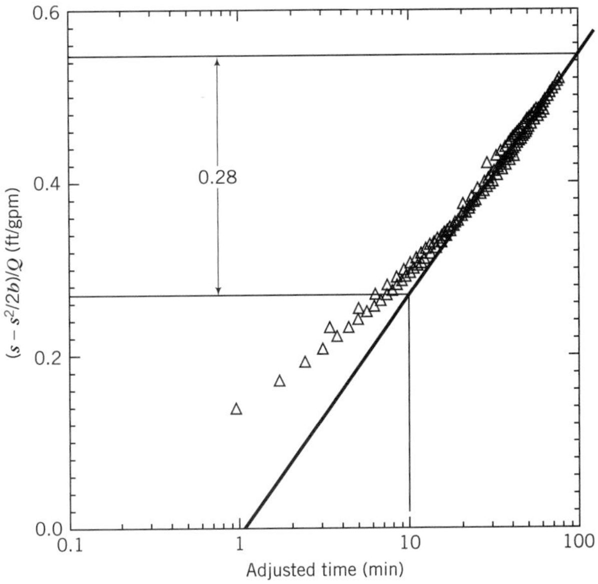

Figure 12.12 Plotting corrected drawdown versus the log of adjusted time to analyze data from a step drawdown test (from Birsoy and Summers, 1980). Reprinted by permission of *Ground Water*. Copyright ©1980. All rights reserved.

well is the summation of head loss in the pumped well and head loss in the aquifer:

$$s_t = BQ + f(Q) \tag{12.20}$$

where B is a function of time for a given aquifer and pumping conditions. The first term BQ represents the drawdown s_a. Different forms of $f(Q)$ were proposed to account for the head loss in the pumped well (Jacob, 1947; Lennox, 1966; Rorabaugh, 1953). According to Jacob (1947), the total head loss (s_t) is

$$s_t = BQ + CQ^2 \tag{12.21}$$

where C is a constant. The step drawdown test is commonly applied to derive constants B and C. Once B and C are obtained, well efficiency may be calculated from such an estimate. The estimated well efficiency is called apparent well efficiency and is defined as

$$E' = \frac{BQ}{BQ + CQ^2} \tag{12.22}$$

For convenience, Equation (12.21) may be rewritten as

$$\frac{s_t}{Q} = B + CQ \tag{12.23}$$

The procedures for determining B and C are summarized as follows:

1. Plot s_t/Q versus Q at the same adjusted time $\beta_n(t)(t - \tau_n)$ in linear scales on graph paper.

2. Approximate the plot by a straight line. B is the intercept at where $Q = 0$ and C is the slope of the plot.

▶ **EXAMPLE 12.5**   Calculate B, C, and apparent well efficiency using the data in Example 12.4.

SOLUTION The pumping rates and drawdowns at adjusted time $\beta_n(t)(t - \tau_n) = 31.38$ min are listed in Table 12.4 and plotted in Figure 12.13. The drawdown in test step 5 is calculated through linear interpolation. The data at step 3 deviates from the straight line. Only data in steps 3, 4, and 5 are utilized. From the graph, we get

$$B = 0.356 \text{ ft/gpm}$$

and

$$C = \frac{0.029 \text{ ft/gpm}}{15 \text{ gpm}} = 0.002 \text{ ft/(gpm)}^2$$

and

$$E' = \frac{BQ}{BQ + CQ^2} = \frac{0.356 \text{ ft/gpm}}{0.356 \text{ ft/gpm} + \left(0.002 \dfrac{\text{ft}}{\text{gpm}^2}\right)(46 \text{ gpm})} = 79\%$$

▶ EXERCISES

12.1. A slug test was carried in a 6-in. well by providing an instantaneous injection of 5.2 ft^3 slug of water. The residual drawdown with time is listed in Table 12.5. Assume that the aquifer is an unconfined aquifer. Calculate hydraulic conductivity using the Hvorslev technique.

12.2. Table 12.6 provides slug-test results in a confined aquifer. The radius of the borehole is 7.6 cm. Calculate the transmissivity of the aquifer using the Cooper–Bredehoeft–Papadopulos slug-test technique. Two template files are located in the directory/chap 12 on the website. The file temp-12.doc

TABLE 12.4 Pumping rates and drawdowns at corrected time $\beta_n(t)(t - \tau_n) = 31.38$ min

Q(gpm)	$[s - s^2/(2b)]/Q$(ft/gpm)
14.7	0.405
32.3	0.401
38.5	0.410
46.0	0.426

Source: From Birsoy and Summers, 1980. Reprinted with permission of *Ground Water*. Copyright ©1980. All rights reserved.

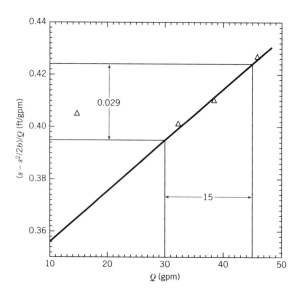

Figure 12.13 Plotting measurement data needed to calculate the apparent well efficiency using step drawdown tests (from Birsoy and Summers, 1980). Reprinted by permission of *Ground Water*. Copyright ©1980. All rights reserved.

TABLE 12.5   Data for a slug injection test at Speedway City, Indiana

Time(min)	Residual Head (ft)	Time(min)	Residual Head (ft)
1.25	0.26	3.87	0.08
1.33	0.25	4.10	0.08
1.50	0.20	4.33	0.08
1.92	0.17	4.52	0.08
2.17	0.16	4.58	0.07
2.30	0.15	4.72	0.07
2.37	0.14	5.17	0.07
2.42	0.14	5.28	0.06
2.67	0.12	5.45	0.06
2.72	0.12	6.10	0.06
2.77	0.12	6.40	0.05
2.92	0.11	6.83	0.05
3.00	0.11	7.17	0.05
3.22	0.10	7.75	0.04
3.28	0.10	8.58	0.04
3.33	0.10	9.37	0.04
3.40	0.09	10.12	0.03
3.47	0.09	11.00	0.03
3.55	0.09	12.5	0.03
3.67	0.09	13.0	0.03
3.77	0.09		

Source: From Ferris and Knowles (1963).

TABLE 12.6   Slug-test results in a confined aquifer

Time(min)	H/H_0	Time(min)	H/H_0
0.010	0.985	0.356	0.803
0.013	0.983	0.452	0.765
0.016	0.980	0.574	0.721
0.020	0.977	0.728	0.671
0.026	0.973	0.924	0.614
0.033	0.967	1.172	0.552
0.042	0.962	1.487	0.485
0.053	0.954	1.887	0.415
0.067	0.945	2.395	0.344
0.085	0.934	3.039	0.275
0.108	0.920	3.857	0.211
0.137	0.905	4.894	0.156
0.174	0.885	6.210	0.111
0.221	0.862	7.880	0.076
0.281	0.835	10.000	0.051

is the type curve for the Cooper–Bredehoeft–Papadopulos slug-test technique. The file temp-12a.doc provides graph paper at the same scale for plotting values of head ratio versus time.

12.3. Calculate hydraulic parameters for slug tests in Tables 12.3 and 12.4 using the Bouwer and Rice technique.

12.4. Table 12.7 gives values of time and drawdown in a pumped well for a step-drawdown test reported by Clark (1977). Calculate the transmissivity and storativity of the aquifer. It is assumed that the effective well radius is 0.3 m for the test.

TABLE 12.7   Values of time and drawdown from a step-drawdown test

Step 1: $Q = 1306$ m³/day		Step 2: $Q = 1693$ m³/day		Step 3: $Q = 2423$ m³/day	
Time (min)	Drawdown (m)	Time (min)	Drawdown (m)	Time (min)	Drawdown (m)
10	3.521	190	5.740	370	8.672
12	3.592	192	5.810	372	8.663
14	3.627	194	5.810	374	8.698
16	3.733	196	5.824	376	8.733
18	3.768	198	5.845	378	8.839
20	3.836	200	5.810	380	8.874
25	3.873	205	5.824	385	8.874
30	4.013	210	5.824	390	8.979
35	3.803	215	5.881	395	8.979
40	4.043	230	6.092	400	8.994
60	4.120	235	6.092	405	9.050
70	4.120	240	6.176	410	9.050
80	4.226	250	6.162	415	9.120
90	4.226	260	6.176	420	9.120
100	4.226	270	6.169	430	9.155
120	4.402	280	6.169	440	9.191
150	4.402	330	6.374	450	9.191
180	4.683	360	6.514	460	9.226
				480	9.261
				510	9.367
				540	9.587

Step 4: $Q = 3261$ m³/day		Step 5: $Q = 4094$ m³/day		Step 6: $Q = 5016$ m³/day	
Time (min)	Drawdown (m)	Time (min)	Drawdown (m)	Time (min)	Drawdown (m)
550	12.325	730	16.093	910	20.917
552	12.360	732	16.198	912	20.952
554	12.395	734	16.268	914	21.022
556	12.430	736	16.304	916	21.128
558	12.430	738	16.374	918	21.163
560	12.501	740	16.409	920	21.198
565	12.508	745	16.586	925	21.304
570	12.606	750	16.621	930	21.375
575	12.712	755	16.691	935	21.480
580	12.747	760	16.726	940	21.551
585	12.783	765	16.776	945	21.619
590	12.818	770	16.797	950	21.656
595	12.853	775	16.938	960	21.663
600	12.853	780	16.973	970	21.691
610	12.888	790	17.079	980	21.762
620	12.923	800	17.079	990	21.832
630	12.994	810	17.114	1000	21.903
640	12.994	820	17.219	1020	22.008
660	13.099	840	17.325	1050	22.184
690	13.205	870	17.395	1080	22.325
720	13.240	900			

Source: From Clark, 1977. *Quart J. Eng. Geol.*, v. 10, no. 2, p. 125–143.

12.5. Calculate the apparent well efficiency using the data in Table 12.7.

▶ CHAPTER
HIGHLIGHTS

1. The slug test is a commonly used approach for measuring the hydraulic conductivity of a geological unit. In a single well or piezometer, water is displaced from an equilibrium level by adding water or withdrawing a slug. The recovery of the water level back to equilibrium is monitored as a function of time. The rate of recovery depends on the well design and the hydraulic conductivity of the unit. Given the construction details of the well, several different mathematical approaches are available for calculating hydraulic conductivity.

2. In general, aquifer tests provide more representative estimates of aquifer parameters than slug tests. Because so little water is actually displaced in a slug test, the test probes only a small volume of the aquifer next to the well. The main attraction of slug tests is their low cost, compared to a full-blown aquifer test with observation wells. Thus, in most studies, hydrologists will have data from a few aquifer tests and many slug tests.

3. The Hvorslev (1951) theory provides a straightforward approach to interpreting data from a slug test. The approach involves calculating the drawdown ratio ($H_{1,2,3,...}$) from the drawdown observations ($s_{1,2,3,...}$), where, for example, $H_1 = s_1/s_0$ and s_0 is the maximum drawdown created when the test began (that is, at time zero). Next, a plot is constructed of $\log H$ versus t. Two points from the best fit line provide the necessary pair of $H - t$ values that are substituted into the working equation (12.1).

4. The Cooper–Bredehoeft–Papodopulos test is a more sophisticated approach for testing single wells. Unlike the Hvorslev approach, it provides estimates of both transmissivity and storativity. This approach involves curve matching, where field data plotted as head ratio versus the log of time are superimposed on a graph of type curves to provide α, β, and t values. These parameters from the curve-matching process are substituted into two simple equations to provide the required T and S values.

5. The Bouwer and Rice (1976) approach is commonly used to interpret slug test data. In principle, the approach is similar to the Hvorslev's method but involves using a set of curves to determine the radius of influence. The governing equation requires parameters that are extracted from a plot of the log water level versus time, details of the well construction, and the radius of influence R_e.

6. As a well is pumped, water moves from the aquifer through the gravel pack and the screen and finally into the casing. Usually, the large volume of water moving toward the well has difficulties in moving through the gravel pack and screen, creating a substantial drop in hydraulic head between the aquifer and the inside of the well. The well efficiency (E) is defined as the ratio of the drawdown (s_a) in the aquifer at the radius just outside of the pumped well to the drawdown (s_t) inside the well, as a percentage.

7. Good practice in well construction is to provide the maximum efficiency. Efficient wells make it possible to use larger pumping rates. Well efficiency is promoted by careful development of the gravel pack around the screen to remove fine-grained materials and by using a screen with a large open area for flow. To keep the water sand-free, the slots in the screen must be properly matched with the gravel pack and formation.

13

SUPERPOSITION AND BOUNDED AQUIFERS

Most well-hydraulics problems are not nearly as simple as the examples presented so far. Complexities arise because more than one well can be pumping at the same time or because aquifers are not infinite in extent. This chapter provides the theoretical basis for accommodating these complexities within the framework of traditional analytical solutions for well-hydraulics problems.

▶ 13.1 MULTIPLE WELLS AND SUPERPOSITION

As a well is pumped, a cone of depression forms and grows with time. When pumping wells are spaced so closely that their cones of depression overlap, they interfere with each other. *Interference* means that there is more drawdown than expected in each of the pumping wells because the water-level decline in a pumping well is due not only to its actual pumping but also to drawdown caused by nearby wells. Heath (1983) illustrates how the cones of depression due to two pumping wells (Figure 13.1*a*) combine to produce a zone of enhanced drawdown (Figure 13.1*b*).

It is simple to analyze this multi-well problem, at least for confined aquifers, using the *principle of superposition*. This principle states that the total drawdown at any location of interest is the sum of the drawdowns due to each well by itself. In practice, you first calculate the drawdown (s_1) at the point of interest (for example, an observation well) due to the first well a distance r_1 away. Next, calculate s_2 for the second well at a distance r_2. The total drawdown at the observation well is the sum of $s_1 + s_2$. The principle of superposition can be applied for confined aquifers because the parameters T and S remain constant. In effect, the differential equations are linear for this case, making the solutions additive. In the case of unconfined aquifers, where T is a function of drawdown, the equations are nonlinear and this approach doesn't hold.

For some problems, both pumping and injection wells may be operating together in the same confined aquifer. The principle of superposition also holds in this case. The total drawdown is the algebraic sum of the drawdowns due to the pumping wells ($+s$) and the buildup due to the injection wells ($-s$). Here is an example that illustrates the principle of superposition for a problem involving a pumping well and an injection well.

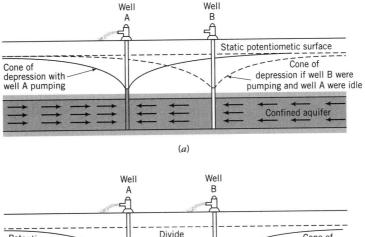

Figure 13.1 Panel (*a*) shows the cones of depression at wells *A* and *B* assuming that one of the wells is idle. Panel (*b*) shows the resulting cone of depression with both wells pumping together. The interference causes significant lowering and the formation of a divide between the two wells (from Heath, 1989).

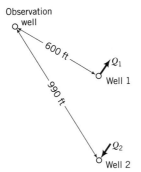

Figure 13.2 Map showing the location of wells in Example 13.1.

▶ **EXAMPLE 13.1**

Figure 13.2 shows the location of a withdrawal well and an injection well in a confined aquifer having a transmissivity of 130 ft²/day and a storativity of 5×10^{-4}. Water is pumped from the withdrawal well (1) at 19,000 ft³/day. Water is injected at 8000 ft³/day at the injection well (2). Calculate the drawdown at the observation well shown in Figure 13.2 after one year of pumping and injection.

SOLUTION Calculate the drawdown at the observation well due to pumping at well 1 and injection at well 2 separately using the Cooper–Jacob approximation.

$$s_1 = \frac{2.3\,Q_1}{4\pi T} \log \frac{2.25\,Tt}{r_1^2 S} = \frac{2.3 \times 19{,}000\,\text{ft}^3/\text{d}}{4 \times 3.14 \times 1500\,\text{ft}^2/\text{d}} \log \frac{2.25 \times 1500\,\text{ft}^2/\text{d} \times 365\,\text{days}}{(600\,\text{ft})^2 \times 0.0005} = 8.9\,\text{ft}$$

$$s_2 = \frac{2.3\,Q_2}{4\pi T} \log \frac{2.25\,Tt}{r_2^2 S} = \frac{2.3 \times (-8000)\,\text{ft}^3/\text{d}}{4 \times 3.14 \times 1500\,\text{ft}^2/\text{d}} \log \frac{2.25 \times 1500\,\text{ft}^2/\text{d} \times 365\,\text{days}}{(990\,\text{ft})^2 \times 0.0005} = -3.3\,\text{ft}$$

Using the principle of superposition, we find that the drawdown at the observation well is the sum of the drawdown due to well 1 and the buildup due to well 2.

$$s_{tot} = s_1 + s_2 = 8.9\,\text{ft} + (-3.3\,\text{ft}) = 5.6\,\text{ft}$$

▶ 13.2 DRAWDOWN SUPERIMPOSED ON A UNIFORM FLOW FIELD

In all of the well-hydraulics problems that we have considered so far, the initial water level in the aquifer is assumed to be a constant. However, for most aquifers, there is a gradient in the hydraulic head because the ground water is flowing. The principle of superposition can be used to determine what the resultant head distribution looks like with two flow elements—the uniform hydraulic-head distribution and the steady-state cone of depression due to a pumping well. Let's illustrate this process conceptually. We start with a uniform flow field that has a regular distribution of equipotential lines (Figure 13.3a). A well, shown on the figure, is pumped for a long time providing a cone of depression (Figure 13.3b). The resulting flow field (Figure 13.3c) is determined by subtracting the drawdown from the original head field.

Though straightforward in principle, this procedure is tedious to do by hand because the drawdown due to the well needs to be calculated at a large number of points. We will therefore consider a steady-state mathematical approach for calculating the effects of pumping superimposed on a uniform flow field. Assume that the regional Darcy velocity is q_0 in the x-direction, that the hydraulic conductivity of the aquifer is K, and that the pumping rate of the well is Q. If the pumped well with a zero initial hydraulic head is located at $x = 0$, the total hydraulic head as shown in Bear (1979) and Javandel et al. (1984) is:

$$h = -\frac{q_0}{K}x + \frac{Q}{2\pi Kb}\ln\frac{r}{r_w} \tag{13.1}$$

where b is the thickness of the aquifer, q_0 is the regional Darcy velocity, and r_w is the radius of the pumped well. The distance from the pumped well to any location in the domain is expressed as

$$r = \sqrt{(x - x_w)^2 + (y - y_w)^2} \tag{13.2}$$

where x and y are coordinates. Looking at Eq. (13.1), we see how the superposition is accomplished. The first term on the RHS defines the hydraulic head as a function of distance, x. The second term describes the steady-state drawdown due to pumping.

FLOWNETz Revisited

In Chapter 8, we introduced the computer program, **FLOWNETz**, to illustrate flownets related to regional flow. This code also provides a capability for calculating hydraulic heads and stream functions for multiple wells in a uniform flow field. The program is located in the directory PROGRAMS/FLOWNETz on the website. The installation instructions are also located in the directory.

We implemented several analytical solutions in **FLOWNETz**. Here we focus on the problem of multiple wells in a uniform flow field, as expressed by Eqs. (13.1) and (13.2). The code is more general than the theory above in that it allows the flow gradient to be at any angle to the x-axis and mathematically calculates streamlines. The steps in using **FLOWNETz** are similar to those presented earlier.

1. Click the **FLOWNETz** on the **Start/Programs** menu to launch **FLOWNETz**.

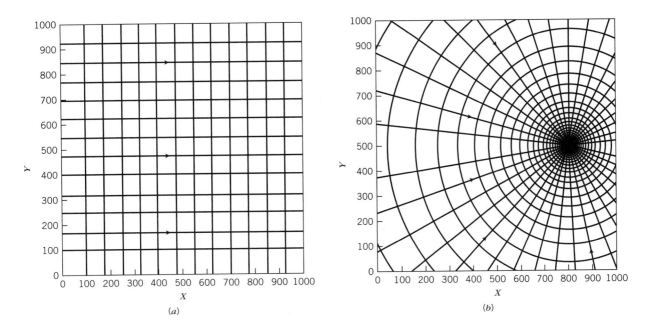

(a)

(b)

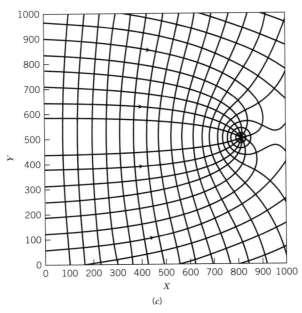

(c)

Figure 13.3 This figure illustrates the principle of superposition that applies to a uniform flow field (Panel *a*) and the steady-state cone of depression due to a pumping well (Panel *b*). The resulting flow field (Panel *c*) shows stream tubes being captured by the well and others bypassing the well.

2. The main menu of the **FLOWNETz** is **Flownet** (Figure 13.4). There are four commands under the **Flownet** menu: **Options, Input, Calculate,** and **Report**. The **Options** command launches a dialog box for users to select the analytical solution for the calculation. Let's choose the option to calculate equipotential lines and flow lines for multiple wells in a uniform flow field. The **Input** command

specifies the Darcy velocity of the uniform flow, the thickness of the aquifer, the angle between the flow direction and the x-axis, and the domain size (Figure 13.5). Well information can be assigned by clicking the **Wells specification** button. A spreadsheet will prompt users to enter the number and, locations of wells, as

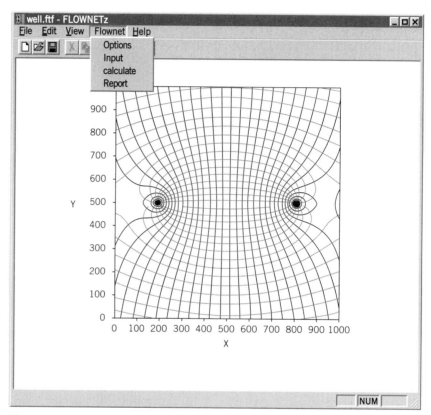

Figure 13.4 Main menu for **FLOWNETz** showing the operations available on the Flownet drop-down menu. The figure shows simulation results for an injection/withdrawal doublet.

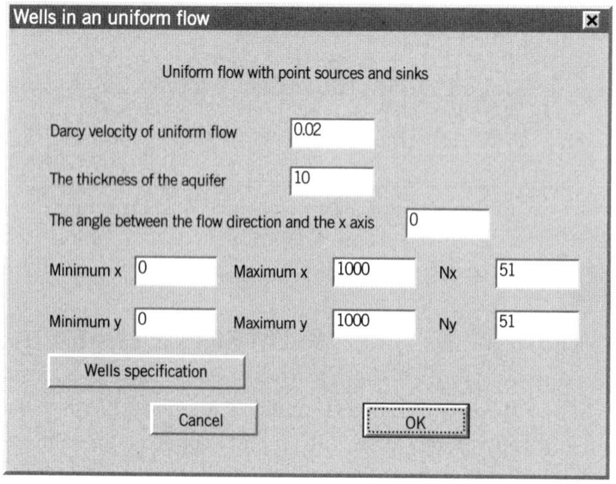

Figure 13.5 Input dialog box for a problem of uniform flow with point sources and sinks.

well as the injection rates of the wells (Figure 13.6). Positive rates are used for injection wells, negative rates for withdrawal wells. Ground water flows in the positive *x*-direction. The **Calculate** command calculates the hydraulic head and stream functions and displays the results on the screen (Figure 13.4). The **Report** command tabulates the input data for the simulation and the results, either in **Notepad** or **Wordpad**.

3. The **View** menu contains **Scale, Display**, and **Contours**. The **Scale** command specifies the vertical scale of the display (Figure 13.7). The **Display** command gives users the options of whether to display axis labels, contours of hydraulic head, and stream functions (Figure 13.8). The **Contours** command specifies the contour parameters of the hydraulic head and stream functions (Figure 13.9).

4. The **FLOWNETz** documents can be opened, saved, and printed using the **Open, Save, Save As, Print, Print Preview**, and **Print Setup** commands in the file menu. Here is an example of the application of **FLOWNETz** to a problem of ambient flow modified by three pumping wells and one injection well.

► **EXAMPLE 13.2**   Four wells, located at points A(200, 200), B(200, 800), C(800, 200), and D(800, 800), are pumped at rates of $-200, 200, -200$, and 200 m^3/days, respectively. The thickness of the aquifer is 10 m. The Darcy velocity for the regional flow field is 0.02 m/day. The flow field is at a 45-degree angle to the *x*-axis. Calculate equipotential and streamlines.

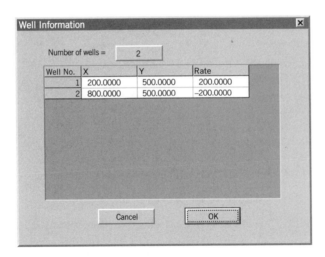

Figure 13.6 Spreadsheet for providing information on the location of individual wells and pumping rates.

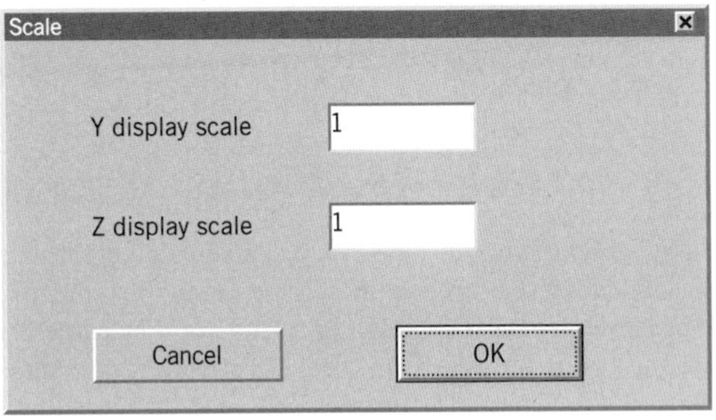

Figure 13.7 Input dialog box for setting the scales for the map of equipotential lines and streamlines.

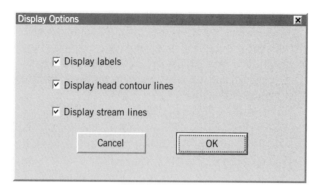

Figure 13.8 Dialog box to control what information is to be plotted on the map.

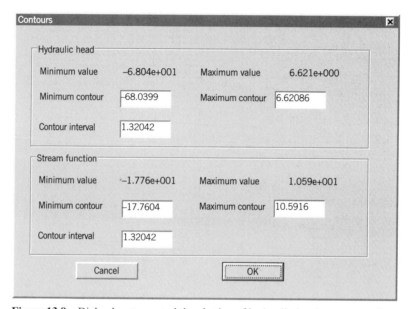

Figure 13.9 Dialog box to control the plotting of hydraulic-head contours and streamlines.

SOLUTION The thickness of the aquifer, the Darcy velocity of the regional flow field, and the angle between the flow field and x-axis are specified in the dialog box (for example, Figure 13.5). Well information is specified in the spreadsheet similar to that in Figure 13.6. The calculated equi-potential and streamlines are shown in Figure 13.10. Well B is the only injection well. Wells A, C, and D are withdrawal wells. Well D collects most of the water injected by well B, whereas wells A and C collect water from the regional flow system. The **Report** command outputs the calculation input and results as shown in Table 13.1.

▶ 13.3 REPLACING A GEOLOGIC BOUNDARY WITH AN IMAGE WELL

The hydraulic theory introduced so far has assumed that the aquifer of interest is infinite in lateral extent. However, this condition is the exception rather than the rule. Aquifers are not infinite because they can be cut by tight faults or they can end abruptly due to changes in geology. These impermeable boundaries effectively halt the spread of the cone of depression and influence the pattern of drawdown related to the well. Similarly, a surface-water body

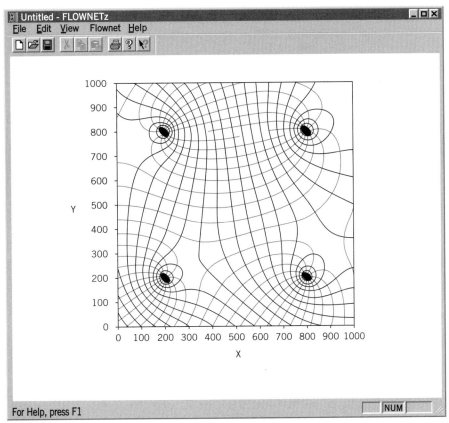

Figure 13.10 Plot of equi-potential lines and streamlines for the four-well problem in Example 13.2.

(for example, a fully penetrating stream or lake) or an adjacent segment of aquifer, having a significantly higher transmissivity or storativity, can halt the spread of a cone of depression by providing a source of recharge to the aquifer. Such recharge boundaries also can influence the pattern of drawdown in the vicinity of the pumping wells.

In general, well-hydraulics theory cannot cope with the presence of one of these aquifer boundaries. It is possible, however, to get rid of the boundary by adding an imaginary well known as an *image well* (Ferris et al., 1962). The following sections illustrate how image-well theory helps both in calculating drawdowns in aquifers containing flow boundaries and in using the results of aquifer tests to estimate the location of boundaries.

Impermeable Boundary

Let's begin by looking at an aquifer with a geologic boundary that forms an impermeable boundary (Figure 13.11a). Figures 13.11b and c illustrate how to replace this boundary by adding an image well. This well is placed across the boundary at the same distance from the boundary as the original well (Figure 13.11c). Adding the pumping well makes the impermeable boundary disappear. Drawdowns in the aquifer can then be calculated using the principle of superposition with two pumping wells. Mathematically, the drawdown due to a single pumping well in a confined aquifer with an impermeable boundary is

$$s = s_r + s_i = \frac{Q}{4\pi T}[W(u_r) + W(u_i)] \tag{13.3}$$

TABLE 13.1   Output from FLOWNETz

Report for the calculation of streamlines and equipotentials for wells in uniform flow

No. of intervals in X: 51
No. of intervals in Y: 51
Minimum X :   0.0 m
Maximum X : 1000 m
Minimum Y :   0.0 m
Maximum Y : 1000 m
Thickness of aquifer: 10.00 m
Number of pumping wells: 4
Pumping rate in cubic meters per day

X	Y	Pumping Rate
200.000	200.000	−200.0
800.000	200.000	−200.0
200.000	800.000	200.0
800.000	800.000	−200.0

X, Y, Hydraulic head

0.000e + 000	0.000e + 000	4.035e + 001
2.000e + 001	0.000e + 000	3.333e + 001
4.000e + 001	0.000e + 000	3.373e + 001
6.000e + 001	0.000e + 000	3.419e + 001

......

X, Y, Stream function value

0.000e + 000	0.000e + 000	−1.844e + 001
2.000e + 001	0.000e + 000	1.108e + 000
4.000e + 001	0.000e + 000	1.871e + 000
6.000e + 001	0.000e + 000	2.545e + 000
8.000e + 001	0.000e + 000	3.065e + 000
1.000e + 002	0.000e + 000	3.318e + 000
1.200e + 002	0.000e + 000	3.371e + 000
1.400e + 002	0.000e + 000	2.919e + 000
1.600e + 002	0.000e + 000	2.471e + 000
1.800e + 002	0.000e + 000	2.477e + 000
2.000e + 002	0.000e + 000	2.874e + 000

where

$$u_r = \frac{r_r^2 S}{4Tt} \tag{13.4}$$

and

$$u_i = \frac{r_i^2 S}{4Tt} \tag{13.5}$$

where r_r and r_i are the distances from the pumped and image wells to an observation well, respectively. For smaller u_r and u_i, the equation can be approximated by

$$s = \frac{2.30Q}{4\pi T}\left[\log\left(\frac{2.25Tt}{r_r^2 S}\right) + \log\left(\frac{2.25Tt}{r_i^2 S}\right)\right] \tag{13.6}$$

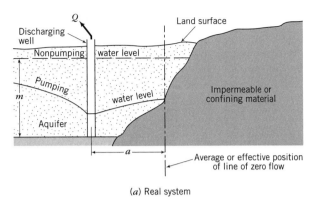

(a) Real system

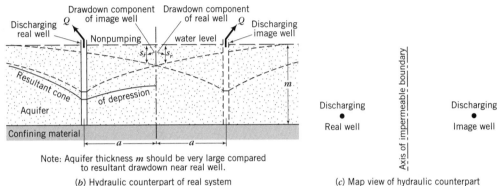

Note: Aquifer thickness m should be very large compared to resultant drawdown near real well.

(b) Hydraulic counterpart of real system

(c) Map view of hydraulic counterpart

Figure 13.11 A cross section illustrating an aquifer with an impermeable boundary. Panel (a) shows the real system; (b) the hydraulic counterpart of the real system (from Ferris et al., 1962); and (c) a map view of the hydraulic counterpart.

In an aquifer with an impermeable boundary, aquifer test data can be interpreted to provide not only hydraulic parameters but also the location of the boundary. Let us consider the simple case of a single straight-line impermeable boundary with a single pumping well. Assume that the observation well where the data have been collected is much closer to the real pumping well than the image well. When drawdown at the observation well is plotted versus the log of time, there should be two distinct slopes. The slope at early times corresponds to the drawdown change caused by the pumped well only. The slope at later times is different because drawdown is due to both the real well and the image well. Drawdown change affected by the combination of the pumped and image wells in one log cycle of time will be twice that of the pumped well alone (Figure 13.12). Following the Cooper–Jacob method, a straight line fitted to the early-time data will yield the aquifer parameters.

Next is an approach to determine the location of the boundary. From Eqs. (13.4) and (13.5), we know that u_r and u_i must be equal in order to generate the same drawdown in an observation well for the pumped well and the image well, respectively. Thus,

$$\frac{r_i^2}{t_i} = \frac{r_r^2}{t_r} \tag{13.7}$$

where t_r and t_i are the times for the pumped wells and the image wells to generate the same amount of drawdowns in an observation well. Equation (13.7) may be rewritten as

$$r_i = r_r \sqrt{\frac{t_i}{t_r}} \tag{13.8}$$

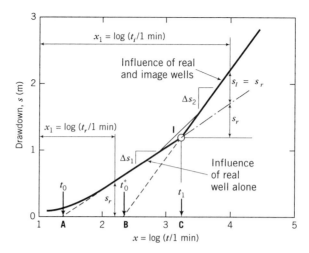

Figure 13.12 Plot of drawdown versus time on a semi-log scale for hydraulic testing near an impermeable boundary in a confined aquifer (from Chapuis, 1994). Reprinted by permission of *Ground Water*. Copyright ©1994. All rights reserved.

Knowing the radial distance from the observation well to the image well begins to define where the boundary is located—except that several observation wells are required. Here are the steps.

1. Install at least four observation wells near the pumped well for the aquifer test (Figure 13.13).

2. Plot the observed drawdown from each of the four wells on a linear scale versus the log of time (Figure 13.14).

3. Determine the aquifer parameters using the Cooper-Jacob straight-line method. Use the early-time drawdown data when the boundary effect is negligible.

4. Calculate and plot the theoretical Theis drawdown for each observation well using the derived aquifer parameters or simply extending the early-time straight line out to the last measurement time (Figure 13.14).

5. Select measured times (t_{i1}, t_{i2}, t_{i3}, and t_{i4}) when the boundary effect is significant at each observation well (Figure 13.14).

6. Calculate the drawdown contribution from the image well only (s_{i1}, s_{i2}, s_{i3}, and s_{i4}) by subtracting the calculated Theis drawdown from the measured drawdown at the selected times (t_{i1}, t_{i2}, t_{i3}, and t_{i4}) (Figure 13.14).

7. Record the times (t_{r1}, t_{r2}, t_{r3}, and t_{r4}) when the Theis drawdown is the same as the drawdown contribution by the image well at each observation well (Figure 13.14).

8. Calculate the distances (r_{i1}, r_{i2}, r_{i3} and r_{i4}) from the image well to all of the observation wells using Eq. (13.8).

9. Draw arcs centered at observation wells with radii of r_{i1}, r_{i2}, r_{i3}, and r_{i4}. The interception of the arcs is the location of the image well. The impervious boundary is located halfway between the image well and the real well (Figure 13.8).

▶ **EXAMPLE 13.3**   The arrangement of wells is shown in Figure 13.15 for a hypothetical aquifer test affected by an impermeable boundary. The pumping rate in the well is 7.52 m³/min. The drawdowns measured at four observation wells are listed in Table 13.2. Calculate the aquifer properties and locate the impermeable boundary.

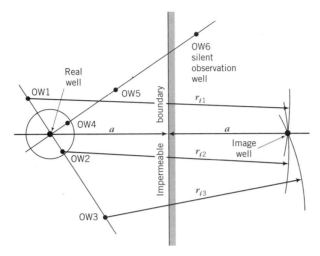

Figure 13.13 Determining the location of an impermeable boundary with at least three observation wells (from Chapuis, 1994). Reprinted by permission of *Ground Water*. Copyright ©1994. All rights reserved.

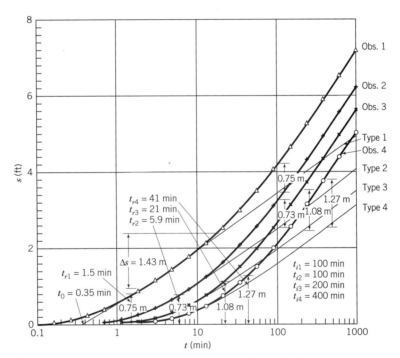

Figure 13.14 An example of drawdown data and Theis fits for observation wells placed near an impervious boundary in a confined aquifer.

SOLUTION Plots of drawdown versus time plots are shown in Figure 13.14 for the four observation wells. Use the early part of the drawdown plot for observation well 1 to determine the aquifer parameters. We recorded $\Delta s = 1.43$ m and $t_0 = 0.35$ min. The transmissivity is obtained by

$$T = \frac{2.3Q}{4\pi \, \Delta s} = \frac{2.3(7.52 \, \text{m}^3/\text{min})}{4\pi(1.43\text{m})} = 0.963 \, \text{m}^2/\text{min} = 1.4 \times 10^3 \, \text{m}^2/\text{day}$$

The storativity is

$$S = \frac{2.25(0.963 \, \text{m}^2/\text{min})(0.35 \, \text{min})}{(50 \, \text{m})^2} = 3.0 \times 10^{-4}$$

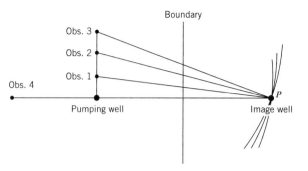

Figure 13.15 Determination of the distances from each of the observation wells to the pumped well provides a basis for locating the image well and the position of the impermeable boundary.

TABLE 13.2 Drawdown measurements in four hypothetical observation wells

Time(min)	s_1(m)	s_2(m)	s_3(m)	s_4(m)
0.10	0.00	0.00	0.00	0.00
0.16	0.03	0.00	0.00	0.00
0.26	0.09	0.00	0.00	0.00
0.43	0.20	0.01	0.00	0.00
0.70	0.37	0.03	0.00	0.00
1.13	0.59	0.10	0.01	0.00
1.83	0.85	0.22	0.05	0.01
2.98	1.14	0.40	0.13	0.04
4.83	1.44	0.63	0.28	0.11
7.85	1.76	0.90	0.48	0.25
12.74	2.11	1.22	0.74	0.44
20.69	2.52	1.60	1.08	0.71
33.60	2.98	2.04	1.50	1.07
54.56	3.50	2.55	1.99	1.50
88.59	4.07	3.11	2.53	2.00
143.84	4.67	3.70	3.11	2.56
233.57	5.29	4.32	3.73	3.15
379.27	5.93	4.95	4.36	3.77
615.85	6.57	5.60	5.00	4.41
1000.00	7.23	6.25	5.65	5.05

Source: Clark, 1977

Theis drawdowns are calculated for each observation well using derived T and S (Figure 13.14). The calculated drawdown is the contribution from the pumped well, assuming the aquifer to be infinite. The drawdown contributions from the image well ($s_{i1} = 0.75$ m, $s_{i2} = 0.73$ m, $s_{i3} = 1.08$ m and $s_{i4} = 1.28$ m) at selected times ($t_{i1} = 100$ min, $t_{i2} = 100$ min, $t_{i3} = 200$ min, and $t_{i4} = 400$ min) are calculated and shown in Figure 13.14. Times are marked on the semi-log drawdown plots for the same contribution from the real and image wells. The calculation of distances from the image well to the various observation wells is summarized in Table 13.3. Four arcs are drawn, centered at the observation wells. The image well is located at point P in Figure 13.14.

Recharge Boundary

The procedure for handling a recharge boundary (for example, a fully penetrating stream; Figure 13.16a) involves replacing the boundary with one or more wells. The image well is located across the boundary, the same distance away as the pumped well

TABLE 13.3 **Distances from the pumped and image wells to observation wells and times to achieve the same drawdowns**

Observation well	No. 1	No. 2	No. 3	No.4
r_r (m)	50.0	100.0	150.0	200.0
t_i (min)	100.0	100.0	200.0	400.0
t_r (min)	1.5	5.9	21.0	41.0
r_i (m)	408.2	411.7	462.9	624.7

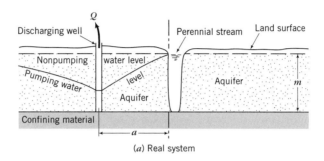

(a) Real system

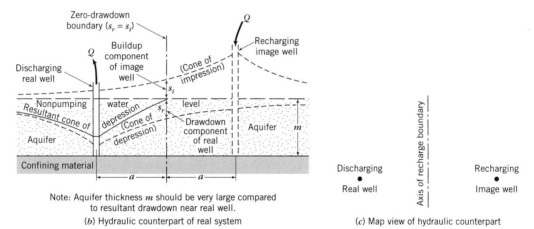

Note: Aquifer thickness m should be very large compared to resultant drawdown near real well.

(b) Hydraulic counterpart of real system

(c) Map view of hydraulic counterpart

Figure 13.16 A cross section illustrating an aquifer with a recharge boundary. Panel (*a*) shows the real system; (*b*) the hydraulic counterpart of the real system (from Ferris et al., 1962); and (*c*) a map view of the hydraulic counterpart.

(Figure 13.16*b, c*). At early time, drawdown in an observation well close to the real well is due only to the real well pumping. Eventually, the total drawdown is the combination of drawdown caused by the pumped well and buildup produced by the image well (Figure 13.17). Along the boundary, the drawdown is zero because the drawdown and the buildup cancel each other. The drawdown in an observation well in this case is expressed as

$$s = s_r - s_i = \frac{Q}{4\pi T}[W(u_r) - W(u_i)] \tag{13.9}$$

As before, time variation in the pattern of drawdown at an observation well can be interpreted to provide hydraulic parameters for the aquifer and the location of the recharge boundary. In calculating aquifer parameters, again use the initial (early-time) slope of the drawdown versus the log of time plot for observation wells. Under the influence of a

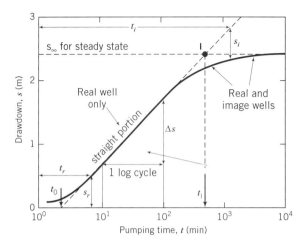

Figure 13.17 A plot of drawdown versus time and the early-time Theis fit for an observation close to a recharge boundary in a confined aquifer (from Chapuis, 1994). Reprinted by permission of *Ground Water*. Copyright ©1994. All rights reserved.

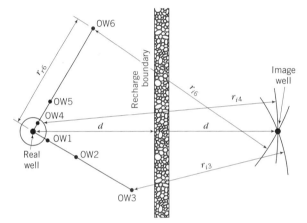

Figure 13.18 Determining the location of a recharge boundary with at least three observation wells (from Chapuis, 1994). Reprinted by permission of *Ground Water*. Copyright ©1994. All rights reserved.

recharge boundary, the buildup of the head attributable to the image well is the difference between the measured drawdown and that due to the calculated Theis drawdown caused by the real well (s_i in Figure 13.17). Equation (13.7) also applies to drawdown produced by real well and buildup due to the image well. In practice, at least four observation wells are necessary to determine the location of the recharge boundary (Figure 13.18).

▷ 13.4 MULTIPLE BOUNDARIES

Image theory is also applicable to the analysis of flow in an aquifer with multiple boundaries. With more than one boundary present, multiple image wells are usually needed to replace the boundaries. In locating the image well, each boundary should be considered to determine the type of image well and where it is located. For example, consider a case where a recharge stream and an impervious boundary intersect at right angles (Figure 13.19a). first, we consider the recharge boundary by adding an injection well (image well 1) and the impermeable boundary by adding a discharge well (image well 2). Although the image well 1 will create zero drawdown along the recharge boundary, it will not provide for no-flow conditions at the impermeable boundary. Another injection well (image well 3) is needed to create no flow at the impermeable boundary. The resultant drawdown will be

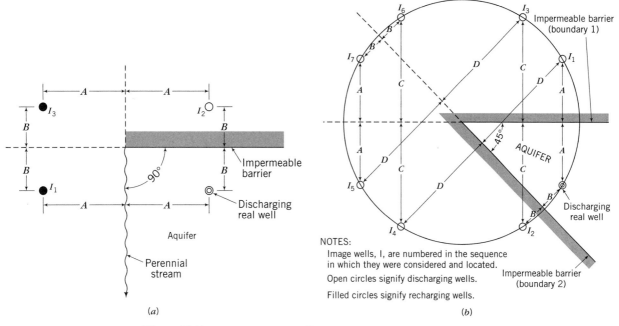

Figure 13.19 The arrangement of image wells for various arrangements of aquifer boundaries. In Panel (a) a fully penetrating stream and an impermeable boundary intersect at right angles. In Panel (b), two impermeable boundaries provide a wedge-shaped aquifer. Dealing with these boundaries for a single withdrawal well requires seven image wells.

$$s = \frac{Q}{4\pi T}[W(u_r) - W(u_{i1}) + W(u_{i2}) - W(u_{i3})] \tag{13.10}$$

For a wedge-shaped boundary for an aquifer pumped by a single discharge well (Figure 13.19b), the number of image wells, as shown by Ferris et al. (1962), is given by

$$n = \frac{360°}{\theta} - 1 \tag{13.11}$$

where θ is the wedge angle.

▶ EXERCISES

13.1. In a hydraulic test near a recharge boundary in a confined aquifer, the pumping rate is 7.52 m³/min. The locations of the observations are shown in Figure 13.20, and the values of time versus drawdown in the four observation wells are listed in Table 13.4. Calculate the hydraulic parameters of the aquifer and determine the recharge boundary.

▶ CHAPTER HIGHLIGHTS

1. When pumping wells are closely spaced, their cones of depression overlap, causing the wells to interfere with one another. There is more drawdown than expected in each of the pumping wells because the water-level decline in a well is due not only to its pumping but also to drawdown caused by nearby wells. For confined aquifers, this problem is handled by the principle of superposition, where the total drawdown at any location of interest is the sum of the drawdowns due to each well by itself.

2. In problems so far, the initial water level in the aquifer is assumed to be a constant. For cases with an initial gradient in the hydraulic head and flowing ground water, the principle of superposition can be used to determine the head distribution. This chapter presents an example with two flow elements—the uniform hydraulic-head distribution due to regional flow in the aquifer and the cone of

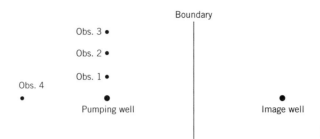

Figure 13.20 A recharge stream and an imprevious boundary intersect at right angles.

TABLE 13.4 **Values of drawdown vs. time for a pumping test in a confined aquifer near a recharge boundary**

Time(min)	$r = 5$m	$r = 100$m	$r = 150$m	$r = 200$m
0.10	0.00	0.00	0.00	0.00
0.16	0.03	0.00	0.00	0.00
0.26	0.09	0.00	0.00	0.00
0.43	0.20	0.01	0.00	0.00
0.70	0.37	0.03	0.00	0.00
1.13	0.59	0.10	0.01	0.00
1.83	0.85	0.22	0.05	0.01
2.98	1.14	0.40	0.13	0.04
4.83	1.44	0.63	0.27	0.11
7.85	1.74	0.88	0.47	0.24
12.74	2.03	1.14	0.68	0.42
20.69	2.27	1.37	0.88	0.61
33.60	2.46	1.55	1.05	0.79
54.56	2.59	1.68	1.18	0.94
88.59	2.69	1.77	1.27	1.05
143.84	2.75	1.83	1.33	1.12
233.57	2.79	1.87	1.37	1.17
379.27	2.81	1.89	1.39	1.20
615.85	2.83	1.91	1.40	1.22
1000.00	2.84	1.92	1.41	1.23

depression due to a pumping well. The resulting flow field (Figure 13.3c) is determined by subtracting the drawdown due to the well from the regional head distribution.

3. The computer program **FLOWNETz** included on the website provides a capability for calculating hydraulic head and stream functions for multiple wells in a regional flow field. The approach allows the flow gradient to be at any angle to the x-axis but requires that the system be at steady state.

4. Hydraulic theory requires that aquifers be infinite in lateral extent. However, this condition is the exception rather than the rule. Aquifers can be cut by tight faults or can end abruptly owing to changes in geology. Such impermeable boundaries halt the spread of the cone of depression and influence the pattern of drawdown related to the well. Similarly, a surface-water body (for example, a fully penetrating stream or lake) or an adjacent segment of aquifer, having a significantly higher transmissivity or storativity, can halt the spread of a cone of depression by providing a source of recharge to the aquifer.

5. Hydraulic theory cannot cope with the presence of one of these aquifer boundaries. However, it is possible to replace a boundary by adding a hypothetical well known as an image well. In practice, this well is placed across the boundary at the same distance from the boundary as the original well. A pumping well stands in for an impermeable boundary, and an injection well stands in for a recharge boundary. Drawdowns at points of interest are calculated using the actual well and the image well.

6. The presence of a boundary adds extra complexity in interpreting aquifer-test results. Generally, one needs to use early-time behavior to estimate T and S before the cone of depression reaches the boundary and drawdown is influenced by the presence of the boundary. A test with four or more observation wells provides sufficient information to estimate the location of the boundary.

14

SOLVING WELL-HYDRAULICS PROBLEMS WITH A PC

One of the most important developments in the study of well hydraulics is PC-based computer codes for evaluating the response of aquifers to pumping. This book provides two computer packages for these calculations. First, we introduce the computer program, **WELLz 2.0,** which facilitates the prediction of drawdown caused by pumping wells in various types of aquifers (for example, confined, leaky-confined, and unconfined aquifers). In particular, **WELLz** is suited for complex situations involving several wells and bounded aquifers. The latter part of this chapter demonstrates the application of **PUMPz,** a companion code that automatically interprets aquifer-test data to provide hydraulic parameters such as T and S.

▶ 14.1 DESCRIPTION OF WELLz

The computer code, **WELLz 2.0,** is designed to estimate drawdown(s) due to one or more pumping and/or injection wells. The present version of the code accommodates calculations involving a fully penetrating well in a confined, leaky-confined, or water-table aquifer. **WELLz 2.0** is programmed in C++ and runs under Windows. This code is an improvement over earlier versions by providing a Laplace transform solution for drawdown in an unconfined aquifer (Moench, 1996) instead of approximating the solution from calculations with the Theis equation. The code is provided in the directory\programs\WELLz2.0 on the website. When the code is run from Windows, the initial dialog box has four main pull-down menus—**File, Editing, Options,** and **View** (Figure 14.1). **File-New** is used when beginning a new problem. A set of default data appears in subsequent dialog boxes, which the user edits. **File-Open** provides the opportunity to load a preexisting data set saved from an earlier run. The user selects the particular file—saved with a *.wzf extension—and the appropriate data appear in the **Input** menu dialog box and subsequent dialog boxes

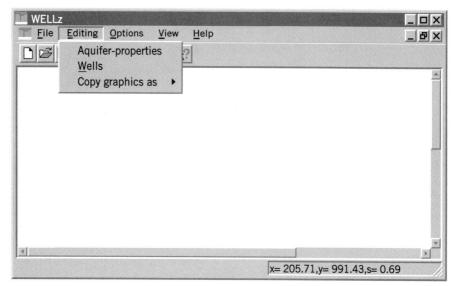

Figure 14.1 The Main Menu of **WELLz 2.0**.

(Figure 14.1). When a preexisting data set is loaded, the first screen that is opened is the contour plot for the problem. To edit aquifer or well parameters for a subsequent run, one would use the **Editing** menu. To return to the Input Menu or first screen, choose **File-New**. In effect, if changes to the **Input** menu are required, the simulation data will likely require major changes, making it a new trial.

Once the **Input** menu dialog box appears, one selects the proper unit parameters, defines the size of the region over which drawdowns are to be contoured, selects the particular aquifer type of interest, and defines the total simulation time (Figure 14.2*a*). For contouring, drawdowns are calculated on an equally spaced row/column grid applied to the region. Normally, users are advised to keep the number of rows and columns in the calculational grid large. The default 50, 50 is recommended, providing drawdown calculations at 2500 separate locations. Note that the simulation region identifies a region in infinite space where drawdowns are to be calculated and contoured. Defining this region has no impact on the calculation, as is the case with a numerical model like MODFLOW. Thus, it is possible to define pumping/injection wells outside of the region, and their impacts may still be felt within the simulation region. Similarly, drawdown may exist outside of the calculation region. The well coordinates need to be defined so that all values are greater than zero.

The next dialog box serves to input the relevant aquifer parameters. The particular dialog box that appears depends on the aquifer type. For example, Figure 14.2*b* illustrates the dialog box for a confined aquifer. The next dialog box (Figure 14.2*c*) contains information on various pumping and observation wells. Use the mouse to add or edit well information. Observation wells are identified by a pumping rate of 0.0. After all of the necessary parameters are provided through the dialog boxes, a window provides the contoured drawdown over the region of interest (Figure 14.2*d*). The drawdown is automatically provided at the designated observation wells. By selecting **Options—Label Contours**, one can use the mouse to label the contour lines or display drawdowns at any designed point on the screen. Simply use the mouse to select the position where you wish the drawdown to be shown and right click. At this point, "File" options similar to those used previously are provided as follows:

New—Rerun the simulation from the beginning.

Open—Open a previously saved data file with a ∗.wzf extension.

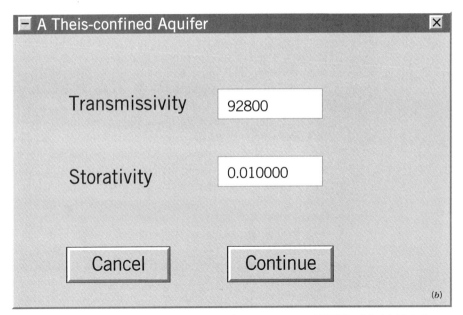

Figure 14.2 Dialog boxes for entering simulation parameters to **WELLz 2.0**. (*a*) Main input menu for **WELLz 2.0**; (*b*) dialog box for the aquifer parameters; (*c*) dialog box for the well parameters; (*d*) graphical display of simulation results.

Save As—Save the input data file with a ∗.wzf extension.

Print—Print the contour map.

Exit—Quit the code.

The number of contour lines and contour intervals are either fixed or specified by the user by selecting **Options—Contour values**. Contour lines are defined by simply dividing the fixed minimum and maximum drawdowns in the field into equal intervals to produce

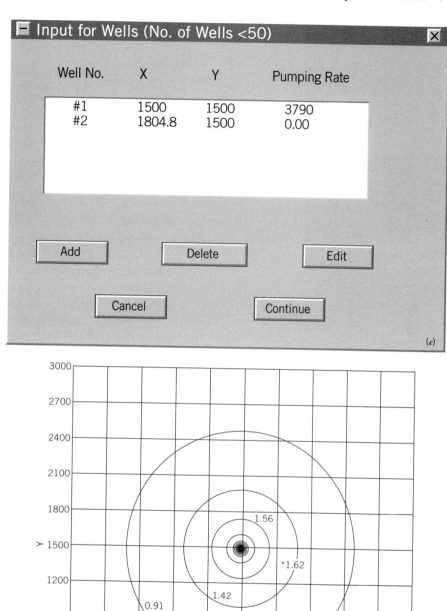

Figure 14.2 Continued

10 contour lines. The simulation results can be viewed and saved by selecting **Report** command in the **View** menu.

▶ 14.2 CODE DEMONSTRATION

The following simple well-hydraulic problem illustrates how the code is used. A well that is pumped at a constant rate of 3790 m^3/day fully penetrates a confined aquifer. The aquifer is effectively infinite with a transmissivity and storativity of 928 m^2/day and 1.0×10^{-2}, respectively. Determine the drawdown in an observation well located 304.8 m from the pumping well at 64 days. We solve the problem using **WELLz,** as follows.

1. Launch WELLz by selecting Start |Programs| WELLz Applications.

2. Select **File-New** because the data set does not already exist.

3. Edit the information that appears in the dialog box as shown in Figure 14.2*a*. The unit parameters for this problem are all meters and days, as shown. The region is set up with the pumping well in the middle having coordinates (1500, 1500) (see Figure 14.3). The calculations will be carried out within a region 3000 m by 3000 m. Note that as required by the code all wells have coordinates greater than zero. For this problem, "Theis confined aquifer" is selected. Having completed the editing of the form, we move to the next dialog box by clicking on "Continue."

4. Edit values of T and S on the next form (Figure 14.2*b*) and continue.

5. Add information on the pumping well and the observation well (Figure 14.2*c*). The coordinates of the observation well are (1804.8, 1500). Click on "Continue" and move to the final dialog box.

6. The drawdown due to the well is contoured in Figure 14.2*d*. At the observation well, the drawdown is indicated to be 1.6 m. By selecting "Option—Label-Contours," one can label the contour lines and other points of interest.

7. Finally, we save the data in the file "ex14-1.wzf" and exit the code.

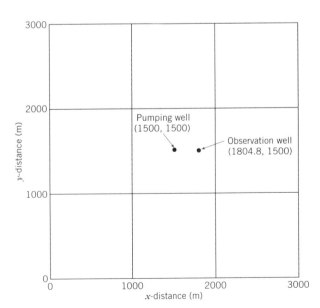

Figure 14.3 A pumping well and an observation well in a confined aquifer.

Bounded Aquifers Revisited

WELLz 2.0 is particularly useful for applications with bounded aquifer systems. Dealing with several wells and image wells together with a relatively large number of observation points makes hand calculations tedious. This example provides another illustration of image-well theory and the application of **WELLz 2.0** to a problem complicated by a relatively large number of wells.

Walton (1962) illustrated the application of image-well theory to the interpretation of an aquifer test by Mikels (1952) at Zion, Illinois. The test involved a single pumping well (A-P), pumped at 99 gpm for 3180 minutes and six observation wells (Figure 14.4). The measurements of drawdown at the observation wells are shown in Figure 14.5 The unconfined gravelly sand aquifer had a saturated thickness of approximately 20 ft. Walton (1962) interpreted the asymmetry in the cone of depression and similarities in water-level responses in the wells and Lake Michigan as indication that the lake was acting as a recharge boundary. He estimated the hydraulic conductivity to be 387.7 ft/day, the specific yield to be 0.01, and the distance to the effective line of recharge to be 206 ft.

Use **WELLz 2.0** to simulate the 3180-minute test at Zion. See how well the simulated drawdowns match the observed drawdowns shown in Figure 14.5.

1. Let us begin by formulating the problem. First, we define a region over which the calculations are to be conducted, and we specify a coordinate system for defining the location of the wells (Figure. 14.4). The exact location of the origin doesn't matter except that the region should be located appropriately. The selected simulation region stops at the recharge boundary 306 ft from the origin. Nevertheless, the image (injection) well is included as part of the input data, and its influence is felt within the simulation domain.

2. We run the code as before, entering the appropriate information. For those more curious, the data set is included as "ex14-2.wzf" in the directory/programs/WELLz on the website and can be loaded through **File-Open**.

3. The simulation results at 3180 minutes are plotted in Figure 14.6. Note how the presence of the recharge boundary leads to reduced drawdowns between the pumping well and Lake Michigan. The simulated drawdowns at the observation wells match the observed drawdowns well.

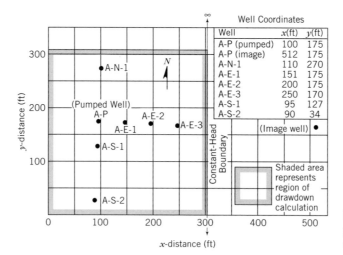

Figure 14.4 Locations of pumping and observation wells for a bounded aquifer.

Well Coordinates		
Well	x(ft)	y(ft)
A-P (pumped)	100	175
A-P (image)	512	175
A-N-1	110	270
A-E-1	151	175
A-E-2	200	175
A-E-3	250	170
A-S-1	95	127
A-S-2	90	34

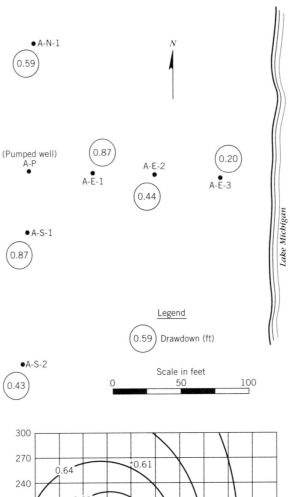

Figure 14.5 Observed drawdowns in a bounded aquifer.

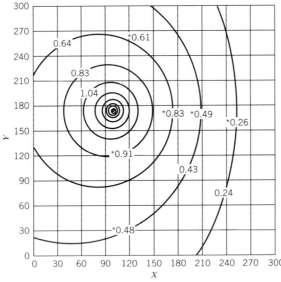

Figure 14.6 Simulated drawdowns in a bounded aquifer.

▶ 14.3 INTERPRETING AQUIFER TESTS

Personal computers are useful for interpreting aquifer tests. The interpretative schemes basically involve minimizing the difference between the measured drawdown in a well and the calculated drawdown in the same well. Eventually, a unique set of aquifer parameters (for example, T and S) will emerge from this analysis. This approach is cast mathematically as minimizing the mean squared error (MSE) between measured and calculated drawdowns

$$\varepsilon = \frac{1}{n} \sum_{i=1}^{n} f_i^2 = \frac{1}{n} \sum_{i=1}^{n} [\, s(t_i, \overline{P}) - s_p(t_i, \overline{P})]^2 \tag{14.1}$$

where ε is the MSE, n is the number of measurement times, s and s_p are measured and predicted drawdowns at time t_i, respectively, and f_i is the difference between measured and predicted drawdowns at time t_i. $\overline{P}$ is a vector of length m with

$$\overline{P} = (\, p_1, p_2, \ldots, p_m) \tag{14.2}$$

where m is the number of parameters to be determined. The process of determining the parameter vector P is called *inversion*. The MSE error is called the *objective function* in an inversion algorithm. The inversion process finds a set of aquifer parameters, which minimize the objective function. There are two categories of inversion techniques to derive the aquifer parameters: (1) methods that do not require the calculation of the derivative of the objective function, such as trial-and-error and downhill simplex methods; and (2) methods that require a calculation of the derivative of the objective function, such as the Marquardt method. The following sections provide information on how these methods can be used to interpret aquifer test results.

Trial-and-Error Method

The trial-and-error method is the simplest scheme to implement and the easiest to understand. You simply begin with an initial guess at a set of hydraulic parameters and use a computer code to predict drawdown with time at each observation well. The root mean squared error (RMSE), defined as $\sqrt{\text{MSE}}$, is calculated for the particular T and S values for the calculated and measured drawdowns using Eq. (14.1). Early in the process, the error will be large. A new set of T, S, and other hydraulic parameters are tested, which hopefully provide a smaller RMSE. The process is continued for other guesses until a best fit is achieved between the predicted and measured drawdowns. With experience, one can usually determine how to change parameters to find the best set of hydraulic parameters. Because of the simplicity in this approach, it can be implemented in a spreadsheet or with any high-level program languages (FORTRAN, C, and C++).

Down-Hill Simplex Method

The down-hill simplex method is a more sophisticated approach. Again, it involves the repetitive calculations of RMSEs with various sets of hydraulic parameters. However, now the code itself figures out how to modify the hydraulic parameters so as to minimize the RMS error. The procedure is straightforward because derivatives of the objective function need not be calculated (Nelder and Mead, 1965; Press et al., 1987). Operationally, the approach involves searching a parameter space (for example, T on the x-axis and S on the y-axis). A family of points are shuffled around on this space to find the minimum RMSE. The term *simplex* defines the geometry of points. Generally, a simplex consists of $m+1$ points and their interconnecting line segments, polygonal faces in m-dimensional space. In two dimensions, it is a triangle, and in three dimensions, it is a tetrahedron.

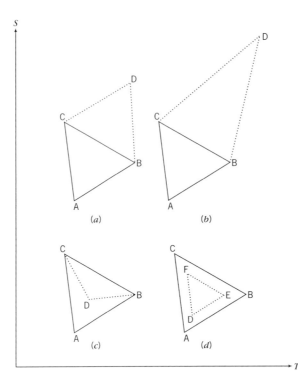

Figure 14.7 Illustration of the downhill simplex method.

The simplex algorithm for minimization takes the set of $m + 1$ points and attempts to move them into a minimum. Let's consider a family of three points on a T and S space. If your initial guess is $\overline{P}_0$, the other points of the simplex can be initialized as

$$\overline{P}_i = \overline{P}_0 + \lambda_i \overline{e}_i \tag{14.3}$$

where $\overline{e}_i$ is a unit vector and λ_i is the characteristic length. Figure 14.7 describes different options as to how the simplex moves around in search of a RMSE minimum. It can: (1) reflect the point with the highest function evaluation (that is, RMSE) in the plane defined by the remaining points (Figures 14.7a,c); or (2) reflect and expand to take larger steps (Figure 14.7b); or (3) contract to shrink the overall volume when it has reached a valley floor with reflections and expansions (Figure 14.7d). The overall effect is for the simplex to crawl around the parameter space, creeping down valleys and shrinking to get to the very bottom of narrow valleys. Eventually, it will discover a best set of T and S values with the minimum RMSE. In some cases, the program could fail to provide a set of convergent parameters for some initial guess of parameter values. The program would need to be restarted in this case with a new initial guess.

Marquardt Method

The Marquardt method (Marquardt, 1963) is the most complicated approach of all. It involves the minimization of Eq. (14.1) in a linearized least-squares sense. We have included the derivation for the interested advanced reader. The measured drawdown can be expanded using Taylor's series at the initial guess $\overline{P}_0$ as

$$s(t_i, \overline{P}) = s_p(t_i, \overline{P}_0) + \sum_{j=1}^{m} \frac{\partial s_p(t_i, \overline{P}_0)}{\partial p_j}(p_j - p_{j0}), (i = 1, \dots, n) \tag{14.4}$$

The above equations may be expressed in matrix format as

$$\Delta \bar{s} = \bar{J} \Delta \bar{P} \tag{14.5}$$

where

$$\overline{\Delta s^T} = [s(t_1) - s_p(t_1), s(t_2) - s_p(t_2), \dots, s(t_n) - s_p(t_n)]^T \tag{14.6}$$

$$\overline{\Delta P^T} = [\, p_1 - p_{10}, p_2 - p_{20}, \dots, p_m - p_{m0}]^T = P^T - P_0^T \tag{14.7}$$

$$\bar{J} = \begin{bmatrix} \dfrac{\partial s_p\,(t_1,\bar{P}_0)}{\partial p_1} & \dfrac{\partial s_p\,(t_1,\bar{P}_0)}{\partial p_2} & \cdots & \dfrac{\partial s_p\,(t_1,\bar{P}_0)}{\partial p_m} \\[2ex] \dfrac{\partial s_p\,(t_2,\bar{P}_0)}{\partial p_1} & \dfrac{\partial s_p\,(t_2,\bar{P}_0)}{\partial p_2} & \cdots & \dfrac{\partial s_p\,(t_2,\bar{P}_0)}{\partial p_m} \\[2ex] \cdots & \cdots & \cdots & \cdots \\[2ex] \dfrac{\partial s_p\,(t_n,\bar{P}_0)}{\partial p_1} & \dfrac{\partial s_p\,(t_n,\bar{P}_0)}{\partial p s_2} & \cdots & \dfrac{\partial s_p\,(t_n,\bar{P}_0)}{\partial p_m} \end{bmatrix} \tag{14.8}$$

where the matrix J is called the Jacobian matrix. To facilitate solution of parameter correction vector P, multiply J^T at both sides of Eq. (14.5). Thus

$$\Delta P = (J^T J)^{-1} J^T \Delta \bar{s} \tag{14.9}$$

One problem encountered is the singularity of matrix $J^T J$ in Eq. (14.9). This can be overcome with Marquardt's approach by writing Eq. (14.9) as

$$\Delta P = (J^T J + \lambda D^2)^{-1} J^T \Delta \bar{s} \tag{14.10}$$

where λ is a parameter known as Marquardt's parameter and D is a unit diagonal matrix. In Marquardt's method, users provide an initial guess for hydraulic parameters, and the computer program calculates the correction term ΔP. The new parameter vector is

$$P = P_0 + \omega\,\Delta P \tag{14.11}$$

where ω is called the relaxation parameter. This process is repeated many times until an acceptable RMSE is achieved. For example, a minimum value of the objective function can be found in the T (transmissivity) $-S$ (Storativity) space after a number of iterations (Figure 14.8). Certain rules are applied in selecting and changing the values of λ and ω, so that the RMSE in the current iteration is less than that in the previous iteration and the aquifer parameters will keep positive during the inversion process.

▶ 14.4 INTRODUCTION TO **PUMP**z (A COMPUTER PROGRAM FOR AQUIFER-TEST ANALYSIS)

A computer code is included on the website for deriving common hydraulic parameters (such as T and S) from aquifer tests. This Windows code is easy to use and ready to install. Go to the directory /programs/pumpz and type setup. The Setup program will guide you through the installation process.

Activate the code by clicking **Start/Program/PUMPz**. There are six menus on the Main Window of the program (Figures 14.9a through 14.9f). The **File** menu contains the

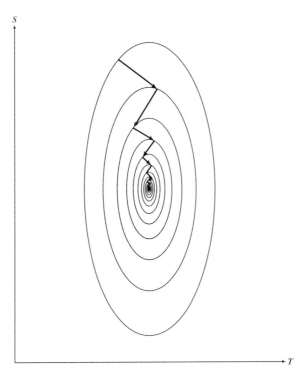

Figure 14.8 Illustration of the convergence path of the objective function in the S-T space.

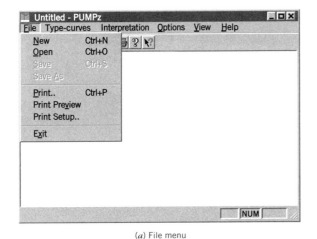

(*a*) File menu

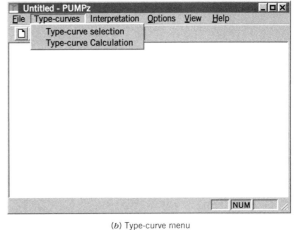

(*b*) Type-curve menu

Figure 14.9 Menus for **PUMPz**. (*a*) File menu; (*b*) Type-curve menu; (*c*) Interpretation menu; (*d*) Options menu; (*e*) View menu; (*f*) Help menu.

typical commands, **Open**, **Save**, **Save as**, **Print**, **Print Preview**, **Print Setup**, and **Exit** (Figure 14.9*a*).

The **Type-curves** menu contains two selections—**Type-curve Selection** and **Type-curve Calculation** commands (Figure 14.9*b*). Clicking the **Type-curve Selection** command opens the dialog box shown in Figure 14.10. There are five categories of type curves: a confined aquifer, a leaky-confined aquifer, a slug test, unconfined aquifers, and fractured aquifers. The **Type-curve Calculation** has the code calculate various types of curves

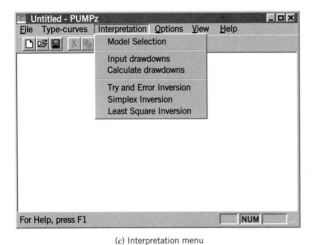

(c) Interpretation menu

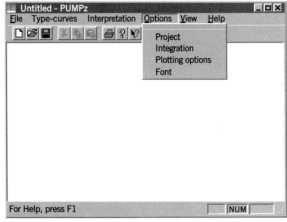

(d) Options menu

Figure 14.9 Continued

(e) View menu

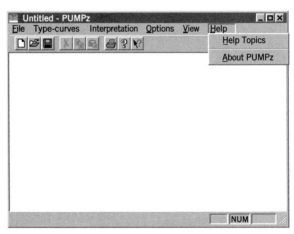

(f) Help menu

Figure 14.9 Continued

(Figure 14.11). The user specifies the *x*- and *y*-dimensions for the curves using a dialog box. It is helpful to look at examples of types of curves in the previous chapters to figure out how many log units are needed.

The **Interpretation** menu contains **Model Selection**, **Input Drawdowns**, **Calculate Drawdowns**, **Trial-and-Error Inversion**, **Simplex Inversion**, and **Least Square Inversion** commands (Figure 14.9c). Choosing **Model Selection** opens the dialog box shown in Figure 14.11. The user then has to decide what type of aquifer setting or test is likely: a confined aquifer, a leaky-confined aquifer, an unconfined aquifer, a fractured aquifer, or a slug test. Next, the measured drawdown data are input to the code. The **Input Drawdowns** command gives users three options: import time/drawdown data from a text file, enter the data in a simple spreadsheet, or select a set of hypothetical drawdown data that you have calculated. The last-named feature is helpful in getting to know the code because you can determine aquifer parameters for a data set that you already know. This hypothetical data set is created using the **Calculate Drawdowns** command and completing the appropriate

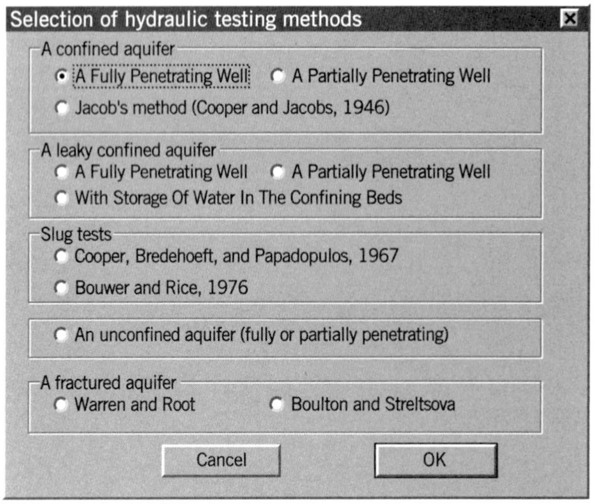

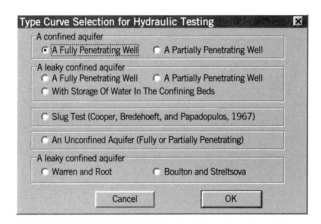

Figure 14.10 Dialog box for the selection of aquifer-test interpretation methods.

Figure 14.11 Dialog box for selecting models of type curves.

dialog boxes. Finally, once you have a set of data to work with, you can choose one of the three techniques—**Trial-and-Error Inversion**, **Simplex Inversion**, or **Least Square Inversion**—to provide the aquifer parameters. When **Simplex inversion** or **Least Square Inversion** is selected, a dialog box opens requesting the maximum number of iterations and the convergence error (Figure 14.12). Before the actual calculation, you need to provide information to one last dialog box concerning basic information about the test (e.g., pumping rate, the distance from a pumping to an observation well). This basic data obviously will change depending on the test technique.

The **Options** menu provides **Integration** and **Plotting** commands (Figure 14.9d). The default values for integration should be used for users without further knowledge of Stehfest Laplace inversion or Gaussian integration schemes. The **Plotting option** command provides ranges and types of horizontal and vertical axes.

The **View** menu contains a very useful command **Report** (Figure 14.9e). The **Report** command provides a summary of the interpreted test results opened by **Notepad**. Users will be able to save the text from **Notepad**. The **Help** menu contains standard commands for **Help Topics** and **About PUMPz** (Figure 14.9f).

All units of the input parameters are in consistent units in **PUMPz.** The term *consistent* means that all lengths and all times are in the same units. The other variables are expressed

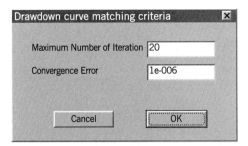

Figure 14.12 Dialog box for specifying the inversion parameters.

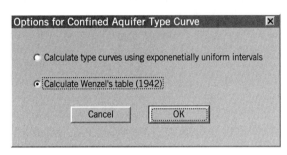

Figure 14.13 Dialog box for type-curve parameters for an aquifer test in a confined aquifer with fully penetrating wells.

in the same units of length and time. For example, if drawdown and the well radius are in meters and time is in minutes, the pumping rate will be in m³/min and the transmissivity will be in m²/min.

► 14.5 APPLICATION OF **PUMP**z

The examples presented next demonstrate how **PUMP**z can be used to help interpret data from aquifer tests.

► **EXAMPLE 14.1**
Creating Type Curves
and Graph Paper

One of the uses for **PUMP**z is in providing type curves for a variety of aquifer-test conditions. For this example, we will create a type curve for an aquifer test in a confined aquifer with a fully penetrating well. First, users select a curve to produce (see list in Figure 14.11) by clicking the **Type-Curve selection** from the **Type-Curves** menu. Next, click the Type-Curve calculation to prompt a dialog box for specifying the type-curve parameters (Figure 14.13). Figure 14.14 is an example of the type curves that can be produced. The type-curve calculation report (Table 14.1) can be viewed, saved, or printed out by clicking on the **Report** in the **View** menu. Values of well functions and dimensionless u are also listed on the report. **PUMP**z can also produce a blank copy of graph paper with the same scale as the type curves (Figure 14.15) for users to plot their measurements. The blank graph paper is created by checking "no" on Horizontal axis->Display label, Vertical axis->Display label, and Data-> Display in the plotting options (Figure 14.16). With the type curve and the proper-sized graph paper, you can use the curve-matching method from Chapter 9.

► **EXAMPLE 14.2**
Interpreting Results
from a Test with a
Confined Aquifer

This example illustrates how aquifer test data can be analyzed automatically using **PUMP**z. It also demonstrates the capability in the code for generating a hypothetical set of aquifer-test data. In other words, given an observation well at distance r, aquifer parameters (for example, T and S), and a pumping rate Q, the code calculates a set of time/drawdown data. Begin Example 2 by generating these time/drawdown data. Let us assume that the test was run for a confined aquifer with a pumping rate of 1 m³/min, and transmissivity and storativity values of 1 m²/min and 1×10^{-5}, respectively. The observation well is assumed to be 50 m away from the pumped well. Here are the procedures to calculate the synthetic set of drawdown data:

1. Select the confined aquifer option with a fully penetrating well by clicking on **Model Selection** in the **Interpretation** menu.

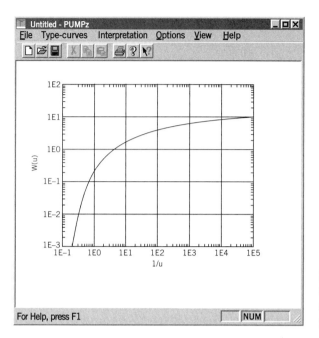

Figure 14.14 Display of the type curve for aquifer tests in a confined aquifer with fully penetrating wells.

TABLE 14.1   Report output from PUMPz for type-curve calculations

REPORT FOR THE CALCULATION OF WELL FUNCTIONS
Operator's Name: John Doe
Project Description: Theis Type Curve
Constant Discharge from a Fully Penetrating Well in a Nonleaky Aquifer
Minimum $1/u = 1.0e - 001$
Minimum $1/u = 1.0e + 004$
Number of Points $= 50$

u	$1/u$	$W(u)$
1.00e + 001	1.00e − 001	4.16e − 006
7.91e + 000	1.26e − 001	4.18e − 005
6.25e + 000	1.60e − 001	2.70e − 004
4.94e + 000	2.02e − 001	1.23e − 003
3.91e + 000	2.56e − 001	4.23e − 003
3.09e + 000	3.24e − 001	1.17e − 002
2.44e + 000	4.09e − 001	2.69e − 002
1.93e + 000	5.18e − 001	5.38e − 002
1.53e + 000	6.55e − 001	9.62e − 002
1.21e + 000	8.29e − 001	1.57e − 001
9.54e − 001	1.05e + 000	2.37e − 001
7.54e − 001	1.33e + 000	3.38e − 001
5.96e − 001	1.68e + 000	4.58e − 001
4.71e − 001	2.12e + 000	5.96e − 001
3.73e − 001	2.68e + 000	7.50e − 001
2.95e − 001	3.39e + 000	9.19e − 001
2.33e − 001	4.29e + 000	1.10e + 000
1.84e − 001	5.43e + 000	1.29e + 000

(Continued)

TABLE 14.1   Continued

1.46e − 001	6.87e + 000	1.49e + 000
1.15e − 001	8.69e + 000	1.70e + 000
9.10e − 002	1.10e + 001	1.91e + 000
7.20e − 002	1.39e + 001	2.13e + 000
5.69e − 002	1.76e + 001	2.35e + 000
4.50e − 002	2.22e + 001	2.57e + 000
3.56e − 002	2.81e + 001	2.79e + 000
2.81e − 002	3.56e + 001	3.02e + 000
2.22e − 002	4.50e + 001	3.25e + 000
1.76e − 002	5.69e + 001	3.48e + 000
1.39e − 002	7.20e + 001	3.71e + 000
1.10e − 002	9.10e + 001	3.94e + 000
8.69e − 003	1.15e + 002	4.18e + 000
6.87e − 003	1.46e + 002	4.41e + 000
5.43e − 003	1.84e + 002	4.64e + 000
4.29e − 003	2.33e + 002	4.88e + 000
3.39e − 003	2.95e + 002	5.11e + 000
2.68e − 003	3.73e + 002	5.35e + 000
2.12e − 003	4.71e + 002	5.58e + 000
1.68e − 003	5.96e + 002	5.82e + 000
1.33e − 003	7.54e + 002	6.05e + 000
1.05e − 003	9.54e + 002	6.28e + 000
8.29e − 004	1.21e + 003	6.52e + 000
6.55e − 004	1.53e + 003	6.75e + 000
5.18e − 004	1.93e + 003	6.99e + 000
4.09e − 004	2.44e + 003	7.22e + 000
3.24e − 004	3.09e + 003	7.46e + 000
2.56e − 004	3.91e + 003	7.69e + 000
2.02e − 004	4.94e + 003	7.93e + 000
1.60e − 004	6.25e + 003	8.16e + 000
1.26e − 004	7.91e + 003	8.40e + 000
1.00e − 004	1.00e + 004	8.63e + 000

2. Click on **Calculate drawdowns** in the **Interpretation** menu.

3. Specify number of times, minimum time, and maximum time in the dialog box shown in Figure 14.17.

4. Specify the pumping rate, the distance from the pumped well to the observation well, transmissivity, and storativity in the dialog box in Figure 14.18.

Here is the procedure for interpreting the aquifer-test results:

1. Click on **Input drawdowns** in the **Interpretation** menu and select **Type values of time and drawdown** in the dialog box in Figure 14.19. The spreadsheet will show values of time, measured drawdown, and calculated drawdown in the second, third, and fourth columns, respectively (Figure 14.20). The measured drawdown is the aquifer-test data to be evaluated. As mentioned, these may be (1) hypothetical draw-down data calculated from selecting **Calculate drawdown** in the **Interpretation** menu, (2) data from an actual test typed in this spreadsheet, or (3) time/drawdown data imported from a file. For this example, we use a set of hypothetical data. Users are allowed to edit values of time and measured drawdown. Clicking on the

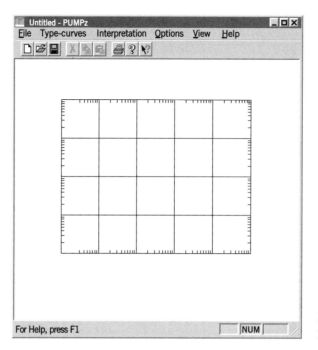

Figure 14.15 Plot of a blank page for type-curve fitting.

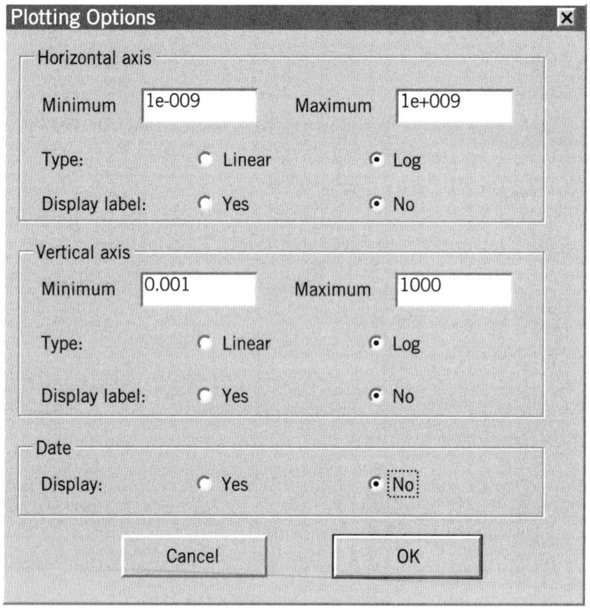

Figure 14.16 Plotting options for **PUMPz**.

button above the spreadsheet changes the number of measurements. Drawdown measurements can be used selectively in the interpretation by specifying the range of data use below the spreadsheet.

2. Select **A confined aquifer** option with **A Fully Penetrating Well** using the dialog box (Figure 14.11) by clicking on **Model Selection** in the **Interpretation** menu.

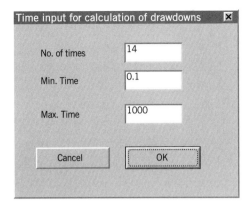

Figure 14.17 Dialog box showing number of times, minimum time, and maximum time.

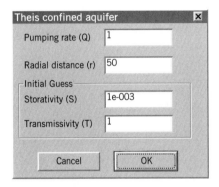

Figure 14.18 Dialog box for specifying an initial guess of aquifers parameters.

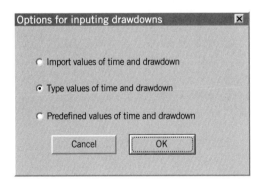

Figure 14.19 Dialog box for selecting options to input measured values of time and drawdown.

3. Click on **Least Square Inversion** in the **Interpretation** menu. In the **Drawdown curve matching criteria** dialog box, specify a maximum number of iterations of 20 and a convergence error of 1×10^{-6} (Figure 14.14). The initial values of the transmissivity and storativity are given as 10 m^2/min and 1×10^{-3}, respectively, in the **Theis confined aquifer** dialog box. The pumping rate and the distance from the pumped well to the observation well are also specified in the box.

4. We are essentially done. The actual and calculated drawdown data for this example are shown in Figure 14.21 after the inversion is finished.

5. The calculation report can be viewed by clicking on **Report** in the **View** menu. The report is opened with Microsoft Notepad or Wordpad. The report for this

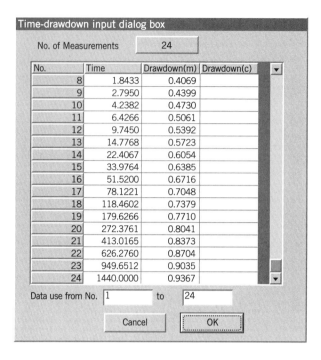

Figure 14.20 Table for inputting measured values of time and drawdown.

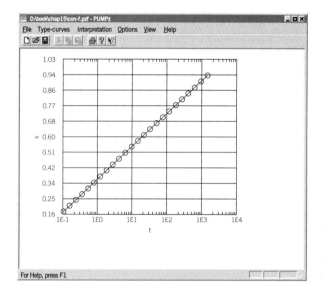

Figure 14.21 Display of measured (small circle) and calculated (solid line) values of drawdown after the inversion.

example is shown in Table 14.2. The last section of the report is the root mean square error (RMSE) for each iteration of the inversion. The inversion will stop when the maximum number of iterations or RMSE is met. Table 14.2 shows that we obtained the correct answer, a T of $1\,m^2/min$ and an S of 1×10^{-5}. The measured and calculated drawdown values are nearly identical because they are both calculated with nearly identical T and S values. With measured drawdown data, the values would obviously not be so similar.

TABLE 14.2   Report output from PUMPz for aquifer parameter interpretations

REPORT OF AQUIFER PARAMETER ESTIMATIONS
Operator's Name: John Doe
Constant Discharge from a Fully Penetrating Well in a Nonleaky Aquifer
Pumping Rate $= 1.00e + 000$
Radial Distance $= 5.00e + 001$
Initial Guess:
Transmissivity $= 1.00e + 001$
Storativity $= 1.00e - 003$

Inversion Results:

Transmissivity $= 1.00e + 000$
Storativity $= 9.99e - 006$

Time	Measured Drawdown	Calculated Drawdown	Measured − Calculated
0.100	0.1795	0.1795	$-3.0e - 008$
0.151	0.2110	0.2110	$-3.0e - 008$
0.229	0.2431	0.2431	$-4.0e - 008$
0.348	0.2755	0.2755	$-3.0e - 008$
0.528	0.3081	0.3081	$-4.0e - 008$
0.801	0.3409	0.3409	$-3.0e - 008$
1.215	0.3738	0.3738	$-4.0e - 008$
1.843	0.4068	0.4068	$-3.0e - 008$
2.795	0.4399	0.4399	$-3.0e - 008$
4.238	0.4729	0.4729	$-4.0e - 008$
6.426	0.5060	0.5060	$-4.0e - 008$
9.745	0.5391	0.5391	$-4.0e - 008$
14.77	0.5722	0.5722	$-4.0e - 008$
22.40	0.6053	0.6053	$-3.0e - 008$
33.97	0.6385	0.6385	$-3.0e - 008$
51.51	0.6716	0.6716	$-3.0e - 008$
78.12	0.7047	0.7047	$-3.0e - 008$
118.46	0.7378	0.7378	$-4.0e - 008$
179.62	0.7710	0.7710	$-3.0e - 008$
272.37	0.8041	0.8041	$-3.0e - 008$
413.01	0.8372	0.8372	$-4.0e - 008$
626.27	0.8703	0.8703	$-3.0e - 008$
949.65	0.9035	0.9035	$-3.0e - 008$
1440.00	0.9366	0.9366	$-3.0e - 008$

User Specified Root Mean Square Error $= 1.00e - 006$
Final Root Mean Square Error $= 3.28e - 008$
User Specified Number of Iterations $= 20$
Final Number of Iterations $= 14$

Iteration No $= 1$	Error $= 5.4e - 001$
Iteration No $= 2$	Error $= 5.1e - 001$
Iteration No $= 3$	Error $= 4.6e - 001$
Iteration No $= 4$	Error $= 3.6e - 001$
Iteration No $= 5$	Error $= 1.7e - 001$
Iteration No $= 6$	Error $= 8.3e - 002$
Iteration No $= 7$	Error $= 6.3e - 002$
Iteration No $= 8$	Error $= 5.1e - 002$
Iteration No $= 9$	Error $= 4.0e - 002$
Iteration No $= 10$	Error $= 3.6e - 002$
Iteration No $= 11$	Error $= 9.9e - 003$
Iteration No $= 12$	Error $= 6.8e - 004$
Iteration No $= 13$	Error $= 6.5e - 006$
Iteration No $= 14$	Error $= 3.2e - 008$

▶ **EXAMPLE 14.3**
Interpreting Results
from a test with
Partially Penetrating
Wells in a Leaky,
Confined Aquifer

Here, **PUMPz** is applied to the interpretation of an aquifer test with partially penetrating wells in a leaky-confined aquifer. A 10-m-thick aquifer is pumped at a constant rate of $1 m^3$/min using a partially penetrating well (screen top, $d/b = 0.5$; screen bottom, $l/b = 0.75$). The leaky confining bed is 21 m thick. Drawdowns are to be measured in an observation well 50 m away. It is also partially penetrating with $d'/b = 0.5$ and $l'/b = 0.75$.

The procedure is similar to that in Example 14.2. The aquifer-test model for the partially penetrating wells in a leaky-confined aquifer is selected from a dialog box as shown in Figure 14.10. Again, we choose to use a hypothetical set of drawdown data, generated in the code assuming that $T = 10 \ m^2$/min, $S = 1 \times 10^{-5}$, and $K' = 1 \times 10^{-3}$. Once the drawdown data are provided to **PUMPz**, click **Least Square Inversion** from the **Interpretation** menu. Complete the dialog box as shown in Figure 14.22. Besides the aquifer-test parameters, this dialog box requires initial guesses for values of T, S, and vertical hydraulic conductivity of the confining bed.

The values of calculated and measured drawdown match perfectly (Figure 14.23). Again this result is no surprise because we used a set of calculated drawdown values. The interpretation report is given in Table 14.3.

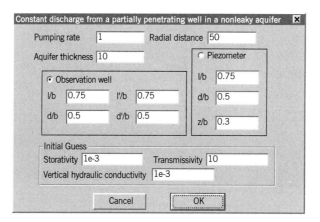

Figure 14.22 Dialog box for specifying the aquifer parameters in a leaky confined aquifer with partially penetrating wells.

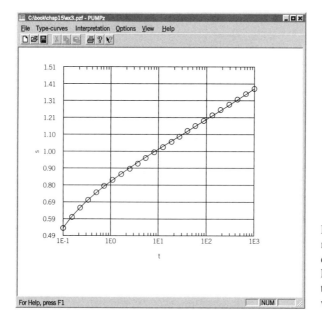

Figure 14.23 Display of measured (small circle) and drawdowns calculated by **PUMPz** (solid line) for an aquifer test in a leaky-confined aquifer with partially penetrating wells.

TABLE 14.3 **Report output from PUMPz for the interpretation of an aquifer test with partially penetrating wells in a unconfined aquifer**

REPORT OF AQUIFER PARAMETER ESTIMATIONS
Operator's Name: Hubao Zhang
Constant Discharge from a Partially Penetrating Well in a Leaky Aquifer
Pumping Rate $= 1.00e+000$
Radius Distance $= 5.00e+001$
Aquifer Thickness $= 1.00e+001$
Thickness of Leaky Bed $= 2.10e + 001$
Screen Top Depth of Pumping well (d/b) $= 5.00e - 001$
Screen Bottom Depth of Pumping Well (l/b) $= 7.50e - 001$
Screen Top Depth of Observation Well (d'/b) $= 5.00e - 001$
Screen Bottom Depth of Observation Well (l'/b) $= 7.50e - 001$
Initial Guess:
Transmissivity $= 1.00e + 001$
Storativity $= 1.00e - 003$
Vertical Hydraulic Conductivity $= 1.00e - 004$
Vertical Hydraulic Conductivity of leaky bed $= 1.00e - 003$
Inversion results: Transmissivity $= 1.00e + 000$
Storativity $= 1.00e - 005$
Vertical Hydraulic Conductivity $= 9.99e - 006$
Vertical Hydraulic Conductivity of Leaky Bed $= 1.00e - 003$

Time	Measured Drawdown	Calculated Drawdown	Measured − Calculated
0.100	0.1909	0.1909	$-1.9e - 009$
0.149	0.2208	0.2208	$-2.3e - 009$
0.222	0.2510	0.2510	$-4.8e - 009$
0.332	0.2813	0.2813	$-7.3e - 009$
0.496	0.3114	0.3114	$-1.7e - 008$
0.740	0.3411	0.3411	$-1.8e - 008$
1.105	0.3700	0.3700	$-2.4e - 008$
1.649	0.3977	0.3977	$-2.4e - 008$
2.462	0.4236	0.4236	$-2.7e - 008$
3.674	0.4471	0.4471	$-3.1e - 008$
5.484	0.4674	0.4674	$-3.4e - 008$
8.185	0.4837	0.4837	$-4.1e - 008$
12.216	0.4954	0.4954	$-4.4e - 008$
18.233	0.5026	0.5026	$-4.7e - 008$
27.213	0.5061	0.5061	$-4.8e - 008$
40.615	0.5073	0.5073	$-5.1e - 008$
60.618	0.5075	0.5075	$-5.2e - 008$
90.473	0.5076	0.5076	$-5.2e - 008$
135.031	0.5076	0.5076	$-5.5e - 008$
201.533	0.5076	0.5076	$-5.3e - 008$
300.788	0.5076	0.5076	$-5.3e - 008$
448.925	0.5076	0.5076	$-5.3e - 008$
670.018	0.5076	0.5076	$-5.3e - 008$
1000.000	0.5076	0.5076	$-5.3e - 008$

User Specified Root Mean Square Error $= 1.00e - 006$
Final Root Mean Square Error $= 4.01e - 008$
User Specified Number of Iterations $= 30$
Final Number of Iterations $= 18$

Iteration No $= 1$ Error $= 3.7e - 001$
Iteration No $= 2$ Error $= 3.3e - 001$

(*Continued*)

▶ **EXAMPLE 14.4**
Interpreting Results
from an Aquifer Test
with Partially
Penetrating Wells in a
Unconfined Aquifer

This last example applies **PUMPz** to the interpretation of an aquifer test with partially penetrating wells in an unconfined aquifer. Details of the test are summarized in the report (Table 14.4). The procedure is again similar to that in Example 14.2. The aquifer-test model for partially penetrating wells in an unconfined aquifer is selected in the dialog box (as shown in Figure 14.10). The drawdown data are generated as before. Click on **Least Square Inversion** from the **Interpretation** menu and supply the appropriate parameters in the dialog box (Figure 14.24). The values of calculated and measured drawdowns match perfectly after 17 iterations (Figure 14.25). The interpretation report is listed in Table 14.4.

TABLE 14.3   Continued

Iteration No = 3	Error = 3.0e − 001
Iteration No = 4	Error = 2.4e − 001
Iteration No = 5	Error = 2.0e − 001
Iteration No = 6	Error = 1.1e − 001
Iteration No = 7	Error = 8.1e − 002
Iteration No = 8	Error = 4.9e − 002
Iteration No = 9	Error = 3.3e − 002
Iteration No = 10	Error = 2.2e − 002
Iteration No = 11	Error = 1.4e − 002
Iteration No = 12	Error = 9.5e − 003
Iteration No = 13	Error = 6.1e − 003
Iteration No = 14	Error = 4.0e − 003
Iteration No = 15	Error = 3.9e − 003
Iteration No = 16	Error = 5.0e − 004
Iteration No = 17	Error = 7.7e − 006
Iteration No = 18	Error = 4.0e − 008

TABLE 14.4   Report output from PUMPz for the interpretation of an aquifer test with partially penetrating wells in a unconfined aquifer

REPORT OF AQUIFER PARAMETER ESTIMATIONS
Operator's Name: Hubao Zhang
Project Description: Anonymous superfund site, X County, Y State
Unconfined Aquifer (Moench, 1993)
Pumping Rate = 1.00e + 000
Radius Distance = 5.00e + 001
Screen Top depth of Pumping Well (d/b) = 0.00e + 000
Screen Bottom Depth of Pumping Well (l/b) = 1.00e + 000
Coordinate of Piezometer (z/b) = 5.00e − 001
Thickness of the Aquifer = 1.00e + 001
Initial Guess:
Transmissivity = 1.00e + 001
Storativity = 1.00e − 003
Vertical Hydraulic Conductivity of the Aquifer = 1.00e − 001
Specific Yield = 2.00e − 001
Inversion results: Transmissivity = 1.00e + 000
Storativity = 1.00e − 005
Vertical Hydraulic Conductivity of the Aquifer = 9.99e − 004
Specific Yield = 2.99e − 001

Time	Measured Drawdown	Calculated Drawdown	Measured − Calculated
0.010	0.0292	0.0292	4.4e − 007
0.014	0.0421	0.0421	5.6e − 007

(Continued)

TABLE 14.4 Continued

Time	Measured Drawdown	Calculated Drawdown	Measured − Calculated
0.021	0.0549	0.0549	6.3e − 007
0.032	0.0667	0.0667	6.4e − 007
0.048	0.0749	0.0749	5.5e − 007
0.071	0.0815	0.0815	4.5e − 007
0.106	0.0847	0.0847	3.3e − 007
0.157	0.0859	0.0859	2.6e − 007
0.233	0.0862	0.0862	2.3e − 007
0.345	0.0863	0.0863	2.2e − 007
0.512	0.0863	0.0863	2.2e − 007
0.760	0.0863	0.0863	2.2e − 007
1.127	0.0864	0.0864	2.2e − 007
1.671	0.0864	0.0864	2.2e − 007
2.477	0.0865	0.0865	2.2e − 007
3.672	0.0866	0.0866	2.2e − 007
5.445	0.0867	0.0867	2.2e − 007
8.072	0.0868	0.0868	2.2e − 007
11.967	0.0870	0.0870	2.1e − 007
17.742	0.0873	0.0873	2.1e − 007
26.303	0.0878	0.0878	2.0e − 007
38.996	0.0885	0.0885	1.9e − 007
57.812	0.0895	0.0895	1.8e − 007
85.708	0.0910	0.0910	1.6e − 007
127.064	0.0931	0.0931	1.3e − 007
188.375	0.0963	0.0963	9.4e − 008
279.270	0.1008	0.1008	3.8e − 008
414.024	0.1073	0.1073	− 1.7e − 008
613.800	0.1162	0.1162	− 9.0e − 008
909.972	0.1284	0.1284	− 1.5e − 007
1349.053	0.1442	0.1442	− 1.8e − 007
2000.000	0.1640	0.1640	− 1.3e − 007

User Specified Root Mean Square Error $= 1.00e − 006$
Final Root Mean Square Error $= 3.01e − 007$
User Specified Number of Iterations $= 20$
Final Number of Iterations $= 17$

Iteration No $=$ 1	Error $= 8.2e − 002$
Iteration No $=$ 2	Error $= 8.0e − 002$
Iteration No $=$ 3	Error $= 7.6e − 002$
Iteration No $=$ 4	Error $= 6.7e − 002$
Iteration No $=$ 5	Error $= 5.3e − 002$
Iteration No $=$ 6	Error $= 5.1e − 002$
Iteration No $=$ 7	Error $= 2.1e − 002$
Iteration No $=$ 8	Error $= 9.5e − 003$
Iteration No $=$ 9	Error $= 6.1e − 003$
Iteration No $=$ 10	Error $= 1.4e − 003$
Iteration No $=$ 11	Error $= 6.8e − 004$
Iteration No $=$ 12	Error $= 3.9e − 004$
Iteration No $=$ 13	Error $= 2.7e − 004$
Iteration No $=$ 14	Error $= 1.5e − 004$
Iteration No $=$ 15	Error $= 1.2e − 004$
Iteration No $=$ 16	Error $= 1.9e − 005$
Iteration No $=$ 17	Error $= 3.0e − 007$

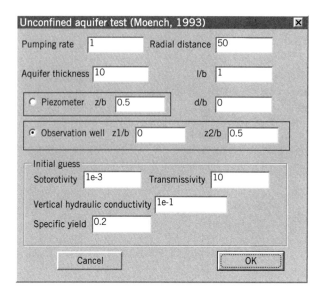

Figure 14.24 Dialog box for specifying the aquifer parameters in an unconfined aquifer with partially penetrating wells.

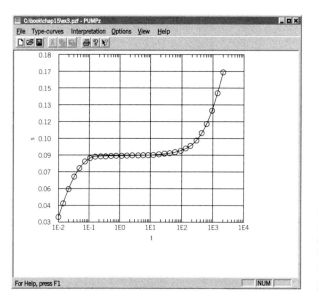

Figure 14.25 Display of measured (small circle) and calculated (solid line) drawdowns by **PUMPz** for an aquifer test in a leaky-confined aquifer with partially penetrating wells.

▶ CHAPTER HIGHLIGHTS

1. Personal computers can help with the calculation of drawdowns in forward calculations and in the interpretation of the results of aquifer tests to provide aquifer parameters. **WELLz** is included on the website. This software helps calculate drawdown due to one or more pumping wells. It accommodates calculations involving a fully penetrating well in a confined, leaky-confined, or water-table aquifer. **PUMPz** is a software package specifically designed for interpreting aquifer tests.

2. **WELLz 2.0** is a Windows code developed in C++. When the code is run, the initial dialog box has four main pull-down menus—**File**, **Editing**, **Options**, and **View**. Users move through a series of pull-down menus to add pertinent information about aquifer parameters, pumping rates, and the location of pumping and observation wells. Drawdowns are calculated on an equally spaced grid

across the simulation domain and at any specified observation wells. The code provides a contour map of the drawdowns and a summary report of the simulation trial.

3. **PUMPz** takes the results of observations from an aquifer test and calculates T, S, and other aquifer parameters for five different types of aquifers or test types: confined aquifer, leaky-confined aquifer, slug test, unconfined aquifer, and fractured aquifer. The interpretative schemes basically involve minimizing the difference between the measured drawdown in a well and the calculated drawdown in the same well. Users can choose from one of three different approaches—the trial-and-error, downhill simplex, or Marquardt method—to do the inversion. The Windows scheme makes it easy to provide the necessary input data to the code.

4. **PUMPz** also gives the user the option of printing paper copies of type curves and blank graph paper at the same scale as the type curves. This feature is very helpful in interpreting aquifer-test data following the classical curve-matching approaches.

15

MANAGEMENT OF GROUND-WATER RESOURCES

This chapter introduces basic concepts in ground-water management: safe yield, artificial recharge, and conjunctive use. It also describes the basic elements of digital ground-water flow models, which are playing a significant role in managing the ground-water resource in large and geologically complex settings. Much of the emphasis here will be on the MOD-FLOW family of codes, which provides an industry-standard tool kit for a wide variety of problems.

▶ 15.1 BASIC CONCEPTS IN MANAGING GROUND-WATER RESOURCES

In the United States, many states depend on ground water as a major water supply (Table 15.1). For other countries, the situation is similar. With the increasing use of ground water, there is a continuing need to manage the resource effectively.

The most fundamental approach to ground-water management is based on water balances within a ground-water basin. The mass balance equation in a ground-water reservoir can simply be written as

$$R_N + Q_i - T - Q_o - Q_p = \Delta S \tag{15.1}$$

where R_N is the recharge to ground water, Q_i is the surface-water inflow to ground-water storage, T is the transpiration, Q_o is the outflow from ground-water storage to surface water, Q_p is the total pumping rate in the basin, and ΔS is the change of storage. Equation (15.1) indicates that for a fixed recharge rate $(R_N + Q_i)$, the increase of the pumping rate (Q_p) eventually will decrease the outflow to surface water (Q_o), the transpiration (T), and storage (S). Decreasing the outflow to surface water could dry up or reduce flow in streams, creeks, lakes, and springs, whereas the decrease of storage will lower the ground-water

TABLE 15.1   Percent of total water withdrawals coming from ground water in 1990

State	Percent	State	Percent	State	Percent
Alabama	5.0%	Kentucky	5.7%	North Dakota	5.3%
Alaska	17.5%	Louisiana	14.3%	Ohio	7.8%
Arizona	41.7%	Maine	7.5%	Oklahoma	54.3%
Arkansas	60.1%	Maryland	3.7%	Oregon	9.1%
California	31.8%	Massachusetts	6.1%	Pennsylvania	10.4%
Colorado	22.0%	Michigan	6.1%	Rhode Island	4.8%
Connecticut	3.4%	Minnesota	24.3%	South Carolina	4.7%
Delaware	6.5%	Mississippi	73.5%	South Dakota	42.4%
D.C.	11.1%	Missouri	10.5%	Tennessee	5.5%
Florida	26.1%	Montana	2.3%	Texas	31.2%
Hawaii	5.2%	Nebraska	53.6%	Utah	21.7%
Georgia	18.6%	Nevada	32.0%	Vermont	7.1%
Idaho	38.5%	New Hampshire	4.9%	Virginia	6.5%
Illinois	5.2%	New Jersey	4.4%	Washington	18.3%
Indiana	6.6%	New Mexico	50.6%	West Virginia	15.9%
Iowa	17.3%	New York	4.4%	Wisconsin	10.5%
Kansas	71.7%	North Carolina	4.9%	Wyoming	5.3%
Puerto Rico	5.3%	Virgin Islands	2.4%		
United States	19.9%				

Source: USGS web, http://www.usgs.gov/

level in aquifers. Exactly how, when, and where these changes will be manifested depends on factors, such as the size of the basin, the geologic setting, and the times involved. Complexity can often arise in real basins because as pumping increases, recharge increases to some extent as well.

Proper management of a ground-water basin requires a capability to estimate how much water is available for development. Historically, one of the earliest approaches to analyzing ground-water yields was built on the concept of safe yield. *Safe yield* is defined as the rate of ground-water extraction from a basin for consumptive use over an indefinite period of time that can be maintained without producing negative effects. The goal of the "safe yield" is to achieve a "long-term balance" between ground-water use and ground-water replacement and to arrest the declines in the water table. The purpose of the safe-yield goal is not to prevent pumping and use of ground water. Rather, it is to limit pumping to the amount of ground water that may be safely "harvested" each year. What is meant by safely harvested is obviously nebulous. Conkling (1946) provides a few rules-of-thumb concerning safe yield. The annual extraction of water should not exceed the average annual recharge. The production of ground water should not lower the water table so that the permissible cost of pumping is exceeded, nor should pumping lead to an unacceptable deterioration in the quality of ground water.

As always, units of analysis need to be clearly defined. In the United States, components of the water budget are often quantified in acre-feet/year. The following equation provides the conversion between the metric unit and the practical U.S. unit.

$$1 \text{ acre-foot/yr} = 226.3 \text{ gallon/min} = 1234 \text{ m}^3/\text{day} = 0.5042 \text{ ft}^3/\text{sec} \qquad (15.2)$$

Here is a case study that looks at how concepts of safe yield are being used to help manage a ground-water supply in Arizona.

SAFE YIELD OF THE PRESCOTT ACTIVE MANAGEMENT AREA

A preliminary safe-yield analysis of the Prescott Active Management Area (AMA) in central Arizona was completed by the Arizona Department of Water Resources in 1997. It used three methods to answer the question as to whether or not the quantity of ground water being produced in the Prescott AMA was below the safe yield for the basin. These methods were: (1) tracking water levels, (2) evaluating annual water budgets, and (3) modeling ground-water flows. The Department examined available water-level data for the Prescott AMA from 1940 to 1994, from 1982 to 1998, and from 1994 to 1998. There was an indication of a gradual, but definite, ongoing decline in water levels in 73 to 75% of wells measured. The water budgets for 1995 to 1997 (Table 15.2) showed that withdrawals of ground water due to municipal, industrial, and agricultural demands were approximately 17,850 acre-feet per year. Natural discharges were approximately 4850 acre-feet per year. The estimated average annual recharge over the area for this period was only 13,900 acre-feet. Thus, ground water in storage decreased an average of 8800 acre-feet per year during this period. This overdraft is causing an ongoing depletion of the amount of ground water in storage in the AMA and producing the decline in the water table.

Beyond the current ground-water utilization, there are significant quantities of ground water in the Prescott AMA that are already committed for new municipal growth but not used. Thus, no ground water is available for allocation to new subdivisions.

The ground-water flow model for the Prescott AMA was constructed and calibrated using the hydrogeological information for the basin. The computer model confirmed observations and conclusions that current extractions exceeded the safe yield. The model also allowed the Department to simulate future conditions and to examine how pumping trends will impact water levels in the AMA. For example, the Department evaluated a scenario for the Prescott AMA using 1992 ground-water pumping and recharge data. In other words, "no growth" values based on 1992 data were used for every year for a period between 1996 and 2025. Given the ongoing growth and increasing water demand in the Prescott AMA, use of 1992 data for every year during that period obviously underestimates future ground-water use and produces a highly optimistic and unlikely scenario. In spite of this optimistic approach, the computer model projections indicate that ground-water levels would continue to decline throughout the AMA during the period through 2025.

Although the "safe-yield" concept is widely used as a ground-water management tool, it has been criticized for not taking surface water into consideration (Sophocleous, 1997). As indicated by Eq. (15.1), overpumping might not only reduce the ground-water level, but also decrease the outflow to surface water. For example, in Kansas, some of the major perennial streams dried up as water levels declined (Figure 15.1). A recent development in the management of water resources tends to treat ground-water and surface-water resources as a single system, so-called conjunctive management. In the arid southwestern states, plants and animals thrive in fragile ecosystems developed along the perennial streams. These systems are particularly at risk when the overdevelopment of ground-water resources lowers water tables in the riparian zones or results in significant water-table fluctuations (Arizona Department of Water Resources, 1994). The Santa Cruz River, near Tucson, Arizona, is one of several rivers where changes in the riparian ecosystems have raised concerns. Along the lower reaches of this river, the extensive pumping of ground water for irrigation and public supply wells has eliminated riparian vegetation. For example, lower ground-water levels along the lower Santa Cruz were likely a major cause of the destruction of mesquite woodlands (Arizona Department of Water Resources, 1994). The State of Arizona is pursuing active programs to maintain and improve the ecological health of riparian ecosystems through the management of ground-water and surface-water resources.

▶ 15.2 MANAGEMENT STRATEGIES

The overexploitation of ground water has also been tied to problems of land subsidence, sinkholes, saltwater intrusion, and costs of pumping. For example, in the city of Shanghai

TABLE 15.2   Prescott Active Management Area 1990–1997 groundwater budget (acre-feet per year)

	1990	1991	1992	1993	1994	1995	1996	1997
GROUNDWATER DEMAND								
MUNICIPAL GROUNDWATER WITHDRAWALS	8289	8667	8756	9595	10044	10303	11635	11594
City of Prescott	5014	5221	5056	5633	5656	5664	6352	6509
Shamrock Water Company	1795	1854	2019	2232	2615	3010	3439	3353
Small Provider	279	335	364	464	493	463	537	521
Exempt Wells	1201	1257	1317	1266	1280	1166	1307	1211
AGRICULTURAL GROUNDWATER WITHDRAWALS	6032	5943	4613	6460	6134	5316	6629	6260
Irrigation Grandfathered Rights	6032	5943	4613	6460	6134	5316	6629	6260
INDUSTRIAL GROUNDWATER WITHDRAWALS	444	486	443	500	533	555	688	628
Turf Facilities	349	399	313	343	357	391	502	434
Non-Turf Facilities	95	67	130	157	176	164	186	192
NATURAL SYSTEM DISCHARGES	4850	4850	4850	4850	4850	4850	4850	4850
Del Rio Springs Underflow from AMA	1500	1500	1500	1500	1500	1500	1500	1500
Del Rio Springs Baseflow from AMA	2100	2100	2100	2100	2100	2100	2100	2100
Upper Agua Fria Baseflow from AMA	1250	1250	1250	1250	1250	1250	1250	1250
GROUNDWATER RECHARGE								
INCIDENTAL RECHARGE	4578	8128	7640	9744	4512	7988	4942	4651
Agricultural Incidental Recharge	2413	3396	2758	3967	2875	4053	2906	2669
CVID Canal Losses	0	2548	2277	3449	1053	3001	635	412
Industrial Incidental Recharge	35	56	48	56	53	59	72	60
Effluent Discharged into Agua Fria	0	0	0	0	531	875	1329	1510
Effluent Recharged with No credits	2131	2128	2559	2272	0	0	0	0
NATURAL SYSTEM RECHARGE	4600	4600	4600	23320	4600	8920	4600	4600
Upper Agua Fria Natural Recharge	2550	2550	2550	2550	2550	2550	2550	2550
Little Chino Natural Recharge	2050	2050	2050	2050	2050	2050	2050	2050
Granite Creek Flood Recharge	0	0	0	18720	0	4320	0	0
CHANGE IN STORAGE								
TOTAL GROUNDWATER DEMAND	19615	19946	18662	21405	21561	21024	23802	23330
TOTAL GROUNDWATER RECHARGE	9178	12728	12240	33064	9112	16908	9542	9251
TOTAL GROUNDWATER OVERDRAFT	−10437	−7218	−6422	11659	−12449	−4116	−14261	−14079
City of Prescott Effluent Recharge Credits	0	0	0	0	1940	2098	1688	2270
CHANGE IN GROUNDWATER STORAGE	−10437	−7218	−6422	11659	−10509	−2018	−12573	−11809

Source: Arizona Department of Water Resources, 1997.

in the People's Republic of China, land subsidence was a serious problem due to the ground-water overextraction. From 1949 to 1959, annual land subsidence increased from 26 mm to more than 98 mm when the net ground-water extraction rate was increased from 10^5 m^3/day to 4×10^5 m^3/day (Figure 15.2).

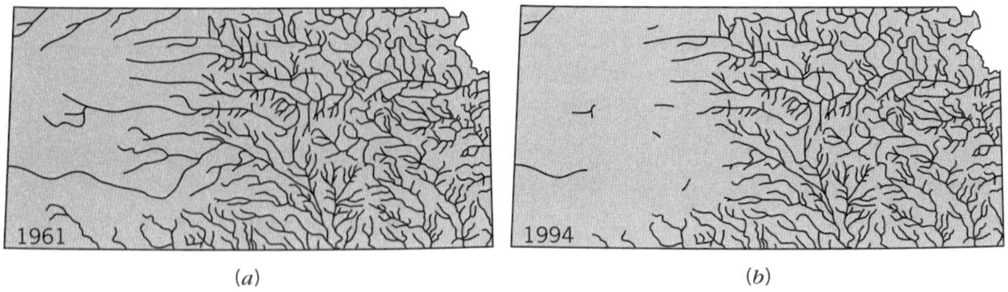

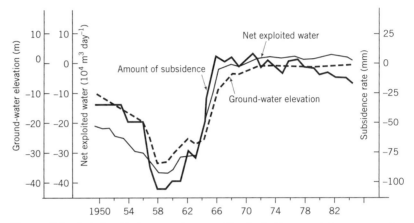

Figure 15.1 Distribution of major perennial streams in Kansas in (*a*) 1962; and (*b*) 1994 (from Sophocleous, 1998).

Figure 15.2 Relationship between the annual land subsidence and the net ground-water extraction rate in Shanghai, China.

The solution to these problems is evident from the water balance equations. One simply can reduce the rate of pumping, but in most cases this remedy is not usually feasible because the demand is fixed. Other possibilities are artificial recharge that increases the quantity of available ground water and conjunctive use, which replaces ground-water supplies by other sources. In Shanghai, the city reduced the net extraction by artificial recharge and other measures. After 1965, the net extraction from the aquifer was close to zero and the land subsidence was controlled. However, artificial recharge will not undo the subsidence that is already manifested. The city had to build floodwalls and pumping stations to control flooding caused by heavy rainfall and tidal invasions in subsiding land along the seacoast. In the following sections, we examine two of the most common strategies for dealing with excessive ground-water withdrawals.

Artificial Recharge

Todd (1980) explains that *artificial recharge* augments the natural infiltration of precipitation or surface water into the ground by some method of construction, the spreading of water, or a change in natural conditions. Artificial recharge is often used (1) to replenish depleted supplies, (2) to prevent or to retard saltwater intrusion, or (3) to store water underground where surface-storage facilities are inadequate to accommodate seasonal demands.

Artificial recharge is commonly classified into two categories: *induced recharge* and *direct recharge*. In induced recharge, wells, spaced at relatively short distances from each other, are installed in a line along a riverbank. These wells pump ground water and induce flow out of the river (Figure 15.3). In some cases, the induced infiltration from the river is significantly limited by the low permeability of the riverbed. In other cases, the well system has to be operated with knowledge that at some times of the year the surface water could be contaminated.

Direct recharge schemes have been developed, including recharge pits, shafts, spreading basins, and wells (Figures 15.4*a* through 15.4*d*). *Spreading basins* are commonly used for recharging shallow aquifers. They can be constructed along streams to capture some

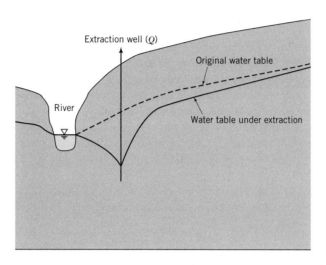

Figure 15.3 Induced infiltration due to a line of closely spaced wells along a river.

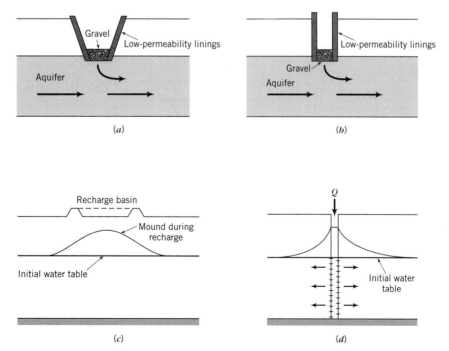

Figure 15.4 Direct recharge schemes. (*a*) recharge pits; (*b*) recharge shafts; (*c*) spreading basins; (*d*) recharge wells.

of the flood flow or in the vicinity of water treatment plants to receive treated water. For deep aquifers, recharge wells are more effective than spreading basins. Pits and shafts are only used in existing stone quarries and gravel pits because the construction of pits and shafts is expensive and they only have a small recharge capacity. Compared with induced infiltration, direct recharge systems can be better managed in terms of quality and quantity. In most cases, pretreatment of the recharged water is necessary to ensure that aquifers will not be contaminated by recharged water.

The most important design issues concerned with artificial recharge are to develop and maintain appropriately high-inflow rates (Todd, 1959). Table 15.3 gives some representative spreading basin recharge rates and some average well recharge rates. Recharge rates associated with spreading basins vary from 15 $m^3/m^2/d$ (m/d) for some gravels to as little as 0.5 m/d in sand and silt (Bear, 1979). Typically, the infiltration rates fall off with time due to clogging of the spreading grounds or recharge wells (Figure 15.5).

In both the surface-spreading and well-injection schemes, clogging of the basins and the recharge wells can greatly reduce the recharge rates and increase pressure heads. The clogging is due to: (1) the existence of suspended silt in the recharge water; (2) the precipitation of solids due to mixing and chemical reaction between native water in an aquifer and recharged water; (3) the growth of algae and bacteria when nutrients are available in the recharged water; and (4) introduction of dissolved gases in the pore of aquifers during recharge.

A spreading basin can be rejuvenated by (1) chemical treatments (polyphosphates prevent precipitation of iron and calcium carbonate), (2) mechanical treatments (plowing,

TABLE 15.3   Representative artificial recharge rates

Spreading Basins		Recharge Wells	
Location	Rate (m/day)	Location	Rate (m^3/day)
California		California	
Los Angeles	0.7–1.9	Fresno	500–2200
Madera	0.3–1.2	Los Angeles	2900
San Gabriel River	0.6–1.6	Manhattan Beach	1000–2400
San Joaquin Valley	0.1–0.5	Orange Cove	1700–2200
Santa Ana River	0.5–2.9	San Fernando Valley	700
Santa Clara Valley	0.4–2.2	Tulare County	300
Tulare County	0.1		
Ventura County	0.4–0.5		
New Jersey		New Jersey	
East Orange	0.1	Newark	1500
Princeton	<0.1		
New York		Texas	
Long Island	0.2–0.9	El Paso	5600
		High Plains	700–2700
Iowa		New York	
Des Moines	0.5	Long Island	500–5400
Washington		Florida	
Richland	2.3	Orlando	500–51,000
Massachusetts		Idaho	
Newton	1.3	Mud Lake	500–2400

Source: Todd (1959). Reprinted from *Groundwater Hydrology*. Copyright © 1959. John Wiley & Sons, Inc. Reprinted by permission of John Wiley & Sons, Inc.

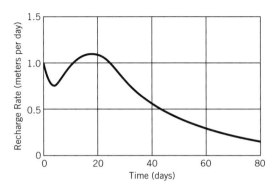

Figure 15.5 Time variation of recharge rate for water spreading on unsaturated soil (from Muckel, 1959).

harrowing to increase porosity; or cyclic operational schemes, such as three weeks of infiltration and one week of drying in summer, three weeks of infiltration, and three weeks of drying in winter; or draining, drying, and scraping procedures), and (3) vegetative treatments (plant native grass cover in the basin to filter suspended colloids).

Diminishing injection rates in wells can be treated either hydraulically or chemically. Backpumping reverses the direction of flow in the well so that clogging materials are pulled back into the well and pumped out. Normally, it takes about 5 to 15 minutes of pumping at a rate equal to the injection rate to reduce by 80% the increase in injection head caused by plugging. For difficult plugging problems, well screens need to be backpumped section by section at a rate greater than the injection rate to remove the clogging materials. Special equipment and long backpumping times (half to one day) are required. When these pumping schemes fail to restore the injection capabilities, chemical treatment is likely the last resort. Chlorine and related compounds are effective in treating plugging by bacteria. Acids are used to dissolve deposits of calcium carbonate. More information on well treatment can be found in the *Ground Water Manual* (Bureau of Reclamation, 1995).

Artificial recharge can degrade ground-water quality when only poor quality water is available. Options to avoid degrading the ground-water resource are to treat the water to remove the most serious contaminants and to ensure that transport in the subsurface is capable of effectively removing the contamination (for example, sorption of phosphate or trace organic compounds).

Conjunctive Use

Conjunctive use involves the coordinated use of surface and ground water to meet some specified water demand in a given area. The objectives of conjunctive use are (1) to maximize net benefits; (2) to minimize costs; and (3) to minimize the degradation of environment.

The net benefit can be expressed as

$$\text{Net benefit} = \text{Revenues} - \text{Costs} \qquad (15.3)$$

For example, Lefkoff and Gorelick (1990a) combined a stream-aquifer model with an economic and agronomic model to study physical processes and economic behavior in irrigated agriculture in Arkansas River Valley in southern Colorado. The economic component of the model involved optimizing production levels for each farm for each season. The agronomic component of the model computes the yield of crops as a function of the quantity and salinity of the irrigation water. The hydrological component of the model determined the volume and salinity of water available for irrigation in space and time in the stream-aquifer system. The results of their simulations were consistent with historical observations. Lefkoff and

Gorelick (1990b) suggested that a water rental market would provide significant economic benefits to all farmers in the area. Their idea is that farmers without ground-water rights buy surface water from farmers with ground-water rights during a dry year, while farmers with ground-water rights increase pumping to replace sold surface water to maximize the profit for the farmers participating in the market.

Importation of Water

Another approach to managing water resources in a basin is to provide alternative sources of water besides ground water. For example, it is technically feasible to move surface water over long distances, as is evident in California. This water can be provided on a seasonal basis and, if appropriate supplies are available, it can change the dynamic of water utilization.

Linking These Water Management Strategies—An Example

There are ground-water basins, especially in the arid southwestern part of the United States, where a basin-wide response to pumping on a massive scale is evident. A case in point is the Upper Los Angeles River Area (ULARA) in southern California, which is comprised of one large ground-water basin, the San Fernando Basin, and three other smaller basins—the Sylmar Basin, the Verdugo Basin, and the Eagle Rock Basin (Figure 15.6). The San Fernando

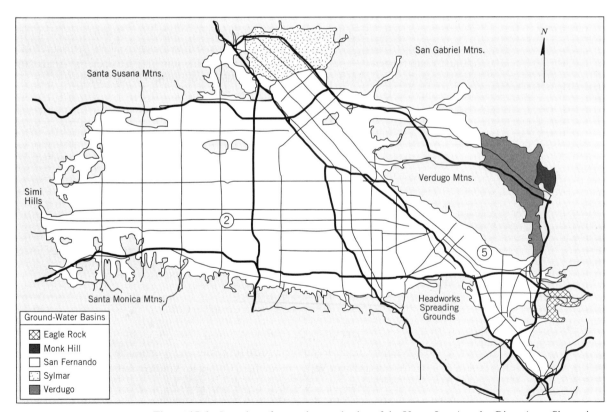

Figure 15.6 Location of ground-water basins of the Upper Los Angeles River Area. Shown in circles is the location of observation wells referred to in the text (modified from Watermaster Report, 1994).

Valley area includes the cities of Los Angeles, Burbank, and Glendale and is home to various Hollywood studios and industrial companies such as Lockheed, Rockwell, and 3M.

The ULARA is bounded by various mountain ranges and hills (Figure 15.6). Ground water is extracted by wells from the thick valley-fill aquifer across much of the San Fernando Valley. Although the valley-fill aquifer is locally productive, rates of recharge are relatively low due mainly to the arid climate and seasonal nature of precipitation. Over the past 100 years, the valley floor received an average of about 14 inches (35.6 cm) of rainfall, most of which falls from December to March (ULARA Watermaster, 1994).

As the population of the San Fernando Valley increased from the 1930s onward, it became inevitable that the valley-fill aquifer could not provide an inexhaustible supply of water. The natural discharge to the rivers was lost initially, followed by a loss of storage in the ground-water system itself. Shown in Figure 15.6 are two observation wells (3700A, located at 2; and 3914H, located at 5). The hydrographs for these wells (Figure 15.7) depict the historical decline of water levels in the valley-fill aquifer from the late 1930s to 1968. As suggested by the hydrographs, water-level declines of the order of 100 to 200 ft (30.4 to 60.8 m) were not uncommon.

Ground-water withdrawals in the basin were limited by court actions in 1968 to approximately 104,000 acre-feet per water year (ULARA Watermaster, 1994). This rate of withdrawal was about two-thirds of the average of the previous six years. This decrease is reflected in the behavior of the hydrographs, which subsequently leveled out or recovered (Figure 15.7). Overall, the basin is managed to keep total ground-water withdrawals within the safe yield of the basin.

Safe-yield concepts were applied to the San Fernando Basin to help allocate the available water. The safe yield of the basin was estimated to be 43,660 acre-feet per year.

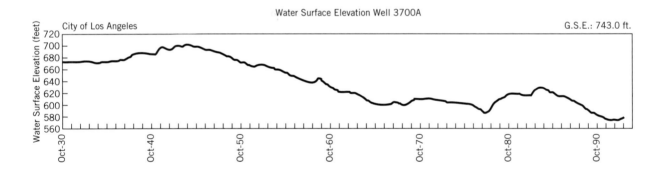

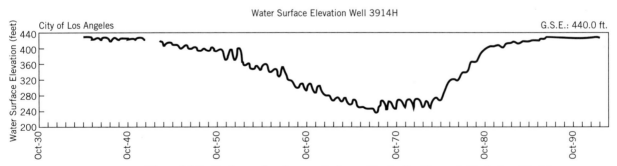

Figure 15.7 Hydrographs of two wells located in the San Fernando Basin (modified from Watermaster Report, 1994).

The entire amount is available to the City of Los Angeles for extraction. This yield represents the long-term average recharge from precipitation. Some of the water that is imported to the basin from other sources ends up as additional sources of recharge to the valley-fill aquifer. This water is also available for pumping by the cities of Los Angeles, Burbank, and Glendale. In the 1993–1994 water year, some 43,900 acre-feet of this imported water was available for pumping (ULARA Watermaster, 1994). Across the San Fernando Basin actually more water ends up in the ground from recharge of imported water than from recharge due to natural precipitation on the basin.

The management scheme for the ULARA also provides the capabilities for Los Angeles, Glendale, and Burbank to store water and extract equivalent amounts in the future. In effect, in years when the pumping for the cities is less than their allocated amount, the water is deemed to be saved and is available for future use in years of high demand.

Within the San Fernando Basin, the ULARA Watermaster Report (1994) outlines plans to enhance the water available for pumping by artificial recharge schemes. The proposed East Valley Water Reclamation Project involves the use of up to 40 million gallons per day of treated sewage from the Donald C. Tillman Water Reclamation Plant to provide water for various purposes (Figure 15.8). A significant proportion of this water will be used for ground-water recharge via two spreading grounds.

Pilot studies have been underway in the San Fernando Basin to examine the water-quality implications of recharging treated sewage. At the Headworks Spreading Ground (see Figure 15.6), water from the Los Angeles River, whose low flow is mainly treated reclaimed water from the Tillman Plant, was spread over an area of approximately 30 acres. The spread water was removed from the ground by a well 1000 ft away. The water removed by pumping was approximately 45% reclaimed water and 55% native ground water. Preliminary indications are that transport through the ground improved the water quality. Reductions were apparent in coliforms, total organic carbon, biological oxygen demand, nitrite, ammonia, and turbidity levels.

Historically, water-management activities like those in the ULARA relied on water balance calculations for broad, basin-scale estimates of inflow and outflow. In the future, there will be greater reliance on calibrated mathematical simulation models at the basin scale that more rigorously account for the inherent variability in material properties, recharge rates, and ground-water withdrawals.

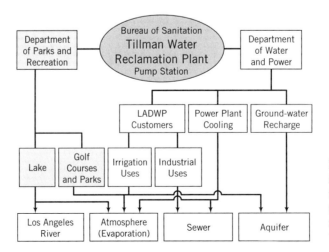

Figure 15.8 Flow diagram illustrating ways proposed for the use of reclaimed water from the Tillman Plant (modified from Watermaster Report, 1994).

► 15.3 INTRODUCTION TO GROUND-WATER MODELING

We have seen how mass balance concepts can be useful in managing ground-water resources in a basin. However, they lack the resolution to provide explicit details on the likely impacts of specific spatial and temporal patterns in water utilization on ground-water and surface-water systems. Managers often want to know in detail how much ground-water levels might be expected to decline in specific areas or in what manner streamflows could be impacted.

Chapters 9 through 11 explained how aquifer models could provide future predictions of the hydraulic-head distribution due to pumping from one or several wells (Chapter 14). Yet, it should be evident that these models were extremely idealized. For example, they could not handle recharge distributed on an areal basis, complicated ground-water/surface-water interactions, and complex heterogeneous aquifer systems. These are the very characteristics common to most large three-dimensional basins that require management.

Analysis of complex problems requires the application of sophisticated numerical models. Unlike the classic analytical solutions to problems of well hydraulics, *numerical models* are based on a fundamentally different mathematical approach. The governing differential equation is approximated by a difference form of the equation that is solved at nodes that comprise a two- or three-dimensional grid. This method is amazingly powerful because you can apply a different set of hydraulic parameters, recharge fluxes, and pumping fluxes to the individual nodes or to the neighborhood of a node. This capability permits the replication of complex hydrogeologic settings with a myriad of pumping wells. Indeed, with these models, the main problem lies not in the capabilities of the code in realistically treating a problem, but in finding sufficient data to describe the hydrogeological problem completely.

Governing Equation

Numerical models provide a solution to a governing ground-water flow equation, subject to boundary and initial conditions. The most general models solve an equation that looks like this

$$\frac{\partial}{\partial x}\left(K_{xx}\frac{\partial h}{\partial x}\right) + \frac{\partial}{\partial y}\left(K_{yy}\frac{\partial h}{\partial y}\right) + \frac{\partial}{\partial z}\left(K_{zz}\frac{\partial h}{\partial z}\right) - W = S_s\frac{\partial h}{\partial t} \qquad (15.4)$$

where K_{xx}, K_{yy}, and K_{zz} are values of hydraulic conductivity along the x-, y-, and z-coordinate axes, h is the hydraulic head, W is a flux term that accounts for pumping, recharge, or other sources and sinks, S_s is the specific storage, x, y, and z are space coordinates, and t is time. Looking at this equation, we can easily appreciate the power of numerical approaches. The solution to Eq. (15.4) provides a transient prediction of hydraulic head in a three-dimensional domain for an anisotropic hydraulic-conductivity field. This form of the equation implies that the principle directions of the hydraulic-conductivity ellipse coincide with the coordinate axes.

One can specify many different types of boundary conditions that vary in space and time. As the model steps forward in time, one can apply seasonally and spatially dependent recharge, turn wells on or off, and change pumping rates. The boundary conditions are appropriately general that the model provides a faithful representation of the actual system.

Numerical Procedures

The main numerical approaches used today for solving ground-water flow equations are finite-difference and finite-element methods (Figures 15.9*a* through 15.9*c*). Both are sophisticated methods that in different ways replace the governing differential equation for ground-water flow by a system of algebraic equations.

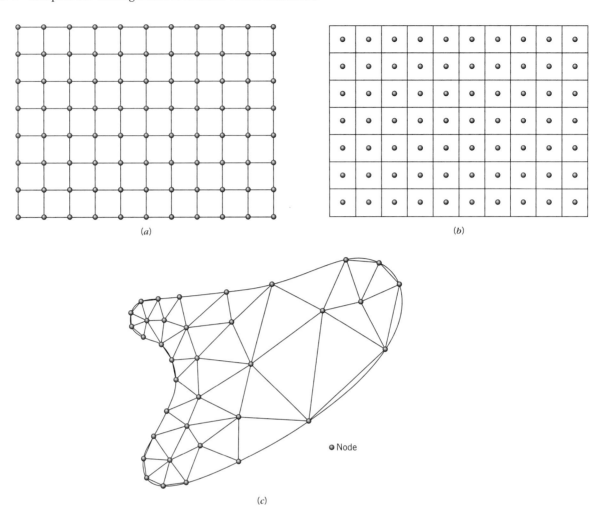

(a)

(b)

(c)

Figure 15.9 Examples of (*a*) mesh-centered and (*b*) block-centered finite-difference grids. Panel (*c*) illustrates the discretization of an irregularly shaped aquifer with linear triangular elements (from Domenico and Schwartz, 1998. *Physical and chemical hydrogeology*). Copyright ©1998 by John Wiley & Sons, Inc. Reprinted by permission of John Wiley & Sons, Inc.

An algebraic equation is written and solved for each node. Thus, both methods require that the region of interest be subdivided using a mesh or grid network. Sometimes, the cumbersome term *discretization* is used to describe this process. The finite-difference approaches utilize a regular discretization, where an aquifer is subdivided into a series of rectangular grid blocks. In a two-dimensional model, each model cell is assumed to have a thickness, m. Thus, each of the grid blocks represents a volume of the aquifer, $m\Delta x\Delta y$. In a three-dimensional model consisting of aquifers and confining beds, individual units are subdivided vertically into cells of a specified thickness.

Operationally, the size of a grid block in the *x-y* plane (that is, a map view) is kept small relative to the overall extent of the aquifer. The spacing between the rows and columns varies, but for simplicity the example grids (Figures 15.9*a* and 15.9*b*) assume a constant spacing (that is, $\Delta x = \Delta y$). Associated with the grid blocks are nodes that represent the points where the unknown hydraulic heads are calculated. Depending on the formulation, node points can be either mesh centered (Figure 15.9*a*) or block centered (Figure 15.9*b*).

The *finite-element method* permits a much more general arrangement of node points. The discretization with triangular elements (Figure 15.9c) illustrates how easy it is to define the boundaries of irregularly shaped aquifers and to ensure that node points coincide with monitoring wells or geographic features (for example, a river). We will leave the detailed discussion of modeling methods to other more specialized texts like Zheng and Bennett (1995) and Huyakorn and Pinder (1983).

In this and following sections, we will develop preliminary concepts of finite-difference methods in relation to the industry-standard code, MODFLOW (McDonald and Harbaugh, 1988).

Description of the Finite-Difference Grid

The finite-difference solution requires that a grid discretize the domain. With MODFLOW, the cells are brick shaped (Figure 15.10). The *grid system* is referenced in terms of a row/column/layer numbering scheme with block-centered nodes (Figure 15.10). As the notation implies (Figure 15.10), the dimensions of each cell can be varied. Thus, a dense system of nodes can be provided around features of interest, and a sparse system can be provided in areas of lesser concern. Overall, the variable grid minimizes the number of nodes in a simulation. However, care must be taken to change cell sizes gradually. The rule-of-thumb is that the dimensions of adjacent cells in a given direction should not differ by more than a factor of 1.5.

The cell sizes in the row and column directions are specified explicitly in the input data. The vertical dimension is specified implicitly by specifying transmissivity values for individual cells. The most common strategy has layers conform to readily identifiable geologic units. Consider the simple example depicted in Figure 15.11a with a coarse sand, a silt, and a sand and gravel layer. Vertical discretization is accomplished using three model layers that coincide with the three stratigraphic units (Figure 15.11b). Changing the

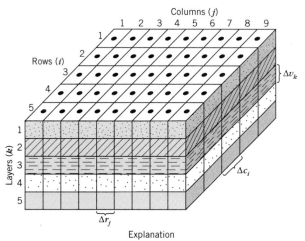

Explanation

• Node

Δr_j Dimension of cell along the row direction. Subscript (j) indicates the number of the column

Δc_i Dimension of cell along the column direction. Subscript (i) indicates the number of the row

Δv_k Dimension of the cell along the vertical direction. Subscript (k) indicates the number of the layer

Figure 15.10 Discretization of a three-dimensional system (modified from McDonald and Harbaugh, 1988).

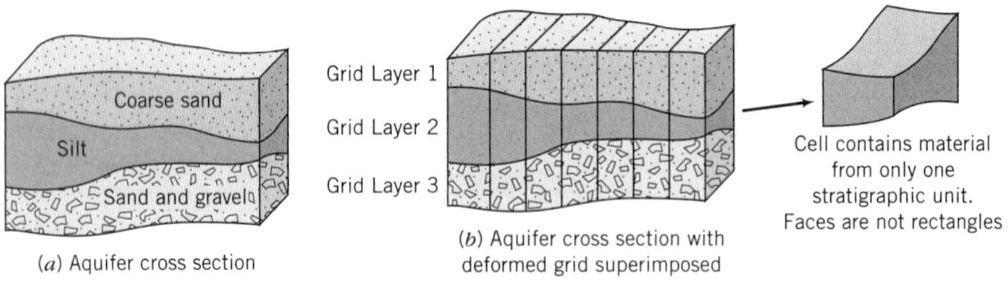
Coarse sand
Silt
Sand and gravel
(a) Aquifer cross section
Grid Layer 1
Grid Layer 2
Grid Layer 3
(b) Aquifer cross section with deformed grid superimposed
Cell contains material from only one stratigraphic unit. Faces are not rectangles

Figure 15.11 One possible approach for representing (*a*) complex layering by (*b*) a deformed grid (modified from McDonald and Harbaugh, 1988).

transmissivity of each cell would accommodate the changing vertical thickness of individual cells, for example, within layer 1.

With the model layers so defined, individual cells are no longer bricks. They have an irregular shape where the cell faces are no longer rectangles. Although this representation of a cell gives rise to errors, they are usually insignificant in relation to errors caused by poor estimates of transmissivity, storativity, and recharge rates.

Derivation of the Finite-Difference Equation

The finite-difference form of the flow equations can be derived in several different ways. One way is to work directly with the governing flow equation (15.4), expressing the derivatives in difference form. The second approach is used with MODFLOW and is discussed here. This approach is based on the equation of continuity which states that the sum of flows into and out of any cell is equal to the time rate of storage plus or minus additions of water from sources or sinks. It is written mathematically as

$$\sum Q_i = S_s \frac{\Delta h}{\Delta t} \Delta V \tag{15.5}$$

where Q_i accounts for ground-water flow into the cell from adjacent cells through six sides and for water added or withdrawn (for example, recharge pumping), S_s is the specific storage, Δh is the change in head over a time interval Δt, and ΔV is the volume of the cell. Q_i in Eq. (15.5) can be expanded using the Darcy equation, written in terms of the gradients between nodes. Doing this substitution and rearranging terms give us a system of equations

$$[A]\{h\} = \{q\} \tag{15.6}$$

where $[A]$ is the coefficient matrix, $\{h\}$ is the vector of unknown head values, and $\{q\}$ is a vector of constant-head terms.

Solving Systems of Finite-Difference Equations

The mathematical solution of the systems of equations (15.6) provides the hydraulic head for the given time step. Procedures for solving systems of algebraic equations can be broadly categorized as direct and iterative. Direct approaches involve rearranging the system of equations to a form that can be easily solved. The methods that most of us learned in high school to solve simple systems of equations are examples of the direct methods. An example of the direct methods is Gaussian elimination. The iterative approaches involve making some initial guess at the unknowns and refining these guesses through a series of repeated calculations until an accurate solution is obtained.

The original MODFLOW contained two iterative schemes. The simpler one is Slice Successive Overrelaxation (SSOR). Instead of solving the entire system of unknowns at the same time, the equations are formulated for a two-dimensional slice (Figure 15.12) with the assumption that the heads in the adjacent two slices are known. The resulting system of equations (actually formulated as the change in hydraulic head) is solved by Gaussian elimination. The slice will usually contain a relatively small number of nodes because in most cases the number of model layers is small. One iteration is complete when all of the slices are processed. After a large number of iterations, the solution converges.

The Strongly Implicit Procedure (SIP) is a more complicated approach. This procedure involves solving the unknowns for the entire grid simultaneously (McDonald and Harbaugh, 1988). More recently, Hill (1990) implemented a preconditioned conjugate gradient procedure (PCG) for use with MODFLOW. In general, this and similar solution techniques are extremely fast and robust, and for these reasons are now often used in solving systems of linear equations.

▶ 15.4 GROUND-WATER MODELING SOFTWARE

MODFLOW has emerged as the de facto standard code for simulating ground-water flow in saturated zones. This code along with other USGS ground-water modeling software is available on the Internet at http://h2o.usgs.gov/software/. The Fortran 66 version of MOD-FLOW was published in 1984 (McDonald and Harbaugh, 1984) and the Fortran 77 version of MODFLOW was published in 1988 (McDonald and Harbaugh, 1988). Major revisions of MODFLOW were released with additions of PCG2 (preconditioned conjugate-gradient solver) (Hill, 1990), BCF3 (version 3 of block-centered flow package) (Goode and Appel, 1992), STR1 (stream package) (Prudic, 1989), HFB1 (horizontal-flow barriers package) (Hsieh and Freckleton, 1993), ISB1 (Interbed-Storage and Time-Variant Specified-Head Packages) (Leake and Prudic, 1988), CHD1 (TIME-VARIANT SPECIFIED-HEAD package), and GFD1 (GENERAL FINITE-DIFFERENCE FLOW PACKAGE) (Harbaugh, 1992) in 1993. TLK1 (transient leakage package) (Leake et al., 1994) and DE45 (direct solution package based on alternating diagonal ordering) (Harbaugh, 1995) were added in 1995. RES1 (reservoir package) was documented in 1996. An updated version of the overall model, which is called MODFLOW-96 was released in 1996. A number of changes were made to make MODFLOW easier to use and easier to enhance. MODFLOW-96 can use existing input data sets and has the same computational methods (Harbaugh and McDonald, 1996a and 1996b). As we complete this book, MODFLOW-2000 is just being released.

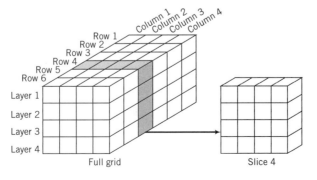

Figure 15.12 Division of the three-dimensional model array into vertical slices for processing in the SSOR package (modified from McDonald and Harbaugh, 1988).

Other stand-alone codes and variants of MODFLOW provide additional capabilities. For example, Pollock (1989) developed MODPATH—a post-processing package that takes output from MODFLOW simulations and computes three-dimensional pathlines. This package has found applications in the simple modeling of contaminant transport and in the location of wells for pump and treat systems for the recovery of dissolved contaminants. Hill (1992) has developed MODFLOWP, a code that includes capabilities for estimating various parameters required in a MODFLOW simulation. This code provides a tool for automatic calibration.

User's Guide to the MODFLOW Software

MODFLOW is built with a modular design that involves a main program and "packages." The packages are groups of independent subroutines, which carry out specific simulation tasks. According to their functions, the packages are divided into five categories: (1) basic package; (2) block-centered flow package; (3) solver packages: SSOR, SIP, and PCG; (4) boundary condition packages: rivers, recharge, wells, evapotranspiration, drain, and general head; and (5) output control package. For each simulation, basic, block-centered flow, and solver packages must be defined. Boundary condition packages are optional depending on the conceptualization of the ground-water flow problems. A default output is used if the output control is not specified. To help students use MODFLOW96, user's guides and examples can be downloaded from www.wiley.com/college/schwartz. For MODFLOW 2000, which provides sensitivity analysis and parameter estimation, please check the USGS website.

Operational Issues

Time-Step Size
Normally, care must be exercised in selecting a time-step size to avoid errors that can occur due to large time steps. These errors are most pronounced in the first few time steps when pumping begins or pumping rates change. The preferred way of controlling errors is to use relatively small time-step sizes. Overall, it is prudent to place less confidence in results coming from the first few time steps after a significant change in withdrawal or injection rates has taken place. MODFLOW lets the user increase the size of the time step as the simulation proceeds. The given time-step size is multiplied by a number greater than 1 between time steps. A time-step multiplier of 1.2 works well in most cases. Thus, an initial time step of 1 day becomes 1.2 days for the second time step and 1.2×1.2 or 1.44 days for the third time step and so on. A variable time-step size will reduce the computational effort compared to a constant-step size and yet preserve the accuracy of the calculation.

Drawdowns at "Pumping" Nodes
In many numerical models, predicting drawdowns at "pumping" nodes is a problem. The head or drawdown at a pumping node does not represent drawdown in the well because the hydraulic head is a "cell average." This value is not the same as that calculated with an analytical model (for example, the Theis equation), which is the exact drawdown at the point of interest. A simple way to reduce the error is to use smaller cell sizes near a well.

Pre- and Post-Processing Software

Several companies market software that assists in the preparation of MODFLOW data sets. Examples include Geraghty & Miller's ModelCad product, Waterloo Hydrogeologic's

Visual MODFLOW, the Department of Defense's Groundwater Modeling System, and Environmental Simulations' Groundwater Vista. Other utilities facilitate the contouring of hydraulic-head fields (SURFER; Golden Software, Inc., and Tecplot, Amtech Engineering, Inc).

▶ 15.5 PRACTICAL CONSIDERATIONS IN THE USE OF GROUND-WATER MODELS

A variety of different tasks are required to construct a ground-water model (Figure 15.13). The main steps include: (1) evaluation of the available data/information, (2) conceptualization of this hydrogeologic setting in a model framework, (3) setup and running of the model, (4) model calibration, and (5) either reevaluation and collection of new data or verification testing, depending on the success of the calibration. Once a model is successfully verified, it can be used productively.

Evaluation of Available Data/Information

In most cases, modeling tasks are performed after a large amount of hydrogeological information is already available. The data should be sufficient to provide: (1) the correct determination of hydraulic conductivity, storativity, and thickness for each hydrogeologic unit; (2) a relatively uniform distribution of hydraulic-head measurements in the simulation domain; and (3) the correct specification of initial and boundary conditions.

Fundamentally, the data required for a ground-water model depends on the modeling objectives. Banks (1993) and Bredehoeft et al. (1995) define three types of exploratory modeling: (1) history matching (adjusts parameters until the measured hydraulic head closely matches the simulated hydraulic head), (2) modeling driven by some policy question (examines plausible scenarios in response to human actions or policy choices), and (3) modeling related to some conceptual problem (compares simulations of different conceptual models). Data requirements can vary for these different modeling purposes. For example, modeling concerned with history matching would require more data than modeling concerned with

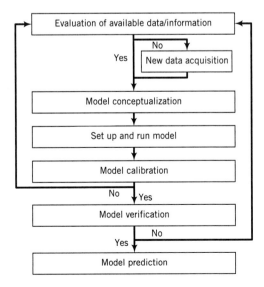

Figure 15.13 Main steps in ground-water modeling.

policy questions or conceptual issues. Most of the ground-water modeling tasks fall into the category of history matching.

Model Conceptualization

Model conceptualization usually involves: (1) defining a simulation domain and the hydrogeological layers; (2) dividing this domain into zones, each of which possesses a unique set of hydraulic properties; (3) collecting values of hydraulic properties (hydraulic conductivity and storativity) for each material zone; (4) determining the outside boundary conditions along six sides of the model domain; (5) determining the internal boundary conditions, such as rivers, wells, recharge, evapotranspiration, drain, and head-dependent fluxes, which are also called stresses; (6) collecting values of measured hydraulic head; and (7) determining cells that are inactive or carry a constant head.

The size of the simulation domain is often larger than the region of interest so that the somewhat artificial conditions along the boundaries don't intrude on the region of interest. Boundary conditions are often difficult to define along the sides, top, and bottom of the domain because hydraulic heads or inflow/outflow rates are often poorly defined. One strategy to overcome this problem is to place model boundaries along "natural" hydrogeologic boundaries or parallel to pathlines. An example of a natural hydrogeologic boundary is a river defined by constant hydraulic-head values equivalent to the stage of the river. Locally, such a boundary condition would provide a zone of recharge or discharge. Another example would be the selection of the major watershed divide as a no-flow boundary. The boundary is no-flow because the gradient is such that flow moves away from a divide. Features of the hydrogeologic setting also help in assigning boundaries. For example, the top of a thick, low-hydraulic conductivity unit can be selected as the bottom of the simulation domain using a no flow boundary. At some depth in almost any system, it should be possible to define a boundary of no flow, implying that deeper circulation is minimal.

Defining boundary conditions along the top of the simulation region is more difficult because often there are active inflows and outflows of water due to recharge and discharge. Experience shows that recharge/discharge rates vary in time and space and are difficult hydrogeologic parameters to measure.

With transient models (that is, models where hydraulic head varies as a function of time), one also needs to define the initial condition for the simulation. In most cases, the initial conditions are so complicated that they would need to be simulated with a steady-state model.

Setting Up and Running Models

Model setup is a process of translating the model conceptualization into the computer input files required by a ground-water model. With the basic version of MODFLOW, the input files must be prepared by strictly following a user's guide. You run the original USGS MODFLOW code by typing the executable file name at the DOS prompt. An example will be given later to illustrate how to set up and run MODFLOW. Commercially available software can help to simplify this process because data can be input easily with drop-down menus. These programs usually contain launching functions to run MODFLOW within the program.

Model Calibration

The hydraulic-head prediction that comes from a flow model is commonly used as the basis for model calibration. Calibration is a process of selecting model parameters to achieve a good match between the predicted and measured hydraulic heads, or other relevant hydrogeologic data like streamflow changes between gaging stations. Most commonly, calibration is accomplished by a trial-and-error adjustment of model parameters. This procedure involves the systematic variation of model parameters like hydraulic conductivity, storativity, flows, or boundary conditions. Automated inverse procedures, like those in MODFLOWP (Hill, 1992), may speed up calibration. Calibration or model tuning is required because ground-water systems are so poorly known.

Each model run provides a predicted hydraulic-head distribution or other flow rates that can be compared to existing field measurements. Commonly, the comparison is not favorable—suggesting problems in the system conceptualization or inappropriate parameter values. The lack of calibration may force a reexamination of how the model is constructed or in some cases may prompt additional data collection studies. After many iterations as shown in Figure 15.13, the predictions of hydraulic heads and various flow rates should match the measured values.

Common practice is to set calibration criteria in advance of the calibration exercise. There are no hard and fast rules for what constitutes a good calibration except that errors should be small relative to the total hydraulic head. Criteria may be based on (1) the mean error, (2) the mean absolute error, or (3) the root mean squared error (Anderson and Woessner, 1992). These measures of the difference between measured and simulated hydraulic-head data are given as

$$\text{Mean Error (ME)}: \quad \frac{1}{n} \sum_{i=1}^{n} (h_m - h_s)_i \tag{15.7}$$

$$\text{Mean Absolute Error (MAE)}: \quad \frac{1}{n} \sum_{i=1}^{n} |(h_m - h_s)_i| \tag{15.8}$$

$$\text{Root Mean Squared Error (RMSE)}: \quad \left[\frac{1}{n} \sum_{i=1}^{n} (h_m - h_s)_i^2 \right]^{0.5} \tag{15.9}$$

where n is the number of points where comparisons are made, h_m is the measured hydraulic head at some point, i, and h_s is the simulated hydraulic head at the same point (Anderson and Woessner, 1992). Of the three error estimates, Anderson and Woessner (1992) point to the root mean squared error as the best quantitative measure if the errors are normally distributed. The mean error is not preferred because large positive and negative errors can cancel each other out. Thus, a small error estimate may hide a poor model calibration. Other requirements are sometimes applied in addition to error estimates such as quantitatively correct flow directions and flow gradients.

Model Verification

Once calibration is complete, a verification test is commonly added to check that the model is a valid representation of the hydrogeologic system. Commonly, model verification involves using the calibrated model to simulate a hydrologic response that is known. For example, one might hold back results from one or more large-scale aquifer tests and examine how well the calibrated model simulates the test. Again, the errors between the observed and simulated hydraulic-head values can be quantified in terms of the error measures. If the model successfully passes this last test, then it can be used for predictive analyses.

Predictions of Ground-Water Flow

Predictions made with simulation models must be interpreted with caution. The "aura of correctness" (Bredehoeft and Konikow, 1993) attached to model calculations often exerts much more influence than is reasonable, given the typically uncertain data on which models are built. Oreskes et al. (1994) believe that "models are representations, useful for guiding further study but not susceptible to proof "(p. 644).

Predictability becomes a problem because ground-water systems are often so poorly characterized. It is usually unrealistic to find data sufficient to describe hydrologic processes in space and time. Thus, the model design depends significantly on the "informed judgment" of its builder rather than on real information. This uncertainty does not disappear simply because a model is constructed.

The calibration-verification process does not lead to a unique description of a hydrogeologic system. For poorly known systems, a large number of different models can be developed without knowing which, if any, is correct. Stated another way, different model developers, given the same hydrogeologic data, will probably develop different conceptualizations of the same system, each of which can be calibrated and verified. An example of the difficulty of calibrating a ground-water flow model is discussed by Freyberg (1988). Different groups using the same set of synthetic data developed quite different predictions concerning system behavior. Sources of variability in the calibration and the resulting prediction were related to (1) the lack of hydrogeologic data, (2) the use of different measures of success in calibration by each group, and (3) differing strategies for calibration (for example, changing local transmissivity values around wells versus changing values over large areas).

The success in model predictions can be examined with *post-audits*. This term describes the process of checking a prediction made by a ground-water model. For example, if one were to make a prediction about the behavior of a system 10 years from now, one could return after 10 years, make the necessary hydraulic-head measurements, and check whether the original model could be validated. Bredehoeft and Konikow (1993), commenting on results from the few available post-audits, indicate that "extrapolations into the future were rarely very accurate." They identified the following problems with the models: "the period of history match (that is, calibration) was too short to capture an important element of the model, or the conceptual model was incomplete, or the parameters were not well defined." They concluded that the record of "validating" models was "not encouraging."

If model predictions are suspect, of what use are they? In addressing this issue, Oreskes et al. (1994) pointed out,

> Models can corroborate a hypothesis by offering evidence to strengthen what may be already partly established through other means. Models can elucidate discrepancies in other models. Models can also be used for sensitivity analysis—for exploring "what if" questions—thereby illuminating which aspects of the system are most in need of further study, and where more empirical data are needed.

In summary, one must use the power provided by computer models carefully. These are useful tools that should be used with full knowledge of their limitations.

▶ 15.6 EXAMPLES OF GROUND-WATER MODELS

Two examples are provided in this section. The first demonstrates how to set up a ground-water flow model using MODFLOW. The second example illustrates the application of MODFLOW to a buried valley aquifer complex near Wooster, Ohio.

Developing a Data Set for a Simple Example

The grid for the first example has 300 cells with two layers, ten rows, and fifteen columns (Figure 15.14a). Two sand aquifers are separated by a layer of silt (Figure 15.14b). Layer 1 represents an unconfined aquifer with a hydraulic conductivity of 50 m/day, and Layer 2 represents a confined aquifer with a transmissivity of 500 m²/day. The confining silt layer is not included as an actual layer but is represented using a vertical conductance between the two sandy aquifers of 0.001 m²/day (see McDonald and Harbaugh, 1988, for calculations). The grid spacings along rows and columns are equal to 500 m.

The problem is transient with two stress periods. The initial condition is a constant initial head of 100 m. At the start of the simulation, it is assumed that continuing irrigation provides a recharge rate of 0.004 m/day and is active throughout both stress periods. River cells are also active along the fifteenth column in the second layer with h_b = 97 m, h_{str} = 100 m, C_{riv} = 10, 000 m²/day (Figure 15.15). These parameters specify the thickness and conductance of the streambed. The first column in the first layer is a constant-head boundary with h = 100 m, while the first column in the second layer is inactive (Figure 15.15). In addition to these boundary conditions, two wells are added in the second stress period at nodes (2,3,4) and (2,3,8) (Figure 15.16).

Before running MODFLOW, a name file needs to be prepared. The name file defines a unit number, a file identification, and a file name for reading or writing information from or to the file. The sequence of input and output files in the name file follows that in Table 15.4. For example, a name file is illustrated in Table 15.5 for the current problem. Files for BAS, BCF, PCG RIV, RCH, WEL, and OC packages are also prepared, as shown in Table 15.4. The first column of Table 15.4 is the element assignment in the array IUNIT (24) for different packages, which indicates that a package is active when IUNIT in the BAS package is greater than zero.

Please refer to the user's guide and examples at www.wiley.com/college/schwartz to understand the input files in Table 15.5. To run this example, follow the following instructions.

1. Create a MODFLOW directory on your hard disk, along with the subdirectory EX1 within the MODFLOW directory.

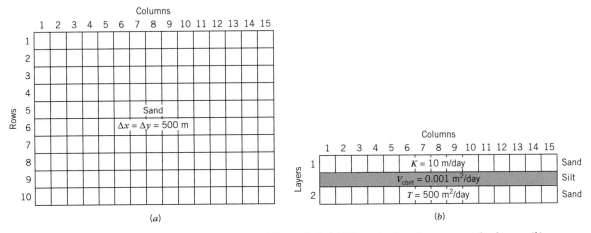

Figure 15.14 Model setup of Example 1. (a) Discretization along rows and columns; (b) hydrogeological layers: two layers of sands separated by a layer of silt.

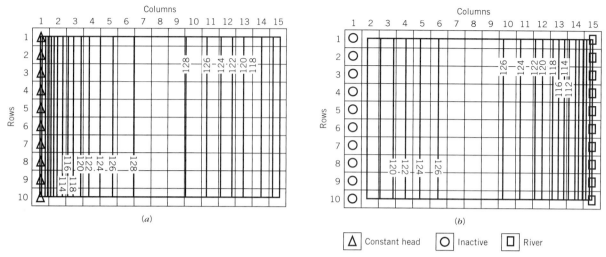

Figure 15.15 Simulation results and boundary conditions for stress period 1. (*a*) Layer 1: constant-head cells along column 1; (*b*) Layer 2: inactive cells along column 1 and river cells along column 15.

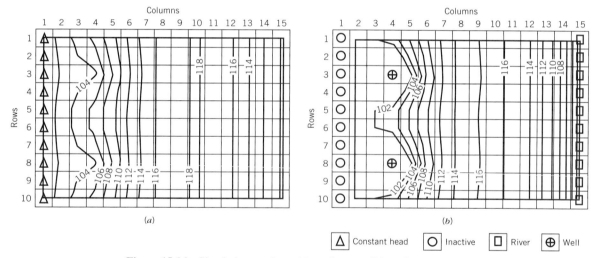

Figure 15.16 Simulation results and boundary conditions for stress period 2. (*a*) Layer 1: constant-head cells along column 1; (*b*) Layer 2: inactive cells along column 1, river cells along column 15, well cells at (3,4) and (8,4).

2. Find the file MODFLOW96.EXE in the directory /PROGRAMS/MODFLOW, and copy it to the MODFLOW directory on your hard drive.

3. Find the subdirectory EX1 in the directory /PROGRAMS/MODFLOW in the programs on the website, and copy all of the files in the EX1 to the EX1 subdirectory in MODFLOW directory in your hard drive.

4. Type "..\MODFLOW96.EXE" in the EX1 subdirectory in your hard disk drive and hit return.

TABLE 15.4   A summary of the input and output files for MODFOW

| Unit | File Identification | | File Description |
	List		Output Listing
	BAS		Basic Package
1	BCF		Block-Centered Flow Package
2	WEL		Well Package
3	DRN		Drain Package
4	RIV		River Package
5	EVT		Evapotranspiration Package
6	TLK		Transient Leakage Package
7	GHB		General Head Boundary Package
8	RCH		Recharge Package
9	SIP		Strongly Implicit Procedure Package
10	DE4		Direct solver
11	SOR		Slice Successive Over-Relaxation Package
12	OC		Output Control Package
13	PCG		Preconditioned Conjugate Gradient Package
14	GFD		General Finite Difference Flow Package
15			
16	HFB		Horizontal Flow Barrier Package
17	RES		Reservoir Package
18	STR		Stream Package
19	IBS		Interbed-Storage Package
20	CHD		Time-Variant Specified-Head Package
21			
22			
23			
24			

Source: Modified from McDonald and Harbaugh, 1988.

5. Type "ex1.nam" and push return.
6. The simulation results will be in file ex1.*l*st.

The output file "ex1.*l*st" contains information about the input, calculated head, and water budget. The calculated head values are shown in Figures 15.14 and 15.15 for stress periods 1 and 2, respectively. The water budget for the stress periods is tabulated in Table 15.5. In stress period 1, water is discharged to the river and the constant-head cells in column 1. In stress period 2, ground-water discharge is due to the wells, the river, and the constant-head cells.

Case Study in the Application of MODFLOW

Model Conceptualization

Springer (1990) and Springer and Bair (1992) applied MODFLOW to a buried valley aquifer complex near Wooster, Ohio. Figure 15.17 shows the geometry of the buried valley, Killbuck Creek and its tributaries, Little Killbuck and Apple Creeks, as well as the location of two major municipal well fields. From 1984 to 1988, the North and South Well fields produced from 3.5 to 4.4 million gallons per day. Today, both well fields may be abandoned due to contamination.

Cross section *A-A'* shows the bedrock valley and the distribution of geologic units. Immediately overlying bedrock is glacial till. This unit is in turn overlain by sand and

TABLE 15.5   Input files of MODFLOW for an aquifer system

Name file: exl.nam

LIST	6	exl.1st
BAS	5	exl.bas
BCF	11	exl.bcf
WEL	12	exl.wel
RIV	14	exl.riv
RCH	18	exl.rch
PCG	19	exl.pcg
OC	20	exl.oc

1. Basic Package: exl.bas

Valley aquifer with 2 sand layers separated	
by silt. Stress period 1 is natural	HEADING(1)
conditions. Stress period 2 adds wells.	HEADING(2)
2 10 15 2 4	NLAY,NROW,NCOL,NPER,ITMUNI
11 12 0 14 0 0 0 18 0 0 0 20 19	IUNIT(24)
0 0	IAPART,ISTRT
5 1 (15I3) 3	IBOUND(NCOL,NROW) for layer 1

```
−1 1 1 1 1 1 1 1 1 1 1 1 1 1 1
−1 1 1 1 1 1 1 1 1 1 1 1 1 1 1
−1 1 1 1 1 1 1 1 1 1 1 1 1 1 1
−1 1 1 1 1 1 1 1 1 1 1 1 1 1 1
−1 1 1 1 1 1 1 1 1 1 1 1 1 1 1
−1 1 1 1 1 1 1 1 1 1 1 1 1 1 1
−1 1 1 1 1 1 1 1 1 1 1 1 1 1 1
−1 1 1 1 1 1 1 1 1 1 1 1 1 1 1
−1 1 1 1 1 1 1 1 1 1 1 1 1 1 1
−1 1 1 1 1 1 1 1 1 1 1 1 1 1 1
```

5 1 (15I3) 3	IBOUND(NCOL,NROW) for layer 2

```
0 1 1 1 1 1 1 1 1 1 1 1 1 1 1
0 1 1 1 1 1 1 1 1 1 1 1 1 1 1
0 1 1 1 1 1 1 1 1 1 1 1 1 1 1
0 1 1 1 1 1 1 1 1 1 1 1 1 1 1
0 1 1 1 1 1 1 1 1 1 1 1 1 1 1
0 1 1 1 1 1 1 1 1 1 1 1 1 1 1
0 1 1 1 1 1 1 1 1 1 1 1 1 1 1
0 1 1 1 1 1 1 1 1 1 1 1 1 1 1
0 1 1 1 1 1 1 1 1 1 1 1 1 1 1
0 1 1 1 1 1 1 1 1 1 1 1 1 1 1
```

999.99	HNOFLO
0 100.	SHEAD(NCOL,NROW) for layer 1
0 100.	SHEAD(NCOL,NROW) for layer 2
1. 1 1.	PERLEN, NSTP, TSMULT for stress period 1
1. 1 1.	PERLEN, NSTP, TSMULT for stress period 2

2. Block-Centered Flow Package: exl.bcf

1 30 777.77 0 1. 0	OISS,IBCFCB,HDRY,IWDFLG,WETFCT,IWETIT,IHDWET
1 0	LAYCON(NLAY)
0 1.	TRPY(NLAY)
0 500.	DELR(NCOL)
0 500.	DELC(NROW)
0 10.	HY(NCOL.NROW) for layer 1
0 50.	BOT(NCOL,NROW) for layer 1

(*continued*)

TABLE 15.5   continued

0	.001					VCONT(NCOL,NROW) for layer 1
0	500.					TRAN (NCOL,NROW) for layer 2

3. Solution package (PCG2): exl.pcg

40	20	1		MXITER,ITER1,NPCOND
	.0010 1000.0 1.0 2 1 1 0			HCLOSE,RCLOSE,RELAX,NBPOL,IPRPCG,MUTPCG

5. Recharge Package: exl.rch

3	30		NRCHOP, IRCHCB
	0		INRECH, INIRCH for stress period 1
	0	.0040	RECH(NCOL, NROW) for stress period 1
	−1		INRECH (recharge for stress period 2 is the same as stress period 1)

4. River Package: exl.riv

10	30					MXRIVR, IRIVCB
	10					ITMP (no. of reaches active in stress period 1)
2	1	15	100.	10000.	97.	LAYER, ROW, COLUMN, STAGE, COND, RBOT
2	2	15	100.	10000.	97.	
2	3	15	100.	10000.	97.	
2	4	15	100.	10000.	97.	
2	5	15	100.	10000.	97.	
2	6	15	100.	10000.	97.	
2	7	15	100.	10000.	97.	
2	8	15	100.	10000.	97.	
2	9	15	100.	10000.	97.	
2	10	15	100.	10000.	97.	
	−1					ITMP (<0, river data from last stress period will be reused)

5. Well Package: exl.wel

2	30			MXWE, IWELCB
	0			ITMP (No. of wells active in stress period 1)
	2			ITMP (No. of wells active in stress period 2)
2	3	4	-35000.	LAYER, ROW, COLUMN, Q
2	8	4	-35000.	

6. Output Control package: exl.oc

−4	−4	50	50	IHEDFM, IDDNFM, IHEDUN, IDDNUN
0	1	1	0	INCODE, IHDDFL, IBUDFL, ICBCFL for stress period 1
1	0	0	0	HDPR, DDPR, HDSV, DDSV for stress period 1
0	1	1	0	INCODE, IHDDFL, IBUDFL, ICBCFL for stress period 2
1	0	0	0	HDPR, DDPR, HDSV, DDSV for stress period 2

7. Partial Listing of Output from MODFLOW: exl.1st

VOLUMETRIC BUDGET FOR ENTIRE MODEL AT END OF TIME STEP 1 IN STRESS PERIOD 1

CUMULATIVE VOLUMES L**3	RATES FOR THIS TIME STEP L**3/AT
IN:	IN:
STORAGE = 0.00000	STORAGE = 0.00000
CONSTANT HEAD = 0.00000	CONSTANT HEAD = 0.00000
WELLS = 0.00000	WELLS = 0.00000
RECHARGE = 0.14000E+06	RECHARGE = 0.14000E+06
RIVER LEAKAGE = 0.00000	RIVER LEAKAGE = 0.00000
TOTAL IN = 0.14000E+06	TOTAL IN = 0.14000E+06

(continued)

TABLE 15.5   Continued

OUT:	OUT:
STORAGE = 0.00000	STORAGE = 0.00000
CONSTANT HEAD = 64437.	CONSTANT HEAD = 64437.
WELLS = 0.00000	WELLS = 0.00000
RECHARGE = 0.00000	RECHARGE = 0.00000
RIVER LEAKAGE = 75563.	RIVER LEAKAGE = 75563.
TOTAL OUT = 0.14000E+06	TOTAL OUT = 0.14000E+06
IN − OUT = −0.20313	IN − OUT = −0.20313
PERCENT DISCREPANCY = 0.00	PERCENT DISCREPANCY = 0.00

VOLUMETRIC BUDGET FOR ENTIRE MODEL AT END OF TIME STEP 1 IN STRESS PERIOD 2

CUMULATIVE VOLUMES L**3	RATES FOR THIS TIME STEP L**3/T
IN:	IN:
STORAGE = 0.00000	STORAGE= 0.00000
CONSTANT HEAD = 0.00000	CONSTANT HEAD = 0.00000
WELLS = 0.00000	WELLS = 0.00000
RECHARGE = 0.28000E+06	RECHARGE = 0.14000E+06
RIVER LEAKAGE = 0.00000	RIVER LEAKAGE = 0.00000
TOTAL IN = 0.28000E+06	TOTAL IN = 0.14000E+06
OUT:	OUT:
STORAGE = 0.00000	STORAGE = 0.00000
CONSTANT HEAD = 78139.	CONSTANT HEAD = 13702.
WELLS = 70000.	WELLS = 70000.
RECHARGE = 0.00000	RECHARGE = 0.00000
RIVER LEAKAGE = 0.13186E+06	RIVER LEAKAGE = 56298.
TOTAL OUT = 0.28000E+06	TOTAL OUT = 0.14000E+06
IN − OUT = −0.40625	IN − OUT = −0.18750
PERCENT DISCREPANCY = 0.00	PERCENT DISCREPANCY = 0.00

gravel outwash, which comprises the aquifer. The outwash unit can be separated by a silt unit up to 21 ft thick (Figure 15.18a). The silt unit is relatively discontinuous; for example, it does not occur in the area of the South Well field. The uppermost geologic unit is comprised of silty-clay lacustrine and floodplain deposits ranging in thickness from 10 to 30 ft. Shown on the map (Figure 15.17) is the location of alluvial fans along Little Killbuck, Clear, and Apple Creeks that directly overlie the outwash aquifer. A variety of hydrogeologic studies and careful stream-discharge measurements have shown that the alluvial fans provide the most important hydraulic connection between surface water and ground water (Breen et al., 1995; Springer and Bair, 1992).

Through the years, a great deal of hydrogeologic data has been collected in relation to this aquifer system (see Springer, 1990). The database includes extensive stream-discharge measurements, hydraulic-conductivity measurements of the streambed, lithologic logs from various test holes, hydraulic information from aquifer tests with a variety of wells, production rates for municipal and industrial wells, and measured hydraulic-head data. The model is supported by extensive hydrogeologic data, which are discussed in detail by Springer (1990). See Breen et al. (1995) for a more recent USGS modeling study of the same system.

Setting Up and Running the Model

The finite difference model is comprised of 28 rows, 46 columns, and 4 layers. The active portion of the grid (Figure 15.19) is determined by the buried valley and is comprised of

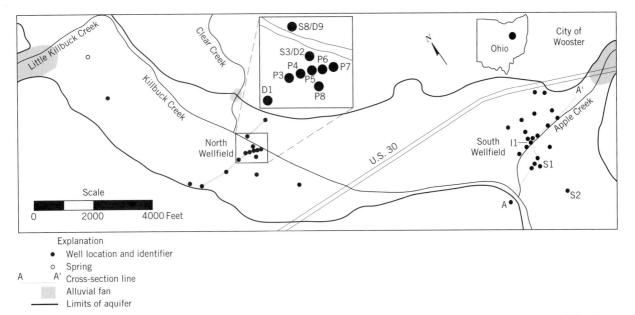

Figure 15.17 Map showing the location of the Wooster study site, the extent of the buried valley aquifer, and the region encompassed by the MODFLOW model (modified from Springer and Bair, 1992). Reprinted by permission of *Ground Water*. Copyright ©1992. All rights reserved.

cells having dimensions $\Delta r_j = 250$ ft and Δc_i ranging between 250 and 500 ft. Figure 15.18b shows how the observed stratigraphy at the site relates to the four layers of the grid. The silty-clay unit that subdivides the outwash along A-A' is not explicitly included as a separate layer. Its hydraulic effects, however, are included by lowering the vertical conductivity between units 2 and 3 in terms of vertical conductance.

Figure 15.19 shows the boundary conditions for layer 1. The various streams are represented with the River Package. Thus, with pumping, induced infiltration from the creeks occurs with the quantity of infiltration determined by the head gradient between the stream and the aquifer, as well as the area and hydraulic conductivity of the streambed. The relatively low-hydraulic conductivity of the lacustrine and floodplain deposits restricts the quantity of induced infiltration. The alluvial fans are important because much more induced infiltration is possible through their higher permeability streambeds as the streams pass over them. Inflows to the alluvial aquifer further up or down valley (that is, drift underflow) are represented by constant fluxes created by adding wells along the end boundaries of the model using the Well Package (Figure 15.19). Inflows to the active cells due to flow through the low-permeability valley walls (that is, bedrock underflow) are simulated using the General-Head Package (Figure 15.19). In effect, the General Head package is a variable-flux boundary condition that lets the inflow through the boundaries increase as the drawdown in the aquifer increases. By being able to specify a conductance (proportional to hydraulic conductivity) between the aquifer and the boundary, the inflow through the boundary can be controlled. In the Wooster example, 29% of the water pumped from the aquifer is thought to originate from the inflow of bedrock water through the side boundaries (Breen, 1995). The boundary conditions for the deeper units (that is, layers 2, 3, and 4) are essentially the same except that the streams and recharge are not included. There are also minor changes around the alluvial fans, but generally the boundary conditions are set up to provide bedrock and drift underflows as was the case with layer 1.

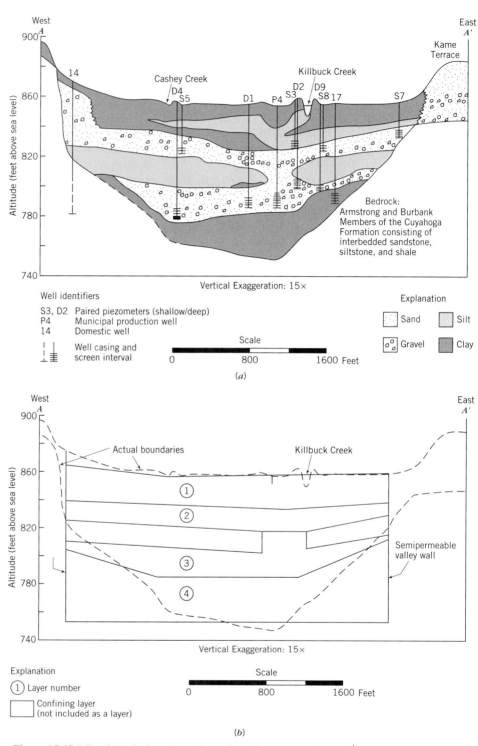

Figure 15.18 Panel (*a*) depicts the geology along the cross section *A–A'* (modified from Springer and Bair, 1992). Reprinted by permission of *Ground Water*. Copyright ©1992. All rights reserved. Panel (*b*) illustrates how the various units are conceptualized for MODFLOW.

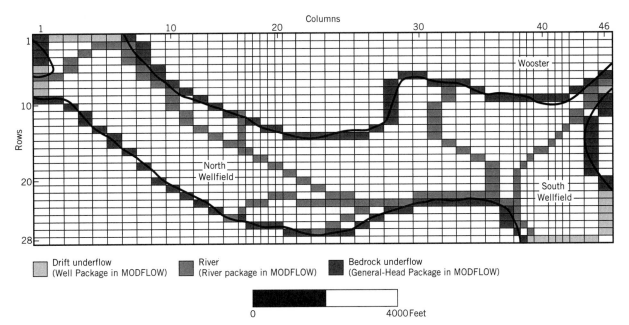

Figure 15.19 Location of active grid cells and a summary of boundary conditions for layer (modified from Springer, 1990).

Overall, the potential sources of inflow to the aquifers are more than capable of balancing the water withdrawn by pumping. Therefore, the flow system will be at steady state. The simulation model then is configured to provide a steady-state result. The drawdown around the North and South Well fields is sufficiently large that layer 1 will desaturate locally and portions of layer 2 will behave as an unconfined aquifer. Thus, layer 1 is considered to be an unconfined aquifer, layer 2 is considered capable of being converted from a confined to an unconfined aquifer, and layers 3 and 4 are considered to be confined.

Through a detailed interpretation of geologic data Springer (1990) developed detailed isopack maps that, together with lithologic and hydraulic test data, provided a full description of transmissivity distributions for the layers representing the outwash aquifer. Thus, transmissivity values for layer 2 varied from 500 to 6000 ft²/d and those for layer 3, from 20 to 90,000 ft²/d. The transmissivity distributions for the lower hydraulic-conductivity units (layers 1 and 4) are less well known and are not represented in detail. Transmissivities of layer 1 range between 0.01 and 200 ft²/d (Springer and Bair, 1992). Layer 4 is assigned a constant transmissivity of 0.0005 ft²/d. Streambed conductance values are based on measured hydraulic conductivities for various reaches of the streambed (Springer, 1990). Recharge rates of 4 in/y for the valley and 10 in/y are assigned to alluvial fans. These values are based on estimates of the U.S. Geological Survey (Breen et al., 1995). Pumping rates for the individual wells are quantified through a synthesis of existing records.

The website contains the complete set of files required to run the Wooster case study in the directory PROGRAMS/MODFLOW/EX2. To run the example, please follow the instructions for Example 1.

Model Calibration

Calibration involved making minor adjustments to the transmissivity field, vertical conductance values (between vertical nodes), streambed hydraulic-conductivity values, and bedrock leakage rates until the steady-state model was successfully calibrated. The calibration criteria involved (1) a good visual comparison between the measured and simulated potentiometric surfaces, (2) a root mean square error between measured and simulated heads of less than 2.5 ft, (3) simulated water losses from streams that compared favorably to measured losses, and (4) simulated leakage from bedrock that agreed with estimates from chemical mixing and isotopic studies (Springer and Bair, 1992).

Shown in Figure 15.20 is the simulated, steady-state potentiometric surface for layer 3 under the imposed conditions of pumping. The cone of depression due to large withdrawals at the North and South Well fields is obvious. The model was validated by removing the pumping stress and successfully simulating the historic, nonpumping potentiometric surface (Springer and Bair, 1992). Figure 15.21 shows a scatter plot of simulated versus measured heads at the North Well field area. The distance between a point and the diagonal line shows the degree of error between the measured and calculated heads. The mean absolute error (MAE) and root mean squared error (RMSE) of the flow model at the North Well field are 3.44 and 5.45 ft, respectively.

▷ **EXERCISE**

15.1. With the help of the MODFLOW code and the data files supplied for the Wooster case study, determine the proportion of water pumped from the system due to natural recharge, leakage from the creeks, drift underflow, and bedrock underflow. To obtain the answer to the problem, it is necessary to plow through the output to find the mass balance summary.

▷ **CHAPTER HIGHLIGHTS**

1. There is an increasing need to manage ground-water resources effectively. The most fundamental approach to ground-water management is based on water balance within a ground-water basin, which is described by the following equation

$$R_N + Q_i - T - Q_o - Q_p = \Delta S$$

where R_N is recharge to ground water, Q_i is surface-water inflow to ground water, T is transpiration, Q_o is outflow from ground water to surface water, Q_p is the total pumping rate in the basin, and ΔS is the change of storage.

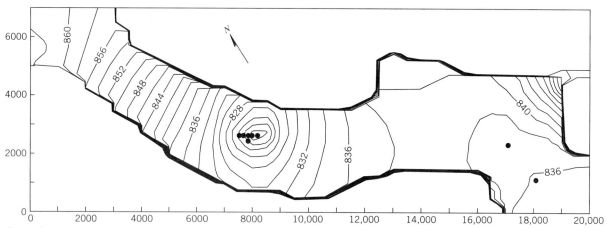

Explanation

Scales in feet

● Pumping well location

−836 — Contour of equal hydraulic head (ft asl)

Contour interval: 2 feet

Figure 15.20 Map of the simulated distribution of hydraulic head for layer 3 (from Springer, 1990).

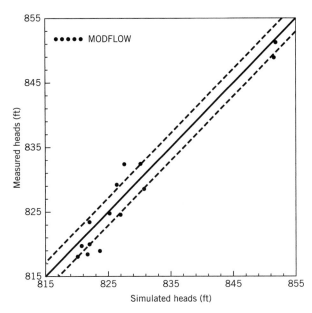

Figure 15.21 A scatter plot of simulated and measured heads (modified from Springer and Bair, 1992). Reprinted by permission of *Ground Water*. Copyright ©1992. All rights reserved.

2. With a relatively fixed quantity of recharge rate ($R_N + Q_i$), an increase in the pumping rate (Q_p) eventually will lead to a decrease in the outflow to surface water (Q_o), the transpiration (T), and storage (S). Reducing Q_o could diminish flow to streams, creeks, lakes, and springs, whereas the decrease of storage will lower ground-water levels. When and where these changes will be manifested depends on the basin size, the geology, and the times involved.

3. Safe yield is the rate of ground-water extraction from a basin for consumptive use over an indefinite period of time that can be maintained without negative effects. The safe yield attempts to prescribe a long-term balance between ground-water use and replacement and to arrest declines in the water table. The goal is not to prevent pumping and use of ground water. Rather, it is to limit pumping to the amount of ground water that can be safely harvested each year.

4. Overexploitation of ground water can create problems of land subsidence, sinkholes, and saltwater intrusion. As implied from the water balance equations, one can reduce the rate of pumping if the demand is flexible. Other possibilities are artificial recharge that increases the quantity of available ground water, conjunctive use, which replaces ground-water supplies by other sources, and water importation.

5. Complex problems can be analyzed with numerical modeling software. Through numerical approaches, the governing differential equation is approximated by a difference form, which is solved at nodes that comprise a two- or three-dimensional grid. This method is amazingly powerful because you can apply a different set of hydraulic parameters, recharge fluxes, and pumping fluxes to the individual nodes or to the neighborhood of a node. One can replicate complex hydrogeologic settings with many wells.

6. The finite-difference approaches involve subdividing the domain into a series of rectangular grid blocks. Operationally, the size of a grid block in a map view is usually kept small relative to the extent of the aquifer. Associated with the grid blocks are nodes where the unknown hydraulic heads are calculated. The finite-element method permits a more general arrangement of nodes. Triangular elements aid in defining the boundaries of irregularly shaped aquifers and in insuring that node points coincide with monitoring wells or geographic features (for example, a river).

7. MODFLOW has emerged as the de facto standard code for simulating ground-water flow in saturated zones. Other stand-alone codes and variants of MODFLOW provide additional capabilities. For example, MODPATH is a post-processing package that takes output from MODFLOW simulations and computes three-dimensional pathlines. Several companies market software that assists in the preparation of MODFLOW data sets.

8. The main steps in constructing a ground-water model include: (a) evaluation of the available data/information, (b) conceptualization of the hydrogeologic setting in a model framework, (c) setup and running of the model, (d) model calibration, and (e) either reevaluation and collection of new data or verification testing, depending on the success of the calibration.

9. Predictions made with simulation models must be interpreted with caution. The "aura of correctness" attached to model calculations often exerts much more influence than is reasonable, given the typically uncertain data from which models are built. Predictability is a problem because ground-water systems are often so poorly characterized. Thus, the model design depends significantly on the "informed judgment" of its builder rather than on real information.

16

DISSOLVED MASS IN GROUND WATER

Many modern-day studies of ground water are concerned with issues of water chemistry. Information on water chemistry is used to make decisions about how water can be used, or whether it contains contaminants that might be hazardous to human or ecological health. This chapter introduces the study of dissolved mass in ground water. It describes the constituents that are found in natural and contaminated ground water, the practical aspects of working with the chemical data, and the way wells are used for chemical sampling.

▶ 16.1 DISSOLVED CONSTITUENTS IN GROUND WATER

Rocks and minerals dissolve in water to form ions. Positively charged species, such as Ca^{2+} or K^+, are *cations*, whereas negatively charged species, such as HCO_3^- or Cl^- are *anions*. Organic substances can also dissolve to form organic cations or anions. However, most organic compounds (for example, trichloroethene, TCE) dissolve as *nonionic* (that is, uncharged) molecules. Most often, ions or organic molecules turn up in water from the dissolution of minerals (for example, halite or calcite), liquids (for example, trichloroethylene, TCE), or gases,

$$
\begin{aligned}
\text{halite dissolution: } & NaCl = Na^+ + Cl^- \\
\text{calcite dissolution: } & CaCO_3 + H^+ = Ca^{2+} + HCO_3^- \\
\text{TCE dissolution: } & TCE = TCE_{aq}
\end{aligned}
$$

Concentration Scales

Concentration measurements tell us how much of a particular ion or molecule is dissolved in the water. In practice, concentrations are reported using these scales.

Molar concentration (M, also mM, μM) represents the number of moles of a species per liter of solution (mol/L). A mole is the formula weight of a substance expressed in grams. For example, a one-liter solution containing 1.42 g of Na_2SO_4 has a (Na_2SO_4) molarity of $1.42/(2 \times 22.99 + 32.06 + 4 \times 16.00)$, or 0.010 M. Because Na_2SO_4 dissociates completely in water as

$$Na_2SO_4 = 2Na^+ + SO_4^{2-}$$

the molar concentrations of Na^+ and SO_4^{2-} are 0.02 and 0.01 M, respectively. In this reaction, one mole of Na_2SO_4 dissolves to produce two moles of Na^+ and one mole of SO_4^{2-}.

Molal concentration represents the number of moles of a species per kilogram of solvent (mol/kg). Typical units are m, mm, or μm. In dilute solutions, this scale for concentrations is almost the same as molar concentrations because a one-liter solution has a mass of approximately 1 kg. For more concentrated solutions, the two scales are increasingly different.

Equivalent charge concentration is the number of equivalent charges of an ion per liter of solution with units such as eq/L or meq/L. The equivalent charge for an ion is equal to the number of moles of an ion multiplied by the absolute value of the charge. For example, with a singly charged species such as Na^+, 1 M Na^+ equals 1 eq/L. With a doubly charged species such as Ca^{2+}, 1 M Ca^{2+} equals 2 eq/L. Equivalent concentrations can also be represented as equivalent charges per unit mass of solution with units such as eq/kg or equivalents per million (epm).

Mass per unit mass concentrations is a scale representing the mass of a species or element per total mass of the system. Many older analyses have been reported using this scale with concentrations in parts per million (ppm) or parts per billion (ppb). More recently, these units of concentration have given way to corresponding concentrations in mg/kg or μg/kg.

Mass per unit volume concentration is the most common scale for concentration. It defines the mass of a solute dissolved in a unit volume of solution. Concentrations are reported in units such as mg/L or μg/L. Again, there is a close correspondence between these last two scales of concentration. For dilute solutions, 1 ppm = 1 mg/kg = 1 mg/L.

Concentration conversions involve a few simple equations. The most common scale change takes the reported values of chemical analyses in mg/L (also mg/kg or ppm) and converts them to molar concentrations

$$\text{molarity} = \frac{\text{mg/L} \times 10^{-3}}{\text{formula weight}} \tag{16.1}$$

Conversion from mg/L to meq/L is sometimes necessary in order to present chemical data graphically or to check chemical analyses. This conversion equation is

$$\text{meq/L} = \frac{\text{mg/L}}{\text{formula wt/charge}} \tag{16.2}$$

As an illustration of the conversion of scales, consider the following example.

▶ **EXAMPLE 16.1**   The concentration of SO_4^{2-} in water is 85.0 mg/L. Express this concentration as molarity and meq/L.

SOLUTION

$$\text{mol/L} = \frac{85 \times 10^{-3}}{32.06 + 4 \times 16.0} = 0.89 \times 10^{-3}$$

$$\text{meq/L} = \frac{85}{(32.06 + 4 \times 16.0)/2} = 1.77$$

▶ 16.2 TYPES OF WATER ANALYSES

Minerals and organic solids, organic liquids, and gases found in the subsurface dissolve in ground water to some extent. Thus, the variety of solutes in ground water (Table 16.1) is not surprising. Inorganic constituents are classified as major constituents with concentrations greater than 5 mg/L (Table 16.1), minor constituents with concentrations ranging from 0.01 to 10 mg/L, and trace elements with concentrations less than 0.01 mg/L (Davis and

TABLE 16.1   The dissolved constituents in groundwater classified according to relative abundance

Major Constituents (greater than 5 mg/L)

Bicarbonate	Silicon
Calcium	Sodium
Chloride	Sulfate
Magnesium	Carbonic acid

Minor Constituents (0.01-10.0 mg/L)

Boron	Nitrate
Carbonate	Potassium
Fluoride	Strontium
Iron	

Trace Constituents (less than 0.1 mg/L)

Aluminum	Molybdenum
Antimony	Nickel
Arsenic	Niobium
Barium	Phosphate
Beryllium	Platinum
Bismuth	Radium
Bromide	Rubidium
Cadmium	Ruthenium
Cerium	Scandium
Cesium	Selenium
Chromium	Silver
Cobalt	Thallium
Copper	Thorium
Gallium	Tin
Germanium	Titanium
Gold	Tungsten
Indium	Uranium
Iodide	Vanadium
Lanthanum	Ytterbium
Lead	Yttrium
Lithium	Zinc
Manganese	Zirconium

Organic Compounds (shallow)

Humic acid	Tannins
Fulvic acid	Lignins
Carbohydrates	Hydrocarbons
Amino acids	

Organic Compounds (deep)

Acetate
Propionate

Source: Modified from Davis and DeWiest, 1966.

DeWiest, 1966). Naturally occurring, dissolved organic compounds in ground water could number in the hundreds. They are typically present in minor or trace quantities. By far, the most abundant organic compounds in shallow ground water are the humic and fulvic acids. However, other organic compounds are also present (Table 16.1). The important gases in ground water are oxygen, carbon dioxide, hydrogen sulfide, and methane.

Routine Water Analyses

It is neither feasible nor necessary to measure the concentration of all constituents that conceivably might occur in water. A *routine* analysis involves measuring the concentration of a standard set of the most abundant constituents. Such a test forms the basis for assessing the suitability of water for human consumption or various industrial and agricultural uses. The routine analysis typically includes the major constituents with the exception of silicon and carbonic acid, and the minor constituents with the exception of boron and strontium. Laboratory results are reported as concentrations in mg/kg or mg/L. The reported concentration for metals (for example, Ca, Mg, and Na) is the total concentration of metals regardless of whether this mass is actually present as the free metal (for example, Ca^{2+}) or metal complexes.

A routine analysis often includes a few other items in addition to concentrations (Table 16.2), such as pH, total dissolved solids (TDS) reported in mg/L, and specific conductance, reported in microsiemens/cm or an older unit micromhos/cm. The *TDS content* is the total quantity of solids when a water sample is evaporated to dryness. *Specific conductance* is a measure of the sample's ability to conduct electricity, and it provides a proxy measure of the total quantity of ions in solution. The measurement is approximate because the specific conductance of a fluid with a given TDS content varies depending on the ions present.

The routine analysis identifies nearly all of the mass dissolved in a sample except if the water is highly contaminated. Unanalyzed ions and organic compounds in natural waters are usually a negligibly small proportion of the total dissolved mass. One simple check on the quality of a routine analysis is to compare the sum of concentrations of cations and anions in milliequivalent per liter. Given that water is electrically neutral, the ratio of the

TABLE 16.2   Example of a routine water analysis

Parameter	mg/L	Parameter	mg/L
pH	7.7	Conductivity	2300
Calcium	1[a]	Magnesium	1
Sodium	550	Potassium	3.5
Iron	8.7	NO_2, NO_3[b]	0.1[a]
Nitrite	0.1[a]	Chloride	45
Sulfate	59	Fluoride	0.25
Bicarbonate	1315	Hardness, T	8
Alkalinity, T	1078		
TDS[c]	1321		
Balance 1.01			

[a]Indicates concentration "less than." Conductivity reported in μS, pH in pH units. All metal parameters expressed as totals. Alkalinity and hardness expressed as calcium carbonate. Nitrate, nitrite, and ammonia expressed as N.
[b]NO_2 = nitrite, NO_3 = nitrate.
[c]Total dissolved solids.

sums should be one. Table 16.3 illustrates this calculation. Conversion factors to change concentrations from mg/L to meq/L are given in Table 16.4. Though not exactly 1, the value is within the $+/-0.05$ range of acceptability used by most laboratories. The *cation/anion ratio* provides one check that concentration determinations are not grossly in error.

Large numbers of routine analyses collected over time provide basic data for research studies in many areas. These data must be used with care. Errors can occur because of a failure to measure rapidly changing parameters in the field, to preserve the samples against deterioration due to long storage, and to assure the quality of the laboratory determinations.

Specialized Analyses

Laboratories can carry out a routine analysis in a rapid and cost-effective manner. Often, there is a need for more specialized analyses of, for example, trace metals (such as Mn, Cr, Cd, Pb, Zn), radioisotopes, organic compounds, various nitrogen-containing species (NO_3^-, NH_4^+), environmental isotopes, or gases. Most of this specialty work is related to problems of ground-water contamination, water-quality assessments, or research and regulatory needs.

At many contaminated sites, the focus of the investigative program is on organic contaminants. Of greatest interest is the identification of what specific organic compounds are present and their concentrations. Dealing with organic contaminants is at first quite

TABLE 16.3 Evaluating the electroneutrality of the example routine analysis

	Cation Concentration			Anion Concentration	
	mg/L	meq/L		mg/L	meq/L
Ca^{2+}	1.0	0.05	HCO_3^-	1315	21.6
Mg^{2+}	1.0	0.08	SO_4^{2-}	59	1.22
Na^+	550	23.9	Cl^-	45	1.27
K^+	3.5	0.09	F^-	0.25	0.01
Fe	8.7	0.31	Total		24.1
	Total	24.4	cation/anion ratio = 1.01		

TABLE 16.4 Factors for converting concentrations in mg/L to meq/L

Ion	Multiply by	Ion	Multiply by
Aluminum (Al^{3+})	0.11119	Iron (Fe^{3+})	0.05372
Barium (Ba^{++})	0.01456	Lead (Pb^{++})	0.00965
Bicarbonate (HCO_3^-)	0.01639	Lithium (Li^+)	0.14409
Bromide (Br^-)	0.01251	Magnesium (Mg^{++})	0.08224
Calcium (Ca^{++})	0.04990	Manganese (Mn^{++})	0.03640
Carbonate (CO_3^-)	0.03333	Manganese (Mn^{4+})	0.07281
Chloride (Cl^-)	0.02820	Nitrate (NO_3^-)	0.01613
Chromium (Cr^{6+})	0.11536	Phosphate (PO_4^{3-})	0.03159
Copper (Cu^{++})	0.03148	Potassium (K^+)	0.02558
Fluoride (F^-)	0.05263	Sodium (Na^+)	0.04350
Hydrogen (H^+)	0.99206	Strontium (Sr^{++})	0.02282
Hydroxide (OH^-)	0.05880	Sulfate (SO_4^{--})	0.02082
Iodide (I^-)	0.00788	Sulfide (S^{--})	0.06237
Iron (Fe^{++})	0.03581	Zinc (Zn^{++})	0.03059

Source: Hem, 1959.

intimidating. Literally thousands of different compounds can be present, many with exotic chemical names. Although it is useful to have a course in organic chemistry, problems can be addressed reasonably well simply by organizing studies around "logical suites of contaminants." Although there are thousands of potential organic contaminants, sites are screened according to what is likely to be there (based on experience from many sites and compounds of regulatory concern). For example, study of most fuel spills involves determining the concentrations of the BTEX compounds (that is, *b*enzene, *t*oluene, *e*thylbenzene, and *x*ylene) in the water and the soil. The reason for this is that fuels such as gasoline and diesel contain these BTEX compounds in addition to hundreds of other compounds. At sites of industrial contamination, one would have water samples analyzed for lists of hazardous, volatile, or nonvolatile substances. To expedite the analyses and to keep their cost down, one screens for likely contaminants that are common for the type of spill. Obviously, if contaminants are known to be present and not on the list, they will be included.

Table 16.5 presents an analysis of this type. It shows that there were three contaminants detected: 1,1,1-trichloroethane at 780 μg/L; trichloroethene at 180 μg/L; and tetrachloroethene at 260 μg/L.

▶ 16.3 WATER-QUALITY STANDARDS

Chemical analyses are commonly used to determine whether water meets various standards for use by humans or for support of the health of aquatic ecosystems. These standards change quite often and vary from country to country. In this section, we provide examples of current standards used in the United States. Interested readers can examine our book's Web site to find links to pages summarizing standards.

In the United States, the National Primary Drinking Water Regulations (NPDWRs) are the legally enforceable standards that apply to public water systems (Table 16.6). They are designed primarily to protect public health by requiring that contaminants or naturally occurring constituents in water (for example, arsenic) be less than certain limits. Families of contaminants and constituents covered by these regulations include microorganisms, disinfection and disinfection byproducts, inorganic chemicals, organic chemicals, and radionuclides. Disinfection and disinfection byproducts relate to chemicals added (or inadvertently produced) to get rid of microorganisms in water.

Standards are presented in terms of MCLs and MCLGs. The *maximum contaminant level* (MCL) is the highest level of a contaminant that is allowed in drinking water. MCLs are enforceable standards that must be met by public drinking water systems. The *maximum contaminant level goal* (MCLG) is the level of a contaminant in drinking water below which there is no known or expected health risk. These nonenforceable goals are targets that should be set in developing water supplies and treatment systems.

Not all water that meets the standards in Table 16.6 is desirable for consumption and home use. There is another set of standards called the National Secondary Drinking Water Regulations (NSDWRs) that help judge the quality of water, especially the natural constituents found in ground water. These nonenforceable guidelines (see Table 16.7) regulate constituents that may cause cosmetic effects (skin or tooth discoloration) or aesthetic effects (taste, odor, or color) in drinking water. The routine water analysis that is often conducted with the domestic water supply provides a check as to how well some of these secondary regulations are being met. Commonly, people will drink water with higher iron concentrations or total dissolved solids contents. Nevertheless, the secondary goals provide simple guidance as to what a good quality drinking water should be like.

There are other important uses for water besides drinking. Water is used for irrigation of agricultural crops or in boilers for making steam. For many of these other uses, water-quality

TABLE 16.5 **Example of organic constituents included in an analysis of volatile organic compounds in ground-water from a contaminated site**

Parameter	Result	Units	Reporting Limit
Chloromethane	ND	μg/L	50
Bromomethane	ND	μg/L	50
Vinyl chloride	ND	μg/L	50
Chloroethane	ND	μg/L	50
Methylene chloride	ND	μg/L	100
Acetone	ND	μg/L	500
Carbon disulfide	ND	μg/L	20
1, 1-Dichloroethene	ND	μg/L	20
1, 1-Dichloroethane	ND	μg/L	20
trans-1, 2-Dichloroethene	ND	μg/L	20
Chloroform	ND	μg/L	20
1, 2-Dichloroethane	ND	μg/L	20
2-Butanone	ND	μg/L	100
1, 1, 1-Trichloroethane	780	μg/L	20
Carbon tetrachloride	ND	μg/L	20
Vinyl acetate	ND	μg/L	100
Bromodichloromethane	ND	μg/L	20
1, 2-Dichloropropane	ND	μg/L	20
trans-1, 3-Dichloropropene	ND	μg/L	20
Trichloroethene	180	μg/L	20
Dibromochloromethane	ND	μg/L	20
1, 1, 2-Trichloroethane	ND	μg/L	20
Benzene	ND	μg/L	20
cis-1, 3-Dichloropropene	ND	μg/L	20
2-Chloroethyl vinyl ether	ND	μg/L	100
Bromoform	ND	μg/L	20
4-Methyl-2-pentanone	ND	μg/L	100
2-Hexanone	ND	μg/L	100
1, 1, 2, 2-Tetrachloroethane	ND	μg/L	20
Tetrachloroethene	260	μg/L	20
Toluene	ND	μg/L	20
Chlorobenzene	ND	μg/L	20
Ethyl benzene	ND	μg/L	20
Styrene	ND	μg/L	20
Total xylenes	ND	μg/L	20

ND means that the compound was not detected in the sample. The reporting limit is the lowest concentration value that can be expected to be reliable.

guidelines apply, and with diligence, one can discover them. Table 16.8 presents guidelines for water use in irrigation. The main issues in the application of irrigation water are requirements to match the salinity of the irrigation water to the salt tolerance of selected crops, to avoid salt buildup in the soil, and to avoid a breakdown of the soil structure and a reduction in permeability by using water high in sodium ions and low in calcium and magnesium ions.

▶ 16.4 EXAMPLES OF DATA COLLECTED IN CHEMICAL SURVEYS

This section provides examples of the kind of data that would be collected in sampling ground water and surface water. Case Study (16–1) is a general study of the ground water

TABLE 16.6   Partial listing of U.S. National Primary Drinking Water Regulations

Family	Contaminant	MCLG (mg/L)	MCL (mg/L)
Inorganic Chemicals	Arsenic	none	0.01
	Barium	2	2
	Cadmium	0.005	0.001
	Chromium (total)	0.1	0.1
	Fluoride	4.0	4.0
	Mercury (inorganic)	0.002	0.002
	Nitrate (as nitrogen)	10	10
	Selenium	0.05	0.05
Organic Chemicals	Atrazine	0.003	0.003
	Benzene	zero	0.005
	Carbon Tetrachloride	zero	0.005
	1,2-Dichloroethane	zero	0.005
	1,1-Dichloroethylene	0.007	0.007
	Ethylbenzene	0.7	0.7
	Polychlorinated biphenyls (PCBs)	zero	0.0005
	Tetrachloroethylene	zero	0.005
	Toluene	1	1
	1,1,1-Trichlorethane	0.2	0.2
	Trichloroethylene	zero	0.005
	Vinyl Chloride	zero	0.002
	Xylenes	10	10
Microorganisms	Total coliforms	zero	5%[a]
Radionuclides	Radium 226 and 228 combined	No MCLG	5 pCi/L
	Uranium	No MCLG	0.030

[a]No more than 5% samples collected in a month can be coliform positive. With fewer than 40 routine samples per month, no more than one sample can be coliform positive.

TABLE 16.7   Secondary standards for drinking water

Contaminant or Constituent	Secondary Standard
Aluminum	0.05 to 0.2 mg/L
Chloride	250 mg/L
Color	15 (color units)
Copper	1.0 mg/L
Corrosivity	noncorrosive
Fluoride	2.0 mg/L
Foaming Agents	0.5 mg/L
Iron	0.3 mg/L
Manganese	0.05 mg/L
Odor	3 threshold odor number
pH	6.5–8.5
Silver	0.10 mg/L
Sulfate	250 mg/L
Total Dissolved Solids	500 mg/L
Zinc	5 mg/L

TABLE 16.8   Guidelines on the suitability of water for irrigation

Constituent	Problem	Water-Quality Hazard		
		Low	Medium	High
Electrical Conductivity (mmhos/cm)	Plant sensitivity to salts and salt buildup	0.00–0.75	0.76–3.00	3.01+
Sodium Adsorption Ratio (SAR)	Soil breakdown	0–6	6–12 extended use on light textured soils 12–18 extended use on all soils	18+
Residual Sodium Carbonate (RSC)	Soil breakdown	0–1.24	1.25–2.50	2.51+

$$SAR = \frac{Na}{\sqrt{[Ca+Mg]/2}}; \text{ all concentrations meq / L}$$

$$RSC = (CO_3 + HCO_3) - (Ca + Mg); \text{ all concentrations meq / L}$$

in a watershed in Canada. It was designed to provide general information on water resources and on the suitability of ground water for drinking water or other uses.

In the mid-1970s, scientists recognized that the careless dumping of organic chemicals had led to severe problems of ground-water contamination—sometimes bad enough to cause deaths. The ground-water quality issue in these cases was to assess potential problems of ground-water contamination and to expedite their cleanup. This new and important area of concern immediately raised the bar as far as chemical knowledge was concerned. Hydrologists now needed to know something about organic contaminants and metals, how they behaved in the subsurface, and what processes worked to modify their concentration.

CASE STUDY 16-1

WATER RESOURCES OF THE UPPER NOTTAWASAGA RIVER BASIN, ONTARIO, CANADA

Sibul and Choo-Ying presented basic water resources information for the Upper Nottawasaga Basin, which is a 470 mi^2 basin in southern Ontario. The basin contains both glacial drift and bedrock aquifers. Ground water comes from unconfined sandy lacustrine deposits and glaciofluvial deposits confined by till and lake clay. Locally, fractured crystalline bedrock is capable of providing a useful water supply.

As is the case with these types of water resources evaluations, emphasis was placed on evaluating the geologic setting and on assessing the productivity of the major aquifer systems. One particular aspect that is germane to this chapter was an evaluation of the water chemistry of drift and bedrock units. Table 16.9 shows several water analyses for samples from drift and bedrock. In these samples, the major ions constitute the bulk of the dissolved mass. Calcium or sodium is the most abundant cations, with bicarbonate the most abundant anion. Further inspection

of the results shows that the water from well 8 (shallow) has unacceptably high concentrations of nitrate. The ground water in the Upper Nottawasaga River Basin generally meets the secondary standards in Table 16.7, although iron concentrations are high in well 7, as are the total dissolved solids in well 8 (Table 16.9). Generally, this ground water should provide an adequate domestic water supply, with confirmatory testing for total coliforms.

Figure 16.1 illustrates how chemical information could be presented in such an evaluation. The circled values indicate samples above the recommended total dissolved solids content at the time of sampling. Sibul and Choo-Ying (1971) provide maps for other key constituents, chloride, iron, and hardness. The sodium adsorption ratios are relatively small for most samples, indicating that the ground water is suitable for irrigation.

Sibul and Choo-Ying, 1971

TABLE 16.9 Chemistry of ground water an indicated by selected analyses of samples from drift and bedrock wells

Source and Number	Geologic unit	pH	Ionic Concentrations (mg/L)										Total Alkalinity as CaCO₃ m (ppm)	Total Hardness as CaCO₃ (ppm)	Total Dissolved Solids (ppm)	Specific Conductance (micromhos) at 25° C
			Calcium	Magnesium	Sodium	Potassium	Bicarbonate	Sulphate	Chloride	Total Iron	Fluoride	Nitrate				
Well 7	Bedrock	7.8	65	33	7	2	263	85	1	0.70	1.3	0.09	216	300	374	570
Well 8	Drift	7.6	129	35	8	2	406	58	28	0.10	0	21	333	470	706	1006
Well 92	Drift	7.4	125	14	10	1.6	390	30	24	0.20	0	1.6	320	370	426	714
Well 119	Bedrock	7.8	28	8	150	13	330	20	115	0.10	0.8	0.41	271	106	502	914

Source: Sibul and Choo-Ying, 1971.

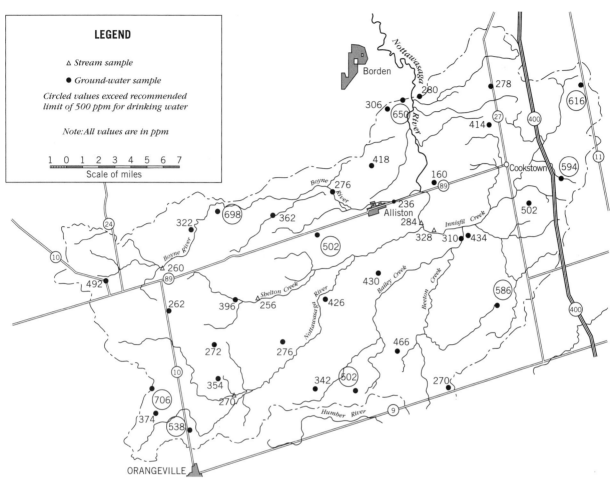

Figure 16.1 Map showing the total dissolved solids content of ground water and surface water in the Upper Nottawasaga River Basin, Ontario (modified from Sibul and Choo-Ying, 1971).

New strategies were needed for sampling contaminants that could easily be lost from bottles by volatilization and for analyzing compounds present at parts per billion levels.

The following case study illustrates a few of the chemical results that came from a problem of industrial contamination. This site in New Hampshire is similar to tens of thousands of such sites found in industrialized countries around the world.

Intensive farming has the potential to contaminate ground water and surface water over large areas. Many of the chemicals used in farming (for example, herbicides, pesticides, and fertilizers) are leached downward out of the soil zone and into the ground waters. Similarly, surface runoff from the cropland can transport these chemicals to nearby streams and rivers. Thus, contaminants can reach streams following ground-water and overland-flow pathways. In the United States, there are indications that large-river systems are having an impact on the Gulf of Mexico and Chesapeake Bay. Thus, small quantities of contamination present across large regions of the country also have a significant impact on ground water and surface waters.

SAVAGE WELL SITE, MILFORD NEW HAMPSHIRE

Milford, a small town in New Hampshire, uses a well system to provide water. In 1983, an analysis of water from one of the town wells detected organic contaminants at high concentrations. The Savage well (Figure 16.2) is located in a thick unconfined aquifer that occurs along the valley of the Souhegan River. This site is downgradient from several industrial plants that used various kinds of organic solvents. Many sampling wells were placed in the aquifer to establish the types and distribution of contaminants. The analyses for contamination involved a typical group of volatile organic compounds (Table 16.10). Results from three wells—16C about 78 ft deep, 17A about 25 ft deep, and 17B about 58 ft deep—indicated the presence of tetrachloroethylene (PCE), trichloroethylene, 1,1,1-trichloroethane, trans-1, 2-dichloroethane, and 1, 1-dichloroethane. This sampling showed that contamination was distributed from top to bottom in this 75-ft-thick aquifer. Concentrations of these contaminants in samples from MW-16C greatly exceeded the drinking water standard of $5\mu g/L$. Figure 16.2 shows the plume of PCE determined using the collection of monitoring well data. You can see how impacted the Savage well was with concentrations approaching $1000 \mu g/L$. The OK Tool Company used tetrachloroethene (PCE) and was thought to be the major source of PCE contamination.

Plumes maps were also developed for the other major contaminants found at the site. This contamination came from chemical dumping on other properties. The complete study of this site required nearly 10 years. Having constructed a slurry wall around the site, ground-water hydrologists are studying new ways to clean up the contamination.

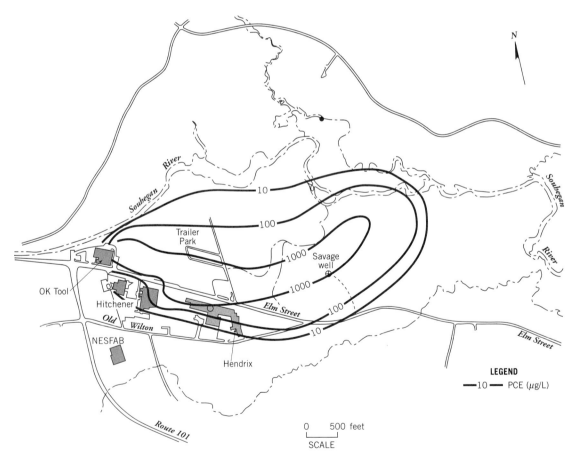

Figure 16.2 Map showing PCE contamination in a shallow drift aquifer along the Souhegan River in Milford, New Hampshire.

TABLE 16.10 **Concentration of volatile organic compounds in three wells downgradient of OK Tools**

Volatile Organic Compounds	CRQL	MW 16-C	MW-17A	MW-17B
			Concentration (μg/L)	
Chloromethane	10μg/l	ND	ND	ND
Bromomethane	10μg/l	ND	ND	ND
Vinyl Chloride	10μg/l	ND	ND	ND
Chloroethane	10μg/l	ND	ND	ND
Methylene Chloride	5μg/l	ND	ND	ND
Trichlorofluoromethane	5μg/l	ND	ND	ND
1,1-Dichloroethylene	5μg/l	ND	ND	ND
1,1-Dichloroethane	5μg/l	ND	8	ND
trans-1,2-Dichloroethene	5μg/l	740	ND	ND
Chloroform	5μg/l	ND	ND	ND
1,2-Dichloroethane	5μg/l	ND	ND	ND
1,1,1-Trichloroethane	5μg/l	75	59	ND
Carbon Tetrachloride	5μg/l	ND	ND	ND
Bromodichloromethane	5μg/l	ND	ND	ND
1,2-Dichloropropane	5μg/l	ND	ND	ND
cis-1,3-Dichloropropene	5μg/l	ND	ND	ND
Trichloroethylene	5μg/l	430	5	TRACE
Benzene	5μg/l	ND	ND	ND
Dibromochloromethane	5μg/l	ND	ND	ND
1,1,2-Trichloroethane	5μg/l	ND	ND	ND
trans-1,3-Dichloropropene	5μg/l	ND	ND	ND
2-Chloroethylvinyl Ether	10μg/l	ND	ND	ND
Bromoform	5μg/l	ND	ND	ND
1,1,2,2-Tetrachloroethane	5μg/l	ND	ND	ND
Tetrachloroethylene	5μg/l	7,900	320	1200
Toluene	5μg/l	ND	ND	ND
Chlorobenzene	5μg/l	ND	ND	ND
Ethylbenzene	5μg/l	ND	ND	ND
Acetone	20μg/l	ND	ND	ND
Carbon Disulfide	10μg/l	ND	ND	ND
2-Butanone	20μg/l	ND	ND	ND
Vinyl Acetate	5μg/l	ND	ND	ND
4-Methyl-2-Pentanone	5μg/l	ND	ND	ND
2-Hexanone	5μg/l	ND	ND	ND
Styrene	5μg/l	ND	ND	ND
Total Xylenes	5μg/l	ND	ND	ND
Dibromoethane (EDB)	5μg/l	ND	ND	ND
Methyl-t-Butylether (MTBE)	10μg/l	ND	46	TRACE

CRQL = Contract Required Quantification Limit.
ND = None Detected (Below Minimum Detectable Level).
TRACE = Confirmed Detection Below MDL.

► 16.5 WORKING WITH CHEMICAL DATA

Describing the concentration or relative abundance of major and minor constituents and the pattern of variability is part of many ground-water investigations. Over time, different graphical and statistical techniques have been developed to assist with this task. Each technique has particular advantages and disadvantages in representing features of the data.

AGRICULTURAL CONTAMINATION ON THE DELMARVA PENINSULA, MARYLAND

The Delmarva Peninsula is located on the eastern shore of Chesapeake Bay near Washington, D.C. This area is intensively farmed, and the ground-water chemistry is showing effects from the agricultural use of crushed dolomite, dolomite, and various nitrogen fertilizers. A factor contributing to ground-water contamination is the relatively high hydraulic conductivity of near-surface deposits. Figure 16.3 illustrates the occurrence of sand and gravel deposits of the fluvial Pensauken Formation and the marine Aquia Formation.

Since the early 1940s, there has been a steady increase in the utilization of nitrogen fertilizers, from about about 2 g/m²/y to 12g/m²/y. This fertilizer usage is responsible for nitrate levels

in ground water exceeding the maximum contaminant level (MCL) of 10 mg/L. This contamination in the shallow ground water is shown in Figure 16.3. Age dates for the ground water in combination with the NO_3^- reflect the more intensive, recent application of fertilizers. Thus, water recharged in the 1960s and 1970s was less contaminated by NO_3^- than water recharged after 1985. Streams receiving ground-water discharge (e.g., Chesterville Branch, Figure 16.3) are being impacted by NO_3^- contamination. In other cases (e.g., Morgan Creek), denitrification reactions reduce NO_3^- concentrations when flow moves into units near or below the bottom the Aquia Formation.

Bohlke and Denver, 1995

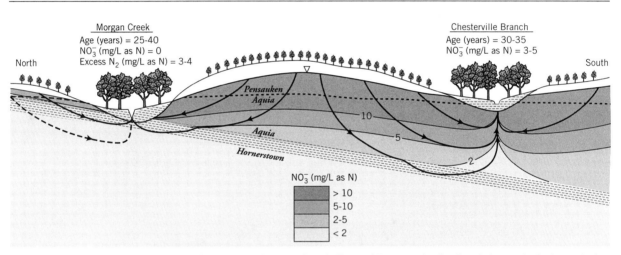

Figure 16.3 Cross section showing shallow NO_3^- contamination in relation to the hydrogeologic setting for a study area on the Delmarva Peninsula, Maryland. More intensive applications of fertilizers in recent years have resulted in NO_3^- concentrations greater than 10 mg/L (modified from Bohlke and Denver, *Water Resour. Res.*, v. 31, no. 9, 2319–2339, 1995). Copyright ©1995 American Geophysical Union.

Thus, in working with a set of chemical data, one should explore alternative ways of presenting results and select the most appropriate.

The methods divide into three major groups. First is a group of graphical approaches for describing abundance or relative abundance. Second are approaches that present patterns of variability in addition to abundance. Third is a group of derived maps that involve various sorts of calculations with the basic data.

Abundance or Relative Abundance

Several different graphical approaches can depict the abundance or relative abundance of ions in individual water samples. The most common approaches are (1) the Collins (1923) bar diagram, (2) the Stiff (1951) pattern diagram, (3) the pie diagram, and (4) the Piper (1944) diagram. Data presented in Table 16.11 will be used to illustrate these approaches. In these plots, concentrations need to be expressed as meq/L or %meq/L. The Collins and Stiff diagrams require absolute concentrations (meq/L), whereas the pie and Piper diagrams require relative concentrations (%meq/L).

Figures 16.4a,b,c, and d illustrate the sample data plotted in different ways. The Collins, Stiff, and pie diagrams are relatively simple to construct. They require only that concentrations be plotted as a bar segment, a point on a line, or a percentage of the pie. The appropriate fields are shaded and possibly labeled in the case of the Collins and Piper diagrams (Figure 16.4). The Stiff diagram can be plotted with or without the labeled axes.

Plotting data for a Piper diagram is complicated because there are three separate diagrams (Figure 16.4d). The relative abundance of cations with the %meq/L of $Na^+ + K^+$, Ca^{2+}, and Mg^{2+} assumed to equal 100% is first plotted on the cation triangle. Similarly, the anion triangle displays the relative abundance of Cl^-, SO_4^{2-} and $HCO_3^- + CO_3^{2-}$. Straight lines projected from the two triangles into the quadrilateral field define a point on the third field (Figure 16.4d). To provide some indication of the absolute quantity of dissolved mass in the sample, the size of the data point is sometimes related to the salinity (TDS). One advantage of all four techniques is that they present the major ion data for a sample on one figure. However, with the exception of the Piper diagram, these approaches are useful only in displaying the results for a few analyses, which are often "type" waters from an area. Presenting a large number of these diagrams together is confusing and not much more helpful than a table of concentration values.

Abundance and Patterns of Change

Graphical/illustrative type diagrams or statistics can define the pattern of spatial change among different geologic units, along a line of section, or along a pathline. One simple way of representing spatial change in a single geological unit is to take the single sample

TABLE 16.11 Sample chemical data used to demonstrate various techniques for plotting chemical data

Chemical Analyses	Sample 1: Tertiary, Czechoslovakia			Sample 2: Upper Cretaceous, Czechoslovakia			Sample 3: Upper Cambrian, Wisconsin		
	mg/L	meq/L	meq (%)	mg/L	meq/L	meq (%)	mg/L	meq/L	meq (%)
Cations									
$Na^+ + K^+$	266.2	10.68	68.24	1913.7	81.54	70.24	7.9	0.34	4.4
Mg^{2+}	21.9	1.80	11.4	132.8	10.95	9.41	43.0	3.54	45.6
Ca^{2+}	61.7	3.08	18.5	468.5	23.38	20.50	78.0	3.89	50.1
Mn^{2+}	Traces	—	—	0.67	0.02	0.01	0.14	0.004	0.05
Fe	2.3	0.08	—	0.15	0.005	0.004	0.11	—	—
Sum	—	15.64	100.0	—	115.89	100.0	—	7.77	100.0
Anions									
Cl^-	11.3	0.32	2.05	850.0	23.98	20.63	17.0	0.48	6.4
NO_3^-	0.0	—	—	0.0	—	—	0.7	0.0	0.1
HCO_3^-	906.1	14.85	95.1	2568.5	42.10	36.23	364.0	5.96	79.5
SO_4^{2-}	21.2	0.44	2.82	2406.5	50.10	43.11	50.0	1.04	13.9
Sum	—	15.61	100.0	—	116.18	100.0	—	7.50	100.0

Source: Zapovozec (1972). Reprinted by permission of *Ground Water*. Copyright ©1972. All rights reserved.

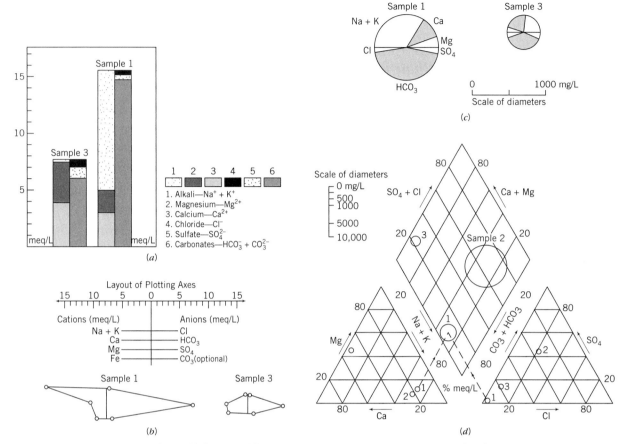

Figure 16.4 Four different ways of plotting major ion data (modified from Zaporozec, 1972). Panel (*a*) is a Collins diagram, Panel (*b*) is a Stiff diagram, Panel (*c*) is a pie diagram, and Panel (*d*) is a Piper diagram. Reprinted by permission of *Ground Water*. Copyright ©1972. All rights reserved.

diagrams (for example, pie or Stiff) and place them on a map. Such maps can convey a sense of how the pattern of ion abundances changes within a unit. Including all of the constituents on a single map can be advantageous. However, making sense out of a collection of geometric shapes is difficult.

When chemical data vary systematically in space, it is often best to plot and contour concentrations (or other data) on maps or cross sections (Figure 16.5). This presentation makes it obvious how individual parameters vary. Any measured or calculated chemical parameter can be represented in this way. One problem is the large number of figures that could be required to describe the chemistry of an area completely. However, given the usefulness of these diagrams, this limitation is not serious.

Another problem arises with "noisy" data, which can result in a complex and cluttered figure. A Piper diagram is preferable to concentration maps when data are noisy. By classifying samples on the Piper diagram, one can identify geologic units with chemically similar water and define the evolution in water chemistry along a flow system (Figure 16.6).

Noisy data can also be smoothed before plotting on a map or cross section. The facies mapping approach (Back, 1961) provides one way of smoothing chemical data. Samples are classified according to facies with two templates for the Piper diagram (Figure 16.7), one for the cations and the other for the anions. The limited number of possibilities for

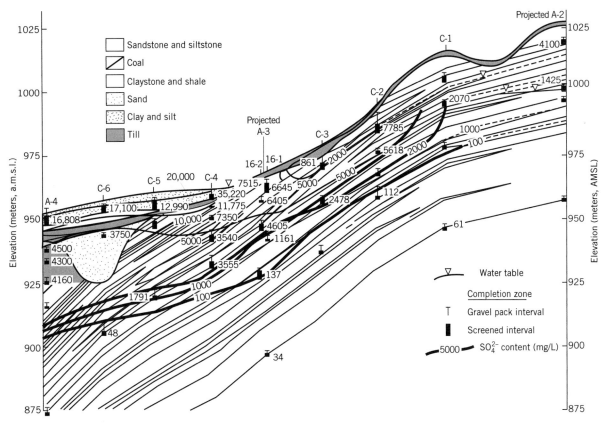

Figure 16.5 Example of how concentration data for an ion can be represented on a cross section. Shown here are SO_4^{2-} data from Blackspring Ridge, Alberta, Canada (from Stein and Schwartz, 1990). Reprinted from *J. Hydrol.*, v. 117, On the origin of saline soils at Blackspring Ridge, Alberta, Canada, p. 99–131, Copyright 1990, with permission from Elsevier Science.

classifying the chemical data effectively eliminates local variability, yet preserves broad trends. The example (Figure 16.8) shows the progressive change in cation chemistry from calcium-magnesium water in the upland part of the Atlantic Coastal Plain to sodium water in deeper units located at the downstream end of the flow system.

▶ 16.6 GROUND-WATER SAMPLING

In most water resource assessments, samples are collected from domestic wells that already exist to provide water to a residence or farm. With a pump and water lines in place, the main challenge is to collect samples without running the water through a large storage tank and plumbing system. The investigator has little control over the well design and the sampling process. This section should help you to become aware of potential problems associated with water samples collected from existing wells.

In contamination studies or research investigations, wells are usually installed to provide samples from specified locations and to assure the quality of the sample. In this case, an investigator must make decisions concerning the location of the wells, their design, and the method of installation. Hopefully, making appropriate decisions in this respect will minimize potential errors.

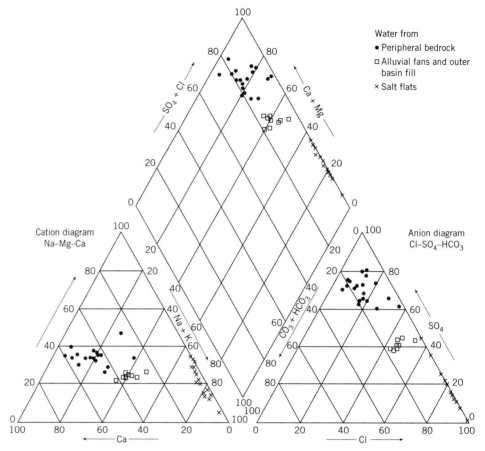

Figure 16.6 Example of the use of a Piper diagram for classifying ground water and defining a pathway of chemical evolution. The diagram shows how samples from the Salt Basin in West Texas can be classified according to the setting, for example, limestone (dots), alluvial fans and basin fills (squares), and the salt flats (crosses). The evolutionary pathway is defined by ground water moving from the limestone, through the alluvium to the salt flats (from Boyd and Kreitler, 1986).

Sources of Error

Collecting water samples involves several steps, and problems can arise in each step. The sampling process begins by taking water from the well using a bailer or a pump. With the water at the surface, a few parameters like pH and specific conductance are measured immediately with portable equipment. Water is also saved in various containers, some of which contain preservation chemicals. These chemicals are added to counter the chemical changes that occur as the water samples age. The samples are packed into ice chests and transported to an appropriate laboratory for analysis. These steps seem simple, yet there are an amazing number of pitfalls and problems that can invalidate the sample results.

Figure 16.9 summarizes the sources of errors in sampling ground water. They run the gamut from improper procedures for installing the well, reactions between the ground water and the well casing or sampler, and poor sample-handling protocols on the surface. As will become clear in later sections, following the recommended procedures in sampling

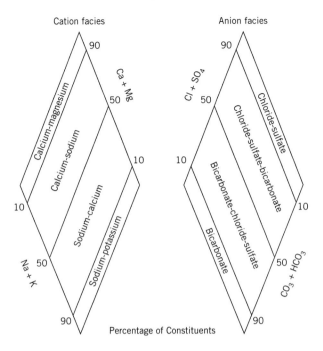

Figure 16.7 Templates for classifying waters into facies for cations and anions (from Back, 1961).

can minimize these problems. Guidance is usually provided in so-called standard operating procedures (SOPs). Most companies and government agencies have developed SOPs from guidance provided by the U.S. Geological Survey, the U.S. Environmental Protection Agency, and other organizations. Use of SOPs is one feature of a comprehensive quality assurance (QA) program that documents the sampling process to provide confidence in the reliability of a data set (Barcelona et al. 1985). Another key element of a QA program is quality control (QC) checks. These tests are built into the sampling process to identify any errors that may have cropped up. We will discuss these QA/QC issues in more detail in a later section.

You might think that problems end when samples are dropped off at the analytical laboratory. Unfortunately, lots of bad things can happen to a water sample in the laboratory as well. These problems encompass errors due to the deterioration of samples and laboratory standards, as well as poor analytical methods. The laboratory problems also require that QC checks be designed to detect errors that originate in the laboratory.

Locating Wells and Piezometers

Often, an investigator has to determine the location of the wells. Among the important considerations in locating wells are (1) the need for sample locations that take into account the character and complexity of chemical patterns and (2) the objectives of the water-quality monitoring program. At a regional scale, changes in water chemistry might reflect broad patterns of chemical evolution along flow systems and the unique geochemical environments provided by specific geologic units. Thus, well locations are selected to provide information keyed to flow paths and hydrogeologic units. Another overall objective in sampling could be to assess the suitability of water for human consumption. Wells should be located so as to provide samples across the entire study area and to include the major aquifers supplying water.

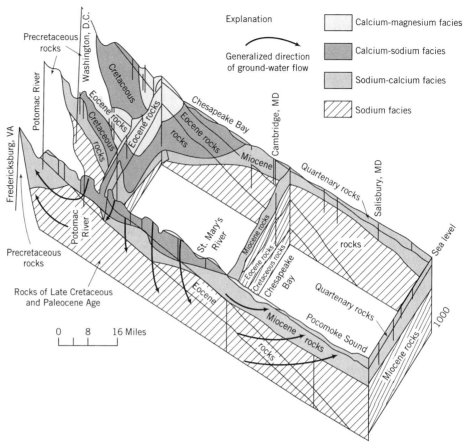

Figure 16.8 Example of hydrochemical facies mapping. Shown on the fence diagram are cation data from part of the north Atlantic Coastal Plain (from Back, 1961).

In the case of contaminant plumes, obviously there are other considerations in locating sampling wells. One of the most important is to define the extent of contamination. Assessing the vertical extent of a plume is difficult because often plumes are vertically quite thin. For determining plume boundaries, wells need to be installed within and closely adjacent to the plume (Figure 16.10). Nests of wells installed along the midline of the plume will establish its overall length and thickness. Lines of wells installed normal to the midline would serve to define the plume's width. This simple design is useful as a starting point but is not universally applicable. For example, in karst systems, patterns of contaminant migration can only be determined by collecting samples from the conduits and springs (Quinlan and Ewers, 1985).

Well Design

At least three well-design issues are pertinent to providing usable chemical data: choosing appropriate screen and casing materials, selecting proper grouts and seals, and designing a screen length appropriate for the monitoring need. Barcelona et al. (1985) recommend that materials for casings and screens be selected on the basis of long-term durability, cleanability, and minimization of secondary effects of sorption or leaching. If the casing materials are able to dissolve, corrode, or interact otherwise with the ground water, the chemical measurements can be affected.

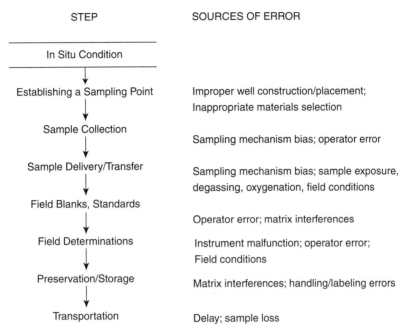

| STEP | SOURCES OF ERROR |

Figure 16.9 Steps in ground-water sampling and sources of error (from Barcelona et al., 1985).

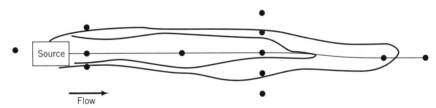

Figure 16.10 Basic plan for locating monitoring wells to define the extent of a dissolved plume of contaminants.

Table 16.12 provides guidance on the selection of materials for monitoring wells. A general rule-of-thumb is to select "plastic"-type materials for monitoring mainly inorganic constituents, and steel for monitoring organic compounds in ground water. Although Teflon® and Stainless Steel 316 are highly recommended, they do not tend to be used in practice because of their higher costs. Flush-threaded PVC (polyvinylchloride) is used most commonly in practice, although there are important restrictions on its use (Table 16.12). The use of PVC cements to couple sections of casing together should be avoided. These cements leach and can contaminate the water samples.

Another important design issue is the choice of sealing and grouting materials. First and foremost, the sealing material must provide a barrier immediately above the well screen. If water leaks down the well bore, then the sample will be a mixture of water from the formation and from formations higher up the hole, which also may have reacted with the grout in the borehole. Barcelona and others (1985) suggest expanding neat cement, bentonite clays, or a mixture of the two as the best sealing materials. All of these materials

TABLE 16.12 Recommendations for rigid materials in sampling applications (in decreasing order of preference)

Material	Recommendations
Teflon® (flush threaded)	Recommended for most monitoring situations with detailed organic analytical needs, particularly for aggressive, organic leachate impacted hydrogeologic conditions. Virtually an ideal material for corrosive situations where inorganic contaminants are of interest.
Stainless Steel 316 (flush threaded)	Recommended for most monitoring situations with detailed organic analytical needs, particularly for aggressive, organic leachate impacted hydrogeologic conditions.
Stainless Steel 304 (flush threaded)	May be prone to slow pitting corrosion in contact with acidic high total dissolved solids aqueous solutions. Corrosion products limited mainly to Fe and possibly Cr and Ni.
PVC (flush threaded) Other noncemented connections, only NSF* approved materials for well casing or potable water applications	Recommended for limited monitoring situations where inorganic contaminants are of interest and it is known that aggressive organic leachate mixtures will not be contacted. Cemented installations have caused documented interferences. The potential for interaction and interferences from PVC well casing in contact with aggressive aqueous organic mixtures is difficult to predict. PVC is not recommended for detailed organic analytical schemes.
	Recommended for monitoring inorganic contaminants in corrosive, acidic inorganic situations. May release Sn or Sb compounds from the original heat stabilizers in the formulation after long exposures.
Low-Carbon Steel Galvanized Steel	May be superior to PVC for exposures to aggressive aqueous organic mixtures. These materials must be very carefully cleaned to remove oily manufacturing residues.
Carbon Steel	Corrosion is likely in high dissolved solids acidic environments, particularly when sulfides are present. Products of corrosion are mainly Fe and Mn, except for galvanized steel which may release Zn and Cd. Weathered steel surfaces present very active adsorption sites for trace organic and inorganic chemical species.

® Trademark of DuPont, Inc.
* National Sanitation Foundation-approved materials carry the NSF logo indicative of the product's certification based on meeting industry standards for performance and formulation purity.
Source: Barcelona et al., 1985.

can impact the water sample if they come in contact. Cement can dissolve and totally wreck the carbonate chemistry (and pH) of the sample. The bentonites can function as ion exchangers. Barcelona et al. (1985) emphasize the importance of adding a layer of fine-grained filter sand near the top of a sand pack to minimize contact of water in the sand pack with the well seal. Our experience is that bentonite seals perform better than cement seals.

The last design variable is screen length. When water levels are measured in piezometers, screen length is not particularly critical. Consider the aquifer in Figure 16.11*a*. As far as hydraulic head is concerned, it makes little difference whether you measure head in a small volume of the aquifer (point sample) or in a larger volume (nonpoint sample). The hydraulic head is about the same in either case (Figure 16.11*a*) because in a permeable unit the gradients in hydraulic head are usually small.

Using the piezometers for chemical sampling, however, can produce dramatically different results. This problem is most evident in sampling plumes of dissolved contaminants that have well-defined upper and lower boundaries (Figure 16.11*b*). For this example, only the point samples provide concentrations suitable for interpreting the contaminant distribution. Estimates of the plume geometry based on the nonpoint samples would be in error because of mixing within the well. Thus, all measurements of contaminant plumes not concerned simply with presence/absence indications should use point sampling.

The problem of screen length is less of an issue in large-scale water resources studies that use existing wells. Within geologic units, there are usually not the same large concentration gradients that one sees with plumes of dissolved contaminants. However, wells screened across more than one geological unit are obviously less useful as monitoring wells.

Drilling and Well Development Methods

Various techniques are available for drilling holes in different kinds of geological materials. Usually, the drilling method selected for a project depends primarily on the geologic setting, the anticipated hole depth, and the cost and availability of equipment. Potential effects on water quality are not often considered (Barcelona et al., 1985). The main problem with several commonly used drilling methods is that they can add large quantities of drilling water and sometimes chemical additives to the subsurface. These fluids usually differ in composition from the native ground water and are difficult to remove completely, once the well is emplaced. The preferred methods either use no fluids or air. Barcelona et al. (1985) provide recommendations for preferred drilling methods in differing geologic settings (Table 16.13).

Once the hole is drilled and the well or piezometers are installed, the well must be *developed* in preparation for sampling. The process of development is designed to remove fine-grained materials from the zone around the screen. It improves the hydraulic connection between the well and the adjacent formation. Development also helps to provide sediment-free samples. Sediment in the water, if left, may react with chemicals added to preserve

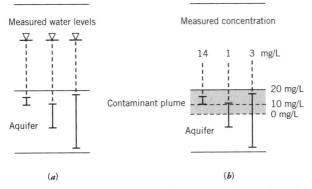

(*a*)   (*b*)

Figure 16.11   (*a*) In many cases, there may not be a substantial difference in hydraulic-head values measured in an aquifer with either point or nonpoint sampling procedures.
(*b*) Concentrations of contaminants can change markedly over relatively small vertical distances. In this example, nonpoint sampling will yield concentrations that are much less than the true concentration (from Domenico and Schwartz, 1998. *Physical and chemical hydrogeology*).
Copyright ©1998 by John Wiley & Sons, Inc. Reprinted by permission of John Wiley & Sons, Inc.

TABLE 16.13 Recommended drilling techniques for various types of geologic settings

Geologic Environment	Recommended Drilling Technique
Glaciated or unconsolidated materials less than 150 ft deep	(1) Hollow-stem continuous-flight auger (2) Solid-stem continuous-flight auger (3) Cable tool
Glaciated or unconsolidated materials greater than 150 ft deep	(1) Cable tool
Consolidated rock formations less than 500 ft deep (minimal or no creviced formations)	(1) Cable tool (2) Air rotary with casing hammer (3) Reverse circulation rotary
Consolidated rock formations less than 500 ft deep (highly creviced formations)	(1) Cable tool (2) Air rotary with casing hammer
Consolidated rock formations greater than 500 ft deep (minimal or no creviced formations)	(1) Air rotary with casing hammer

Source: Barcelona et al., 1985.

the sample. It also can increase processing times if the filters keep clogging (Barcelona et al., 1985).

Development is accomplished by surging water in and out of the well through the screen. In low-permeability units, this surging action is created by circulating clean water, which is not likely to enter the formation (Barcelona et al., 1985). In more conductive formations, you can use a jetting tool that sends compressed air through the screen and into the filter pack. Adding compressed air also removes water from the well, which helps with the development. Other development approaches include the use of a surge block or a bailer.

▶ 16.7 STEPS IN WATER SAMPLING

With the well developed, it is ready for sampling. This section explains the steps, procedures, and essential elements important in sampling. The summary of steps in Figure 16.12 comes from Barcelona et al. (1985).

Well Inspection

Well inspection, the first step, provides (1) a check on the integrity of the well and casing protector and (2) an initial water-level measurement before any water is removed from the well. The integrity check looks for signs of vandalism or tampering with the well. With the checks out of the way, a drop cloth is laid down around the well to help keep the contamination away from the open casing and equipment.

Typically, water levels are measured using an electric tape, with the top of the casing surveyed as a measurement point. In most cases, the same tape is sent down different monitoring wells. Thus, care must be taken to decontaminate the electric tape in moving from one hole to another. Most SOPs for water-level measurements provide specific guidance on how to clean the tape after making a measurement. Because cross-contamination could even destroy wells for water-quality monitoring purposes, personnel must always avoid sending "dirty" equipment down a borehole.

Well Purging

Well purging is designed to get rid of water in the well casing and to bring new water from the aquifer (or geologic unit) into the well for sampling. Water that resides in the casing

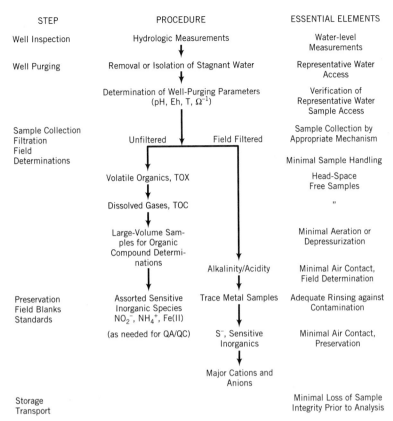

Figure 16.12 Generalized flow diagram showing the steps and essential elements of ground-water sampling (from Barcelona et al., 1985).

for a long time can lose key dissolved gases or react with the casing and seal, causing the chemistry to change. Stimulating this flow of "new" water to the well yields a water sample with a chemical composition representative of *in situ* conditions. Because water purged from the well might be contaminated, it must be collected at the surface and sent for appropriate disposal.

Purging the well, and sampling for that matter, require some capability of getting water out of the well. One common method uses a bailer. A *bailer* is a piece of pipe with a one-way valve at the bottom (Figure 16.13*a*). It is lowered down the hole until it fills with water, and then it is pulled back up the hole and emptied. Although used commonly, Boulding and Barcelona (1991) don't recommend bailers for purging because this approach is often slow to remove water and keeps mixing the old and new water in the well. Again, care must be taken to avoid cross-contaminating wells by failing to clean the bailer and the rope in between sampling stops at different wells. Disposable bailers can help in this respect.

Pumping the well is another way to bring water up to the surface. A variety of pumping techniques can be used. Barcelona et al. (1985) suggest bladder pumps as the most appropriate for sampling. A *bladder pump* (Figure 16.13*b*) works through squeezing action much like the human heart. Gas pressure compresses the sample chamber in the pump and moves the water up the sample line through a one-way valve. When the air pressure is released, water from the well refills the sample chamber, and the pumping cycle is repeated. This method uses positive pressures (not suction) to move the water, and the water moves up the sample line without contacting atmospheric gases.

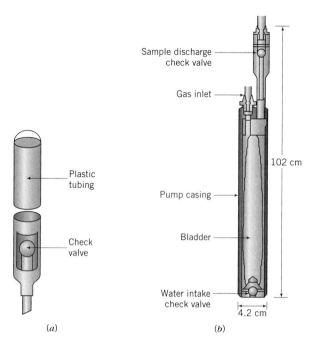

Figure 16.13 Two common approaches for removing water from a well with purging and sampling. Panel (*a*) illustrates a plastic, disposable bailer with a one-way valve on the bottom. Panel (*b*) is a schematic of a bladder pump that uses compressed gas pulses to move water under pressure up the pump line (from Pohlmann et al.,1990).

An *electrical submersible* pump is also useful for purging and sampling wells. This pump combines a waterproof electric motor and a turbine pump that are installed down in the well. This kind of pump is great for purging a well but not quite as desirable for collecting the actual sample. Boulding and Barcelona (1991) think that this pump could impact concentrations of gases or volatile organic compounds in the water.

If the purging pump is moved from well to well, then great care must be taken to clean the pump and sampling line after each use. This cleaning can be tedious and time consuming. In some cases, it is more cost effective to dedicate a pump and a sample line to each monitoring well.

When water levels are close to the ground surface, you might consider purging and sampling using a *vacuum pump*. This pumping system works like a straw in a glass of milk—the vacuum sucks the fluid up a rigid sampling line. This method is inexpensive and easy to implement. In sampling applications, however, it can cause gases and volatiles to be lost. For this reason, it is not commonly recommended (Boulding and Barcelona, 1991).

SOPs for purging usually recommend that from three to five casing volumes of water be removed from the well before a sample is collected. Boulding and Barcelona (1991) consider this guidance as simply a starting point. They recommend a well-specific procedure that uses indicator parameters (for example, pH and specific conductance) that are monitored while purging. When the indicators stabilize, they indicate that new water has been drawn into the well.

Sample Collection, Filtration, and Preservation

The approaches used in purging are also useful in providing water samples for analysis. Bladder pumps and bailers usually provide the most reliable approaches for collecting water samples. Care must be exercised in selecting materials (for the pump and sample line) that do not interact with water to an appreciable extent (see Barcelona et al., 1985, for guidance).

Figure 16.12 points out that it is necessary to field filter some of the samples. If samples aren't filtered, the acid added to preserve the metals, for example, can react with suspended solids and produce anomalously high concentrations. A variety of containers are used at the surface to hold the samples. These containers often contain the chemicals required to preserve the samples. For collecting water containing volatile organic compounds, you need specialized sample vials that have nearly gas-tight lids and that can be filled completely (no airspace, also called headspace). It is beyond the scope of this book to document the various protocols for sample preservation and container selection. Barcelona et al. (1985) and Boulding and Barcelona (1991) provide a useful starting point.

The easiest approach to filtering a sample is to attach the filter to the sample line and filter the flowing water using pump pressure. Other filtering schemes are possible, but vacuum filtering should be avoided. Typically, most field filtering involves a 0.45 μm filter. Barcelona et al. (1985) recommend using glass fiber or Teflon® filters. Other filters (such as cellulose nitrate, cellulose acetate, or polycarbonate) are not recommended because they may react with organic contaminants in the water.

Field Determinations

The values of some parameters and the concentrations of some chemical constituents can change rapidly upon sampling. It is difficult to preserve samples to avoid these changes. For example, oxygen added to an anoxic sample results in a rapid change in E_H that can cause dissolved trace metals to precipitate as solids. Loss of CO_2 can raise the pH and decrease the HCO_3^- concentration and in some cases cause $CaCO_3$ to precipitate. The extent of the change in composition depends in many respects on the chemistry of the particular sample.

Obtaining valid measurements for these parameters requires that measurements be made in the field when the water is sampled. The most reliable measurements are made by using a temperature-controlled flow-through cell with electrodes inserted (Jackson and Inch, 1980). Water being pumped to the surface is routed through the cell, thereby avoiding contact with the atmosphere. It is often convenient to measure specific conductance and turbidity in the field.

The instruments used in the field need to be properly maintained and calibrated using standards. In most cases, SOPs let the user check the operation and calibration of the instrument.

► 16.8 MAINTAINING THE QUALITY OF CHEMICAL DATA

The discussion so far has emphasized the myriad of problems that can frustrate the collection of chemical data. Most government and industrial organizations have well-defined operational procedures to minimize potential sampling problems. These approaches are embodied in what is known as a *sampling plan*. Usually, a sampling plan is created on a project-by-project basis to set down formally how the sampling will take place. One purpose of a plan is to define standard operating procedures (SOPs) for the various activities. The SOPs would be available by regulators or other interested parties to review during the study. The sampling plan also describes necessary activities (QA) to document the quality of the sampling process (e.g., field blanks, trip blanks, field duplicates). Finally, the study plan makes sure that the legal requirements often associated with sampling (for example, chain of custody requirements) are maintained. Taken together, the sampling plan provides a management approach that minimizes errors by applying consistent and scientifically proven methods in the field, a system of checks that might identify problems early, and sample-handling approaches that minimize the possibilities of sample tampering and shipping problems.

SOPs are detailed, time-tested approaches to conducting field operations. They are typically confidential assets that can provide an important business advantage. For example, if my SOPs are more rigorous and more thoroughly tested and implemented than yours, it might give me a competitive advantage in bidding for work. Thus, not all SOPs are the same; they might vary from company to company. SOPs cover activities like water-level measurements, well purging, sample collection, sample filtration and preservation, bottle labeling, and on-site measurements. For exotic sampling, you might have to create a new SOP.

Fortunately, not all guidance that might let you create an SOP is locked in corporate vaults. The U.S. EPA provides information in its *Compendium of Superfund Field Operations Methods* (U.S. EPA, 1987). The American Society of Testing and Materials (ASTM) establishes standard methods and publishes them in book form. The U.S. Geological Survey has a report series called Techniques in Water Resources Investigations, which is also useful.

A *quality assurance (QA) program* is a system of documented checks that validate the reliability of a data set (Barcelona et al., 1985). The implementation of a QA program is through quality control (QC) checks at different stages of the overall sampling process. Following are examples of some QC procedures that could be conducted as part of the sampling program to reduce problems and to provide data on the overall quality of the sampling effort.

1. Field (or cleaning) blanks establish whether there are problems due to field operations and transport. Field blanks are samples of deionized, contaminant-free water that are sent through decontaminated sampling devices and filtering systems. They are collected periodically, handled just like any other ground-water sample, and sent to the laboratory for analysis. Usually, every 10 to 20 regular water samples are accompanied by a field blank.

2. Trip blanks are usually used with samples containing volatile organic compounds (VOCs). These samples start out from the laboratory as a sample bottle filled with analyte-free water. They are taken into the field in shipping chests and returned to the laboratory without being opened. The trip blank measures the possibility of cross contamination among the sample bottles during transport. A trip blank is normally collected for each 10 to 20 regular samples.

3. Occasionally, two samples from the same well are sent to the laboratory. Field duplicates measure the analytical precision of the laboratory. *Precision* in analysis is the ability to get the same answer by analyzing the same sample over and over again. However, just because an analysis is precise doesn't mean it's accurate. Duplicates are sent in every batch of 10 or 20 samples.

4. A spiked sample provides a way of quantifying the accuracy of a laboratory. It is prepared by splitting a field sample and adding a known quantity of a particular constituent to one of the splits. If the laboratory is accurate, it should report back the known or spiked value. Usually, one spiked sample is sent with every group of samples.

Because the various QC checks require the laboratory to analyze samples, the process is expensive. In other words, many of the samples that are sent to the laboratory in the sampling program are not actual ground-water samples but "check" samples. One might be concerned about the cost (or effort if you are analyzing the samples). However, the expense is often worthwhile insurance because potential problems in the chemical results can be identified at an early stage.

For regulatory and legal problems, one normally has to maintain and document official custody of samples from when they are collected to when they are discarded. In many respects, the concern is similar to airline personnel who ask whether your bags have been in your possession since you packed them.

► EXERCISES

16.1. Routine analysis of a water sample provides the following concentrations (as mg/L). Calculate the concentrations of the ions in terms of molarity and milliequivalents per liter (meq/L).

Concentrations in mg/L: Ca−120; Mg−27; Na−53; K−4.4; Fe−0.5;

HCO$_3$ − 546; SO$_4$ − 10; Cl −26; NO$_3$ − 0.1; pH 7.6

16.2. Assess in a preliminary way the quality of the analytical results in Exercise 16.1 by determining the cation/anion balance.

16.3 Develop a relatively general Excel spreadsheet that will take ion concentrations in mg/L and convert them to mol/L and meq/L. Plan for each water sample to be processed on a separate sheet. Add labels to make it clear what you are doing. Save your program on a disk. Show that your spreadsheet works by repeating the calculations in Exercises 16.1 and 16.2.

16.4 This question and others in later chapters relate to a study of the Alliston Aquifer in Ontario, Canada, conducted by Aravena et al. (1995). The Alliston aquifer is a regionally confined aquifer system located north of Toronto, Canada, in the Upper Nottawasaga Basin. The aquifer is composed of sand and gravel lenses and is confined above by thick glacial till and below by Paleozoic bedrock or

TABLE 16.14   Water chemistry data for overburden and bedrock wells

Well	pH	Ca, mg/L	Mg, mg/L	Na, mg/L	K, mg/L	Fe, mg/L	HCO$_3$, mg/L	SO$_4$, mg/L	Cl, mg/L	NO$_3$, mg/L
OV2	7.43	86.8	29.3	5.87	2.62	0.07	336	46.5	12	10.2
OV6	7.88	71.5	16.6	3.02	2.15	1.84	242	33.7	1.02	<0.01
OV7	7.78	45.9	18.9	84.7	5.09	4.06	339	0.23	81.1	<0.01
OV9	7.7	46.1	17.5	24.9	3.1	3.18	247	0.3	27.3	<0.1
OV15	7.36	102	25.4	75.3	2.94	0.16	426	20.3	103	43.2
OV17	7.5	48.6	19.0	32.5	2.84	1.59	319	<0.01	2.41	<0.01
OV19	8.1	4.28	3.62	63.8	0.34	0.2	169	1.96	66.3	0.67
OV21	8.01	23.0	17.5	99.3	2.43	0.64	283	<0.01	89.5	<0.01
OV23	7.69	42.8	20.1	239	3.17	4.54	428	<0.01	309	<0.01
OV29	7.75	65.4	18.7	4.1	2.5	0.3	253	32.5	2.11	<0.01
OV32		53.5	21.6	37.7	1.22	1.83	395	2.28	7.04	5.69
OV34	7.85	2.56	5.21	68.0	0.78	1.34	234	0.1	2.58	3.89
OV43	7.78	40.0	19.2	47.7	5.86	2.13	342	0.64	7.2	0.26
OV45	7.85	43.7	10.1	3.36	1.04	2.16	186	20.7	1.93	<0.01
OV49	7.64	54	18	38.8	6.08	2.14	391	0.28	6.31	<0.01
BR2	7.84	37.6	22.3	177	22.5	1.6	306	2.92	239	<0.01
BR3	7.65	25.8	13.7	148	14.4	0.71	532	<0.01	38.8	<0.01
BR7	7.7	56.8	28.3	29.9	1.54	1.07	310	10.5	30.9	<0.01
BR14	7.8	1.8	18.2	168	2.9	1.66	185	0.01	226	<0.01
BR18	7.98	23.7	16.9	30.0	3.3	0.16	203	1.64	18.9	<0.01

Overburden BR = bedrock.
Source: Modified from Aravena et al.,1995.

glacial till. The aquifer is part of a unit extending north toward Georgian Bay and south toward Lake Ontario. Shown in Table 16.14 are data from water samples collected in the aquifer and from bedrock underlying the aquifer. Process all of the chemical data through your spreadsheet from Exercise 16.3 and use the cation/anion balances to evaluate the quality of the chemical data.

16.5 Represent the following chemical data graphically using (a) a Collins diagram and (b) a Piper diagram. (c) Classify the water using the concept of hydrochemical facies. (as mg/L): $Ca^{2+} - 93.9$; $Mg^{2+} - 22.9$; $Na^+ - 19.1$; $HCO_3^- - 334.$; $SO_4^{2-} - 85.0$; $Cl^- - 9.0$ and a pH of 7.20.

▶ CHAPTER
HIGHLIGHTS

1. Rocks and minerals dissolve in water to form ions. Positively charged species, such as Ca^{2+} or K^+, are cations; negatively charged species, such as HCO_3^- or Cl^-, are anions. However, most organic compounds (for example, trichloroethene, TCE) dissolve as nonionic or uncharged molecules.

2. Concentration measurements indicate how much of a partuclar ion or molecule is dissolved in water. Mass per unit volume concentration is the most common scale for concentration. It defines the mass of a solute dissolved in a unit volume of solution, for example, mg/L or μg/L.

3. It is neither feasible nor necessary to measure the concentration of all constituents that conceivably might occur in water. A routine analysis involves measuring the concentration of a standard set of the most abundant constituents and parameters like pH, total dissolved solids, and specific conductance. Such a test forms the basis for assessing the suitability of water for human consumption or various industrial and agricultural uses.

4. In the United States, the National Primary Drinking Water Regulations (NPDWRs) are the legally enforceable standards that are designed primarily to protect public health. Contaminants and constituents covered by these regulations include microorganisms, disinfection and disinfection byproducts, inorganic chemicals, organic chemicals, and radionuclides. Standards are presented in terms of MCLs and MCLGs, the maximum contaminant level (MCL), and the maximum contaminant level goal (MCLG).

5. Another set of standards, the National Secondary Drinking Water Regulations (NSDWRs), helps judge the quality of water, especially the natural constituents found in ground water. These nonenforceable guidelines apply to constituents that may cause cosmetic effects (skin or tooth discoloration) or aesthetic effects (taste, odor, or color) in drinking water.

6. Several different graphical approaches can depict the abundance or relative abundance of ions in individual water samples. The most common are (a) the Collins (1923) bar diagram, (b) the Stiff (1951) pattern diagram, (c) the pie diagram, and (d) the Piper (1944) diagram.

7. The sources of errors in sampling ground water run the gamut from improper procedures for installing the well, reactions between the ground water and the well casing or sampler, and poor sample-handling protocols on the surface. The use of standard operating procedures (SOPs) can minimize these problems.

8. Important considerations in deciding where to locate wells include the need for sample locations that take into account the character and complexity of chemical patterns and the objectives of the monitoring program. At a regional scale, changes in water chemistry might reflect the chemical evolution of water along flow systems or the unique geochemical environments provided by a specific geologic unit. In the case of contaminant plumes, wells are located to define the extent of contamination.

9. At least three well-design issues are pertinent to providing usable chemical data: choosing appropriate screen and casing materials, selecting proper grouts and seals, and using a screen length appropriate for the monitoring need.

10. The key steps in water sampling include (a) well inspection, (b) well purging, (c) sample collection, filtration, and preservation, and (d) field determinations. Usually, a sampling plan is created on a project-by-project basis to set down formally how the sampling will take place. One purpose of a plan is to define SOPs. for the various activities. The SOPs would be available by regulators or other interested parties to review during the study.

17

KINETIC AND EQUILIBRIUM REACTIONS

▼ 17.1 LAW OF MASS ACTION AND CHEMICAL EQUILIBRIUM

▼ 17.2 DEVIATIONS FROM EQUILIBRIUM

▼ 17.3 KINETIC REACTIONS

In natural waters, chemical reactions operate to increase or decrease the abundance of dissolved constituents in water. For example, mass may be removed from the aqueous phase as ions are incorporated into a mineral that is precipitating or as minerals or pieces of organic matter sorb ions. We begin this exploration of chemical reactions in ground water with an introduction of the equilibrium and kinetic framework that is essential for describing chemical and biological reactions. This chapter also illustrates how mineral saturation is used to describe the saturation state of ground water with respect to key minerals.

▶ 17.1 LAW OF MASS ACTION AND CHEMICAL EQUILIBRIUM

Equilibrium and kinetic concepts provide a basis for treating chemical reactions. The starting point for exploring these concepts is the *law of mass action*. Consider the following reaction where ions C and D react to produce ions Y and Z

$$cC + dD = yY + zZ \qquad (17.1)$$

with c, d, y, and z representing the number of moles of these constituents. For a dilute solution, the law of mass action describes the equilibrium distribution of mass between reactants and products as

$$K = \frac{(Y)^y (Z)^z}{(C)^c (D)^d} \qquad (17.2)$$

where K is the *equilibrium constant* and (Y), (Z), (C), and (D) are the molal (or molar) concentrations for reactants and products. Equation (17.2) provides a good way to think about chemical equilibrium. When a reaction is at equilibrium in a closed system, there is no chemical energy available to alter the relative distribution of mass between the reactants and products in a reaction (for example, Eq. 17.2). Away from equilibrium, energy is available to spontaneously drive a system toward equilibrium by allowing the reaction to progress.

Clearly, the quantitative description of chemical reactions requires actual numerical values for the equilibrium constant in Eq. (17.2). Such values can be calculated using thermodynamic principles or from tabulations in geochemical textbooks. Equilibrium constants, expressed as a function of temperature, are also included in the databases of computer codes for modeling aqueous systems such as SOLMNEQ (Kharaka and Barnes, 1973), EQ3/6 (Wolery, 1979), and PHREEQE (Parkhurst et al., 1980).

Activity Models

Simple application of equilibrium equations like (17.2) is frustrated by the fact that once a solution is no longer dilute, one needs to use "thermodynamically effective concentrations" or activities instead of actual concentrations. *Activity models* provide the way to calculate these activities and incorporate the nonideal behavior of the solution. One writes the mass action equations in terms of activities instead of molal concentrations. For the case of species Y, the following equation relates activity to molar concentration:

$$[Y] = \gamma_Y(Y) \tag{17.3}$$

where γ_Y is the *activity coefficient*. Typically, γ is close to 1 in dilute solutions and decreases as salinity increases. Activities of species with more than one charge usually are smaller than those with a single charge. At relatively high salinities, the activity coefficients in many cases begin to increase and even exceed one.

We can calculate γ for ions from one of several different activity models. The simplest is the Debye-Hückel equation

$$\log \gamma_i = -A z_i^2 (I)^{0.5} \tag{17.4}$$

where A (Table 17.1) is a constant that is a function of temperature, z_i is the ion charge, and I is the ionic strength of the solution. Mathematically, the ionic strength in mols/L is

$$I = 0.5 \sum M_i z_i^2 \tag{17.5}$$

TABLE 17.1 Parameters used in the extended Debye–Hückel equation at 1 atmosphere pressure

Temperature (C°)	A	B (×10⁸)	$a_i \times 10^{-8}$ cm	Ion
0	0.4883	0.3241	2.5	Rb^+, Cs^+, NH_4^+, Ag^+,
5	0.4921	0.3249	3.0	$K^+, Cl^-, Br^-, I^-, NO_3^-$
10	0.4960	0.3258	3.5	OH^-, F^-, HS^-, BrO_3^-,
15	0.5000	0.3262	4.0–4.5	$Na^+, HCO_3^-, H_2PO_4^-$,
20	0.5042	0.3273		$HSO_3^-, Hg_2^{2+}, SO_4^{2-}$,
25	0.5085	0.3281		HPO_4^{2-}, PO_4^{3-}
30	0.5130	0.3290	4.5	$Pb^{2+}, CO_3^{2-}, MoO_4^{2-}$,
35	0.5175	0.3297	5.0	$Sr^{2+}, Ba^{2+}, Ra^{2+}, Cd^{2+}$,
40	0.5221	0.3305		$Hg^{2+}, S^{2-}, WO_4^{2-}$
50	0.5319	0.3321	6	$Li^+, Ca^{2+}, Cu^{2+}, Zn^{2+}$,
60	0.5425	0.3338		$Sn^{2+}, Mn^{2+}, Fe^{2+}, Ni^{2+}$,
				Co^{2+}
			8	Mg^{2+}, Be^{2+}
			9	H^+, Al^{3+}, Cr^{3+}, trivalent
				rare earths
			11	$Th^{4+}, Zr^{4+}, Ce^{4+}, Sn^{4+}$

Source: Reprinted with permission from Manov, G. G. et al., *J. Am. Chem. Soc.*, v. 65, 1943, p. 1764–1767. Copyright © 1943 American Chemical Society.

with (M_i) as the molar concentration of species i having a charge z. *Ionic strength* is a measure of the total concentration of ions that emphasizes the increased contribution of species with charges greater than one to solution nonideality.

Use of the Debye-Hückel equation is limited to solutions with ionic strengths less than 0.005 M, which is fresh, potable ground water. It is inappropriate for use with more saline water.

The extended Debye-Hückel equation provides estimates of γ_i to a maximum ionic strength of about 0.1 M or a total dissolved solids content of approximately 5000 mg/L (Langmuir and Mahoney, 1984). This equation is

$$\log \gamma_i = \frac{-A z_i^2 (I)^{0.5}}{1 + B a_i (I)^{0.5}} \tag{17.6}$$

where A and B are temperature-dependent constants and a_i is the radius of the hydrated ion in centimeters (Table 17.1).

Geochemical studies of basinal brines require other activity models to deal with the higher salinity water. The simplest model takes an equation like (17.6) and adds additional correction or curve-fitting parameters. The Davies equation is an example. Establishing activities in even more saline water requires the use of ion interaction models of a type proposed by Pitzer and Kim (1974). This sophisticated activity model is more difficult to use than the Debye-Hückel or Davies-type equations but provides accurate estimates of activity up to 20 M (Langmuir and Mahoney, 1984).

On an activity versus ionic strength plot for Cl^- and Ca^{2+}, note the typical curve shapes for monovalent and divalent ions (Figure 17.1). We have shown activities calculated using the Pitzer model, but they are not strictly comparable to the other three. At low ionic strengths, all four methods give approximately the same activity coefficients. However, as the ionic strength increases (Figure 17.1), estimates of activity from the Debye-Hückel, extended Debye-Hückel, and Davies equations differ noticeably.

Because the extended Debye-Hückel equation or the Davies equation is applicable to charged ionic species, they cannot determine γ's for neutral species that exist as ion pairs or other complexes. For solutions with ionic strengths less than about 0.1 M, activity coefficients are assigned a value of 1. In solutions with higher ionic strengths, activity

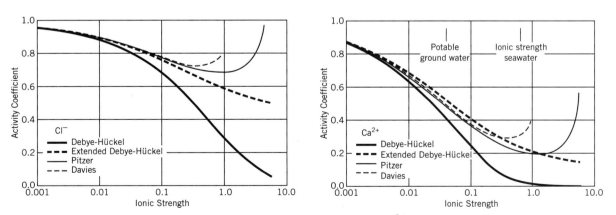

Figure 17.1 Comparison of activities of Cl^- and Ca^{2+} as a function of ionic strength, calculated using the Debye-Hückel, extended Debye-Hückel, Davies, and Pitzer equations (unpublished data from A. S. Crowe).

estimates are difficult. Possible approaches include making activities a proportion of the ionic strength or equivalent to the activity of CO_2.

Typically for ground waters, activity coefficients are calculated using the extended Debye-Hückel equation (Eq. 17.6) or the Davies equation. The following example illustrates the use of the extended Debye-Hückel equation.

▶ **EXAMPLE 17.1**   Water contains ions in the following molar concentrations Ca^{2+} 3.25 × 10^{-3}, Na^+ 0.96 × 10^{-3}, HCO_3^- 5.75 × 10^{-3}, and SO_4^{2-} 0.89×10^{-3}. Calculate activity coefficients for Na^+ and Ca^{2+} at 25°C using Eq. (17.6).

SOLUTION   First, calculate the ionic strength from Eq. (17.5) or

$$I = 0.5 \times (3.25 \times 2^2 + 0.96 \times 1^2 + 5.75 \times 1^2 + 0.89 \times 2^2) \times 10^{-3} = 0.012 \, \text{M}$$

This value of I is at the upper end of the range of applicability of the extended Debye-Hückel equation. Now, determine the γ values

$$\log \gamma_{Na^+} = -0.508 \times 1^2 (0.012)^{0.5}/(1 + 0.328 \times 10^{+8} \times 4.0 \times 10^{-8}(0.012)^{0.5})$$

$$= -0.049$$

$$\gamma_{Na^+} = 10^{-0.49} = 0.89$$

$$\log \gamma_{Ca^{2+}} = -0.508 \times 2^2 (0.012)^{0.5}/(1 + 0.328 \times 10^{+8} \times 6.0 \times 10^{-8}(0.012)^{0.5})$$

$$= -0.183$$

$$\gamma_{Ca^{2+}} = 10^{-0.183} = 0.66$$

▶ 17.2 DEVIATIONS FROM EQUILIBRIUM

Equilibrium calculations are most commonly used to assess whether the ground water is in equilibrium with respect to one or more minerals. For example, if we added gypsum crystals to distilled water, the solution would be initially undersaturated with respect to gypsum. Gypsum would dissolve until the dissolution reaction reached equilibrium, at which point the dissolution would stop. The accumulation of dissolved constituents in ground water can be viewed as a reaction of rainwater with the assemblage of minerals comprising the porous medium. Some of these minerals would dissolve rapidly, and the ground water would rapidly achieve equilibrium with respect to these minerals. In other cases, the ground water may not have reached equilibrium with respect to others. The term *partial equilibrium system* describes a complex mineral–water system in which reactions may not be at equilibrium (for example, dissolution or precipitation reactions). Hydrogeologists use water chemistry data to sort out the equilibrium status with respect to various minerals. These calculations in a regional water quality investigation help to explain why the major and minor ion chemistry is changing due to mineral dissolution and precipitation reactions.

The departure of a reaction from equilibrium is determined as the ratio of the ion activity product to the equilibrium constant (*IAP/K*). The ion activity product is calculated by substituting sample activity values in the mass law expression for a reaction. For example, given a ground water with known activities of $[C]$, $[D]$, $[Y]$, and $[Z]$, the ion activity product (*IAP*) for the reaction in Eq. (17.1) is

$$IAP = \frac{[Y]^y [Z]^z}{[C]^c [D]^d} \qquad (17.7)$$

If the $IAP > K$, the reaction is progressing from right to left reducing $[Y]$ and $[Z]$ and increasing $[C]$ and $[D]$. If $IAP < K$, the reaction is proceeding from left to right. The IAP at equilibrium is equal to the equilibrium constant.

This theory provides the saturation state of a ground water with respect to one or more mineral phases. When $IAP/K < 1$, the ground water is *undersaturated* with respect to the given mineral. When $IAP/K = 1$, the ground water is in chemical equilibrium with the mineral, and when $IAP/K > 1$, the ground water is *supersaturated*. Undersaturation with respect to a mineral results in the net dissolution, provided the mineral is present. Supersaturation results in the net precipitation of the mineral, should suitable nuclei be present.

The saturation state is often expressed in terms of a *saturation index* (SI), defined as $\log (IAP/K)$. When a mineral is in equilibrium with respect to a solution, the SI is zero. Undersaturation is indicated by a negative SI and supersaturation by a positive SI. Example 17.2 illustrates how to calculate the state of saturation with respect to a mineral.

▶ **EXAMPLE 17.2**

Given a ground water with the molar composition as follows, calculate the saturation state with respect to calcite and dolomite. The activity coefficients for Ca^{2+}, Mg^{2+}, and CO_3^{2-} are 0.57, 0.59, and 0.56, respectively. The equilibrium constants defining the solubility of calcite and dolomite are 4.9×10^{-9} and 2.7×10^{-17}, respectively.

$$(Ca^{2+}) = 3.74 \times 10^{-4}: (Mg^{2+}) = 4.11 \times 10^{-6}: (Na^+) = 2.02 \times 10^{-2}:$$
$$(K^+) = 6.14 \times 10^{-5}: (H^+) = 10^{-7.9}$$
$$(HCO_3^-) = 1.83 \times 10^{-2}: (CO_3^{2-}) = 5.50 \times 10^{-5}:$$
$$(SO_4^{2-}) = 1.24 \times 10^{-3}: (Cl^-) = 1.19 \times 10^{-3}$$

SOLUTION

$$
\begin{aligned}
IAP_{cal} &= [Ca^{2+}][CO_3^{2-}] \\
&= 0.57 \times 3.74 \times 10^{-4} \times 0.56 \times 5.50 \times 10^{-5} \\
&= 6.56 \times 10^{-9}
\end{aligned}
$$

$$\{IAP/K\}_{cal} = 6.56 \times 10^{-9}/4.90 \times 10^{-9} = 1.34$$

The sample is slightly oversaturated with respect to calcite.

$$IAP_{dol} = [Ca^{2+}][Mg^{2+}][CO_3^{2-}]^2 = 4.89 \times 10^{-19}$$

$$\{IAP\}_{dol} = 4.89 \times 10^{-19}/2.7 \times 10^{-17} = 0.018$$

The sample is strongly undersaturated with respect to dolomite.

Saturation calculations with respect to selected minerals provide a useful way of exploring how the pore fluid has interacted with the minerals making up the porous medium and of understanding the pattern of chemical evolution along the flow system. Example 17.2 showed how these calculations can be carried out by hand. There is a problem, however, in that this simple approach doesn't take account of the presence of complexes.

Let's consider the Ca ion as an example. A water analysis gives the total calcium ion (Ca_T) present in a sample. In simple equilibrium calculations, one assumes that (Ca^{2+}) is the same as determined from the water analysis (Ca_T). In potable waters, this approximation is good with (Ca_T) approximately the same as (Ca^{2+}). Thus, the hand calculations are approximately correct. In more saline waters, calcium forms complexes with, for example, organic compounds and anions. Thus, (Ca_T) is greater than (Ca^{2+}) and the assumption breaks down. One needs to come up with an accurate estimate of (Ca^{2+}) that accounts for the complexes likely to form.

The best way to do this speciation calculation and the equilibrium calculations is to utilize a computer code like WATEQF (Plummer et al., 1976), MINTEQ (Felmy et al., 1983), or PHREEQE (Parkhurst et al., 1980). The advantages of using one of these codes are that the burden of calculations is carried by the computer and the consistency of the database of equilibrium constants facilitates the comparison of results from various studies.

▶ 17.3 KINETIC REACTIONS

We have just considered the theory applicable to reactions at equilibrium. However, this theory is limited in that it provides no information about the time required to reach equilibrium or the reaction pathways involved. A different, more general approach is required if one is interested in quantitatively describing those features of a system. Kinetic theory provides a useful framework for studying reactions in relation to time and reaction pathways. A kinetic description is applicable to any reaction but is required for irreversible reactions or reversible reactions that are slow in relation to mass transport. An irreversible reaction proceeds only in one direction. For example, the reaction that turns a shiny steel nail to rust is irreversible. There is no way the rusty nail will turn back to shiny steel.

First-Order Kinetic Reactions

Radioactive decay is an example of an irreversible reaction. For example, carbon-14 decays to nitrogen and an electron with an equation of the following form

$$^{14}C \rightarrow {}^{14}N + e \qquad (17.8)$$

Notice that this reaction is written with an arrow and not an equal sign, signifying that it is irreversible, proceeding in only one direction. The kinetic equation for ^{14}C in this reaction

---

CASE STUDY 17-1

SATURATION CALCULATIONS

In Indiana, carbonate rocks of Silurian and Devonian age are important aquifers. Some of these aquifers are formed from thick carbonate reefs, which have a porosity of 10 to 15% and relatively high hydraulic conductivities. Nonreef aquifers are less porous and permeable and contain more shale units. The study involved drilling a well at three different sites and testing various zones using a packer system. Following in Table 17.2 are 12 water analyses from Sites 1, 2, and 3. At Site 1, the aquifer is a nonreef type. At Sites 2 and 3, the aquifer is a reef. This example will examine the water chemistry in the two types of aquifer (reef and nonreef). Fortunately, the results of the study by Schnoebelen and Krothe (1999) reduce the work involved.

A cursory inspection of the chemical data in Table 17.2 suggests that there is a significant difference in the chemistry of water from the two different aquifer settings. Ground water from

Site 1 is a $Na-HCO_3$ type water. Water from Sites 2 and 3 is a $Ca-Mg-HCO_3$ type. A Piper diagram (Figure 17.2) is useful in illustrating the differences in the relative ion abundances in the two types of aquifer. Clearly, waters from all three sites are different in chemistry, but Site 1, the nonreef aquifer, is distinctly different from the other two waters.

The code WATEQF was used in the study to characterize saturation of the ground water with respect to various mineral species. The results of these calculations (Table 17.3) show that Site 1 tends to be somewhat undersaturated in terms of carbonate minerals, as compared to Sites 2 and 3. Site 1 is closer to saturation with respect to fluorite as compared to samples from the other two sites. Thus, the combination of graphical analysis and saturation determinations was useful in differentiating the chemical features of ground water from the two reef environments.

From Schnoebelen and Krothe, 1999.

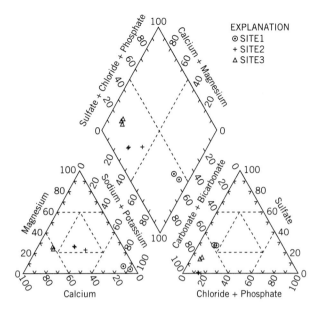

Figure 17.2 Trilinear diagram showing results of water analyses from sites 1, 2, and 3 (from Schnoebelen and Krothe, 1999). Reprinted by permission of *Ground Water*. Copyright ©1999. All rights reserved.

TABLE 17.2   Ground water chemistry data for Sites 1, 2, and 3 in the study area

Site and Depth	Field Values			Concentration of Ions (mg/L)									
(Meters below land surface)	T °C	SpC µs/ cm	pH	Ca	Mg	Na	K	Si	F	CO₃	HCO₃	Cl	SO₄
Site 1													
56–59	11.6	1114	8.01	20	9.1	210	7.2	7.5	2.4	0	404	48	150
67–70	12.1	1279	8.02	11	6.3	260	7.1	6.7	3.1	10	449	64	170
Site 2													
16–19	12.6	693	7.51	60	23	47	4.6	11	0.50	0	369	32	2.8
38–41	12.1	632	7.40	58	22	46	4.7	11	0.50	0	376	31	1.9
78–81	12.5	631	7.42	59	22	48	4.5	11	0.50	0	366	33	1.0
102–105	12.6	704	7.33	59	23	50	4.5	11	0.40	0	383	39	0.90
127–183	12.6	867	7.34	62	25	87	5.6	10	0.50	0	388	81	2.1
Site 3													
20–23	11.7	522	7.70	60	14	16	1.3	12	0.50	0	254	13	35
56–59	11.9	556	7.70	67	17	15	1.4	12	0.40	0	290	17	41
107–110	12.0	559	7.65	68	17	16	1.5	12	0.40	0	263	18	42
130–133	12.0	552	7.60	66	17	15	1.4	12	0.40	0	259	16	41
130–183	12.0	552	7.60	65	15	15	1.4	12	0.50	0	259	16	41

T, water temperature; °C, degrees Celsius; SpC, specific conductance; µg/cm, microsiemens per centimeter; Alk., alkalinity; *, laboratory alkalinity, µg/L, micrograms —, not available.

Source: Schnoebelen and Krothe, 1999. Reprinted with permission of *Ground Water*. Copyright © 1999. All rights reserved.

has the following form:

$$\frac{d(^{14}C)}{dt} = -k_1(^{14}C)^1 \qquad (17.9)$$

TABLE 17.3 **Saturation indices for selected minerals in the study area**

Site and Depth (meters below top of land surface)	SI Dolomite	SI Calcite	SI Aragonite	SI Gypsum	SI Fluorite	SI FCO$_3$ Apatite
Site 1						
56–59	+ 0.14	+ 0.16	+ 0.01	− 2.00	− 0.69	—
67–70	− 0.14	− 0.31	− 0.18	− 2.23	− 0.76	+ 6.61
Site 2						
16–19	+ 0.14	+ 0.19	+ 0.04	− 3.20	− 1.52	+ 7.63
38–41	− 0.12	+ 0.07	− 0.08	− 3.38	− 1.53	+ 5.03
78–81	− 0.06	+ 0.09	− 0.06	− 3.65	− 1.53	+ 4.37
102–105	− 0.20	+ 0.02	− 0.13	− 3.70	− 1.73	—
127–183	− 0.19	+ 0.02	− 0.13	− 3.36	− 1.54	+ 5.32
Site 3						
20–23	− 0.03	+ 0.22	+ 0.07	− 2.07	− 1.48	+ 10.15
56–59	+ 0.20	+ 0.32	+ 0.16	− 1.98	− 1.65	+ 9.66
107–110	+ 0.03	+ 0.24	+ 0.83	− 1.96	− 1.64	+ 6.85
130–133	− 0.09	+ 0.17	+ 0.02	− 1.98	− 1.46	+ 8.42
130–183	− 0.15	+ 0.17	+ 0.01	− 1.98	− 1.46	—

SI, saturation indice; − , not calculated.
Source: Schnoebolen and Krothe, 1999. Reprinted with permission of *Ground Water.* Copyright © 1999. All rights reserved.

where (^{14}C) is the activity (for simplicity concentration) of carbon-14, t is time, and k_1 is the rate constant for the reaction. The negative sign in front of k_1 indicates that the concentration of carbon-14 is being reduced in the reaction. Notice the superscript 1 that is added to the term (^{14}C). In most cases, this superscript is omitted when writing the equation because a value of 1 can be assumed. Nevertheless, we need to keep track of the superscripts on the right-hand side of kinetic equations because these will tell us the *order* of the reaction. As we will learn shortly, Eq. (17.9) is an example of a first-order kinetic reaction with the general form

$$\frac{dC}{dt} = -k\,C \tag{17.10}$$

In fact, many of the important reactions that we deal with are first-order reactions.

Equation (17.10) is an ordinary differential equation that can be solved given an initial condition such as $C(0) = C_0$. In words, this condition states that at time equals zero there is some initial quantity of a constituent C_0. The solution to Eq. (17.10) lets us determine the concentration of C for all times greater than zero. This is the wonderful property of kinetic equations that lets us track the decrease or increase of dissolved constituents in aqueous solution with time. Mathematically, the solution has the following form:

$$C = C_0 e^{-kt} \tag{17.11}$$

In words, this equation states that some initial concentration of carbon-14 will be reduced exponentially as a function of time and the rate constant k. The following calculation illustrates these points.

▶ **EXAMPLE 17.3**   A chemical constituent in ground water disappears following a first-order kinetic rate law. Given some initial concentration of 100 mg/L and a rate constant k_1 of 0.00693 days^{-1}, calculate how much of the constituent will remain after 100 days.

SOLUTION   The solution comes simply by substituting the numbers into Eq. (17.10).

$$C = C_0 e^{-k_1 t}$$

$$C = 100 \, mg/L \times e^{-0.00693 \times 100}$$

$$C = 100 \, mg/L \times 0.5 = 50 \, mg/L$$

Thus, at the end of 100 days there is 50 mg/L of the constituent remaining.

Plot the function $C = 100 e^{-0.00693 t}$ using both arithmetic and semi-log scales. As expected, the arithmetic plot (Figure 17.3a) shows the typical exponential loss of the constituent as a function of time. Plotting the equation as a semi-log plot (Figure 17.3b), however, yields a straight line. This result also can be shown mathematically as follows:

$$C = 100 e^{-0.00693 t}$$

$$\ln C = \ln 100 + (-0.00693 t)$$

$$2.3 \log C = 2.3 \log 100 - 0.00693 t$$

$$\log C = \log 100 - 0.003 t$$

$$\log C = 2 - 0.003 t$$

Commonly, the tendency for kinetic data to fall along a straight line when plotted in this manner is taken as evidence of a first-order kinetic equation.

The rate constant in a kinetic equation can also be expressed by an equivalent parameter termed the half life ($t_{1/2}$). The half life is the time required to reduce the concentration by one-half. The relationship between $t_{1/2}$ and k can easily be determined starting from

$$C = C_0 e^{-kt}$$

$$\frac{C}{C_0} = \frac{1}{2} = e^{-kt_{1/2}}$$

$$\ln \frac{1}{2} = -kt_{1/2}$$

$$t_{1/2} = -\frac{\ln 1/2}{k} = \frac{0.693}{k}$$

For our example problem, the half life was 100 days, so that after 100 days the initial 100 mg/L would shrink to 50 mg/L.

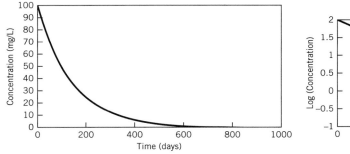
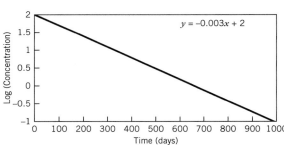

Figure 17.3   Panel (a) shows the exponential loss of a constituent as a function of time. A plot of the log of concentration versus time in Panel (b) yields a straight line typical of a first-order kinetic reaction.

General Theory of Kinetic Reactions

Not all reactions are the simple, first-order type considered in the previous section. Here, we present a more general theory available to analyze more complicated reactions. Consider a case in which several reactants and products are involved in the reaction as follows:

$$aA + bB + \ldots \text{(other reacts)} \underset{k_1}{\overset{k_2}{\rightleftharpoons}} rR + sS + \ldots \text{(other prods)} \tag{17.12}$$

where k_1 and k_2 are the rate constants for the forward and reverse reactions, respectively. Each constituent in Eq. (17.12) has a reaction rate (for example, r_A) that describes the rate of change of concentration as a function of time $\{d(A)/dt\}$. Because of the stoichiometry, these rate expressions are related

$$-r_A/a = -r_B/b = \ldots = r_R/r = r_S/s = \ldots \tag{17.13}$$

where a, b, r, and s are the stoichiometric coefficients.

The rate law for a component (say A) in Eq. (17.12) has this form

$$r_A = d(A)/dt = -k_1(A)^{n1}(B)^{n2} \ldots \text{(other reactants)} \tag{17.14}$$

$$+k_2(R)^{m1}(S)^{m2} \ldots \text{(other products)}$$

where $n1$, $n2$, ... and $m1$, $m2$, ... are empirical coefficients. This equation expresses the rate of change in (A) as the difference between the rate at which the component is being used in the forward reaction and generated in the reverse reaction. Let's consider an irreversible reaction involving the oxidation of ferrous iron between pH 2.2 and 3.5:

$$Fe^{2+} + 1/4\,O_2 + 1/2\,H_2O \rightarrow FeOH^{2+} \tag{17.15}$$

The rate law for Fe^{2+} depends both on the concentrations of Fe^{2+} and O_2, and as shown in Langmuir and Mahoney (1984) is written as

$$d(Fe^{2+})/dt = -k_1(Fe^{2+})(P_{O2}) \tag{17.16}$$

Kinetic reactions are described according to the order of the reaction. For the reaction,

$$2A + B \rightarrow C \tag{17.17}$$

the rate law in terms of (A) is

$$d(A)/dt = -2d(B)/dt = -2d(C)/dt = -k_1(A)^{n1}(B)^{n2} \tag{17.18}$$

The coefficients $n1$ and $n2$ define the order of the reaction in A and B. The order of the overall reaction is the sum of all the n's. Thus, Eq. (17.18) is of the order $n1 + n2$. Applying this scheme to Eq. (17.15), we see that this equation is second order.

As Eq. (17.14) illustrates, the rate law for an equilibrium reaction depends not only on how fast a constituent is being consumed in the forward reaction but also on how fast it is being created in the reverse reaction. For this reaction (see Langmuir and Mahoney, 1984):

$$Fe^{3+} + SO_4^{2-} \underset{k_1}{\overset{k_2}{\rightleftharpoons}} FeSO_4^+ \tag{17.19}$$

rate laws for Fe^{3+} and SO_4^{2-} in the forward direction are of second order and are written

$$d(Fe^{3+})/dt = d(SO_4^{2-})/dt = -k_1(Fe^{3+})(SO_4^{2-}) \tag{17.20}$$

The reverse reaction has the following rate law:

$$d(FeSO_4^+)/dt = k_2(FeSO_4^+) \tag{17.21}$$

The change in (Fe^{3+}) with time for the overall reaction is

$$d(Fe^{3+})/dt = -k_1(Fe^{3+})(SO_4^{2-}) + k_2(FeSO_4^+) \tag{17.22}$$

Rate laws are even more complex for reactions that are in parallel or series and that cannot be written as a single overall reaction.

Solutions to rate equations like Eq. (17.22) describe how the concentrations of various components change from some initial condition. This quantitative approach is useful mainly for characterizing reactions in laboratory experiments, but not in modeling chemical changes in ground-water systems. In practice, kinetic equations like Eq. (17.22) must be integrated into a complete equation for mass transport, which includes all of the other processes.

► EXERCISES

17.1. The routine analysis of a water sample provides the following concentrations (as mg/L): $Ca^{2+} - 93.9$; $Mg^{2+} - 22.9$; $Na^+ - 19.1$; $HCO_3^- - 334.$; $SO_4^{2-} - 85.0$; $Cl^- - 9.0$ and a pH of 7.20.

    a. Calculate the concentrations of Ca^{2+}, Na^+, HCO_3^-, and Cl^- in terms of molarity and milliequivalents per liter (meq/L).

    b. What is the ionic strength of the sample?

    c. Calculate the activity coefficients for Ca^{2+}, and HCO_3^- using the extended Debye-Hückel equation.

17.2. Write mass law expressions for the following equilibrium reactions.

    a. $CaCO_3 = Ca^{2+} + CO_3^{2-}$

    b. $CO_2(g) = CO_2(aq)$

    c. $Mn^{2+} + Cl^- = MnCl^+$

17.3. For the water sample in Exercise 17.1, determine the saturation index for calcite at 25°C given that $[CO_3^{2-}] = 0.34 \times 10^{-5}$ and that the equilibrium constant for calcite dissolution is 4.9×10^{-9}. What does the saturation index indicate about the state of saturation with respect to calcite?

17.4. Determine the order of the following kinetic expressions:

    a. degradation of organic matter (org): $d(org)/dt = -k_1(org)$

    b. oxidation of pyrite: $d(Fe^{3+})/dt = -k_1 \sum(Fe^{3+})$, where $\sum$ is the surface area of reacting mineral

    c. general reaction: $d(A)/dt = -k_1$

    d. general reaction: $d(C)/dt = -k_1(A)(B)^2$

17.5. Often (CO_3^{2-}) is not reported in water-quality analyses because the concentration is too low to measure directly. Develop an expression from the appropriate mass laws to calculate the concentration of (CO_3^{2-}) given pH and (HCO_3^-).

17.6. A water sample has the following ionic composition at 25°C.

Ionic Concentration (M) $\times 10^{-3}$

Ca	Mg	Na	HCO₃	SO₄	Cl	pH
4.17	1.32	11.7	12.0	5.49	0.08	7.60

Given that the activity coefficients for Ca^{2+}, Mg^{2+}, and HCO_3^- are 0.52, 0.55, and 0.85, respectively, answer the following questions.

    a.   What is the partial pressure of CO_2 in equilibrium with the sample?

    b.   Assume that the concentration reported for calcite represents $(Ca)_T$ and that three significant ion pairs form $CaCO_3^0$, $CaHCO_3^+$, and $CaSO_4^0$. Develop an equation to calculate the concentration of free Ca^{2+} in the sample, assuming the solution is ideal.

▶ CHAPTER
HIGHLIGHTS

1. Equilibrium and kinetic concepts provide a basis for treating chemical reactions. Consider this reaction where ions C and D react to produce ions Y and Z

$$cC + dD = yY + zZ$$

with c, d, y, and z representing the number of moles of these constituents. The law of mass action describes the equilibrium distribution of mass between reactants and products as

$$K = \frac{(Y)^y (Z)^z}{(C)^c (D)^d}$$

where K is the equilibrium constant and (Y), (Z), (C), and (D) are the molal (or molar) concentrations for reactants and products.

2. Application of mass law expressions is more complicated for nondilute solutions. Thermodynamically effective concentrations or activities are used instead of actual concentrations. Thus, mass action equations are written in terms of activities instead of molal concentrations. For species Y, the following equation relates activity to molar concentration:

$$[Y] = \gamma_Y (Y)$$

where γ_y is the activity coefficient. Typically, γ is close to 1 in dilute solutions and decreases as salinity increases.

3. The simplest activity model is the Debye-Huckel equation

$$\log \gamma_i = -Az_i^2 (I)^{0.5}$$

where A is a constant that is a function of temperature, z_i is the ion charge, and I is the ionic strength of the solution. Ionic strength is a measure of the total concentration of ions that emphasizes the increased contribution of species with charges greater than 1 to solution nonideality.

4. The departure of a reaction from equilibrium is determined as the ratio of the ion activity product to the equilibrium constant (IAP/K). The ion activity product is calculated by substituting sample activity values in the mass law expression for a reaction. When $IAP/K < 1$, the ground water is undersaturated with respect to the given mineral. When $IAP/K = 1$, the ground water is at chemical equilibrium, and when $IAP/K > 1$, the ground water is supersaturated.

5. Kinetic theory provides a useful framework for studying reactions in relation to time and reaction pathways. A kinetic description is applicable to any reaction but is required for irreversible reactions or reversible reactions that are slow in relation to mass transport. An irreversible reaction proceeds in only one direction.

6. Many of the important reactions that we deal with are first order. Following is an example of a first-order kinetic reaction with the general form

$$\frac{dC}{dt} = -k\,C$$

where C is concentration, k is the rate constant for the reaction, and t is time. The negative sign indicates that the concentration of C is being reduced in the reaction.

7. The rate constant in a kinetic equation can also be expressed by an equivalent parameter termed the half life $(t_{1/2})$. The half life is the time required to reduce the concentration by one-half.

18

KEY REACTIONS INFLUENCING GROUND-WATER CHEMISTRY

A number of chemical reactions operate in ground-water systems to transfer mass among ions or phases. These reactions in various ways give rise to the diversity in the chemistry of natural waters and control the mobility of contaminants in ground water.

▶ 18.1 ACID-BASE REACTIONS

Acid-base reactions involve the transfer of hydrogen ions (H^+) among the ions present in the aqueous phase. The concentration of H^+ determines the pH of the solution, which is a key factor controlling many of the other processes that follow. For example, at low pH waters are capable of carrying a large load of dissolved metals. At high pH, the metals tend to precipitate as solids. pH is defined formally as the negative logarithm of the hydrogen ion activity, or $pH = -\log[H^+]$. Water is acidic when pH <7, neutral when pH $=7$, or basic when pH >7.

Acid-base reactions include

- dissociation of water: $H_2O = H^+ + OH^-$

- dissolution of CO_2 in water: $CO_2(g) + H_2O = H_2CO_3$

- two-stage ionization of carbonic acid: $H_2CO_3 = HCO_3^- + H^+$

$$HCO_3^- = CO_3^{2-} + H^+$$

TABLE 18.1   Important weak acid-base reactions in natural water systems

Reaction	Mass Law Equation	Eq.	$-\log$ $K(25°C)$
$H_2O = H^+ + OH^-$	$K_w = (H^+)(OH^-)$	18.1	14.0
$CO_2(g) + H_2O = H_2CO_3^*$	$K_{CO_2} = \dfrac{(H_2CO_3^*)}{P_{CO_2}(H_2O)}$	18.2	1.46
$H_2CO_3^* = HCO_3^- + H^+$	$K_1 = \dfrac{(HCO_3^-)(H^+)}{(H_2CO_3^*)}$	18.3	6.35
$HCO_3^- = CO_3^{2-} + H^+$	$K_2 = \dfrac{(CO_3^{2-})(H^+)}{(HCO_3^-)}$	18.4	10.33

The last three reactions show that when CO_2 gas is added to water it partitions among three other dissolved constituents: carbonic acid (H_2CO_3), bicarbonate ion (HCO_3^-), and carbonate ion (CO_3^{2-}).

Table 18.1 shows equilibrium constants for these important acid-base reactions. Given this information, one can understand how mass is distributed among the carbonate species H_2CO_3, HCO_3^- and CO_3^{2-}, as a function of changing pH. Following is an example illustrating this calculation.

▶ **EXAMPLE 18.1**   Assume that CO_2 is dissolved in water so that $(CO_2)_T = 10^{-3}$ M and it is distributed to some extent among the three carbonate species H_2CO_3, HCO_3^-, and CO_3^{2-}. What is the concentration of the carbonate species at a pH of 6.35? We chose this particular pH to simplify the calculation.

SOLUTION

$$10^{-3} = (H_2CO_3) + (HCO_3^-) + (CO_3^{2-}) \tag{18.5}$$

Start by finding the mass law expression in Table 18.1 for

$$H_2CO_3 = HCO_3^- + H^+$$

or

$$K_1 = \frac{(HCO_3^-)(H^+)}{(H_2CO_3)} \tag{18.6}$$

Substitution of the known pH into Eq. (18.6) gives

$$10^{-6.35} = (HCO_3^-)(10^{-6.36})/(H_2CO_3) \tag{18.7}$$

Simplification gives $(HCO_3^-) = (H_2CO_3)$. Substitution into Eq. (18.4) provides

$$10^{-10.33} = (CO_3^{2-})(10^{-6.35})/HCO_3^-$$

and shows that $(HCO_3^-) \gg (CO_3^{2-})$. Assuming for the moment that (CO_3^{2-}) is negligible, a solution to Eq. (18.5) gives $(HCO_3^-) = (H_2CO_3^*) = 10^{-3.3}$ M. Substitution into Eq. (18.4) gives $(CO_3^{2-}) = 10^{-7.28}$ M.

Thus, at the specified pH, (6.35), the concentrations of H_2CO_3 and HCO_3^- are the same and CO_3^{2-} is much lower.

We could repeat this calculation over a broad range of pH. Interestingly, it would show that as the pH changes, the dominance among the carbonate species changes as well. A

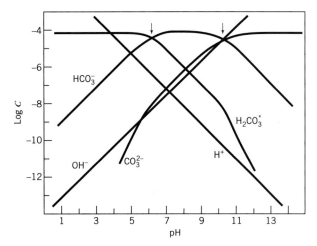

Figure 18.1 Log C-pH diagram for the carbonate system with $(CO_3)_T = 10^{-4}$ M. (from Morel and Hering, 1998, *Principles and applications of aquatic chemistry*). Copyright ©1998 by John Wiley & Sons, Inc. Reprinted by permission of John Wiley & Sons, Inc.

useful way of illustrating this relationship is by means of a plot of the logarithm of activities (or concentrations) of various species versus pH. A log C-pH diagram for the carbonate system is shown in Figure 18.1 for $(CO_2)_T = 10^{-4}$. The arrows on the figure depict the crossover points, where at pH 6.35 $(H_2CO_3) = (HCO_3^-)$ and at 10.33 $(HCO_3^-) = (CO_3^{2-})$. These are known as the pK_a values for aqueous CO_2 (i.e., $pK_1 = 6.35$ and $pK_2 = 10.33$). Below pH 5, H_2CO_3 is the dominant carbonate species (Figure 18.1). In the range from approximately pH 7 to 9, HCO_3^- is the most abundant species. The carbonate ion (CO_3^{2-}) is dominant when pH is above 10. Over a pH range of natural ground water (5–8.5), HCO_3^- is the most abundant species.

Another important group of acid-base reactions involves the reaction of minerals with hydrogen ion. These reactions are particularly important because they produce the load of dissolved mass (that is, ions) found in natural ground waters. Here are examples of these reactions:

- dissolution/precipitation of calcite: $CaCO_3 + H^+ = Ca^{2+} + HCO_3^-$
- dissolution/precipitation of silicate minerals: silicate $+ H^+ =$ cations $+ H_2SiO_3$

The spectacular landforms and cave features that are seen around the world are generated as ground water dissolves calcite $(CaCO_3)$ and adds Ca^{2+} and HCO_3^- to the ground water.

▶ 18.2 DISSOLUTION AND PRECIPITATION REACTIONS INVOLVING SALTS AND LIQUIDS

The previous section introduced acid-base reactions involving minerals. However, not all dissolution reactions are acid-base reactions. It is these additional reactions that we consider in this section. First, there are cases where mineral salts like halite, gypsum, and anhydrite dissolve or precipitate to influence both mineral and ion distributions in solution. Here is what one of these reactions looks like.

- dissolution and precipitation of halite: $NaCl = Na^+ + Cl^-$

In a later chapter, we'll illustrate how this and other reactions can lead to the formation of salt-karst in ground water or the formation of saline soils.

Organic Compounds in Water

In the subsurface, there can be other liquids like naturally occurring oils, or contaminants, like gasoline or industrial solvents. These liquids can migrate as a separate liquid phase or be dissolved in ground water. It is this latter process that is of interest here.

Organic compounds differ widely in their overall solubility (Mackay et al., 1986). Some solutes like methanol are extremely soluble, while others such as PCBs or DDT are sparingly soluble. As a general rule, the most soluble organic compounds are charged species or those that contain oxygen or nitrogen. Examples of this latter group are alcohols or carboxylic acids. Most organic compounds have a relatively low solubility because they commonly are relatively polar and uncharged compounds.

Let's examine this question of *polarity* in more detail. A water molecule is an example of a polar compound. It is comprised of two hydrogen atoms and one oxygen atom. Shared electrons form covalent bonds between atoms. Hydrogen atoms are located asymmetrically with respect to the oxygen atom (Figure 18.2a). Because of this structure and the covalent bonding, the electrical center of the negative charges (electrons) has a different location than that of the positive charges. It is this local separation of charge centers which makes the molecule polar. When a molecule is perfectly symmetrical (e.g., carbon tetrachloride, benzene, and other hydrocarbons), there is no separation of charge centers and the molecule is *nonpolar*.

The water molecules are joined by a series of hydrogen bonds (Figure 18.2b). These bonds form from the electrostatic interaction between a hydrogen and an oxygen atom. These interactions produce an irregular tetrahedral network (Figure 18.2c). This molecular structure explains the unique characteristics of water such as freezing and boiling points.

This tendency for the water molecule to be polar also explains why minerals, which dissolve to produce ions, can be so soluble. Charged species (for example, ions or molecules) can pop right into the structure of water. In effect, the charged species forces a local rearrangement in the polar molecules (Figure 18.2d). Water molecules immediately adjacent bind to the ions. Those nearby reorient themselves because of the charge.

In the case of an uncharged species, such as many uncharged, nonpolar organic molecules, there is much less interaction with the water molecule. The organic molecule in effect has to find a space for itself in the water structure. These effects contribute to the relative insolubility of nonpolar molecules in water. However, some organic molecules can

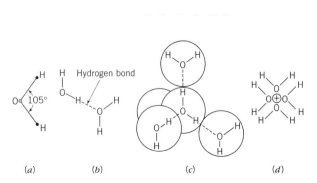

Figure 18.2 Essential features of the structure of water and adjustments due to the presence of an ion (adapted from Pytkowicz, 1983): (*a*) the position of the hydrogen and oxygen atoms in the polar water molecule; (*b*) hydrogen bonding between water molecules, giving rise to (*c*) the tetrahedral structure of water; (*d*) water molecules rearrange themselves around a cation due to the polar nature of the water.

form a bond with the water molecule (for example, various alcohols). These compounds can be extremely soluble in water.

▶ 18.3 COMPLEXATION REACTIONS

A complex is an ion that forms by combining simpler cations, anions, and sometimes molecules. The cation or central atom is typically one of the large number of metals making up the periodic table. The anions, often called ligands, include many of the common inorganic species found in ground water such as Cl^-, F^-, Br^-, SO_4^{2-}, PO_4^{3-} and CO_3^{2-}. The ligand might also comprise various organic molecules such as amino acids. Complexation is an important chemical process because it facilitates the transport of metals. Naturally occurring metals like Pb and Zn, when transported in natural ground waters, for a long time can give rise to the formation of ore deposits. Complexes can also help potentially toxic metals such as cadmium, chromium, copper, uranium, or plutonium migrate away from sites of contamination. Metals that occur as a free metal ion (e.g., Zn^{2+}) are typically not very mobile because they tend to be sorbed onto clay minerals.

Here are examples of complexation reactions:

> • simple combination of a metal and a ligand: $Mn^{2+} + Cl^- = MnCl^+$
>
> • complexes formed when complexes combine with ligands:
>
> $$Cr^{3+} + OH^- = Cr(OH)^{2+}$$
>
> $$Cr(OH)^{2+} + OH^- = Cr(OH)_2^+$$
>
> $$Cr(OH)_2^+ + OH^- = Cr(OH)_3^0 \ldots$$

Stability of Complexes and Speciation Modeling

Most inorganic reactions involving complexes are kinetically fast. Thus, they can be examined quantitatively using equilibrium concepts. For example, the mass law equations for the examples are

$$K_{MnCl^+} = \frac{[MnCl^+]}{[Mn^{2+}][Cl^-]} \tag{18.8}$$

and

$$K_{Cr(OH)^{2+}} = \frac{[Cr(OH)^{2+}]}{[Cr^{3+}][OH^-]} \tag{18.9}$$

$$K_{Cr(OH)_2^+} = \frac{[Cr(OH)_2^+]}{[Cr(OH)^{2+}][OH^-]} \tag{18.10}$$

and so on.

When a metal, for example, Cr^{3+}, is distributed among several complexes, one can solve a series of mass law equations to determine the concentration of Cr^{3+} in each of the complex species. These calculations are simpler with the complexation reactions written in terms of *association reactions*. Let's look at an example of what these association reactions might look like. Shown below on the left are the series of complexation reactions involving Cr^{3+}, which we presented earlier. On the right are the equivalent set of association reactions:

$$Cr^{3+} + OH^- = Cr(OH)^{2+} \qquad Cr^{3+} + OH^- = Cr(OH)^{2+}$$

$$Cr(OH)^{2+} + OH^- = Cr(OH)_2^+ \qquad Cr^{3+} + 2OH^- = Cr(OH)_2^+$$

$$Cr(OH)_2^+ + OH^- = Cr(OH)_3^0 \qquad Cr^{3+} + 3OH^- = Cr(OH)_3^0$$

An association reaction, then, is simply another way of writing a complexation reaction as the combination of a metal and an appropriate number of ligands. The equilibrium constant for an association reaction is given the name *stability constant*, usually represented as β. For example, here is an example mass law equation for the third of the association reactions:

$$\beta_3 = \frac{(Cr[OH]_3^0)}{(Cr^{3+})(OH^-)^3} \tag{18.11}$$

Large values of β_i are associated with stronger or more stable complexes. Morel and Hering (1993, p. 332) tabulate stability constants for a host of metal-ligand reactions.

The stability constants for a series of metal ligand reactions provide the basic information necessary to determine how the total concentration of a metal in solution $(M)_T$ is distributed as a metal ion and various complexes. The following example illustrates the speciation calculation when Cr hydrolyzes.

▶ **EXAMPLE 18.2**

A solution with a pH of 5.0 contains a trace quantity of Cr $\{(Cr)_T = 10^{-5} \text{ M}\}$. Determine the speciation for Cr among the various hydroxy complexes and Cr^{3+}. Stability constants (as base 10 logarithms) are 10.0, 18.3, and 24.0 for $Cr(OH)^{2+}$, $Cr(OH)_2^+$ and $Cr(OH)_3^0$ respectively. Assume that the pH does not change with the addition of the metal, no solids form, and the solution behaves ideally.

SOLUTION   The mole balance equation for $(Cr)_T$ is

$$(Cr)_T = (Cr^{3+}) + (Cr[OH]^{2+}) + (Cr[OH]_2^+) + (Cr[OH]_3^0)$$

Substitution of the appropriate mass law equations for the association reactions into this equation gives

$$(Cr)_T = (Cr^{3+}) + \beta_1(Cr^{3+})(OH^-) + \beta_2(Cr^{3+})(OH^-)^2 + \beta_3(Cr^{3+})(OH^-)^3$$

$$(Cr)_T = (Cr)^{3+}\{1 + \beta_1(OH^-) + \beta_2(OH^-)^2 + \beta_3(OH^-)^3\}$$

Because pH is fixed, $(OH^-) = 10^{-9}$ M, and all the terms in the brackets are known. Substitution of the known values solving for (Cr^{3+}) gives:

$$(Cr^{3+}) = 10^{-5}/\{1 + 10^{10}(10^{-9}) + 10^{18.3}(10^{-18})10^{24}(10^{-27})\}$$

$$= 10^{-5}/\{1 + 10 + 10^{0.3} + 10^{-3}\}$$

$$= 10^{-6.12} \text{ M}$$

The concentration of the complexes is calculated by substituting known values of (Cr^{3+}) and (OH^-) in the mass law equations:

$$(Cr[OH]^{2+}) = \beta_1(Cr^{3+})(OH^-)$$

$$= 10^{10.0}(10^{-6.12})(10^{-9})$$

$$= 10^{-5.12}\text{M}$$

$$(Cr[OH]_2^+) = 10^{-5.82}\text{ M}$$

$$(Cr[OH]_3^0) = 10^{-9.12}\text{ M}$$

Thus, most of the Cr occurs as $Cr(OH)^{2+}$ and $Cr(OH)_2^+$. Only about 7.5% of $(Cr)_T$ is the free ion Cr^{3+}.

In Example 18.2, the speciation of Cr is controlled by solution pH. The log C-pH plot for $(Cr)_T = 10^{-5}$ (Figure 18.3) shows that at a pH below 4, Cr^{3+} is the dominant Cr species. As the pH increases, the various hydroxy complexes dominate.

This use of stability constants to determine the speciation of metals can be extended to a mixed group of mononuclear complexes. For example, if trace quantities of Pb are present in a ground water, a mixed group of chloride, hydroxyl, and carbonate complexes can form. Speciation is determined by taking the appropriate mole balance equation

$$(Pb)_{Total} = (Pb^{2+}) + (PbCl_2^0) + (PbCl_3^-) + (PbOH^+) + (PbCO_3^0) \qquad (18.12)$$

and substituting the appropriate mass law equations. This approach requires that metal concentration in the ground water be sufficiently small so as not to influence the major ion chemistry or pH of the solution. At some point, however, adding large quantities of metal affects the pH and carbonate chemistry. The speciation becomes much more difficult to determine in this case because the system of mole balance and mass law equations is larger and more complex. An example of this more complex situation is the case where the chemistry of a ground water changes when the most abundant ions, Ca^{2+}, Mg^{2+}, Na^+, K^+, and H^+ complex with the common ligands SO_4^{2-}, Cl^- HCO_3^-, and OH^-.

Major Ion Complexation and Equilibrium Calculations

The simplest calculations of mineral saturations that we have examined so far assumed that the major ions in a water analysis (Ca_{Total}, Mg_{Total}) are present as free ions (for example,

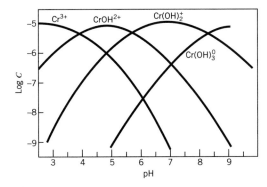

Figure 18.3 Log C-pH diagram for chromium hydroxide complexes (from Domenico and Schwartz 1998. *Physical and chemical hydrogeology*). Copyright ©1998 by John Wiley & Sons, Inc. Reprinted by permission of John Wiley & Sons, Inc.

Ca^{2+}, Mg^{2+}). However, these same Ca and Mg will be tied up in complexes. Thus, the free ion (or ligand) concentration will be less than the total concentration. For this reason, the calculated IAP/K ratios for some minerals will be less than what was calculated, depending on the extent of complex formation. Let us reexamine Example 17.2 to illustrate this point. Now, the molar concentrations for the ions in Example 17.2 are total concentrations, not free ion calculations. Using a computer code, it is possible to calculate the concentrations of the free species and the complexes. Table 18.2 presents the concentrations of the major complexes. Although the concentrations of the complexes are relatively small, the reduction in $[Ca^{2+}]$ and $[CO_3^{2-}]$ due to complexation reduces the IAP/K ratio for calcite from 1.34 to 1.03. For dolomite, the ratio decreases from 0.018 to 0.011. This example shows that even in relatively fresh ground water (ionic strength = 0.023 M), the error in determining mineral saturation can be substantial when complexation is not considered. This error can become even larger as the water becomes more saline. In seawater for example, only approximately 40% of $(SO_4)_T$ exists as SO_4^{2-}, while the remainder exists as $NaSO_4^-$ (37%) and $MgSO_4^0$ (19%) (Morel and Hering, 1993).

Enhancing the Mobility of Metals

Complexation reactions are also important because they can enhance the mobility of metals in ground water. These metals can occur naturally, and over time they form ore deposits; or they can be contaminants and produce health problems. In general, metals in ground water are most mobile at a low pH where most of the mass occurs as a charged metal ion. Ignoring for a moment the effects of sorption, mobility begins to decline once pH increases to the point where saturation is reached with respect to a solid phase, usually a metal-hydroxide, metal-carbonate, or metal-sulfide. At this point, equilibrium with the solid determines that most of the metal will be associated with the solid phase.

Over the pH range common to most natural ground waters (5–8), the concentration of most metals is small, reflecting this control. Complexes in these circumstances can enhance the solubility of metals. The transport of uranium in ground water is a good example. Figure 18.4 shows the relative abundance of various uranyl complexes in ground water with a composition typical of the Wind River Formation in Wyoming (Langmuir, 1978). Across the entire range of pH represented on the figure, a significant proportion of the total uranium is complexed. Below a pH of about 4, a uranyl-fluoride complex is the most dominant species (Figure 18.4) In the range of typical ground water for this formation (pH 6.6–8.3), both phosphate and carbonate complexes are important. This extensive complexing especially over the neutral to alkaline pH range increases the solubility of some uranium minerals by several orders-of-magnitude. This increase makes possible the formation of uranium ore deposits.

TABLE 18.2 Calculated molar concentrations of complexes formed from major ion species (based on data from Example 17.2)

Metal	Ligand		
	HCO_3^-	CO_3^{2-}	SO_4^{2-}
Ca^{2+}	5.46×10^{-5}	8.03×10^{-6}	2.16×10^{-5}
Mg^{2+}	4.98×10^{-7}	1.46×10^{-7}	2.78×10^{-7}
Na^+	1.54×10^{-4}	1.13×10^{-5}	6.66×10^{-5}

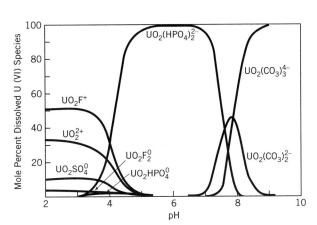

Figure 18.4 Distribution of uranyl complexes versus pH for typical ligand concentrations in ground water of the Wind River Formation at 25°C. $P_{CO_2} = 10^{-2.5}$ atm, $\Sigma F = 0.3$ ppm, $\Sigma Cl = 10$ ppm, $\Sigma SO_4 = 100$ ppm, $\Sigma PO_4 = 0.1$ ppm, $\Sigma siO_2 = 30$ ppm (from Langmuir, 1978). Reprinted from *Geochim. Cosmochim. Acta*, v. 42, Uranium solution-mineral equilibria at low temperatures with applications to sedimentary ore deposits, p. 547–569, Copyright © 1978, with permission from Elsevier Science.

Metal complexation can also involve organic ligands that are present naturally in ground water or added as contaminants. For example at a site near Ottawa, Canada, Killey et al. (1984) showed how up to 80% of the ^{60}Co present as a contaminant in ground water occurred as weakly anionic complexes. Thus, with the formation of organic complexes, metals can be made much more mobile.

► 18.4 SURFACE REACTIONS

Surface reactions can cause wholesale changes to the geochemistry of both natural and contaminated ground water. They involve reactions between ions in solution and the grains or crystals in a rock or sediment. For example, water in clay-rich marine deposits can experience a loss of Ca and Mg and a gain of Na due to cation exchange processes with clay minerals. These reactions naturally soften the water in much the same way as a commercial softener that people install in their houses. The tendency for some contaminants to be sorbed on aquifer materials also reduces the spread of contaminants.

Sorption Isotherms

When water containing a trace constituent with a concentration C_i is mixed with granular solids and allowed to equilibrate, mass often partitions between the solution and the solid. The following equation represents this process

$$S = \frac{(C_i - C)(\text{solution volume})}{sm} \tag{18.13}$$

where the equilibrium concentration, C, and the starting concentration, C_i, have units such as mg/L or μg/L; sm is the sediment mass (for example, g); and S is the quantity of mass sorbed on the surface (for example, mg/g or μg/g). Such an experiment provides the single point a on the S versus C plot (Figure 18.5). By repeating the procedure at the same temperature (hence, isotherm) with different values of C_i, the family of points forms a sorption isotherm (Figure 18.5). This experiment is known as a batch test.

Real isotherms have no prescribed shape. They can be linear, concave, convex, or a complex combination of all these shapes. Sorption is modeled by fitting an experimentally derived isotherm to theoretical equations. Two of the most common relationships are the

Freundlich: $\quad S = KC^n \tag{18.14}$

and Langmuir isotherms: $$S = \frac{Q^0 KC}{1 + KC} \qquad (18.15)$$

where K is a *partition coefficient* reflecting the extent of sorption, n is a constant usually ranging between 0.7 and 1.2, and Q^0 is the maximum sorptive capacity for the surface.

Figure 18.6 depicts Freundlich isotherms calculated with a K value of 1.5 and values of n of 0.5, 1.0, and 1.5, as well as Langmuir isotherms with values of K of 0.5 and 1.5 and a Q^0 value of 30 mg/g. These examples illustrate the range in curve shapes that can be fit with these two equations. If the fit is not satisfactory, there are other equations.

A Freundlich isotherm with $n = 1$ is a special case because this linear isotherm is easy to incorporate into mass transport models. The following equation relates S to C

$$S = K_d C \qquad (18.16)$$

where K_d with units like cm^3/g is the *distribution coefficient* (i.e., the slope of the linear sorption isotherm). Large K_d values are indicative of a greater tendency for sorption.

Sorption of Organic Compounds

Hydrophobic (or water-hating) organic molecules, which are nonionic (i.e., they are neither cations nor anions), tend to partition preferentially into solid organic matter (for example,

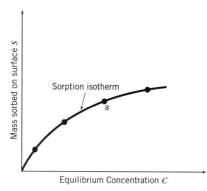

Figure 18.5 Example of a simple sorption isotherm constructed using data points from batch tests (from Domenico and Schwartz, 1998. *Physical and chemical hydrogeology*). Copyright ©1998 by John Wiley & Sons, Inc. Reprinted by permission of John Wiley & Sons, Inc.

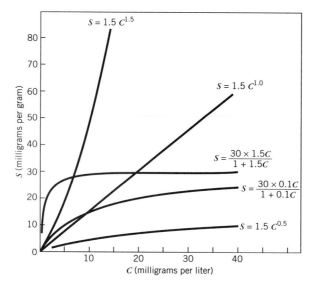

Figure 18.6 Examples of Langmuir and Freundlich isotherms (from Domenico and Schwartz, 1998. *Physical and chemical hydrogeology*). Copyright ©1998 by John Wiley & Sons, Inc. Reprinted by permission of John Wiley & Sons, Inc.

humic substances and kerogen). This solid matter occurs in small amounts in the porous media as discrete solids, as films on individual grains, or as stringers of organic material in grains. Nonpolar organic molecules essentially hate water and prefer organic matter. Overall, the more hydrophobic a compound is, the greater is its tendency to partition into a solid phase. Generally, the larger the size of the molecule, the greater is the tendency to partition out of the water.

The following example illustrates the tendency for an organic compound to be sorbed by aquifer materials. Also, at relatively small concentrations, this sorption can be described by a linear isotherm.

EXAMPLE 18.3

The following table summarizes data collected from batch sorption experiments with pyrene using 25 mg of porous media in 50 mL of solution. Construct a sorption isotherm and calculate the distribution coefficient in cm^3/g.

Test No.	$C_i(\mu g/L)$	$C(\mu g/L)$	$S(\mu g/g)$
1	19.4	5.9	27.0
2	39.7	9.5	60.4
3	60.9	14.1	93.6
4	79.1	20.7	117.8
5	100.3	26.1	148.2

SOLUTION

$$S = \frac{(C_i - C) \times 0.050}{0.025}, \qquad \text{units} \rightarrow \frac{\mu g/L \cdot L}{g}$$

S is plotted versus C, as shown in Figure 18.7 to establish the isotherm. K_d is the slope of the isotherm. The calculation illustrates a strong affinity for the solid by pyrene.

Several studies (for example Karickhoff et al., 1979; Schwarzenbach and Westall, 1981) have shown how the distribution coefficient can be expressed as the product of constants describing the contaminant and the porous medium, or

$$K_d = K_{oc} f_{oc} \qquad (18.17)$$

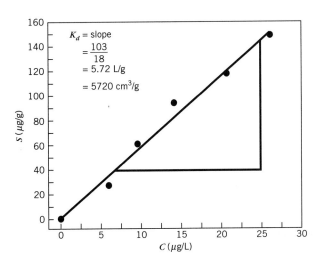

K_d = slope
$= \frac{103}{18}$
= 5.72 L/g
= 5720 cm^3/g

Figure 18.7 Calculated isotherm for the sorption of pyrene for Example 18.3 (from Domenico and Schwartz, 1998. *Physical and chemical hydrogeology*). Copyright ©1998 by John Wiley & Sons, Inc. Reprinted by permission of John Wiley & Sons, Inc.

where K_{oc} is the partition coefficient of a compound between organic carbon and water with typical units such as cm^3/g and f_{oc} is the weight fraction of organic carbon (dimensionless), defined as g_{oc}/g_s, or grams of solid organic carbon to grams of total aquifer solids.

When values of f_{oc} and K_{oc} are known, the K_d value determined from Eq. (18.17) can be used to predict the extent of partitioning. Organic carbon contents can be measured in the laboratory on porous-medium samples. However, this parameter is not well characterized. The range of values reported for geologic materials is extremely variable (see Table 18.3). In estimating K_{oc}, it is fortunate that a good correlation exists between $\log K_{oc}$ and $\log K_{ow}$, the *octanol-water partition coefficient*. This correlation is expected because the partitioning of an organic compound between water and organic carbon is not much different from that between water and octanol.

Regression equations in practice describe the relationship between K_{oc} and K_{ow} (Griffin and Roy, 1985)

$$\text{Karickhoff et al., 1979:} \qquad \log K_{oc} = -0.21 + \log K_{ow} \qquad (18.18)$$

$$\text{Schwarzenbach and Westall, 1981:} \quad \log K_{oc} = 0.49 + 0.72 \log K_{ow} \qquad (18.19)$$

$$\text{Hassett et al., 1983:} \qquad \log K_{oc} = 0.088 + 0.909 \log K_{ow} \qquad (18.20)$$

where K_{ow} is a tabulated constant (dimensionless) and K_{oc} has units of cm^3/g. The similarity in these equations is evident in Figure 18.8. Overall, they provide a preliminary estimate of K_{oc} when other information is lacking. In field studies, K_{oc} values estimated from these equations need to be refined with field or laboratory experiments. The following example illustrates the application of the empirical equations in estimating a distribution coefficient.

TABLE 18.3   **A synthesis of some data on the organic carbon content of sediments**

Site Name	Type of Deposit	Texture	Organic Carbon Content
Borden, Ontario[a]	Glaciofluvial	Fine-medium sand	0.0002
Gloucester, Ontario[b]	Glaciofluvial	Sands and gravels	0.0006
North Bay, Ontario[c]	Glaciofluvial	Medium sand	0.00017
Woolwich, Ontario[c]	Glaciofluvial	Fine-medium sand	0.00023
Chalk River, Ontario[c]	Glaciofluvial	Fine sand	0.00026
Cambridge, Ontario[c]	Glaciofluvial	Medium sand	0.00065
Rodney, Ontario[c]	Glaciofluvial	Fine sand	0.00102
Wildwood, Ontario[c]	Lacustrine	Silt	0.00108
Palo Alto, Baylands[d]	?	Silty sand	0.01
River Glatt, Switzerland[e]	Glaciofluvial	Sand, gravel	<0.000–0.01
Oconee River, Georgia[f]	River sediment	Sand,	0.0057
		coarse silt,	0.029
		medium silt,	0.02
		fine silt	0.0226

[a] Mackay et al. (1986).
[b] Jackson (personal communication, 1989).
[c] J. Baker, University of Waterloo (personal communication, 1987).
[d] Mackay and Vogel (1985).
[e] Schwarzenbach and Giger (1985).
[f] Karickhoff (1981).

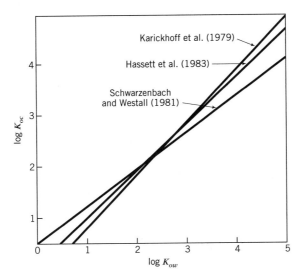

Figure 18.8 Correlation between the log octanol-water partition coefficient (K_{ow}) and the log organic carbon/water partition coefficient as determined by three different studies (Modified from Griffin and Roy, 1985). Reprinted by permission of the Environmental Institute for Waste Management Studies, The University of Alabama.

▶ **EXAMPLE 18.4**

An aquifer has an f_{oc} of 0.01. Estimate the K_d value characterizing the sorption of 1,2-dichloroethane having a log $K_{ow} = 1.48$.

SOLUTION Starting with the basic equation $K_d = f_{oc} K_{oc}$ and taking logs of both sides gives

$$\log K_d = \log f_{oc} + \log K_{oc}$$

Substitution for log K_{oc} with the Schwarzenbach and Westall (1981) equation yields

$$\log K_d = \log 0.01 + 0.49 + 0.72(1.48) = -0.444$$

The K_d value is $10^{-0.444}$ or 0.36 cm^3/g.

It is tempting to extend this model of organic sorption to less hydrophobic (hydrophilic) compounds (e.g., methanol). Such compounds have large aqueous solubilities and relatively low values of log K_{ow} (e.g., −0.66 for methanol). In contrast to the nonpolar compounds, these are preferentially partitioned to the aqueous phase, with much less affinity for organic carbon. Karickhoff (1984) recommends that the model of hydrophobic sorption should only be used for compounds with solubilities less than 10^{-3} M. In addition, this model of sorption begins to break down when f_{oc} becomes small or the concentrations of the organic contaminant become large.

Ion Exchange

Mineral surfaces often bind cations and anions. Because these ions are held "loosely," there is an opportunity for them to exchange with ions in solution. This kind of exchange can result in a dramatic change in the chemistry of the water. A case in point is the natural water-softening reactions that exchange Ca and Mg ions in the water with Na ions sorbed onto clay minerals. Relatively high concentrations of Ca + Mg make the ground water hard. Hard water tastes better than soft water but doesn't clean as well because soaps have difficulty in lathering.

Here is how these reactions are written chemically

$$Ca^{2+} + Na\text{-clay} = 2Na^+ + Ca\text{-clay}$$
$$Mg^{2+} + Na\text{-clay} = 2Na^+ + Mg\text{-clay}$$

The exchange process effectively replaces the Ca + Mg with Na, which "softens" the water. In contamination problems, ion exchange processes can reduce the mobility of hazardous ions (for example, radionuclides or trace metals).

Surfaces bind cations and anions in several different ways. For example, ion sorption occurs because of imperfections or elemental substitutions in the crystal lattice of minerals. The result is a surface with a net negative or positive charge, which is balanced by ions (called counter ions) attracted on to the surface. The most common situation is that in which negative charges in the mineral structure are balanced by cations (Figure 18.9).

Clay minerals are the most important group of minerals carrying a significant fixed negative surface charge. Cations bound to the clays are available to exchange with cations in the ground water. The negative charge in the clays is due to substitutions of cations of a lower valence (for example, Al^{3+} for Si^{4+} in the lattice) and to a lesser extent because of broken bonds at the edges of the mineral.

The *cation exchange capacity (CEC)* of an exchanger describes how good the exchanger really is. CEC is defined as the quantity of cations bound to the mineral surface, which are available for exchange with ions in solution. Because electrical charges are involved, equivalent units of concentration are the most convenient. CEC is expressed as the number of milliequivalents that can be exchanged at pH 7 in a sample with a dry mass of 100 g. Thus, the greater the CEC, the more exchange that is possible.

Table 18.4 presents a summary of the key properties of the common clay minerals: kaolinite, illite, chlorite, vermiculite, and montmorillonite. In most cases, the clay minerals have fixed charges or mostly fixed charges. In other words, the surface charge doesn't change as the chemistry of the pore fluid changes. Of the list, the chlorites, vermiculites, and montmorillonites have the greatest exchange capacity.

Clay minerals exhibit a preference as to which ion occupies an exchange site. For example, clay minerals would rather have Ca and Mg on the exchange sites than Na. Attempts have been made over the years to establish a selectivity sequence, where equivalent amounts of cations are arranged according to their relative affinity for an exchange site. Many variations of this sequence have been published, such as the following one provided by Yong (1985)

$$Li^+ < Na^+ < H^+ < K^+ < NH_4^+ < Mg^{2+} < Ca^{2+} < Al^{3+}$$

In general, cations with larger charges have a greater affinity for an exchange site. The idea of an affinity sequence, though useful for addressing exchange preference, is too simplified to be applied rigorously.

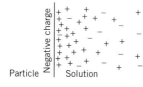

Figure 18.9 Schematic of the surface of a typical clay mineral where the negative charge in the mineral is balanced by cations.

TABLE 18.4   Properties of the common clay minerals

Clay Mineral	Lattice Description	CEC (meq/100 g)	Surface Area (m^2/g)	Source of Charge	Charge Characteristics
Kaolinites		5–15	15	Edges, broken bonds (hydroxylated edges)	Variable charge
Illites	K+	25	80	Isomorphous substitution, some broken bonds at edges	Mostly fixed charge
Chlorites		10–40	80	Isomorphous substitution	Fixed charge
Vermiculites	H_2O Ex H_2O	100–150	n.d.	Isomorphous substitution	Fixed charge
Montmorillonites	H_2O Ex H_2O	80–100	800	Isomorphous substitution, some broken bonds at edges	Mostly fixed charge

n.d.—not determined.
Ex-Exchangeable cations.

Source: Yong (1985). Reprinted by permission of Second Canadian/American Conference on Hydrogeology. Copyright ©1985. All rights reserved.

Clay Minerals in Geologic Materials

Clay minerals are abundant in fine-grained sediments (e.g., glacial till, lake clays) and sedimentary rocks of all kinds (e.g., shale, mudstone). Often, coarser grained sediments (e.g., sand and gravel) and sedimentary rocks (sandstone) contain small but important quantities of clay minerals.

An indirect way of quantifying the abundance of clay minerals in the sample is with a grain size (granulometric) analysis. This standard laboratory analysis provides a measure of the weight percent of a sample in sand, silt, and clay-sized fractions. The sizes are determined according to the diameter of the grains making up the sediment. Following is the nomenclature for a commonly used classification scheme:

2 mm	0.062 mm	0.004 mm	
Gravel	Sand	Silt	Clay

This wording choice is unfortunate because the "clay" size fraction can include tiny fragments of other minerals (e.g., quartz). Commonly, however, the clay-size fraction is made up mostly of clay minerals. The silt and sand-size fractions usually include the common rock-forming minerals other than clay minerals.

Not surprisingly, fine-grained sediments include a broad distribution of particle sizes. Figure 18.10 shows the textural composition of various glacial tills and glaciolacustrine/marine clays from across Canada (Grisak and Jackson, 1978). In the case of glacial tills, the clay-size fraction is often less than half of the total sample mass. Glaciolacustrine and marine clays (Figure 18.10c) deposits have a significant clay-size fraction. However, other size fractions are represented as well.

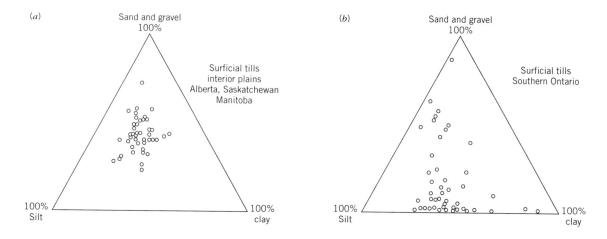

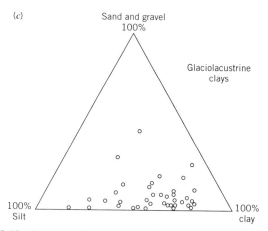

Figure 18.10 Summary diagrams depicting the abundance of sand, silt, and clay fractions in glacial tills (Panels *a* and *b*) and lacustrine clay (Panel *c*) (from Grisak and Jackson, 1978). Reproduced with the permission of the Minister of Public Works and Government Services Canada, 2001 and courtesy of Environment Canada.

A variety of factors come into play in determining exactly what clay minerals will be present in a particular fine-grained deposit. Modern clay-rich sediments being transported to oceans consist of weathering products such as montmorillonite and kaolinite. With time and burial, these minerals are progressively converted to illite and chlorite. Thus, old sedimentary rocks (for example, Paleozoic) contain mostly illite.

Glacial sediments on the continents have a diverse clay mineralogy that depends in large measure on the rocks that were eroded to form glacial till, from which other glacial deposits (e.g., outwash sands, glaciolacustrine deposits) developed. Let's look at examples provided by Grisak and Jackson (1978) for a few areas of North America. Glacial tills in Illinois (USA) and Ontario (Canada) are illitic in character, reflecting the dominant contribution of materials from Paleozoic marine deposits (Figure 18.11*a*). Glacial tills from the Interior Plains are comprised mainly of montmorillonite from the erosion of young Mesozoic sediments containing bentonites (Figure 18.11*b*).

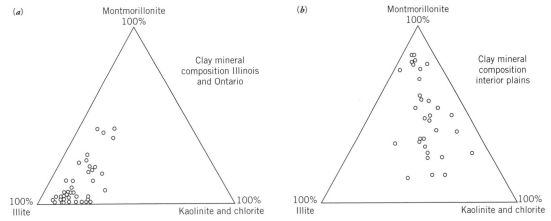

Figure 18.11 Clay mineral composition of glacial tills in Illinois and Ontario (Panel *a*) and the Interior Plains of the U.S. and Canada (Panel *b*) (from Grisak and Jackson, 1978). Reproduced with the permission of the Minister of Public Works and Government Services Canada, 2001 and courtesy of Environment Canada.

Reactions on Oxide and Oxyhydroxide Surfaces

Another important type of sorption reaction involves solids whose surface charges change as a function of ground-water composition. Examples of such solids include kaolinite, metal oxides (for example, SiO_2, Al_2O_3), and metal oxyhydroxides (for example, $Al(OH)_3$, $Si(OH)_4$). The metal oxides and oxyhydroxides typically form a thin coating on sand grains. The surface reaction involves hydroxyl groups and chemisorbed water on the surface. These species protonate or dissociate to surface forms like XOH, XO^-, and XOH_2^+, where XOH represents the hydroxylated surface and XO^- the ionized surface site (Morel and Hering, 1993). These species are free to react with metals or metal complexes. For example, a metal ion (M^{2+}) can be sorbed to the surface by a reaction such as

$$M^{2+} + XO^- = XOM^+ \tag{18.21}$$

The distribution of surface species is controlled by a set of equilibrium reactions that look like this

$$XOH = H^+ + XO^- \quad XOH + H^+ = XOH_2^+$$

The presence of H^+ means that the character of the surface changes depending on the pH of the ground water.

Typically, the surface has an abundance of XOH_2^+ at low pH's, which favors the sorption of anions. However, at higher pH's, the surface species changes to XO^-, favoring the sorption of cations. There is a pH in between, known as the *zero point of charge* (pH_{zpc}), where the surface has zero net charge. The term *isoelectric point* defines the pH_{zpc} when the binding and dissociation of protons (H^+) are the only reactions affecting surface charge. These surface-charge relationships are unique for given solids and solutions, and usually must be determined experimentally. Compiled in Table 18.5 are estimates of the isoelectric points for various solids.

Stollenwerk (1991) studied the sorption of molybdate (MoO_4^{2-}) on ferrihydrite, which coats the quartz grains of an outwash aquifer. A variety of reactions involving the {S} $FeOH^0$ complexation site cause the sorptive properties of the aquifer to change as a function of pH.

TABLE 18.5 **Examples of the isoelectric points (pH) for various solids**

Solid	pH
Quartz α-SiO_2	2–3.5
Albite $NaAlSi_3O_8$	2.0
Kaolinite $Al(Si_4O_{10})(OH)_8$	<2–4.6
Montmorillonite	$\leq$2.5
Hematite Fe_2O_3	5–9
Magnetite Fe_3O_4	6.5
Goethite $FeOOH$	6–7
Corundum Al_2O_3	9.1
Gibbsite $Al(OH)_3$	~9

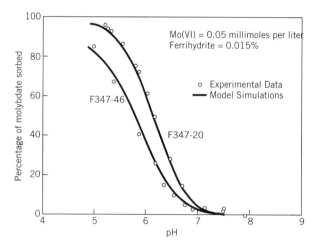

Figure 18.12 Variation in molybdate sorption as a function of pH. The circles present data from actual measurements. The line presents the results of model simulations. The difference in the results is due to the increased competition caused by elevated SO_4^{2-}, and PO_4^{3-} present from the deeper well F347-46 (from Stollenwerk, 1991).

At pH's below about 7, the ferrihydrite coatings are capable of sorbing molybdate (Figure 18.12). In this pH range, below the pH_{zpc} for ferrihydrite, the surface has a net positive charge. At very lower pH's, almost 100% of the total molybdate ends up sorbed on the surface. Once the pH climbs above the pH_{zpc}, the surface has a net positive charge and loses its capability of sorbing molybdate.

▶ 18.5 OXIDATION-REDUCTION REACTIONS

Oxidation-reduction or redox reactions are enormously important in controlling the geochemistry of natural waters. They are unlike other reactions so far because they involve the transfer of electrons and are mediated by microorganisms. The microorganisms act as catalysts, speeding up what otherwise are extremely sluggish reactions.

Redox reactions involve elements (for example, C, O, N, Fe, Mn), that can gain or lose electrons. These transfers of electrons imply that these elements have different oxidation states. The oxidation number refers to charges that are assigned to an element. For example, in the case of CO_3^{2-}, the oxidation number of carbon is (+IV) and for oxygen (−II), giving the molecule a net negative charge of −2. By convention, oxidation numbers are written as roman numerals.

Oxidation is the removal of electrons from an atom, forcing a change in the oxidation number of an element. For example, the oxidation of Fe^{2+} can be written as

$$Fe^{2+} = Fe^{3+} + e^- \tag{18.22}$$

where e^- is an electron. The oxidation number for iron changes from (+II) to (+III).

Reduction refers to the addition of an electron to lower the oxidation number

$$Fe^{3+} + e^- = Fe^{2+} \tag{18.23}$$

All redox reactions transfer electrons and involve elements with more than one oxidation number. Listed in Table 18.6 are some important elements, their typical oxidation states, and a few of the ions and solids that form. A host of trace metals not included in the table also have variable oxidation numbers.

Redox reactions involve a transfer of electrons from a *reductant* (electron donor) to an *oxidant* (electron acceptor). In any reaction, both oxidation and reduction reactions occur (that is, a pair of redox reactions). Thus, no free electrons result in a redox reaction. Redox reactions have this general form

$$Ox_1 + Red_2 = Red_1 + Ox_2 \tag{18.24}$$

TABLE 18.6   Some elements found with more than one oxidation state and examples of ions or solids formed from those elements

Element and Oxidation State in Parentheses	Example
C(+IV)	HCO_3^-, CO_3^{2-}
C(O)	CH_2O, C
C(−IV)	CH_4
Cr(+VI)	CrO_4^{2-}, $Cr_2O_7^{2-}$
Cr(+III)	Cr^{3+}, $Cr(OH)_3$
Fe(+III)	Fe^{3+}, $Fe(OH)_3$
Fe(+II)	Fe^{2+}
N(+V)	NO_3^-
N(+III)	NO_2^-
N(0)	N_2
N(−III)	NH_4^+, NH_3
S(+VI)	SO_4^{2-}
S(+V)	$S_2O_6^{2-}$
S(+II)	$S_2O_3^{2-}$
S(−II)	H_2S, HS^-

Source: Morel and Hering, 1993. *Principles and applications of aquatic chemistry.* Copyright ©1993 by John Wiley & Sons, Inc. Reprinted by permission of John Wiley & Sons, Inc.

where Ox and Red refer to oxidants and reductants, respectively. The more specific example makes this point about electron transfers more obvious. Electrons generated by the oxidation of Fe^{2+} are accepted by the O, which is reduced in the reaction

$$O_2 + 4Fe^{2+} + 4H^+ = 2H_2O + 4Fe^{3+} \tag{18.25}$$

$$O(0) \quad Fe(+II) \qquad O(-II) \quad Fe(+III)$$

$$Ox_1 \quad Red_2 \qquad Red_1 \quad Ox_2$$

The overall reaction can be represented as two separate *half-redox reactions* Ox_1-Red_1 and Red_2-Ox_2

$$O_2 + 4H^+ + 4e^- = 2H_2O \tag{18.26}$$

$$4Fe^{2+} = 4Fe^{3+} + 4e^-$$

Adding these two reactions together gives the original reaction Eq. (18.25). Half reactions are unique in that electrons are reactants or products. Otherwise, these reactions are similar to other equilibrium reactions.

Mass law equations can be written in terms of the concentration or activities of reactants and products (including the electrons), and an appropriate equilibrium constant. For example, the mass law expression for the half reaction

$$Ox + ne^- = Red$$

is

$$K = \frac{[Red]}{[Ox][e^-]^n} \tag{18.27}$$

Rearrangement of Eq. (18.27) gives the electron activity $[e^-]$ for a half reaction

$$[e^-] = \left\{ \frac{[Red]}{[Ox]K} \right\}^{1/n} \tag{18.28}$$

In the same way that pH defines $[H^+]$, *pe* defines electron activity, $[e^-]$. Rewriting Eq. (18.28) by taking the logarithm of both sides gives

$$pe = -\log[e^-] = \frac{1}{n} \left\{ \log K - \log \frac{[Red]}{[Ox]} \right\}$$

When a half reaction is written in terms of a single electron or $n = 1$, the log K term is written as pe^0 so that

$$pe = pe^0 - \log \frac{[Red]}{[Ox]} \tag{18.29}$$

Morel and Hering (1993, p. 430) and Pytkowicz (1983, p. 249) tabulate values of pe^0 or log K. Some of the important half reactions together with values of pe^0 are presented in Table 18.7.

TABLE 18.7 Some Important Half-Redox Reactions

	$pe° = \log K$
Hydrogen	
$H^+ + e^- = \frac{1}{2}H_2(g)$	0.0
Oxygen	
$\frac{1}{2}O_3(g) + H^+ + e^- = \frac{1}{2}O_2(g) + \frac{1}{2}H_2O$	$+35.1$
$\frac{1}{4}O_2(g) + H^+ + e^- = \frac{1}{2}H_2O$	$+20.75$
$\frac{1}{2}H_2O_2 + H^+ + e^- = H_2O$	$+30.0$
(Note also $HO_2^- + H^+ = H_2O$; $\log K = 11.6$)	
Nitrogen	
$NO_3^- + 2H^+ + e^- = \frac{1}{2}N_2O_4(g) + H_2O$	$+13.6$
(Note also $N_2O_4(g) = 2NO_2(g)$; $\log K = -0.47$)	
$\frac{1}{2}NO_3^- + H^+ + e^- = \frac{1}{2}NO_2^- + \frac{1}{2}H_2O$	$+14.15$
(Note also $NO_2^- + H^+ = HNO_2$; $\log K = 3.35$)	
$\frac{1}{3}NO_3^- + \frac{4}{3}H^+ + e^- = \frac{1}{3}NO(g) + \frac{2}{3}H_2O$	$+16.15$
$\frac{1}{4}NO_3^- + \frac{5}{4}H^+ + e^- = \frac{1}{8}N_2O(g) + \frac{5}{8}H_2O$	$+18.9$
$\frac{1}{5}NO_3^- + \frac{6}{5}H^+ + e^- = \frac{1}{10}N_2(g) + \frac{3}{5}H_2O$	$+21.05$
$\frac{1}{8}NO_3^- + \frac{5}{4}H^+ + e^- = \frac{1}{8}NH_4^+ + \frac{3}{8}H_2O$	$+14.9$
Sulfur	
$\frac{1}{2}SO_4^{2-} + H^+ + e^- = \frac{1}{2}SO_3^{2-} + \frac{1}{2}H_2O$	-1.65
(Note also $SO_3^{2-} + H^+ = HSO_3^-$; $\log K \simeq 7$)	
$\frac{1}{4}SO_4^{2-} + \frac{5}{4}H^+ + e^- = \frac{1}{8}S_2O_3^{2-} + \frac{5}{8}H_2O$	$+4.85$
$\frac{1}{6}SO_4^{2-} + \frac{4}{3}H^+ + e^- = \frac{1}{48}S_8^0 \text{ (s. ort.)} + \frac{2}{3}H_2O$	$+6.03$
(Note also S_8^0 (s. ort.) $= S_8^0$ (s. col.); $\log K = -0.6$)	
$\frac{3}{19}SO_4^{2-} + \frac{24}{19}H^+ + e^- = \frac{1}{38}S_6^{2-} + \frac{12}{19}H_2O$	$+5.40$
$\frac{5}{32}SO_4^{2-} + \frac{5}{4}H^+ + e^- = \frac{1}{32}S_5^{2-} + \frac{5}{8}H_2O$	$+5.88$
(Note also $S_5^{2-} + H^+ = HS_5^-$; $\log K = 6.1$ and	
$HS_5^- + H^+ = H_2S_5$; $\log K = 3.5$)	
$\frac{2}{13}SO_4^{2-} + \frac{16}{13}H^+ + e^- = \frac{1}{26}S_4^{2-} + \frac{8}{13}H_2O$	$+5.12$
(Note also $S_4^{2-} + H^+ = HS_4^-$; $\log K = 7.0$ and	
$HS_4^- + H^+ = H_2S_4$; $\log K = 3.8$)	
$\frac{1}{8}SO_4^{2-} + \frac{5}{4}H^+ + e^- = \frac{1}{8}H_2S(aq) + \frac{1}{2}H_2O$	$+5.13$
(Note also $H_2S(g) = H_2S(aq)$; $\log K_H = -1.0$, and other	
acid-base, coordination, and precipitation reactions)	

(*continued*)

TABLE 18.7   continued

	$pe° = \log K$
Trace Metals	

Cr

$$\tfrac{1}{3}HCrO_4^- + \tfrac{7}{3}H^+ + e^- = \tfrac{1}{3}Cr^{3+} + \tfrac{4}{3}H_2O \qquad\qquad +20.2$$

> (Note also $HCrO_4^- = \tfrac{1}{2}Cr_2O_7^{2-} + \tfrac{1}{2}H_2O$, $\log K = -1.5$;
> $HCrO_4^- = H^+ + CrO_4^{2-}$, $\log K = -6.5$; and various
> Cr(III) precipitation and coordination reactions.)

Mn

$$\tfrac{1}{5}MnO_4^- + \tfrac{8}{5}H^+ + e^- = \tfrac{1}{5}Mn^{2+} + \tfrac{4}{5}H_2O \qquad\qquad +25.5$$

$$\tfrac{1}{2}MnO_2(s) + 2H^+ + e^- = \tfrac{1}{2}Mn^{2+} + H_2O \qquad\qquad +20.8$$

Fe

$$Fe^{3+} + e^- = Fe^{2+} \qquad\qquad +13.0$$

$$\tfrac{1}{2}Fe^{2+} + e^- = \tfrac{1}{2}Fe(s) \qquad\qquad -7.5$$

$$\tfrac{1}{2}Fe_3O_4(s) + 4H^+ + e^- = \tfrac{3}{2}Fe^{2+} + 2H_2O \qquad\qquad +16.6$$

Co

$$Co(OH)_3(s) + 3H^+ + e^- = Co^{2+} + 3H_2O \qquad\qquad +29.5$$

$$\tfrac{1}{2}Co_3O_4(s) + 4H^+ + e^- = \tfrac{3}{2}Co^{2+} + 2H_2O \qquad\qquad +31.4$$

Cu

$$Cu^{2+} + e^- = Cu^+ \qquad\qquad +2.6$$

$$\tfrac{1}{2}Cu^{2+} + e^- = \tfrac{1}{2}Cu(s) \qquad\qquad +5.7$$

Se

$$\tfrac{1}{2}SeO_4^{2-} + 2H^+ + e^- = \tfrac{1}{2}H_2SeO_3 + \tfrac{1}{2}H_2O \qquad\qquad +19.4$$

$$\tfrac{1}{4}H_2SeO_3 + H^+ + e^- = \tfrac{1}{4}Se(s) + \tfrac{3}{4}H_2O \qquad\qquad +12.5$$

$$\tfrac{1}{2}Se(s) + H^+ + e^- = \tfrac{1}{2}H_2Se \qquad\qquad -6.7$$

> (Note also $H_2Se = H^+ + HSe^-$, $\log K = -3.9$;
> $H_2SeO_3 = H^+ + HSeO_3^-$, $\log K = -2.4$;
> $HSeO_3^- = H^+ + SeO_3^{2-}$, $\log K = -7.9$;
> $SeO_4^{2-} + H^+ = HSeO_4^-$, $\log K = +1.7$)

Ag

$$AgCl(s) + e^- = Ag(s) + Cl^- \qquad\qquad +3.76$$

$$Ag^+ + e^- = Ag(s) \qquad\qquad +13.5$$

Hg

$$\tfrac{1}{2}Hg^{2+} + e^- = \tfrac{1}{2}Hg(1) \qquad\qquad +14.4$$

$$Hg^{2+} + e^- = \tfrac{1}{2}Hg_2^{2+} \qquad\qquad +15.4$$

Pb

$$\tfrac{1}{2}PbO_2 + 2H^+ + e^- = \tfrac{1}{2}Pb^{2+} + H_2O \qquad\qquad +24.6$$

> (Note many other reactions for Mn, Fe, Co, Cu, Se, Ag,
> Hg, and Pb)

Source: Morel and Hering, 1993. *Principles and applications of aquatic chemistry*. Copyright ©1993 by John Wiley & Sons, Inc. Reprinted by permission of John Wiley & Sons, Inc.

The redox state of soils and water can be expressed in terms of E_H, the redox potential. Values of E_H have units of volts, acknowledging that a redox reaction involves the transfer of electrons. E_H is related to pe by the following equation:

$$E_H = \frac{2.3\,RT}{F}pe \qquad (18.30)$$

where F is the Faraday constant defined as the electrical charge of one mole of electrons (96,500 Coulombs) with $2.3\,RT/F$ equal to 0.059 V at 25°C. In practice, pe is taken to be the calculated value of electron activity, and E_H is the measured electrode potential for an electrochemical cell. In other words, pe is a calculated quantity and E_H is a measured one. The following example illustrates how the pe of a ground water is calculated assuming redox equilibrium.

▶ **EXAMPLE 18.5** A ground water has $(Fe^{2+}) = 10^{-3.3}$ M and $(Fe^{3+}) = 10^{-5.9}$ M. Calculate the pe at 25°C assuming that the activities of Fe species are equal to their concentrations. What should be the measured E_H of this solution?

SOLUTION From Table 18.7, the half reaction for the reduction of Fe^{3+} to Fe^{2+} is

$$Fe^{3+} + e^- = Fe^{2+} \quad \text{with } pe^0 = 13.0$$

Substitution into Eq. (18.29) gives

$$pe = pe^0 - \log\{(Fe^{2+})/(Fe^{3+})\}$$

$$pe = 13.0 - \log\{10^{-3.3}/10^{-5.9}\}$$

$$pe = 13.0 - 2.6 = 10.4$$

$$E_H = \frac{2.3\,RT}{F}pe = 0.059 \times 10.4 = 0.61 \text{ V}$$

Kinetics and Dominant Couples

So far, we have presented redox phenomena as equilibrium processes, implying that the half-redox reactions are in equilibrium. However, redox reactions are not usually at equilibrium because the number of active microorganisms could be small or microorganisms do not metabolize some of the reactants easily. Also, redox reactions with extremely large equilibrium constants are essentially irreversible. Take, for example, a reaction where an excess of dissolved oxygen oxidizes organic matter (as CH_2O)

$$CH_2O + O_2 \rightarrow CO_2 + H_2O$$

The reaction progresses to the right until all of the organic matter disappears. This example raises an important question—what good are equilibrium concepts involving pe or E_H if the reactions are not at equilibrium? Fortunately, a constant value of pe or E_H exists when the concentration of one of the couples, O_2/H_2O in this case, is much greater than that of the other. Because utilization of all the organic matter will require a negligibly small proportion of the oxygen, the redox potential defined by the O_2/H_2O couple does not change as a function of the reaction progress.

At one time it was thought that the less dominant couples would be controlled by the dominant couples. For example, if the O_2/H_2O couple is dominant, the other couples like Fe^{2+}/Fe^{3+} would adjust their concentrations according to the pe defined by the dominant couple. However, research has shown that redox equilibrium among all the couples does not occur (Lindberg and Runnells, 1984). Thus, while the concept of some master *pe* is alluring, each half reaction has a different *pe* (Lindberg and Runnells, 1984). The main use for *pe* calculations is to provide indications of the direction in which a system is evolving (Stumm and Morgan, 1981).

Control on the Mobility of Metals

Redox reactions influence the mobility of metal ions in solution. One example is the behavior of metals in sulfur systems involving the sulfate/sulfide couple. Assume that the total sulfur in a system (S_T) exists as four possible species, SO_4^{2-}, H_2S, HS^-, and S^{2-}, where in the first sulfur occurs as S(+VI) and in the last three as S(−II). The relationship between H_2S, HS^-, and S^{2-} is fixed by equilibrium relationships, much like the carbonate system. In oxic environments, SO_4^{2-} is the dominant species of the couple with essentially none of the S(−II) species present. Sulfate ion is reduced in an anoxic environment, and the other three species dominate.

As the sulfur species change with changing redox conditions, so do the solids that precipitate. When SO_4^{2-} dominates, metal concentrations can be relatively high because there are no solubility constraints to force the removal of solids from solution. Once the system becomes reducing, almost all of the metal and sulfur is removed from the system as metal sulfides, which are usually relatively insoluble. Thus, in one case, the total concentration of the metal is found as a metal ion, and in the other as a solid.

With a more complex system, the pH could change, other ions could be present, and one of several solids could form. For such a system, *E_H-pH or pe-pH diagrams* provide a useful way of presenting the essential features of the redox and solid-phase chemistry. Shown in Figure 18.13 is an E_H-pH diagram for iron. The upper and lower lines on the

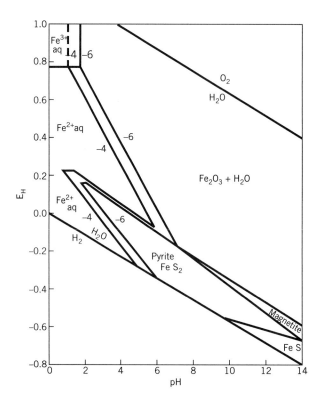

Figure 18.13 Example of an E_H-pH diagram showing the stability relationships of iron oxides and sulfides (from Macumber, 1984). Reprinted by permission First Canadian/American Conference on Hydrogeology. Copyright ©1984. All rights reserved.

field represent the oxidation of water to O_2 and the reduction of water to H_2. These two half reactions define the upper and lower limits for oxidation or reduction. The other lines are boundaries for what are known as *stability fields*. These fields are labeled with the one dominant ion or solid for the specified E_H and pH. When an ion dominates (for example, Fe^{2+}), the element is mobile because a relatively small proportion of the metal occurs as a solid. In the stability fields for solid phases (for example, FeS_2 and Fe_2O_3), nearly all of the iron will exist as that phase, making the element essentially immobile.

Biotransformation of Organic Compounds

A large number of half reactions involve the oxidation of organic compounds into simpler inorganic forms such as CO_2 and H_2O. These reactions are referred to as *biodegradation or biotransformation reactions* because they are microbially catalyzed. In some cases, oxygen acts as the electron acceptor

$$\tfrac{1}{4}CH_2O + \tfrac{1}{4}O_2(g) = \tfrac{1}{4}CO_2(g) + \tfrac{1}{4}H_2O$$

However, when O_2 is unavailable, other species, for example, NO_3^-, Fe(+III), Mn(+IV), SO_4^{2-}, and CO_2, accept electrons. A few of the most important of these biotransformation reactions include

> Fe(III) reduction: $\tfrac{1}{4}CH_2O + Fe(OH)_3 + 2H^+ = \tfrac{1}{4}CO_2(g) + Fe^{2+} + \tfrac{11}{4}H_2O$
>
> denitrification: $CH_2O + \tfrac{4}{5}NO_3^- + \tfrac{4}{5}H^+ = CO_2(g) + \tfrac{2}{5}N_2 + \tfrac{7}{5}H_2O$
>
> sulfate reduction: $CH_2O + \tfrac{1}{2}SO_4^{2-} + \tfrac{1}{2}H^+ = \tfrac{1}{2}HS^- + H_2O + CO_2(g)$
>
> methane formation: $CH_2O + \tfrac{1}{2}CO_2(g) = \tfrac{1}{2}CH_4 + CO_2(g)$

where CH_2O is a "type" of organic compound.

Biotransformation reactions are important when the organic compound is a ground-water contaminant. If biotransformation occurs rapidly, the reaction will reduce contaminant concentrations and perhaps alter the major ion chemistry.

► 18.6 MICROORGANISMS IN GROUND WATER

Microorganisms occur ubiquitously in the subsurface. Although bacteria predominate, protozoa and fungi are also common. *Bacteria* are small single-celled organisms. In order to live, they metabolize dissolved organic matter that is present naturally or that is due to contamination. *Protozoa* are somewhat larger, single-celled organisms that can feed on bacteria. *Fungi* live by degrading other organic matter and effectively recycling plant and animal debris (Chapelle, 1993).

Bacteria occur in the subsurface in different ways. Some are adapted to move freely in the water with flagella that propel them. The direction of motion can be either random or directed in response to concentrations of organic compounds (*chemotactic*). This property of motility can be important for the local redistribution of bacteria and may influence larger scale transport. Most commonly, bacteria are *immotile*, bound to the surfaces of the aquifer solids. The attached population forms what is called a *biofilm*, which consists of bacteria held together to the particles by extracellular polymers (McCarty et al., 1984).

In pristine ground-water systems, the bulk of the bacterial population is attached to the particle surface. Theoretical work suggests that when the concentration of metabolizable organic compounds (i.e., the substrate for growth) is low, immotile bacteria have a competitive advantage over motile forms (Kelly et al., 1988). The proportion of motile bacteria may be larger in contaminated systems containing metabolizable organic compounds (Ghiorse and Wilson, 1988). Thus, with increasing substrate concentrations, motile bacteria may become more competitive.

Biofilms

Biofilms are the site for transformation reactions that are important in reducing the concentrations of some organic contaminants. They are commonly conceptualized in terms of a simple layer, with molecular diffusion operating to bring metabolizable organic compounds from the bulk pore fluid to clusters of bacteria cells in the biofilm (Figure 18.14). Real biofilms are more complex. For example, in cases where organic substrate concentrations are relatively low, patchy or discontinuous biofilms form (Rittmann, 1993). Furthermore, detailed microscopic studies of model biofilms demonstrated fluid flow within biofilm channels (Stoodley et al., 1994). These channels provide a delivery system for nutrients into the biofilm that is more efficient than diffusive transport.

Quantifying Microbial Abundances

A variety of techniques have been developed to enumerate the abundance of microorganisms in aquatic systems. Depending on the purpose of the study, one might collect either water samples or soil material. Water samples can be collected from conventional monitoring wells using sterile techniques (e.g., Harvey et al., 1984; NcNabb and Mallard, 1984). However, soil cores are most often collected because most of the microbial population is attached on grain surfaces. Again, care must be exercised in keeping the sampling equipment and core tubes sterile. Readers interested in the details of these sampling approaches can refer to Armstrong et al., (1988) and McNabb and Mallard (1984).

Once the sample of water or solid is collected, techniques are available in the laboratory to characterize the size of microbial population. In the case of a soil sample, however, an additional step is required to remove bacteria from the solid materials. The most common procedure involves shaking sediment-distilled water slurry. Actual microbial numbers are determined via a plate count or direct counting procedures.

A *plate count* involves smearing a sample (water or sediment/water slurry) containing the bacteria on a growth medium, incubating the plate for some time, and counting the bacterial colonies. The number of bacteria can be expressed, for example, as colony-forming units per milliliter (CFU/mL) in the case of water samples, or CFU/[g dry weight of sediment]. A slightly different plating procedure, the most-probable-number (MPN) approach, determines how much a sediment slurry or water sample can be diluted before no growth takes place. The dilution information provides a basis for estimating the number of cells per volume (or mass) of the original sample. Direct counting procedures use a

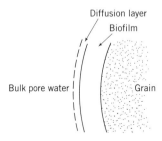

Figure 18.14 Conceptualization of a biofilm at a pore scale (from Domenico and Schwartz, 1998. *Physical and chemical hydrogeology*). Copyright ©1998 by John Wiley & Sons, Inc. Reprinted by permission of John Wiley & Sons, Inc.

microscope to count stained bacteria. This procedure provides an estimate of the number of cells per gram of material.

Microbial Ecology of the Subsurface

The subsurface is home to a diverse set of microorganisms. In pristine ground-water systems, population densities range between 10^5 and 10^7 cells per gram dry weight (gdw) (Ghiorse and Wilson, 1988). Generally, these population densities are small relative to those found in the soil zone and the unsaturated zone. This difference in the size of microbial populations is caused by a decrease in the concentration of organic matter as recharging ground water moves through the soil.

The distribution of microorganisms depends in a complex way on the residence time of water in the system, and the geochemistry of organic matter transported along the flow system. As Ghiorse and Wilson (1988) suggest, the quantity of organic matter and nutrients passed along a pristine ground-water system is relatively meager. Thus, microbial populations appear adapted for growth and survival in nutrient-poor (*oligotrophic*) conditions (Balkwill et al., 1989). Theoretically, in a system closed to organic carbon, the concentration of organic matter should decline until it is no longer possible to support the microbial population (Ghiorse and Wilson, 1988). However, this end point has not been discovered in large flow systems.

Results from several studies show that even down to several hundred meters, microbial abundance does not diminish with depth. However, there can be considerable spatial variability both vertically and laterally. For example, in holes drilled as part of the Deep Probe Project of the U.S. Department of Energy, Levine and Ghiorse (1990) found a highly significant correlation between bacterial abundance and hydraulic conductivity. Bacteria were more abundant in sandy sediments and much less abundant in clayey sediments. Similar distributions are observed in shallow environments as well. This pattern may reflect the much greater difficulty in originally colonizing the fine-grained units because of straining filtration of migrating organisms (Zachara, 1990). In addition, the larger fluxes of water through the more permeable units would likely make significantly more dissolved organic matter available. Fredrickson et al. (1990) found significant diversity in the ability of bacteria to utilize various types of organic compounds even when nearby samples were compared. Their data indicate a great diversity in bacteria at the population and organism levels even within the same geologic unit.

Biodegradation Reactions

There are two different modes in which organic substrates participate in biodegradation reactions. With *primary utilization*, a dissolved organic compound provides the main source of energy and carbon for the microorganisms. In effect, the microorganisms produce energy through the electron transfer involved with redox reactions. In other cases, an organic compound or substrate of interest is *cometabolized* along with some other primary substrate. The secondary substrate by itself is not capable of providing sufficient energy to maintain the bacterial population.

TABLE 18.8   Examples of primary substrates

Aerobic and anaerobic	Glucose, acetone, isopropanol, acetate, benzoate, phenol
Aerobic primarily	Alkanes, benzene, toluene, xylene, vinyl chloride, methane, propane

A variety of organic compounds can serve as the primary substrate for microbial metabolism (National Research Council, 1993). Table 18.8 lists examples of primary substrates. Following are examples of half reactions involving the primary utilization of organic contaminants.

benzene oxidation: $12H_2O + C_6H_6 = 6CO_2 + 3OH^- + 30e^-$

toluene oxidation: $14H_2O + C_6H_5CH_3 = 7CO_2 + 36H^+ + 36e^-$

Some chlorinated solvents, like trichloroethene (TCE), dichloroethene (DCE), dichloroethane (DCA), and vinyl chloride (VC), are relatively oxidized compounds and resist biotransformation in aerobic settings because they do not provide much energy. However, they can be cometabolized with other organic compounds. For example, Roberts et al. (1989) demonstrate how methanogenic bacteria are stimulated through the addition of methane and oxygen. The oxidation of methane provides the energy and carbon necessary for bacterial growth and leads to the production of enzymes required to degrade the secondary substrates like TCE and VC. Chemically, both TCE and VC end up being mineralized to CO_2. This possibility for cometabolizing chlorinated solvents holds promise as a remedial approach for aquifers contaminated by TCE. In anaerobic settings, these compounds can be biotically transformed in reductive dehalogenation reactions. Here are a few examples of these redox reactions:

PCE reductive dehalogenation: $C_2Cl_4 + H^+ + 2e^- = C_2HCl_3 + Cl^-$

TCE reductive dehalogenation: $C_2HCl_3 + H^+ + 2e^- = C_2H_2Cl_2 + Cl^-$

Degradation of Common Contaminants

Biotransformation reactions are important because they influence the distribution of organic contaminants in ground water. The last section in this chapter will examine the extent to which biotransformation reactions influence the most important families of contaminants—hydrocarbons and chlorinated solvents. Commonly, biotransformation processes are examined in relation to two prototypical settings. In *aerobic settings*, oxygen is the major electron acceptor for the various biotransformation reactions. In *anaerobic settings*, oxygen is absent, and other compounds, like nitrate, sulfate, and carbon dioxide, function as electron acceptors.

Hydrocarbons and Derivatives

Hydrocarbons and related compounds constitute one of the most important contaminant families in ground water. In aerobic ground-water systems, aromatic hydrocarbons with up to two benzene rings are mineralized relatively rapidly with minimal lag times. For example, Nielsen and Christensen (1994a) found that benzene, toluene, and o-xylene degraded with short lag times (about four days). Initial concentrations of approximately 150 μg/L of benzene and toluene decreased to < 2 μg/L in less than one month. The degradation of o-xylene required about three months and was substantially less complete. These laboratory findings of relatively rapid degradation rates are also corroborated by field tests. Studies of an oil spill at Bimidji, Minnesota (Cozzarelli et al., 1988) showed that hydrocarbons degrade in anaerobic systems as well.

Other components of fuels (or coal-tar derivatives) are polyaromatic hydrocarbons (PAHs) such as naphthalene, anthracene, or pyrene. These compounds are comprised of a series of benzene rings. In general, PAHs are biodegradable in aerobic settings, with the rate decreasing as the molecular weight of the compound increases. In anaerobic systems, PAHs may be somewhat more resistant to degradation. However, Lyngkilde and Christensen (1992) report the reduction of naphthalene concentrations under ferrogenic conditions.

Phenolic compounds are also found as components in fuels or as industrial contaminants. Phenols are characterized by an aromatic ring with an attached hydroxyl group. Examples of these compounds include phenol, nitrobenze, o-cresol, and o-nitrophenol. Microcosm studies by Nielsen and Christensen (1994b) show that these compounds are mineralized in aerobic systems. Indications are that they biodegrade relatively rapidly with broad variation in the rates. p-, m-, and o-cresol also degrade under anaerobic conditions (Smolenski and Sulflita, 1987). The remaining hydrocarbons in Table 18.9, alcohols and ketones, are readily degradable in ground water.

Halogenated aliphatic compounds include the common industrial solvents such as carbon tetrachloride (CT), tetrachloroethylene (PCE), trichloroethylene (TCE), trichloroethane (TCA), vinyl chloride (VC), and dichloroethylene (DCE). These compounds are formed from carbon atoms joined together in chains with attached atoms. In aerobic systems, they resist degradation and are extremely persistent. There are no known bacteria that can oxidize these compounds as a primary substrate because the reaction yields so little energy (Chapelle, 1993). The exception to this general behavior occurs when the aerobic system contains methane either naturally or applied in a bioremediation scheme. Compounds like TCE may be metabolized as a secondary substrate when methane is the primary substrate.

Under anaerobic conditions (methanogenic), halogenated compounds are commonly biotransformed. The reactions involve a sequential loss of chloride atoms in a process called reductive dehalogenation. Figure 18.15 illustrates reductive dechlorination reactions beginning with PCE. The last step from vinyl chloride to CO_2 is quite slow relative to the other dechlorination reactions, so that vinyl chloride often tends to accumulate.

▶ EXERCISES

18.1. Often (CO_3^{2-}) is not reported in water-quality analyses because the concentration is too low to measure directly. Develop an expression from the appropriate mass laws to calculate the concentration of (CO_3^{2-}) given pH and (HCO_3^-).

TABLE 18.9   Important families of organic contaminants found in ground water

Chemical Family	Examples of Compounds
Hydrocarbons and derivatives	
Fuels	Benezene, toluene, *o*-xylene, butane, phenol
PAHs	Anthracene, phenanthrene
Alcohols	Methanol, glycerol
Creosote	*m*-cresol, *o*-cresol
Ketones	Acetone
Halogenated aliphatics	Tetrachloroethene, trichloroethene, dichloromethane
Halogenated aromatics	Chlorobenzene, dichlorobenzene
Polychlorinated biphenyls	2,4′-PCB, 4,4′-PCB

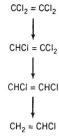

$CCl_2 = CCl_2$

$CHCl = CCl_2$

$CHCl = CHCl$

$CH_2 = CHCl$

Figure 18.15 Reductive dehalogenation from PCE to TCE to DCE and VC (from Domenico and Schwartz, 1998. *Physical and chemical hydrogeology*). Copyright ©1998 by John Wiley & Sons, Inc. Reprinted by permission of John Wiley & Sons, Inc.

18.2. Following are the results of a batch sorption experiment in which two grams of porous media were mixed with 10 ml of a solution containing various concentrations of Cd.

Initial (Cd)	**Equilibrium** (Cd)
Concentration (mg/ml)	Concentration (mg/ml) $\times 10^{-4}$
0.0005	0.040
0.001	0.093
0.010	1.77
0.050	39.3
0.096	205.2

a. Find the equation for the Freundlich isotherm that best fits these data. (*Hint*: Transform the basic equation and data in terms of the logarithm of concentration.)

b. Assuming that only the lowest concentrations are of interest, fit the same set of data with a linear Freundlich isotherm (i.e., $n = 1$) to obtain a distribution coefficient (K_d). Be sure to indicate the units for K_d.

18.3. At a site, the following contaminants are found (i) parathion (log $K_{ow} = 3.80$), (ii) chlorobenzene (log $K_{ow} = 2.71$), and (iii) DDT (log $K_{ow} = 6.19$).

a. Using the equation of Hassett et al. (1983) in Eq. (18.20), estimate the partition coefficient between solid organic carbon and water (K_{oc}).

b. With an f_{oc} of 0.01, estimate the distribution coefficients (K_d) for these compounds.

c. On the basis of sorption alone, which of these compounds is the most mobile and which is the least mobile?

18.4. Compare the range in variability of the distribution coefficients for carbon tetrachloride (log K_{ow} 2.83) estimated using the three different regression equations (18.18, 18.19, and 18.20) and an f_{oc} of 0.01.

18.5. The properties of six organic compounds are as follows:

Compound	**log K_{ow}**
#1	2.04
#2	0.89
#3	0.46
#4	0.27
#5	2.83
#6	1.48

List the organics in order of decreasing mobility.

18.6. Consider the following half-redox reaction involving NO_3^- with a concentration of 10^{-5} M and ammonium with a concentration of 10^{-3} in ground water with a pH of 8.

$$\tfrac{1}{8} NO_3^- + \tfrac{5}{4} H^+ + e^- = \tfrac{1}{8} NH_4^+ + \tfrac{3}{8} H_2O$$

What is the *pe* and E_H of the system at 25°C assuming redox equilibrium?

18.7. The couple O_2/H_2O is found to be the dominant couple, producing a *pe* of 12.0. By assuming redox equilibrium, determine how $(Fe)_T = 10^{-5}$ M would be distributed as Fe^{2+} and Fe^{3+}.

► CHAPTER
HIGHLIGHTS

1. A variety of chemical reactions operate in ground-water systems to transfer mass among ions or phases. These reactions give rise to the diversity in the chemistry of natural waters and control the mobility of contaminants in ground water.

2. Acid-base reactions transfer hydrogen ions (H^+) among the ions present in the aqueous phase. The concentration of H^+ determines the pH of the solution, which is a key factor in controlling many of the other processes that follow. pH is defined as the negative logarithm of the hydrogen ion activity, or pH $= -\log[H^+]$. Water is acidic when pH <7, neutral when pH $= 7$, or basic when pH >7. Here are examples of acid-base reactions:

- dissociation of water: $H_2O = H^+ + OH^-$
- dissolution of CO_2 in water: $CO_2(g) + H_2O = H_2CO_3$
- two-satge ionization of carbonic acid: $H_2CO_3 = HCO_3^- + H^+$
$$HCO_3^- = CO_3^{2-} + H^+$$

3. Minerals and liquids (for example, oil/gasoline) dissolve in the subsurface and add ions and/or organic constituents to ground water. Solubility depends on how well the dissolved phase fits into the tetrahedral structure of water. The most soluble organic compounds are charged species or those containing oxygen or nitrogen (e.g., alcohols or carboxylic acids), which fit well in the water structure. Most organic compounds have a low solubility because they don't fit well into the water structure.

4. A complex is an ion that forms by combining simpler cations, anions, and sometimes molecules. The cation or central atom is typically a metal. The anions, often called ligands, can include Cl^-, F^-, Br^-, SO_4^{2-}, PO_4^{3-}, and CO_3^{2-}. The ligand might also comprise various organic molecules such as amino acids. Here are examples.

- simple combination of a metal and a ligand: $Mn^{2+} + Cl^- = MnCl^+$
- complexes can also form by combining with ligands: $Cr^{3+} + OH^- = Cr(OH)^{2+}$
$$Cr(OH)^{2+} + OH^- = Cr(OH)^{2+}$$
$$Cr(OH)_2^+ + OH^- = Cr(OH)_3^0..$$

5. Surface reactions involve the ions in solution and the grains or crystals in a rock, or grain coatings. For example, water in clay-rich deposits can experience a loss of Ca and Mg and a gain of Na due to cation exchange processes with clay minerals. Organic contaminants can partition into solid organic carbon.

6. Sorption is usually studied by batch experiments. Sorption isotherms are mathematical equations that are fit to experimental data. The simplest model characterizes sorption in terms of a distribution coefficient or K_d. The K_d-based approach has had good success in predicting the mobility of organic contaminants in ground water.

7. Mineral surfaces, particularly clays, often have structural imperfections that produce a net negative charge. With cation exchange, cations that are attracted to the surface can exchange with ions in solution. Another important sorption reaction involves solids, for example, kaolinite, metal oxides, and metal oxyhydroxides, whose surface charges change as a function of ground-water composition.

In this reaction, ions in solution react chemically with hydroxyl groups and chemisorbed water on the surface.

8. Redox reactions involve elements (for example, C, O, N, Fe, Mn), which have different oxidation states. Oxidation is the removal of electrons from an atom, forcing a change in the oxidation number of an element. Reduction refers to the addition of an electron to lower the oxidation number. All redox reactions transfer electrons and involve elements with more than one oxidation number.

9. In an oxidation reaction, no free electrons are produced. Both oxidation and reduction reactions occur together. An overall reaction can then be represented as two half-redox reactions. These half-redox reactions can be treated using equilibrium concepts. Interestingly, an electron activity appears in the expression. The redox state of soils and water can be expressed in terms of E_H, the redox potential with units of volts.

10. Microorganisms occur ubiquitously in the subsurface. Bacteria are small single-celled organisms. In order to live, they metabolize dissolved organic matter present naturally or due to contamination. In pristine ground-water systems, the bulk of the bacterial population is attached to the particle surface as biofilms. Molecular diffusion brings metabolizable organic compounds from the pore fluid to bacteria in the biofilm.

11. Biodegradation reactions are important because they influence the distribution of organic contaminants in ground water. In aerobic settings, oxygen is the major electron acceptor for biotransformation reactions. Contaminants that degrade there include hydrocarbons and fuels, and PAHs. In anaerobic settings, nitrate, sulfate, and carbon dioxide accept electrons. There, industrial solvents like carbon tetrachloride, tetrachloroethylene, trichloroethylene, trichloroethane, and vinyl chloride degrade.

CHAPTER

19

BASICS OF MASS TRANSPORT, ADVECTION, AND DISPERSION

► 19.1 ADVECTION

► 19.2 DIFFUSION

► 19.3 DISPERSION

► 19.4 DISPERSION COEFFICIENTS AT MICROSCOPIC
AND MACROSCOPIC SCALES

► 19.5 STATISTICAL PATTERNS OF MASS SPREADING

► 19.6 A GEOSTATISTICAL MODEL OF DISPERSION

► 19.7 MIXING IN FRACTURED MEDIA

► 19.8 TRACERS AND TRACER TESTS

So far we have discussed examples of how chemical reactions influence the geochemistry of ground water. This section looks at dissolved mineral mass within a more general theory of mass transport. This theory provides a rigorous description of how dissolved mass in ground water originates and is added to the water, how it is moved or redistributed, and how it interacts chemically with other dissolved mass in ground water or the porous medium. As the cartoon suggests (Figure 19.1), *mass transport* is simply a delivery/distribution problem. Like the fruits and vegetables being moved down the conveyor belt, dissolved mass (that is, ions and molecules) is being moved along by the flowing ground water. Fruits and vegetables can be eaten or lost from the belt and end up on the ground. Mass in flowing ground water can also be eaten—in this case by bacteria that consume various organic compounds—or lost from the flow by being sorbed onto the aquifer grains.

With this theory of mass transport, the question of whether the system is natural or contaminated is relevant only to where the initial mass originated. For example, does dissolved iron come from the weathering of an iron silicate mineral or an old car in a landfill? Once in the ground water, dissolved iron will behave similarly regardless of the source.

It is difficult to explain the transport processes in a simple way because so much is happening at the same time. The *physical processes* determine the way mass is moved from one point to another—the system of ground-water conveyance and mixing. The *chemical and biological processes* that were presented in Chapter 18 redistribute mass among different chemical forms, or into and out of the aqueous system.

Figure 19.2 summarizes key processes that contribute to mass transport along a flow system. By now, we should know something about the chemical and biological processes

443

Figure 19.1 Key elements of mass transfer. Unpublished cartoon provided by T. Davis.

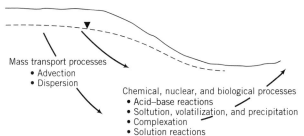

Mass transport processes
• Advection
• Dispersion

Chemical, nuclear, and biological processes
• Acid–base reactions
• Soltution, volatilization, and precipitation
• Complexation
• Solution reactions
• Oxidation–reduction reactions
• Hydrolysis reactions
• Isotropic reactions

Figure 19.2 Physical and chemical mass transport processess operating in a ground-water flow system (from Domenico and Schwartz, 1998. *Physical and chemical hydrogeology*). Copyright ©1998 by John Wiley & Sons, Inc. Reprinted by permission of John Wiley & Sons, Inc.

(Figure 19.2). What remains is to describe the physical mass transport processes, advection, and dispersion.

▶ 19.1 ADVECTION

Advection is mass transport due simply to the flow of water in which the mass is dissolved. It is the main process conveying dissolved mass from one point to another. The direction and rate of transport coincide with the ground water. Thus, knowing something about the pattern of ground-water flow automatically implies knowledge about advection. For example, in the case of topographically driven flow systems, the factors that influence ground-water flow patterns—water-table configuration, pattern of geologic layering, size of the ground-water basin, pumping, or injection—also control the direction and rate of mass transport. Thus, the background from the earlier treatment of regional ground-water flow allows us to consider advection in just a few pages.

The two flownets in Figure 19.3 illustrate basic ideas about advection, which hold in systems with a constant fluid density. First, when only advection is operating, dissolved mass in one or more stream tubes will remain in those stream tubes (Figure 19.3a). It takes other processes to move dissolved mass between stream tubes (Figure 19.3b). Second, pathlines define the direction of mass spreading in steady-state systems, even with relatively complex flow patterns.

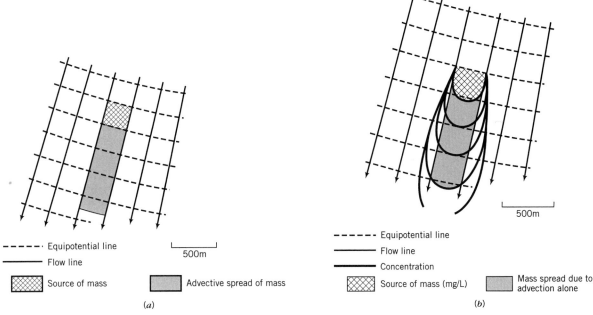

Figure 19.3 Mass spreading in a shallow unconfined aquifer due to (*a*) advection alone and (*b*) advection and dispersion (from Domenico and Schwartz, 1998. *Physical and chemical hydrogeology*). Copyright ©1998 by John Wiley & Sons, Inc. Reprinted by permission of John Wiley & Sons, Inc.

For most practical problems, ground water and dissolved mass will move at the same rate (in the absence of other processes) and in the same direction. Accordingly, the Darcy equation developed in Chapter 3 describes the velocity of advective transport

$$v = (v_x, v_y, v_z) = \left(-\frac{K_{xx}}{n} \frac{\partial h}{\partial x}, -\frac{K_{yy}}{n} \frac{\partial h}{\partial y}, -\frac{K_{zz}}{n} \frac{\partial h}{\partial z} \right) = -\frac{K}{n} \text{grad } h \qquad (19.1)$$

where v is the linear ground-water velocity, K_{xx}, K_{yy}, and K_{zz} are the hydraulic conductivity along x-, y-, and z-directions, n is the effective porosity, and $-\partial h/\partial x$, $-\partial h/\partial y$, and $-\partial h/\partial z$ are the hydraulic gradients along the x-, y-, and z-directions. Also, as is evident from Eq. (19.1), the linear ground-water velocity and, therefore, the velocity of advective transport increase with decreasing effective porosity. This relationship is particularly important in fractured rocks, where the effective porosity can be much less than the total porosity, often as low as 1×10^{-4} or 1×10^{-5}.

► **EXAMPLE 19.1** A small volume of tracer is added to an unconfined aquifer that has a hydraulic conductivity of 1 m/d and a porosity of 0.35. The hydraulic gradient is 0.07. Calculate how far the center of mass of the tracer will move in one year.

SOLUTION Assuming that transport is due mainly to advection, calculate the advective velocity using Eq. (19.1).

$$v = -\frac{K}{n} \text{grad } h$$

$$= \frac{1 \text{m/d}}{0.35} \times 0.07 = 0.2 \text{ m/d}$$

After one year the tracer will move a distance equal to vt, or

$$d = v \times t$$

$$= 0.2 \text{ m/d} \times 365 \text{ d} = 73 \text{ m}$$

Thus, the center of mass will move 73 down the flow system in one year.

The term *mass flux* is used to describe the quantity of mass being carried by ground water. For systems where advection is the main transport mechanism, the advective mass transport flux is written as

$$J_{adv} = nCv \tag{19.2}$$

where n is the porosity of the medium, C is the concentration of the contaminant, and v is the ground-water velocity vector. The advective mass transport flux is defined as mass transported through a unit area per time. For example, if the concentrations are expressed in μg/L, distances in meters, and time in days, the mass flux has units of g/m^2/day.

▷ **EXAMPLE 19.2**

The porosity of a geological medium is 0.4. The specific discharge of ground water is 0.02 m/day. The concentration of a contaminant in a control volume is 100 μg/L. Calculate the contaminant mass flux in the control volume.

SOLUTION   The mass flux is calculated using Eq. (19.2):

$$J_{adv} = (0.4)(100\,\mu\text{g/L})\frac{(0.02 \text{ m/day})}{0.4} = 2 \times 10^{-3} \text{ g/m}^2\text{/day}$$

▷ 19.2 DIFFUSION

In Chapter 3, Darcy's equation was introduced as a linear law relating fluid flow q to the gradient in the hydraulic head. For mass, there is a similar law that describes the chemical mass flux as proportional to the gradient in concentration. This law is known as Fick's law, and for a simple one-dimensional aqueous (nonporous) system it is:

$$J_{\text{dif}} = -D_m \frac{dC}{dx} \tag{19.3}$$

In this formulation, J_{dif} is a chemical mass flux, and $-dC/dx$ is the concentration gradient with the negative sign indicating that the transport is in the direction of decreasing concentration. The proportionality constant D_m is termed the molecular diffusion coefficient of a chemical in a fluid environment. If J is expressed in units of moles/L^2/T, and the concentration is in moles/L^3, the proportionality constant has the units L^2/T in that the gradient operator $d(\ )/dx$ has the units L^{-1}.

Molecular diffusion is mixing caused by random molecular motions due to the thermal kinetic energy of the solute. Because of molecular spacing, the coefficient describing this scattering is larger in gases than in liquids and larger in liquids than in solids. The diffusion coefficient in a porous medium is smaller than in pure liquids primarily because collision with the solids of the medium hinders diffusion. Figure 19.4 provides a range in values for diffusion coefficients for various fluid environments. Table 19.1 cites molecular diffusion coefficients in water for common cations and anions. With the exception of H$^+$ and OH$^-$, values of D_m range from 5×10^{-6} to 20×10^{-6} cm^2/s, with the smallest values associated with ions having the greatest charge. Table 19.2 lists the diffusion coefficients of some organic compounds in air and water. Note how the diffusion coefficients in air are much

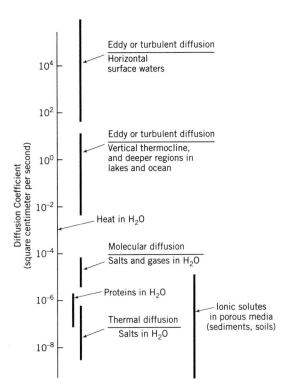

Figure 19.4 Diffusion coefficients characteristic of various environments. (Reprinted with permission from A. Lerman, 1971, in *Non-equilibrium systems in water chemistry*. J. D., Hem, ed., ACS Advances in Chemistry Series 106. Copyright ©1971 American Chemical Society).

TABLE 19.1 Diffusion coefficients in water for some ions at 25°C

Cation	$D_d(10^{-6}$ cm^2/s)	Anion	$D_d(10^{-6}$ cm^2/s)
H^+	93.1	OH^-	52.7
Na^+	13.3	F^-	14.6
K^+	19.6	Cl^-	20.3
Rb^+	20.6	Br^-	20.1
Cs^+	20.7	HS^-	17.3
		HCO_3^-	11.8
Mg^{2+}	7.05		
Ca^{2+}	7.93	CO_3^{2-}	9.55
Sr^{2+}	7.94	SO_4^{2-}	10.7
Ba^{2+}	8.48	PO_4^{3-}	6.12
Ra^{2+}	8.89		
Mn^{2+}	6.88		
Fe^{2+}	7.19		
Cr^{3+}	5.94		
Fe^{3+}	6.07		

Source: Li and Gregory, 1974.

larger than those in water. In sediments, these values are reduced by a factor of about 4 to 100. In unfractured crystalline, effective diffusion coefficients are several orders of magnitude smaller than free-water values.

In a porous medium, diffusion takes place in the liquid phase, containing granular solids. In practice, Eq. (19.3) is modified to reflect the presence of the porous medium.

TABLE 19.2 Diffusion coefficients for organic compounds in air and water

CAS No.	Compound	$D_{I,a}(cm^2/s)$	$D_{I,w}(cm^2/s)$
67-64-1	Acetone	1.24×10^{-1}	1.14×10^{-5}
309-00-2	Aldrin	1.32×10^{-2}	4.86×10^{-6}
120-12-7	Anthracene	3.24×10^{-2}	7.74×10^{-6}
71-43-2	Benzene	8.80×10^{-2}	9.80×10^{-6}
65-85-0	Benzoic acid	5.36×10^{-2}	7.97×10^{-6}
75-25-2	Bromoform	1.49×10^{-2}	1.03×10^{-5}
75-15-0	Carbon disulfide	1.04×10^{-1}	1.00×10^{-5}
56-23-5	Carbon tetrachloride	7.80×10^{-2}	8.80×10^{-6}
57-74-9	Chlordane	1.18×10^{-2}	4.37×10^{-6}
108-90-7	Chlorobenzene	7.30×10^{-2}	8.70×10^{-6}
67-66-3	Chloroform	1.04×10^{-1}	1.00×10^{-5}
95-57-8	2-Chlorophenol	5.01×10^{-2}	9.46×10^{-6}
72-54-8	DDD	1.69×10^{-2a}	4.76×10^{-6a}
72-55-9	DDE	1.44×10^{-2}	5.87×10^{-6}
50-29-3	DDT	1.37×10^{-2}	4.95×10^{-6}
95-50-1	1,2-Dichlorobenzene	6.90×10^{-2}	7.90×10^{-6}
106-46-7	1,4-Dichlorobenzene	6.90×10^{-2}	7.90×10^{-6}
75-34-3	1,1-Dichloroethane	7.42×10^{-2}	1.05×10^{-5}
107-06-2	1,2-Dichloroethane	1.04×10^{-1}	9.90×10^{-6}
75-35-4	1,1-Dichloroethylene	9.00×10^{-2}	1.04×10^{-5}
156-59-2	cis-1,2-Dichloroethylene	7.36×10^{-2}	1.13×10^{-5}
156-60-5	trans-1,2-Dichloroethylene	7.07×10^{-2}	1.19×10^{-5}
60-57-1	Dieldrin	1.25×10^{-2}	4.74×10^{-6}
115-29-7	Endosulfan	1.15×10^{-2}	4.55×10^{-6}
72-20-8	Endrin	1.25×10^{-2}	4.74×10^{-6}
100-41-4	Ethylbenzene	7.50×10^{-2}	7.80×10^{-6}
76-44-8	Heptachlor	1.12×10^{-2}	5.69×10^{-6}
58-89-9	γ-HCH (Lindane)	1.42×10^{-2}	7.34×10^{-6}
67-72-1	Hexachloroethane	2.50×10^{-3}	6.80×10^{-6}
74-83-9	Methyl bromide	7.28×10^{-2}	1.21×10^{-5}
75-09-2	Methylene chloride	1.01×10^{-1}	1.17×10^{-5}
91-20-3	Naphthalene	5.90×10^{-2}	7.50×10^{-6}
98-95-3	Nitrobenzene	7.60×10^{-2}	8.60×10^{-6}
87-86-5	Pentachlorophenol	5.60×10^{-2}	6.10×10^{-6}
108-95-2	Phenol	8.20×10^{-2}	9.10×10^{-6}
129-00-0	Pyrene	2.72×10^{-2a}	7.24×10^{-6a}
100-42-5	Styrene	7.10×10^{-2}	8.00×10^{-6}
79-34-5	1,1,2,2-Tetrachloroethane	7.10×10^{-2}	7.90×10^{-6}
127-18-4	Tetrachloroethylene	7.20×10^{-2}	8.20×10^{-6}
108-88-3	Toluene	8.70×10^{-2}	8.60×10^{-6}
120-82-1	1,2,4-Trichlorobenzene	3.00×10^{-2}	8.23×10^{-6}
71-55-6	1,1,1-Trichloroethane	7.80×10^{-2}	8.80×10^{-6}
79-00-5	1,1,2-Trichloroethane	7.80×10^{-2}	8.80×10^{-6}
79-01-6	Trichloroethylene	7.90×10^{-2}	9.10×10^{-6}
75-01-4	Vinyl chloride	1.06×10^{-1}	1.23×10^{-6}
108-38-3	m-Xylene	7.00×10^{-2}	7.80×10^{-6}
95-47-6	o-Xylene	8.70×10^{-2}	1.00×10^{-5}
106-42-3	p-Xylene	7.69×10^{-2}	8.44×10^{-6}

CAS = Chemical Abstracts Service.
[a]Estimated using correlations in WATER8 model.

Fick's law for diffusion in sediments may be rewritten as

$$J_{\text{dif}} = -n\,\tau D_m\frac{dC}{dx} \tag{19.4}$$

where x is the coordinate, n is the effective porosity, and τ is the tortuosity of the medium. The tortuosity can be estimated from an empirical equation developed by Millington and Quirk (1961)

$$\tau = \frac{n_w^{7/3}}{n^2} \tag{19.5}$$

where n_w is the water-filled porosity and n is the porosity of the material. In an unsaturated medium, n_w is the water saturation times the porosity ($S_w n$). For vapor-phase diffusion processes, the tortuosity in a medium is defined as

$$\tau = \frac{n_a^{7/3}}{n^2} \tag{19.6}$$

where n_a is the air-filled porosity of the medium.

In the saturated zone, the tortuosity of the medium is estimated as

$$\tau = n^{1/3} \tag{19.7}$$

Unfortunately, the literature on diffusion is complicated by various definitions applied to porous media. For example, the effective diffusion coefficient in a porous medium is

$$D' = n\,\tau D_m = nD^* \tag{19.8}$$

where the D' is the effective diffusion coefficient, and $D^*(=\tau D_m)$ is the bulk diffusion coefficient. The effective diffusion coefficient in porous media increases with increasing porosity and decreases as the pathways for diffusion around the grains become more tortuous. For water-filled sediments, Fick's law in one dimension is written as

$$J_{dif} = -D'\frac{dC}{dx} = -nD^*\frac{dC}{dx} \tag{19.9}$$

Various experimental techniques are available to measure the effective diffusion coefficients in sediments. These typically involve separating solutions of differing chemistry by a wafer of porous media and observing the rates of diffusive mass transfer through the medium.

► **EXAMPLE 19.3**   In the unsaturated zone at a contaminated site, the molecular diffusion coefficients of the contaminant in air and water are 1.0×10^{-6} and 1.0×10^{-10} m²/s, respectively. The water saturation is 30% in the medium, with a porosity of 0.40. Calculate the tortuosity and the bulk diffusion coefficients for both the vapor and aqueous phases. If the contaminant concentration gradients in the vapor and the aqueous phases are 0.1 mg/L/m and 0.0001 mg/L/m, respectively, calculate the diffusive fluxes.

In the saturated zone, the porosity of the medium is also 0.4. Calculate the tortuosity and the bulk diffusion coefficient of the contaminant in the saturated zone. If the concentration gradient of the contaminant is 0.0002 mg/L/m in the saturated zone, calculate the diffusive flux.

SOLUTION   The tortuosity of the medium for vapor-phase diffusion processes in the unsaturated zone is

$$\tau = \frac{[(1 - 0.3)(0.4)]^{7/3}}{0.40^2} = 0.32$$

The tortuosity of the medium for aqueous-phase diffusion processes in the unsaturated zone is

$$\tau = \frac{[(0.3)(0.4)]^{7/3}}{0.40^2} = 0.044$$

The bulk diffusion coefficient of the medium for vapor-phase diffusion processes in the unsaturated zone is

$$D^* = \tau D_m = (0.32)(1.0 \times 10^{-6} \, \text{m}^2/\text{s}) = 3.2 \times 10^{-7} \, \text{m}^2/\text{s}$$

The bulk diffusion coefficient of the medium for aqueous-phase diffusion processes in the unsaturated zone is

$$D^* = \tau D_m = (0.044)(1.0 \times 10^{-10} \text{m}^2/\text{s}) = 4.4 \times 10^{-12} \text{m}^2/\text{s}$$

The contaminant vapor-phase mass flux in the control volume of the unsaturated zone is

$$J_{\text{dif}} = -n_a D \times \frac{dC}{dx} = (0.4 \times 0.7)(3.21 \times 10^{-7} \text{m}^2/\text{s})(0.1 \, \text{mg/L/m}) = 9.0 \times 10^{-9} \text{g/m}^2/\text{s}$$

The contaminant aqueous-phase mass flux in the control volume of the unsaturated zone is

$$J_{\text{dif}} = -n_w D \times \frac{dC}{dx} = -(0.4 \times 0.3)(4.4 \times 10^{-12} \text{m}^2/\text{s})(0.0001 \, \text{mg/L/m})$$
$$= 5.3 \times 10^{-17} \text{g/m}^2/\text{s}$$

The tortuosity of the medium in the saturated zone is

$$\tau = n^{1/3} = 0.40^{1/3} = 0.74$$

The bulk diffusion coefficient of the medium in the saturated zone is

$$D^* = \tau D_m = (0.74)(1.0 \times 10^{-10} \, \text{m}^2/\text{s}) = 7.37 \times 10^{-11} \, \text{m}^2/\text{s}$$

The contaminant mass flux in the control volume of the saturated zone is

$$J_{\text{dif}} = -n D^* \frac{dC}{dx} = (0.4)(7.4 \times 10^{-11} \, \text{m}^2/\text{s})(0.0002 \, \text{mg/L/m})$$
$$= 5.9 \times 10^{-15} \, \text{g/m}^2/\text{s} = 5.1 \times 10^{-10} \, \text{g/m}^2/\text{day}$$

▶ 19.3 DISPERSION

Dispersion is a process of mixing that causes a zone of mixing to develop between a fluid of one composition that is adjacent to or being displaced by a fluid with a different composition. Thinking again about flow tubes in a steady-state ground-water system is a good place to begin examining the concept of dispersion. Dispersion spreads mass beyond the region it normally would occupy due to advection alone. This idea can be illustrated with a simple column apparatus (Figure 19.5a). The apparatus is similar to a Darcy column, with additional plumbing to permit the controlled and continuous addition of a tracer. Initially, a steady flow of water is set up through the column. The test begins with a dissolved tracer at a relative concentration $C/C_0 = 1$ added across the entire cross section of the column on a continuous basis. Monitoring the outflow from the column establishes the relative concentration of the tracer as a function of time.

The relative concentration of tracer varies in water going into and coming out of the column (Figure 19.5a). Such relative concentration versus time plots is known as the *source-loading curve* and the *breakthrough curve*, respectively. The breakthrough curve does not

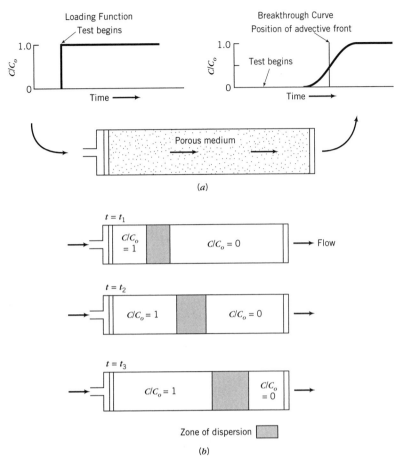

Figure 19.5 Experimental apparatus to illustrate dispersion in a column: (*a*) The test begins with a continuous input of tracer $C/C_0 =$ at the inflow end. The relative concentration versus time function at the outflow characterizes dispersion in the column; (*b*) Schematic representation of dispersion within the porous medium at three different times. A progressively larger zone of mixing forms between the two fluids ($C/C_0 = 1$ and $C/C_0 = 0$) displacing one another (from Domenico and Schwartz, 1998. *Physical and chemical hydrogeology*). Copyright ©1998 by John Wiley & Sons, Inc. Reprinted by permission of John Wiley & Sons, Inc.

have the same "step" shape as the source loading function. Dispersion creates a zone of mixing between the displacing fluid and the fluid being displaced. Some of the dissolved tracer leaves the column in advance of the advective front, which is defined as the product of the seepage velocity and time since displacement first started. The position of the advective front at breakthrough corresponds to a C/C_0 value of 0.5 (Figure 19.4*a*). In the absence of dispersion or other processes, the shapes of the loading function and the breakthrough curve are identical.

A zone of mixing gradually develops around the advective front (Figure 19.5*b*). Dispersion moves some tracer from behind to in front of the advective front. The size of the zone of mixing increases as the advective front moves further from the source. This column experiment is an example of one-dimensional transport involving advection and dispersion. Similar mixing will occur in both two and three dimensions. Again, dispersion spreads some of the mass beyond the region it would occupy due to advection alone. There is spreading both ahead of the advective front in the same flow tube and laterally into adjacent flow tubes. This dispersion is referred to as longitudinal and transverse dispersion, respectively. Dispersion in three dimensions involves spreading in two transverse directions as well as longitudinally.

Dispersion is recognizable in real ground-water systems when the mixing water bodies exhibit a difference in chemical composition. For example, in a problem of ground-water contamination, the contaminant(s) serves as a tracer to mark the body of contaminating fluid (Figure 19.6). Dispersion will also be manifest in a regional flow-system context when some change occurs in the chemistry of the recharging ground water. For example, nuclear weapons testing in the 1950s resulted in an abrupt jump in ^{36}Cl and tritium levels in recharge all around the world. Although dispersion occurs in both contaminated and natural ground-water systems, it is studied mostly with contamination problems.

Dispersion occurs in ground water because of two processes—diffusion and mechanical dispersion. *Diffusion* is a process of mass transport in response to a concentration gradient. Many examples of this process occur in nature. Suppose we brought a skunk into a classroom and the skunk hated hydrogeology. Assuming no air circulation, those sitting in the class would know the skunk was there because the odor would diffuse from a zone of high concentration around the skunk to zones of lower concentration in the vicinity of where you were sitting. The same transport would occur if, within a porous medium, waters having a high concentration were placed identically next to waters of low concentration. Solute molecules would move from the body of water with a high concentration into the body of low concentration. Thus, diffusion is a transport process because the dissolved mass occupies a larger volume than before.

Mechanical dispersion is a process of mixing that occurs because of local differences around some mean velocity of flow. We will illustrate this concept with a "thought" experiment. Think about what will happen when a large cluster of "rubber duckies" is released in the river shown in Figure 19.7. Will the duckies stay in a cluster or will they spread out? Obviously, the duckies do in fact spread out. Even though the river has some mean velocity, there is variability in the local velocities about the mean. Thus, some duckies will be fortunate and will generally find the high-velocity pathways. Others will fall behind as they encounter stretches of slower moving water. These same patterns of variability develop within a porous medium as a result of various nonidealities that will become clear shortly.

Mechanical dispersion is mixing that is caused by local variations in velocity around some mean velocity of flow. Thus, mechanical dispersion is an advective process and not a chemical one. With time, mass occupying some volume becomes gradually more dispersed as different fractions of mass are transported in these varying velocity regimes. Variability in the direction and rate of transport is caused by nonidealities in the porous medium. The most important variable in this respect is hydraulic conductivity. Table 19.3 lists some of

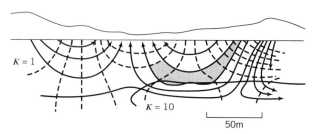

Figure 19.6 Transport of a tracer in a ground-water flow system (from Domenico and Schwartz, 1998. *Physical and chemical hydrogeology*). Copyright ©1998 by John Wiley & Sons, Inc. Reprinted by permission of John Wiley & Sons, Inc.

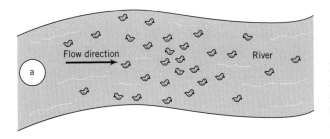

Figure 19.7 "Rubber duckies" released in a river from the circle at point "a" will end up highly dispersed due to local variability in the flow velocity.

TABLE 19.3 Geological features contributing to nonidealities in a porous medium

A. Microscopic heterogeneity: pore to pore
  1. Pore size distribution
  2. Pore geometry
  3. Dead-end pore space
B. Macroscopic heterogeneity: well to well or intraformational
  1. Stratification characteristics
    a. Nonuniform stratification
    b. Stratification contrasts
    c. Stratification continuity
    d. Insulation to cross-flow
  2. Permeability characteristics
    a. Nonuniform permeability
    b. Permeability trends
    c. Directional permeability
C. Megascopic heterogeneity: formational (either fieldwide or regional)
  1. Reservoir geometry
    a. Overall structural framework: faults, dipping strata, etc.
    b. Overall stratigraphic framework: bar, blanket, channel fill, etc.
  2. Hyperpermeability-oriented natural fracture systems

Source: Alpay, 1972. Copyright ©1972, Society of Petroleum Engineers Inc., JPT (July 1972).

the geological features that produce nonidealities. Nonidealities exist at a variety of scales ranging from microscopic to megascopic to an even larger scale (not included in the table) involving groups of formations.

Nonidealities at all of these scales are responsible for mechanical dispersion. For example, variability in velocity at the microscopic scale develops because of differing flow regimes across individual pore throats or variability in the tortuosity of the flow channels (Figure 19.8a). The heterogeneities shown in Figure 19.8b produce dispersion by creating variability in flow at a macroscopic scale. This example, adapted from Sudicky (1986), represents the plotted results of a large number of hydraulic conductivity measurements for the shallow unconfined aquifer at Canadian Forces Base Borden.

The fluid flow velocity and grain size are the main controls on mechanical dispersion in a column. The results of a large number of column experiments determined that longitudinal mechanical dispersion is approximately proportional to velocity. Two- and three-dimensional experiments in the laboratory also showed that mass spreading could occur transverse to the direction of mean ground-water flow. In other words, if the flow is in the x-direction, mass could spread in the y- and z-directions as well. One component of transverse spreading occurs in the horizontal plane, and the second occurs in the vertical

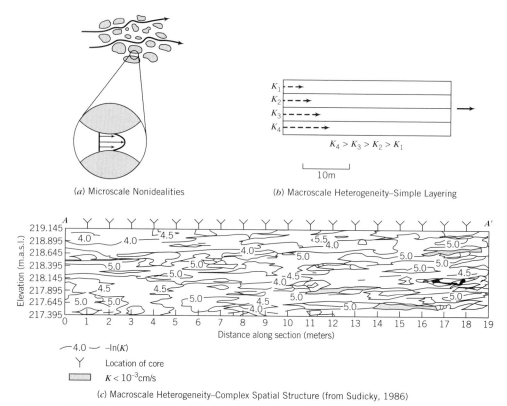

(a) Microscale Nonidealities

(b) Macroscale Heterogeneity–Simple Layering

—4.0— −ln(K)

Y   Location of core

▭   K < 10⁻³cm/s

(c) Macroscale Heterogeneity–Complex Spatial Structure (from Sudicky, 1986)

Figure 19.8 Examples of nonidealities at different scales giving rise to mechanical dispersion. Panel (c) (from Sudicky, *Water Resources Res.*, v. 22, p. 2069–2082. 1986). Copyright by American Geophysical Union.

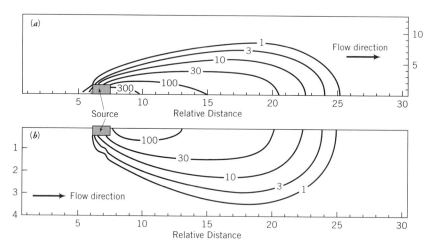

Figure 19.9 Illustrations of transverse dispersions: (a) horizontal transverse dispersion; (b) vertical transverse dispersion.

plane. This point is illustrated with the plume shown in Figure 19.9. Along the horizontal plane of the plume, notice how the concentrations decrease away from the midline of the plume toward the edges. The slice across the nose of the plume shows this same pattern in the downward or z-direction.

The Mathematical Theory of Dispersion

The mechanical dispersive and diffusive rates can be combined as a total dispersive flux

$$J_{dis} = J_{mdis} + J_{dif} = -n(D_{disp} + \tau D_m)\nabla C = -nD\nabla C \tag{19.10}$$

where J_{dis} is the dispersive mass flux and J_{mdis} is the mechanical dispersive mass flux. The dispersion coefficient tensor D is a function of velocity vector, diffusion coefficient, and dispersivities (Burnett and Frind, 1987).

$$D = \begin{pmatrix} D_x & D_{xy} & D_{xz} \\ D_{yx} & D_y & D_{yz} \\ D_{zx} & D_{zy} & D_z \end{pmatrix} \tag{19.11}$$

where

$$D_x = \alpha_L \frac{v_x^2}{|v|} + \alpha_{TH} \frac{v_y^2}{|v|} + \alpha_{TV} \frac{v_z^2}{|v|} + \tau D_m \tag{19.12}$$

$$D_y = \alpha_L \frac{v_y^2}{|v|} + \alpha_{TH} \frac{v_x^2}{|v|} + \alpha_{TV} \frac{v_z^2}{|v|} + \tau D_m \tag{19.13}$$

$$D_z = \alpha_L \frac{v_z^2}{|v|} + \alpha_{TV} \frac{v_y^2}{|v|} + \alpha_{TV} \frac{v_x^2}{|v|} + \tau D_m \tag{19.14}$$

$$D_{xy} = D_{yx} = (\alpha_L - \alpha_{TH})\frac{v_x v_y}{|v|} \tag{19.15}$$

$$D_{xz} = D_{zx} = (\alpha_L - \alpha_{TV})\frac{v_x v_z}{|v|} \tag{19.16}$$

$$D_{yz} = D_{zy} = (\alpha_L - \alpha_{TV})\frac{v_y v_z}{|v|} \tag{19.17}$$

$$|v| = \sqrt{v_x^2 + v_y^2 + v_z^2} \tag{19.18}$$

where α_L is the longitudinal dispersivity, α_{TH} is the horizontal transverse dispersivity, and α_{TV} is the vertical transverse dispersivity.

Dispersivities have units of length, and like hydraulic conductivity they are characteristic properties of a medium. Clearly, the dispersion coefficients will be zero if the ground-water flow velocity and dispersivities are zero. The longitudinal dispersivity represents the effect of heterogeneity on the spreading of dissolved mass in the horizontal flow direction, the horizontal transverse dispersivity quantifies the spreading in the horizontal direction perpendicular to the horizontal flow, and the vertical dispersivity is the effect of heterogeneity on the spreading in the vertical direction. The magnitude of the spreading is also proportional to the mean ground-water flow vector. In some literature, the horizontal (α_{TH}) and vertical (α_{TV}) dispersivities are assumed to be the same (Bear, 1972). The mass transport rate representing advection and dispersion is written as mass flux

$$J = J_x i + J_y j + J_z k = -nD\nabla C + nvC \tag{19.19}$$

where

$$J_x = nv_xC - nD_x\frac{\partial C}{\partial x} - nD_{xy}\frac{\partial C}{\partial y} - nD_{xz}\frac{\partial C}{\partial z} \tag{19.20}$$

$$J_y = nV_yC - nD_{yx}\frac{\partial C}{\partial x} - nD_y\frac{\partial C}{\partial y} - nD_{yz}\frac{\partial C}{\partial z} \tag{19.21}$$

$$J_z = nV_zC - nD_{zx}\frac{\partial C}{\partial x} - nD_{zy}\frac{\partial C}{\partial y} - nD_z\frac{\partial C}{\partial z} \tag{19.22}$$

where J_x, J_y, and J_z are mass fluxes along the x-, y-, and z-directions, respectively.

For a unidirectional flow where $v_x = v_L, v_y = 0$, and $v_z = 0$, the components of dispersion coefficient are simplified as shown in Scheidegger, 1961:

$$D_x = D_L = \alpha_L v_x + \tau D_m \tag{19.23}$$

$$D_y = D_{TH} = \alpha_{TH} v_x + \tau D_m \tag{19.24}$$

$$D_z = D_{TV} = \alpha_{TV} v_x + \tau D_m \tag{19.25}$$

$$D_{xy} = D_{yx} = D_{xz} = D_{zx} = D_{yz} = D_{zy} = 0 \tag{19.26}$$

$$J_x = nv_xC - nD_x\frac{\partial C}{\partial x} \tag{19.27}$$

$$J_y = -nD_y\frac{\partial C}{\partial y} \tag{19.28}$$

$$J_z = -nD_z\frac{\partial C}{\partial z} \tag{19.29}$$

▶ **EXAMPLE 19.4**   The advective flux was calculated in Example 19.2, and the diffusive flux was calculated in Example 19.3. Calculate (1) the dispersive mass flux for a geological medium with a longitudinal dispersivity of 1.0, a horizontal transverse dispersivity of 0.1, and a vertical transverse dispersivity of 0.01; (2) the total mass flux in the control volume of the saturated zone. We assume that concentration gradients across the volume in x-, y-, and z- directions are 0.0002 mg/L/m.

SOLUTION   The longitudinal dispersion coefficient is calculated by Eq. (19.23):

$$D_x = \alpha_L v_x = (1\,\text{m})\frac{(0.02\,\text{m/day})}{0.4} = 0.05\,\text{m}^2/\text{day}$$

The horizontal transverse dispersion coefficient is calculated by Eq. (19.24):

$$D_y = \alpha_{TH} v_x = (0.1\,\text{m})\frac{(0.02\,\text{m/day})}{0.4} = 0.005\,\text{m}^2/\text{day}$$

The vertical transverse dispersion coefficient is calculated by Eq. (19.25):

$$D_z = \alpha_{TV} v_x = (0.01\,\text{m})\frac{(0.02\,\text{m/day})}{0.4} = 0.0005\,\text{m}^2/\text{day}$$

The longitudinal dispersive flux is

$$J^x_{mdis} = -nD_x\frac{\partial C}{\partial x} = (0.4)(0.05\,\text{m}^2/\text{day})(0.0002\,\text{mg/L/m}) = 4\times 10^{-6}\,\text{g/m}^2/\text{day}$$

The horizontal dispersive flux is

$$J^y_{mdis} = -nD_y\frac{\partial C}{\partial y} = (0.4)(0.005\,\text{m}^2/\text{day})(0.0002\,\text{mg/L/m}) = 4\times 10^{-7}\,\text{g/m}^2/\text{day}$$

The vertical transverse dispersive flux is

$$J^z_{mdis} = -nD_z\frac{\partial C}{\partial z} = (0.4)(0.0005\,\text{m}^2/\text{day})(0.0002\,\text{mg/L/m}) = 4\times 10^{-8}\,\text{g/m}^2/\text{day}$$

The total longitudinal mass flux is

$$J_x = J^x_{adv}+J^x_{mdis}+J^x_{dif} = (2\times 10^{-3}+4\times 10^{-6}+5.1\times 10^{-10})(\text{g/m}^2/\text{day}) = 2.004\times 10^{-3}\,\text{g/m}^2/\text{day}$$

The total horizontal dispersive mass flux is

$$J_y = J^y_{adv}+J^y_{mdis}+J^y_{dif} = (0+4\times 10^{-7}+5.1\times 10^{-10})(\text{g/m}^2/\text{day}) = 4.005\times 10^{-7}\,\text{g/m}^2/\text{day}$$

The total vertical transverse mass flux is

$$J_z = J^z_{adv}+J^z_{mdis}+J^z_{dif} = (0+4\times 10^{-8}+5.1\times 10^{-10})(\text{g/m}^2/\text{day}) = 4.05\times 10^{-8}\,\text{g/m}^2/\text{day}$$

For most mass transport problems that involve the migration of dissolved contaminants in aquifers, the effects of mechanical dispersion are much greater than the diffusive component. Thus, the overall dispersion is described in terms of aquifer dispersivity or α values. Only in units with low-hydraulic conductivity and small advective velocity will diffusion become important.

► 19.4 DISPERSION COEFFICIENTS AT MICROSCOPIC AND MACROSCOPIC SCALES

Dispersion coefficients are also a function of scale. In small, microscopic-scale systems, like a laboratory column, mixing at the pore scale produces mechanical dispersion. A typical range of longitudinal dispersivities is from 10^{-2} to 1 cm. Moving up in scale, dispersivity values become much larger.

Gelhar et al. (1992) undertook a critical review of field experiments at 59 sites around the world. Tests yielded some 106 longitudinal dispersivity values ranging from 0.01 m to 5500 m at scales of 0.75 m to 100 km. It appears that longitudinal dispersivities increase indefinitely with scale (Figure 19.10). However, not all of the values are reliable (Figure 19.10). Looking at the most reliable results, however, we find that the trend of increasing longitudinal dispersivity as a function of distance was more apparent than real (Gelhar et al., 1992). The most reliable dispersivity values are all at the low end of the range.

From such tests, a consistent view about macroscopic dispersion has emerged. Heterogeneity at the macroscopic scale contributes significantly to dispersion because it creates local-scale variability in velocity. Values of macroscopic dispersivity are in general two or more orders of magnitude larger than those from column experiments. In field experiments, values range from approximately 0.1 to 2 m over relatively short transport distances. At any given scale, longitudinal dispersivities can range over two to three orders of magnitude depending on the variability in hydraulic conductivity. Although no reliable, large-scale studies have been carried out, longitudinal dispersivity values in excess of 10 m probably

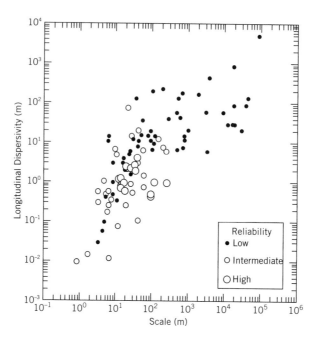

Figure 19.10 Longitudinal dispersion versus scale data classified by reliability (from Gelhar et al., *Water Resources Res.*, v. 28, p. 1955–1974, 1992). Copyright © by American Geophysical Union.

exist. Values, however, are not likely as large as estimates determined using contaminant plumes and environmental isotopes, which are less reliable (Gelhar et al., 1992).

Experiments in the field commonly show that dispersivity values increase as a tracer moves away from a source. Eventually, however, the dispersivity values become constant. Gelhar et al. (1979) refer to this constant macroscale dispersivity as the *asymptotic dispersivity*. For aquifers, a tracer may have to spread tens or hundreds of meters before the asymptotic dispersivity is obtained. This behavior is observed because of aquifer heterogeneity, which is the main reason for dispersivity at the field scale. When a tracer spreads a few meters, it encounters little of the heterogeneity present in the aquifer. Thus, at small displacements from the source, there is relatively limited dispersion—in most cases resembling dispersion in a column. When a tracer spreads farther away, it will encounter the large-scale heterogeneity (see Figure 19.8c), which is much more important in causing dispersion. In effect, there is a transition zone between the microscale and macroscsale heterogeneity in the medium.

▶ 19.5 STATISTICAL PATTERNS OF MASS SPREADING

Both laboratory and field experiments show that mass spreading tends to have a Gaussian distribution. Figure 19.11 shows this feature for both one-dimensional and two-dimensional transport. The position of the mean of the concentrations distribution represents transport at the linear ground-water velocity. The variance of the distribution (σ_L^2) is proportional to the dispersion in the system. The two-dimensional spread of a tracer in a unidirectional flow field results in an elliptically shaped concentration distribution (Figure 19.11b) that is normally distributed in both the longitudinal and transverse directions. Typically, longitudinal dispersion is greater than transverse dispersion (Figure 19.11b). Concentration distributions in three dimensions form ellipsoids of revolution (football shapes) when the two components of transverse spreading are the same and longitudinal dispersion is larger.

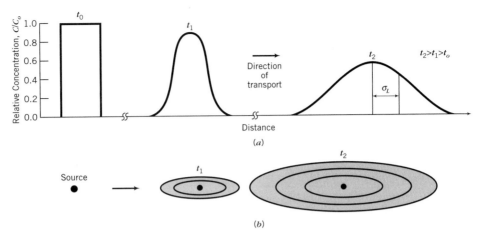

Figure 19.11 Variations in concentration of a tracer spreading in (a) one or (b) two-dimensional constant-velocity flow system (from Domenico and Schwartz, 1998. *Physical and chemical hydrogeology*). Copyright ©1998 by John Wiley & Sons, Inc. Reprinted by permission of John Wiley & Sons, Inc.

TABLE 19.4   Properties of the normal distribution

$\dfrac{x}{\sigma}$	Values from the Normal Distribution	Percentage of the Mass Contained (Calculated as Two Times the Value of the Normal Distribution)
0	0	0
0.1	0.0398	8
0.2	0.0793	15.8
0.3	0.1197	24
0.4	0.1554	31
0.5	0.1915	38
0.6	0.2257	45
0.7	0.2580	52
0.8	0.2881	57.6
0.9	0.3159	63.2
1.0	0.3413	68.3
1.2	0.3849	77
1.4	0.4192	83.8
1.6	0.4452	89
1.8	0.4641	92.8
2.0	0.4773	95.4
2.5	0.4938	98.7
3.0	0.4987	99.74
4.0	0.49996	99.99
infinity	0.5	100.00

When the vertical transverse dispersivity is small, as is often the case, plumes take on a surfboard shape.

Before proceeding, let us examine some features of a normal distribution. The standard deviation σ and the variance σ^2 are measures of the spread in the data about the mean. Thus, from Table 19.4, a spread of 4σ ($x/\sigma = 2$) incorporates about 99.9% of the mass or area under the concentration distribution. One standard deviation contains about 68.3% of the total mass. The standard deviation of a concentration distribution is related to the dispersion coefficient as

$$\sigma = (2Dt)^2 \qquad (19.30)$$

where t is time and D is the dispersion coefficient. Thus, for a one-dimensional column,

$$D_L = \sigma_L^2/2t \qquad (19.31)$$

where D_L is the longitudinal dispersion coefficient.

When mass spreads in two or three dimensions, the distributions of mass sampled normal to the direction of flow are also normally distributed, with variances that increase in proportion to $2t$. Thus, coefficients of transverse dispersion are defined as

$$D_T = \sigma_T^2/2t \qquad (19.32)$$

► **EXAMPLE 19.5**
Analysis of the Borden Data

A field experiment at the Canadian Forces Base Borden illustrates how this theory can be applied in the analysis of concentration data (Freyberg, 1986; Mackay et al., 1986). In this experiment, water containing Cl^- at 892 mg/L was injected into a shallow, unconfined sand aquifer. What started out as a roughly lenticular volume of tracer, approximately 6 m in diameter and 1 m thick, was transported down the flow system. A dense network of monitoring wells was used to track this tracer for almost three years and 60 m.

As expected, dispersion worked to spread this lens. Figure 19.12 shows this original Cl^- plume after transport for 462 days and a distance of 42 m from the source. It is now elliptical in shape, some 30 m long and 10 m wide. The maximum Cl^- concentration is about 65 mg/L. Dispersion has been especially effective in elongating the plume in the longitudinal direction.

Given the standard deviation of the concentration, calculate the variance and use Eq. (19.13) to calculate the dispersion coefficient and eventually the dispersivity.

SOLUTION A simple graphical way to estimate the standard deviation in the longitudinal concentration distribution (σ_L) is to find the half-width of the distribution in the longitudinal direction (Γ_L), as shown in Figure 19.13.

Let's apply this simple approach to the Cl^- distribution in Figure 19.12. The maximum Cl^- concentration is about 65 mg/L. Thus, Γ_L would be the distance along the curving longitudinal axis

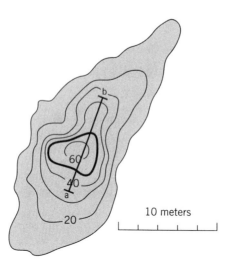

10 meters

Figure 19.12 Map view of the Cl^- ion distribution in the tracer test at Borden at 462 days. (modified from Mackay et al., *Water Resour. Res.*, v. 22, p. 2017–2029, 1986. Copyright by American Geophysical Union).

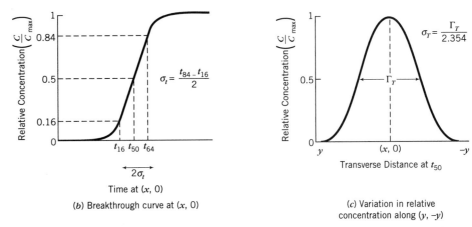

(a) Spreading of a tracer from a point source

(b) Breakthrough curve at $(x, 0)$

(c) Variation in relative concentration along $(y, -y)$

Figure 19.13 Graphical procedure for estimating the standard deviation of a normal distribution (from Robbins, 1983).

from concentration 32.5 mg/L to 32.5 mg/L, or from point a to point b on Figure 19.12. Γ_L is about 10 m, which provides

$$\sigma_L = \frac{10}{2.354} = 4.25 \, \text{m}$$

The dispersion coefficient (D_L) is given as

$$D_L = \frac{\sigma^2}{2t} = \frac{4.25 \times 4.25}{2 \times 462} = 0.020 \, \text{m}^2/\text{d}$$

We also know that $D_L = \alpha_L \, v$, with v given as the distance of travel (x) over the travel time (t). Given that $d = 42$ m and $t = 462$ days, the longitudinal dispersivity is calculated as

$$\alpha_L = \frac{D_L}{v} = \frac{D_L \times t}{x} = \frac{0.020 \times 462}{42} = 0.22 \, \text{m}$$

The dispersivity calculated here is approximately the same as that calculated by Freyberg et al. (1986). It is a relatively small value because the Borden sand aquifer is not very heterogeneous.

► 19.6 A GEOSTATISTICAL MODEL OF DISPERSION

Exciting new contributions to the study of dispersion are quantitative approaches that relate asymptotic dispersivities to geostatistical models of hydraulic conductivity. Eventually, these approaches will form the basis for a family of practical field techniques for characterizing dispersivity. It is beyond the scope of this text to treat the theory in detail.

The heterogeneous character of aquifers can be characterized in the simplest case by three statistical parameters: the mean hydraulic conductivity, the variance in hydraulic conductivity, and the correlation length. Assume that we have hydraulic-conductivity measurements from a large number of boreholes along a two-dimensional section (Figure 19.14). For the entire set of measurements, we could find the mean and variance using simple statistics. Note that because hydraulic-conductivity distributions, are normal as the log or ln of the measured value, the statistics are done on the log-transformed data.

The *correlation length* is a measure of the spatial persistence in some property of the medium. For example, if we measured the hydraulic conductivity at two points close to one another, we would expect the values to be very similar to one another. If we moved one of the measurement points farther away, the values might be somewhat similar. Finally, if one of the points was moved far away, there would be no similarity in the values. In effect, as the separation distance increases, the correlation between the values decays. The correlation length scale measures the distance at which the correlation decays to some prescribed value.

The correlation lengths can be different in each of the three coordinate directions. In most cases, the vertical correlation length scale is much smaller than the horizontal length scale. For example, this was the case with the hydraulic-conductivity field depicted in Figure 19.14. Several authors have enough hydraulic-conductivity measurements at sites to estimate the geostatistical parameters for a site.

The beauty of this approach to geostatistical characterization of the medium is that it provides an indirect way to calculate asymptotic dispersivity value related to the heterogeneity of the medium. Gelhar and Axness (1983) provide the following theoretical equation for this purpose:

$$A_L = \sigma_Y^2 \lambda / \gamma^2 \tag{19.33}$$

where A_L is the asymptotic longitudinal dispersivity, σ_Y^2 is the variance of the log-transformed hydraulic conductivity [*i.e.*, $Y = \ln K$], λ is the correlation length in the mean direction of flow, and γ is a flow factor that Dagan (1982) considers equal to 1.

An example in the estimation of longitudinal asymptotic dispersivity is provided with data collected by Sudicky (1986) and shown in Figure 19.14. This section is oriented parallel to the direction of tracer migration at Canadian Forces Base Borden (Freyberg,

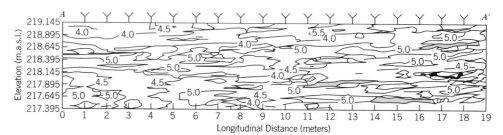

Figure 19.14 Variations in hydraulic conductivity (as − ln K) along the cross section A–A' at Canadian Forces Base Borden (modified from Sudicky, *Water Resources Res.*, v. 22. p. 2069–2082, 1986). Copyright © by American Geophysical Union.

1986; Mackay et al., 1986). The variance in hydraulic conductivity (σ_Y^2) has a value of 0.29, and horizontal and vertical correlation length scales (α_L and α_T) of 2.8 and 0.12 m, respectively (Sudicky, 1986). These values confirm what is apparent from looking at the contoured field, namely, that the correlation structure is anisotropic.

The contribution of heterogeneity to the asymptotic longitudinal dispersivity can be calculated from Eq. (19.15) as

$$A_L = \sigma_L^2 \lambda_L / \gamma^2 = 0.29 \times 2.8/1^2 = 0.81 \text{ m}$$

where the flow factor γ is taken to be 1. With the contribution from diffusion essentially negligible and the local-scale dispersivity (α_L) estimated to be 0.05 m, most of the longitudinal dispersion arises because of heterogeneity in the hydraulic-conductivity field.

An attractive feature of this method is that it does not require that dissolved mass pass through the test volume. Thus, it has the potential to be applied to large-scale systems—systems that are too big to be tested using more conventional methods. One limitation is the relatively large number of hydraulic-conductivity values that are required to define the spatial structure. The only practical way to address this limitation is through indirect determination of hydraulic conductivity using various kinds of borehole logs or a borehole flowmeter.

▶ 19.7 MIXING IN FRACTURED MEDIA

Many of the concepts of mixing discussed in the previous sections apply to fractured media. Nevertheless, there is sufficient added complexity to warrant treating these media separately. The present discussion focuses on the only type of fractured medium that has been studied in detail. This is a case where fractures are the only permeable pathways through the rock. The hydraulic conductivity of the unfractured part of the medium is so small that fluid flow, and hence advective transport, are negligible. However, mass can diffuse into the unfractured rock matrix in response to concentration gradients.

Mixing in a fracture system is the result of both mechanical mixing and diffusion. Concentrations in the fractures will be affected by diffusion into the matrix, a chemical process (Figure 19.15a); variability within individual fractures caused by asperities, an advective dispersive process (Figure 19.15b); fluid mixing at the fracture intersections, a diluting or possibly diffusive process (Figure 19.15c); and variability in velocity caused by differing scales of fracturing (Figure 19.15d) or by variations in fracture density (Figure 19.15e). Some of these processes are coupled in a complex fashion and compete.

Diffusion into the matrix provides an important process to attenuate the transport of contaminants. This result contradicts experience with porous media, where the contribution of diffusion to dispersion is generally swamped by mechanical dispersion. Diffusion is more important in fractured media because localizing mass in fractures provides the opportunity for large-concentration gradients to develop. Theoretical studies by Grisak and Pickens (1980) and Tang et al. (1981) explored this process using numerical and analytical models of a single fracture bounded by an infinite porous matrix. Dispersion in a single fracture is caused by the variability in fracture aperture. This variability develops because of the roughness of the fracture walls and the precipitation of secondary minerals. At many locations, the fracture may be closed to flow and transport. Such aperture topology (see Figure 19.15b) gives rise to *channeling* (Neretnieks, 1985), where mass moves predominantly along networks of irregularly shaped pathways in the plane of a fracture.

This channelization model has developed from tracer tests carried out in fractured rocks. For example, studies at the Stripa mine (Neretnieks, 1985) showed how the inflow

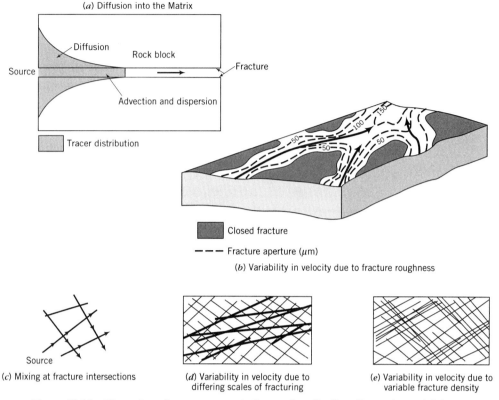

Figure 19.15 Illustration of mass transport in fractured media (from Domenico and Schwartz, 1998. *Physical and chemical hydrogeology*). Copyright ©1998 by John Wiley & Sons, Inc. Reprinted by permission of John Wiley & Sons, Inc.

of water to tunnels was extremely localized. Approximately one-third of the flow entered from approximately 2% of the fractured rocks. These results are strongly indicative of localized channel flow within individual fractures.

Channels can be so poorly interconnected that they may not interact with one another over appreciable distances (Neretnieks, 1985). This behavior is termed *pure channeling*. From this discussion, it is easy to understand why the smooth, parallel plate model used by some investigators is really a simplified representation of a fracture.

Dispersion at the next larger scale occurs when the geometry of the three-dimensional network begins to influence mass transport. A tracer moving along a fracture to an intersection (Figure 19.15c) partitions into two or more fractures. Because the water is also partitioned and because concentration is the mass per unit volume of solution, partitioning by itself does not affect the concentrations. However, concentrations will be affected if there is mixing with water that does not contain the tracer. This dilution depends directly on the type of mixing process in the intersection and on the quantity of water flowing along the fractures.

The dispersion in Figures 19.15d and e is analogous to that caused by heterogeneities in porous media. Differing scales of fracturing or spatial variability in fracture density create variability in local velocity. Even larger scale dispersion could develop if individual discontinuities exist on a regional scale or the variability in fracture density includes several units.

► 19.8 TRACERS AND TRACER TESTS

Tracers and tracer tests are useful in understanding dispersion and advection and in estimating various transport parameters (for example, ground-water velocity and dispersivity). The most important tracers include: (1) ions that occur naturally in a ground-water system such as Br^- or Cl^-, (2) environmental isotopes such as ^2H, ^3H, or ^{18}O, (3) contaminants of all kinds that enter a flow system, or (4) chemicals added to a flow system as part of an experiment. This latter group could include radioisotopes such as ^3H, ^{131}I, and ^{82}Br; ions such as Cl^-, Br^-, and I^-; and organic compounds such as rhodamine WT, fluorescein, and uranine. Many of these ions or compounds do not react to any appreciable extent with other ions in solution and the porous medium. They are considered *ideal tracers*. Reactive tracers are used more specifically to define the nature of reactions.

Given this variety of tracers and testing strategies, confusion may arise as to which to use in site studies. In most cases, the choice is linked inexorably to the scale of the study or the presence of a contaminant plume. For example, tracing flow and dispersion in a unit of regional extent (10's to 100's of km) will involve either naturally occurring ions or environmental isotopes. Conducting a tracer experiment on such a large scale is simply not feasible because of the long times required for tracers to spread regionally. On a more localized or site-specific basis (several kilometers), the presence of a contaminant plume that has spread over a long time automatically makes it the tracer of choice. Again, time is insufficient (except perhaps in karst) to run an experiment at the scale of interest. Only for small systems (for example, some fraction of a kilometer) is there a possibility of running a field tracer experiment. It is for these experiments that a selection must be made among the various tracers.

Tracers are transported in two kinds of flow environments. In a natural gradient system, tracers move as a result of the natural flow of ground water. In a system stressed by injection and/or pumping, transport occurs in response to hydraulic gradients that are typically much larger than those in natural systems. A small-scale tracer experiment can involve either of these flow conditions.

Success in these tests depends on adequately characterizing concentration distributions in space and time. Typically, a large three-dimensional network of monitoring points is necessary to define the plume accurately. Point sampling is essential to avoid concentration averaging within the well bore. For example, estimated values of dispersivity tend to be larger when the number of individual sampling points is small or samples are collected over relatively large vertical intervals. This excess dispersion represents error due to inadequacies in the sampling network.

The *natural gradient test* involves monitoring a small volume of tracer as it moves down the flow system. Keeping the quantity of tracer small minimizes the disturbance of flow conditions. The resulting concentration distributions provide the data necessary to determine advective velocities, dispersivities, and occasionally equilibrium and kinetic parameters. Such an experiment typically requires a dense network of sampling points. This chapter concludes with a brief description of some of the important tracer tests of this type run over the last decade.

The *single-well pulse test* (Figure 19.16a) involves injecting first a tracer and then water into an aquifer at a constant rate. After an appropriate period of injection, the aquifer is pumped at the same rate. The concentration of tracer in water being withdrawn is monitored with time or total volume of water pumped. These concentration-time data provide a basis for estimating the longitudinal dispersivity and chemical parameters like K_d within a few meters of the well (for example, Pickens et al., 1981).

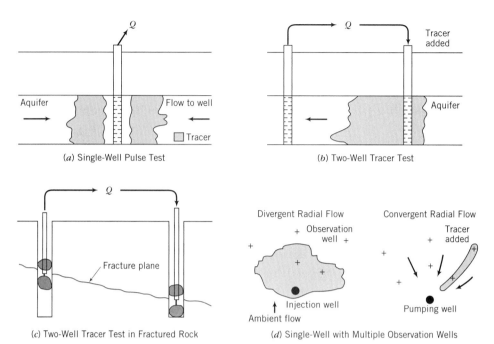

Figure 19.16 Several types of tracer tests (from Domenico and Schwartz, 1998. *Physical and chemical hydrogeology*). Copyright ©1998 by John Wiley & Sons, Inc. Reprinted by permission of John Wiley & Sons, Inc.

Though historically of some interest, this test has limited applicability in estimating values of dispersivity. It is generally not possible to scale up dispersivity measurements made at such a small scale to the larger scales of interest in most problems. This test holds more promise for evaluating geochemical processes.

In the *two-well test*, water is pumped from one well and injected into another at the same rate to create a steady-state flow regime (Figure 19.16*b*). The tracer is added continuously at a constant concentration at the injection well and monitored in the withdrawal well. The simplest way of running this test is by recirculating the pumped water back to the injection well, as is shown on the figure. However, the tracer concentration will begin to increase at the injection well once breakthrough occurs at the pumped well. The test can also be run without recirculation by providing the water for the injection well from an alternative source. The resulting concentration versus time data are interpreted in terms of processes and parameters. These tests can be conducted over several hundred meters in highly permeable systems. However, with only a single monitoring point, the test provides at best only a crude estimate of dispersivity. Estimates can be improved by adding more observation wells between the pumping/injection doublet. Interpretive techniques for two-well tracer tests are described by Grove and Beetem (1971), Grove (1971), Güven et al. (1986), and Huyakorn et al. (1986a).

The final type of tracer test that we illustrate is a *single-well injection or withdrawal with multiple observation wells*. This test involves creating a transient radial flow field by injection or withdrawal. The radially divergent test (Figure 19.16*d*) involves monitoring the tracer as it moves away from the well. The radially convergent test involves adding the tracer at one of the observation wells and monitoring as it moves toward the pumped well. Parameters can be estimated at scales of practical interest with reasonable accuracy, given a reasonable number of observation wells. Between these two tests, the divergent flow test is preferable (Gelhar et al., 1985). The converging flow field (Figure 19.16*d*) counteracts spreading due to dispersion and is thought to be less useful.

Estimates of dispersivities at scales greater than several hundred meters are based on mixing patterns caused by the presence of a dissolved contaminant plume or temporal variations in the chemistry of natural recharge. These latter changes can be caused by the atmospheric testing of nuclear weapons, which impacts tritium and chlorine-36 levels in rainfall, or by longer term changes in the climate, which impact oxygen-18 and deuterium (for example, Pleistocene to present). The detailed examination and interpretation of data from contaminant plumes and environmental isotopes will be reserved for several later chapters.

Estimates of advection and dispersion based on data from plumes or environmental measurements are less reliable than field tracer tests. There is uncertainty in defining the loading function for the constituent of interest, and often, there is an inadequate number of sampling points (Gelhar et al., 1992).

Examples of Tracer Tests

This section illustrates the usefulness of tracer tests by looking at three natural gradient tests that are important because of the important knowledge gained from them. The tests, run at Borden, Cape Cod, and Columbus, are unique because between 5000 and 10,000 monitoring points provided precise three-dimensional pictures of the evolving tracer plumes over travel times of approximately two to three years. These tests were run to learn more about the manifestations of physical and chemical transport processes at large scale in shallow unconfined aquifers, and to validate the geostatistical concepts of dispersion. The study of chemical processes involved the use of reactive tracers in addition to conservative tracers. The test at Borden used chloride and bromide as the conservative (that, is nonreactive) tracers and five organic compounds as reactive tracers (Table 19.5). The test at Cape Cod used bromide as the nonreactive tracer, and lithium (Li^+) and molybdate (MoO_4^{2-}) as reactive tracers.

These tests are important scientifically because investigators were able to show that the dispersion observed in the field during the tests was comparable to values predicted from estimates of the correlation lengths and the variance in ln K (Table 19.6). Thus, the relatively small longitudinal dispersivities observed at Borden and Cape Cod are due to the relative homogeneity of the deposits there (that is, small S_y^2). The Columbus site is much more heterogeneous in terms of the hydraulic conductivity (that is, large S_y^2). As a consequence, there is more significant dispersion than at the other two sites.

▶ EXERCISES

19.1. The hydrogeologic cross section on the figure illustrates the pattern of ground-water flow along a local flow system. Assume that the source of contamination develops with advection as the only

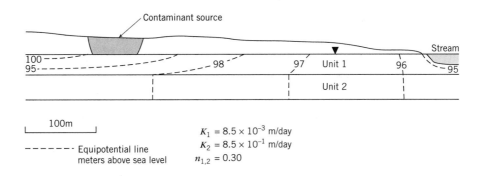

TABLE 19.5   Information on massively instrumented tracer tests

Site	Aquifer Material	Test Scale (m)	References
Canadian Forces Base Borden, Ontario	Glaciofluvial sand	90	Mackay et al., 1986 Freyberg, 1986 Sudicky, 1986
Cape Cod, Massachusetts	Sand and gravel	250	LeBlanc et al., 1991 Garabedian and LeBlanc, 1991 Hess et al., 1992
Columbus, Mississippi	Sandy gravel/gravelly sand	280	Boggs et al., 1992 Adams and Gelhar, 1992 Rehfeldt et al., 1992

TABLE 19.6   Geostatistical description of aquifers at the three sites and comparison of dispersivity values estimated from field tracer tests and stochastic theory

Site	Aquifer Properties			Longitudinal Dispersivity (m)	
	λ_b	λ_v	S_y^2	Theory[a]	Test
Borden	2.8	0.12	0.29	0.45–0.6	0.36
Cape Cod	5.1	0.26	0.26	0.35–0.78	0.96
Columbus	5.3	0.7	2.8	1.5	5–10

[a]These calculations involve a variety of stochastic approaches, and the values are not the same as those that would come from the simple equation we presented.

λ_b = correlation length in the horizontal direction
λ_v = correlation length in the vertical direction
S_y^2 = sample estimate of the variance in ln K

CASE STUDY 19-1

CAPE COD TRACER EXPERIMENT

The U.S. Geological Survey conducted a large-scale natural gradient tracer test in a stratified sand and gravel aquifer on Cape Cod, Massachusetts. Several reactive and nonreactive tracers were utilized, although we will focus on Br^-, which is assumed to be nonreactive. The term *nonreactive* means that the tracer does not sorb or otherwise interact with the medium and is transported with the same velocity as the ground water.

The test was carried out in the bottom of an abandoned sand and gravel pit (Figure 19.17); 7.6 m^3 of tracer solution was injected into the aquifer. The tracers were transported southward and were monitored using a network of 656 multilevel wells (Figure 19.18). Each well had up to 15 different depths. Sampling was expedited through the use of a cart, which let all of the multilevel samplers at a site be sampled at the same time.

As the plume of Br^- was transported along the flow system, it gradually dispersed. Longitudinal dispersion was most evident and produced significant concentration reductions in Br^- concentrations. Figure 19.19 provides three "snapshots" in time of the plume and clearly shows longitudinal spreading with time. Presenting the results in cross section demonstrates how that vertical transverse dispersion is limited (Figure 19.20). In other words, the plume does not tend to spread vertically. These patterns of behavior were similar to those observed at Borden. The extent of dispersion was appropriate to the aquifer heterogeneity. Overall this test helped to confirm the geostatistical model of dispersion.

LeBlanc et al., 1991

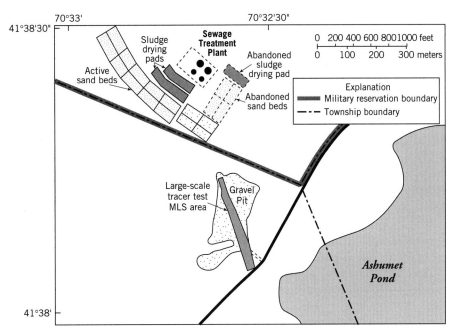

Figure 19.17 Location of the Cape Cod natural-gradient tracer test (from Rea et al., 1994).

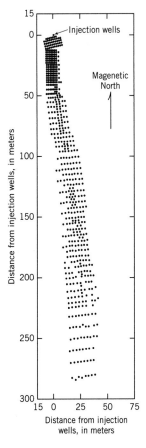

Figure 19.18 Location of monitoring wells used in the Cape Cod tracer experiment (from LeBlanc et al., *Water Resour. Res.*, v. 27, no. 5, p. 895–910, 1991). Copyright by American Geophysical Union.

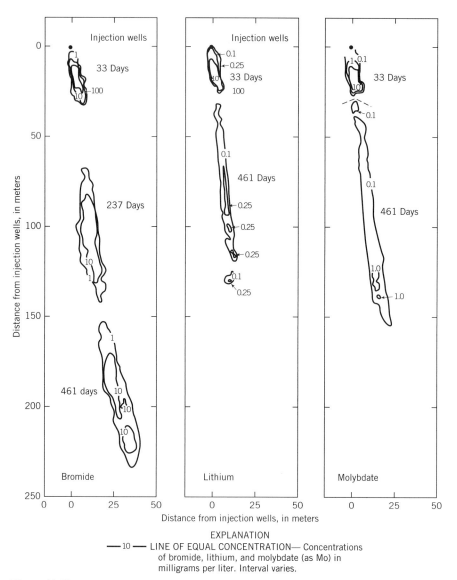

Figure 19.19 Areal distribution of bromide, lithium, and molybdate at several transport times (from LeBlanc et al., *Water Resour. Res.*, v. 27, no. 5, p. 895–910, 1991). Copyright by American Geophysical Union.

operative transport process. Describe the pathway for contaminant migration and estimate at what time in the future the plume will reach the stream.

19.2. Ground water flows through the left face of a cube of sandstone (1 m on a side) and out of the right face with a linear ground-water velocity of 10^{-5} m/s. The porosity of the sandstone is 0.10, and the effective diffusion coefficient is 10^{-10} m²/s. Assume that a tracer has concentrations of 120, 100, and 80 mg/l (1.2, 1.0, and 0.8×10^5 mg/m³) at the inflow face, the middle of the block, and the outflow face, respectively. Calculate the mass flux through the central plane due to advection and diffusion.

19.3. A contaminant is added as a point source to ground water flowing with a constant velocity of 4×10^{-6} m/s. Assuming longitudinal and two transverse dispersivities (*y*- and *z*-directions) of 1.0,

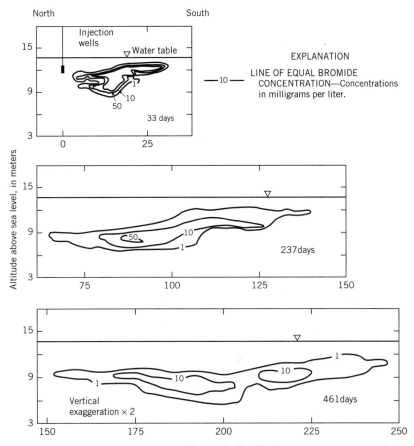

Figure 19.20 Cross-sectional view of bromide distribution at several transport times (from LeBlanc et al., *Water Resour. Res.*, v. 27, no. 5, p. 895–910, 1991). Copyright by American Geophysical Union.

0.1, and 0.01 m, determine the spatial standard deviations ($\sigma_{x,y,z}$) in the plume size after 400 m of transport.

19.4. A series of hydraulic-conductivity measurements for an unconfined aquifer provide a mean hydraulic conductivity (Y) of 0.004 m/s, where $Y = \ln K$, a variance in the log-transformed hydraulic conductivity (σ_Y^2) of 1.0, and a correlation length in the direction of the mean flow (σ_L) of 10.0 m. Estimate the asymptotic macroscale dispersivity for the aquifer.

19.5. In Figure 19.19, look at the pattern of evolution in the Br$^-$ concentration as a function of travel distance. Assume that at 237 days, Γ_L is equal to 24 m. Calculate the dispersion coefficient and the dispersivity for the aquifer.

▶ CHAPTER
HIGHLIGHTS

1. Mass transport theory explains how dissolved mass in ground water originates and is added to the water, how it is moved or redistributed, and how it interacts chemically with the water and the porous medium. The physical processes determine the way mass is moved from one point to another—the system of ground-water conveyance and mixing. The chemical and biological processes redistribute mass among different chemical forms, or into and out of the aqueous system.

2. Advection is mass transport due to the flow of water in which the mass is dissolved. It conveys dissolved mass from one point to another. The direction and rate of transport coincide with the ground water. Thus, the factors that influence ground-water flow patterns—water-table configuration, pattern

of geologic layering, size of the ground-water basin, pumping, or injection—also control the direction and rate of mass transport.

3. When only advection is operating, dissolved mass in one or more stream tubes remains in those stream tubes. Other processes are required to move dissolved mass between stream tubes. In addition, pathlines define the direction of mass spreading in steady-state systems, even with relatively complex flow patterns.

4. Dispersion is a process that causes a zone of mixing to develop between a fluid of one composition that is adjacent to or being displaced by a fluid with a different composition. Longitudinal dispersion occurs in the direction of flow; transverse dispersion occurs perpendicular to the direction of flow.

5. Dispersion is due to diffusion and mechanical dispersion. Diffusion is a process of mass transport in response to a concentration gradient. Mechanical dispersion is a process of mixing that occurs because of local differences around some mean velocity of flow. This variability in the direction and rate of transport is caused mostly by point-to-point variations in hydraulic conductivity.

6. For unidirectional flow, the coefficient of mechanical dispersion (D') in the longitudinal and transverse directions depends on the linear ground-water velocity (v) and dispersivity (α)

$$D'_L = \alpha_L v \quad D'_{TH} = \alpha_{TH} v \quad D'_{TV} = \alpha_{TV} v$$

where α_L is longitudinal dispersivity, and α_{TH} and α_{TV} are transverse dispersivities in the horizontal and vertical directions, respectively. Dispersivities, like hydraulic conductivity, have units of length and are characteristic properties of a medium. Large α values imply large dispersion along the flow system.

7. Quantitative approaches define the relationship between dispersivity values and geostatistical models of hydraulic conductivity. A mean, a variance, and a correlation length can characterize a hydraulic-conductivity field. The correlation length is a measure of the spatial persistence in some property of the medium in a given direction. This information can yield estimates of the extent of mechanical dispersion in the system.

8. Many of the porous-medium concepts of mixing apply to fractured media. Mixing in a fractured medium comes from mechanical mixing and diffusion. In particular, diffusion into the matrix has an important influence on transport in many fractured-rock systems.

9. Tracers and tracer tests are useful in estimating various transport parameters. Tracers include: (a) ions that occur naturally in a ground-water system, (b) environmental isotopes such as ^{2}H, ^{3}H, or ^{18}O, (c) contaminants of all kinds that enter a flow system, or (d) chemicals added as part of an experiment.

10. Natural gradient tracer tests run at Borden, Cape Cod, and Columbus have contributed knowledge about physical and chemical transport processes in shallow unconfined aquifers, and have validated geostatistical concepts of dispersion. These experiments are unique because many monitoring points provided precise three-dimensional pictures of the tracer distribution over two to three years.

20

ISOTOPES AND AGE DATING

Isotopes are widely used in research and practice in ground-water hydrology. They can be divided broadly into stable and radiogenic varieties. The stable isotopes are used mainly for flow system tracing and climate reconstruction; the radiogenic isotopes are used to date ground water. This chapter is designed to provide a basic understanding of how isotopes are used and the basic interpretive techniques that are involved. The first part of this chapter presents basic definitions and units of measurements. Next, we look in detail at the stable isotopes, oxygen-18 (^{18}O) and deuterium (D), which are used most commonly. The last part deals with approaches in dating ground water, which have applications to problems of geologic waste disposal.

▶ 20.1 STABLE AND RADIOGENIC ISOTOPES

Isotopes are atoms of the same element that differ in terms of their mass. For example, hydrogen with an atomic number of 1 has three isotopes, $^{1}_{1}$H, $^{2}_{1}$H, and $^{3}_{1}$H, with mass numbers (superscripts) of 1, 2, and 3, respectively. The first of these isotopes is stable, whereas the last $^{3}_{1}$H (usually written ^{3}H) decays radioactively to $^{3}_{2}$He. Thus, radioactive decay is one of the important processes that involve isotopes. In radioactive decay, atoms of a particular isotope change spontaneously to a new, more stable isotope. Isotopic concentrations also change due to processes like evaporation, condensation, or water/rock interactions. Typically, these processes favor one of the isotopes of a given element over others, producing fractionation.

Stable Isotopes

The isotopes of hydrogen, oxygen, carbon, and sulfur are useful in studying chemical processes. Table 20.1 lists seven isotopes of environmental significance. Commonly, ground-water studies involve ^{2}H or D, ^{18}O, ^{13}C, and ^{34}S.

Typically, one isotope dominates the rest in terms of relative abundances (Table 20.1). Thus, changes due to fractionation are too small to measure accurately, requiring that isotopic abundances be reported as positive or negative deviations of isotope ratios away

TABLE 20.1   Isotopes of environmental significance and their relative abundances

Element	Isotopes	Average Abundance % of Stable Isotopes
Hydrogen	^1_1H	99.984
	$^2_1H^a$	0.015
	$^3_1H^b$ radioactive	10^{-14} to 10^{-16}
Oxygen	$^{16}_8O$	99.76
	$^{17}_8O$	0.037
	$^{18}_8O$	0.10
Carbon	$^{12}_6C$	98.89
	$^{13}_6C$	1.11
	$^{14}_6C$ radioactive	$\sim 10^{-10}$
Sulfur	$^{32}_{16}S$	95.02
	$^{33}_{16}S$	0.75
	$^{34}_{16}S$	4.21
	$^{36}_{16}S$	0.02

[a]Deuterium also noted as D.
[b]Tritium, often referred to by T.

from a standard. This convention is represented in the following general equation, derived from Fritz and Fontes (1980):

$$\delta = \frac{R_{sample} - R_{standard}}{R_{standard}} \times 1000 \qquad (20.1)$$

where δ, reported as permil (‰), represents the deviation from the standard and R is the particular isotopic ratio (e.g., $^{18}O/^{16}O$) for the sample and the standard. For example, we would express sulfur-isotope ratios as

$$\delta^{34}S_{sample} = \frac{\left(^{34}S/^{32}S\right)_{sample} - \left(^{34}S/^{32}S\right)_{standard}}{\left(^{34}S/^{32}S\right)_{standard}} \times 1000$$

A $\delta^{34}S$ value of -20‰ means that the sample is depleted in ^{34}S by 20‰, or 2% relative to the standard.

This way of expressing isotopic compositions takes advantage of the ability of mass spectrometers to measure isotopic ratios accurately. The errors involved in determining δ values are typically a small proportion of the possible range of values.

Deuterium and oxygen-18 compositions of water are usually measured with respect to the SMOW (Standard Mean Ocean Water) standard (Fritz and Fontes, 1980). This choice of standard is particularly appropriate because precipitation that recharges ground water originated from the evaporation of ocean water. However, the isotopic composition of rain or snow in most areas is not the same as ocean water. Evaporation and subsequent cycles of condensation significantly change the isotopic composition of water vapor in the atmosphere.

Radioactive Decay

Radioactive decay occurs mainly by the emission of an α particle ($^4_2$He) or a β particle (electron_$^0_{-1}$e). Often accompanying the emission of these particles is γ radiation, which is electromagnetic energy of short wavelength. This radiation forms when nuclides produced in an excited state (noted by *) revert to a so-called groundstate. As the following examples illustrate, α-decay changes both the mass number and the atomic number, β-decay changes the atomic number only, and γ-emission changes neither.

$$\alpha\text{-decay}: \quad ^{232}_{90}\text{Th} \rightarrow ^{228}_{88}\text{Ra} + ^4_2\text{He}$$

$$\beta\text{-decay}: \quad ^{228}_{88}\text{Ra} \rightarrow ^{228}_{89}\text{Ac} + ^0_{-1}\text{e}$$

$$\gamma\text{-emission}: \quad ^{236}_{92}\text{U}^* \rightarrow ^{236}_{92}\text{U} + \gamma$$

The arrows in these equations indicate that radioactive decay is irreversible. The quantity of the reacting or parent isotope continually decreases, while the product or daughter isotope increases. Reactions become more complex when the daughter itself decays through a series of other products until a stable form is finally created. We refer to these reactions as decay chains or disintegration series (for example, $^{232}_{90}\text{Th}$, $^{235}_{92}\text{U}$, and $^{238}_{92}\text{U}$ series).

The decay of radioactive isotopes is independent of temperature and follows a first-order rate law. The rate of decrease in the activity of a radioactive substance is commonly expressed in terms of a radioactive half-life ($t_{1/2}$), which is the time required to reduce the number of parent atoms by one-half. Listed in Table 20.2 are radioactive half-lives for many radionuclides of interest in ground-water investigations.

Radioactive decay is important for two reasons. When radioactive isotopes occur as contaminants, decay through several half-lives can reduce the hazard when the residence time in a flow system is much greater than the half-life for decay. This capacity to attenuate radioactivity provides the rationale to support the subsurface disposal of some radioactive contaminants. Radioactive decay also forms the basis for various techniques used to age-date ground water. The most important radioactive isotopes used for this purpose are ^{3}H (tritium), ^{14}C, and ^{36}Cl.

The concentration of these important radiogenic isotopes is measured in slightly different ways. Tritium [T] is measured in terms of its relative abundance in the water molecule as compared to hydrogen of mass 1 [H]. *One tritium unit* (TU) corresponds to one atom of ^{3}H in 10^{18} atoms of ^{1}H (Fontes, 1980). Tritium occurs naturally in precipitation at concentrations less than 20 TU. However, the main use of tritium as a tracer comes from elevated levels that arose from thermonuclear testing of bombs in the atmosphere. Tritium levels in precipitation in the early 1960s exceeded 1000 TU.

Measurements of ^{14}C are reported as *percent modern carbon* (pmc). This measure is the ratio of the ^{14}C activity in the sample to that of an international standard, expressed as a percentage. Using appropriate models, pmc can be expressed as a water age. Measurements of ^{36}Cl are reported as the ratio of ^{36}Cl/Cl. Typical values for meteoric water fall in a range of 100 to 500×10^{-15}. We can interpret these ratios in terms of water ages, or as a relative indication of bomb pulse tritium. Global fallout from high-yield nuclear weapons elevated ^{36}Cl levels several orders of magnitude.

TABLE 20.2 Radioactive half-lives for elements of interest in age dating and contaminant studies

Nuclide	$t_{1/2}(\text{Yr})$	Nuclide	$t_{1/2}$ (Yr)
^{3}H	12.3	^{231}Pa	3×10^4
^{14}C	5×10^3	^{241}Am	432
^{36}Cl	3.1×10^5	^{243}Am	7×10^3
^{63}Ni	100	^{79}Se	6.5×10^4
^{90}Sr	29	^{93}Mo	3.5×10^3
^{93}Zr	1.5×10^6	^{99}Tc	2×10^5
^{94}Nb	2×10^4		
^{107}Pd	7×10^6	^{126}Sn	1×10^5
^{129}I	2×10^7	^{151}Sm	90
^{135}Cs	3×10^6	^{147}Sm	1.3×10^{11}
^{137}Cs	30	^{106}Ru	1.0
^{154}Eu	8.2	^{235}U	7×10^8
^{210}Pb	22	^{238}U	4.5×10^9
^{226}Ra	1.6×10^3	^{237}Np	2×10^6
^{227}Ac	22		
^{230}Th	8×10^4		
^{232}Th	1.4×10^{10}		

Source: Modified from Moody, 1982.

▷ 20.2 ^{18}O AND D IN THE HYDROLOGIC CYCLE

When water changes from a gas to a liquid or from a solid to a liquid, isotopic fractionation occurs. Fractionation causes the relative abundance of the isotopes to change because the heavier isotopes (^{18}O and D) tend to be more abundant in the condensed phase. For example, as water evaporates from the ocean, the heavier molecules H_2 ^{18}O and HDO are more abundant in the water phase than in the vapor phase.

The isotopic composition of rain or snow in most areas is not the same as ocean water. Evaporation and subsequent cycles of condensation significantly change the isotopic composition of water vapor in the atmosphere. Observation shows that water vapor in equilibrium with ocean water (δD and δ^{18}O $\sim$ 0‰) has δD $= -80$‰ and δ^{18}O $= -10$‰. We can check this observation by using measured fractionation factors (α) and the following equation

$$\alpha = \frac{\left(^{18}\text{O}/^{16}\text{O}\right)_{\text{water}}}{\left(^{18}\text{O}/^{16}\text{O}\right)_{\text{vapor}}} \tag{20.2}$$

or in terms of del notation

$$\alpha = \frac{1000 + \delta^{18}\text{O}_{\text{water}}}{1000 + \delta^{18}\text{O}_{\text{vapor}}} \tag{20.3}$$

The fractionation factors for liquid-vapor transformations are 1.0098 for ^{18}O and 1.084 for D at 20°C. Knowing that δ^{18}O$_{\text{water}}$ is 0‰, we can rearrange Eq. (20.3) to provide

$$\delta^{18}\text{O}_{\text{vapor}} = \frac{1000 + \delta^{18}\text{O}_{\text{water}}}{\alpha} - 1000$$

Substituting the known values gives

$$\delta^{18}O_{vapor} = \frac{1000 + 0}{1.0098} - 1000 = -9.7‰$$

A similar calculation for δD gives a value of $-77.4‰$.

Several additional points need to be made about this calculation. First, the difference in fractionation factors for δD and $\delta^{18}O$ causes an $8\times$ greater fractionation in D as compared to ^{18}O. Second, the fractionation factors are temperature dependent. At $0°C$, the fractionation factors for liquid-vapor transformations are 1.0117 for ^{18}O and 1.111 for D. Because these numbers are bigger, the vapor will be even more depleted in the two isotopes. The following calculation illustrates this point.

► **EXAMPLE 20.1**   Calculate the isotopic composition of a vapor in equilibrium with seawater at $0°C$. The fractionation factors for ^{18}O and D are 1.0117 and 1.111, respectively.

SOLUTION

$$\delta^{18}O_{vapor} = \frac{1000 + 0}{1.0117} - 1000 = -11.6‰$$

$$\delta D_{vapor} = \frac{1000 + 0}{1.111} - 1000 = -99.9‰$$

As expected, the vapor is more depleted than the water with respect to ^{18}O and D. The lower temperature has enhanced the fractionation. The temperature control on fractionation also explains why winter precipitation is more depleted in D and ^{18}O.

These simple equilibrium models help to explain what happens as fractionation occurs. However, natural processes are more complex, and isotopic ratios of water vapor in air masses associated with oceans are not in equilibrium with the water. In general, vapor is more depleted than expected. Rainfall sampled in coastal areas near the equator has an isotopic composition most like seawater, but it is still slightly depleted.

With evaporation and condensation processes, both ^{18}O and D are fractionated in a consistent way, although the extent of fractionation is different in both cases. Thus, water circulating from oceans into the atmosphere and falling as rain has $\delta^{18}O$ and δD values that are correlated. In other words, if you know one of the values, you can predict the other. Isotope chemists discovered this fact in the early 1950s when they started analyzing samples of rainfall from sites all around the world. When measured δD and $\delta^{18}O$ values for samples of meteoric water are plotted together, they lie along a straight line (Figure 20.1), known as the *meteoric water line* (Craig, 1961). The equation for this line is approximately $\delta D = 8\delta^{18}O + 10‰$. The slope of this line, 8, reflects the difference in the fractionation behavior between ^{18}O and D, which we discussed using Eq. (20.3).

The usual way of interpreting the results of ^{18}O and D analyses is to take the data and plot δD versus $\delta^{18}O$. Where the data fall in relation to the meteoric water line provides clues as to what processes have occurred. Water with an isotopic composition falling on the meteoric water line is assumed to have originated from condensation of water vapor and to be unaffected by other isotopic processes. The actual location on the meteoric water can vary because of temperature and other effects, which we will examine in more detail. Deviations from the meteoric water line result from other isotopic processes. In most cases, these processes affect the relationship between δD and $\delta^{18}O$ in such a unique way that the position of the data points can help to identify a process. Figure 20.2 illustrates how various processes push the composition of water away from the meteoric water line.

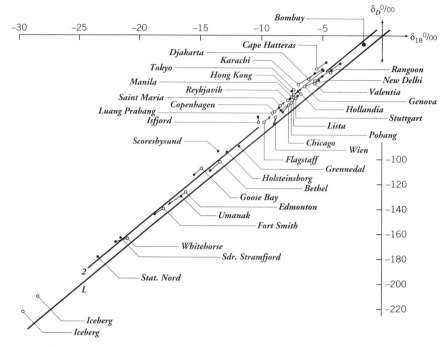

Figure 20.1 Weighted (open circles) and unweighted (dots) means for Northern Hemisphere continental stations, disregarding Africa and the Near East. (Reproduced from Dansgaard, 1964. *Stable isotopes in precipitation*. Tellus, v. 16, p. 436–438. Copyright Blackwell Science, Ltd.)

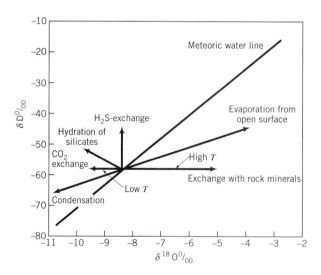

Figure 20.2 Deviations in isotopic compositions away from the meteoric water line as a consequence of various processes.

Behavior of D and ^{18}O in Rain

The particular δD and $\delta^{18}O$ values measured for rainfall can be explained in terms of temperature and the transport history of the air mass. In the previous section, we saw how temperature controls fractionation. At colder temperatures, there is more fractionation between phases. This temperature effect is important as air masses move onto continents because as rain continues to fall it depletes the air mass with respect to the heavier isotopes.

If the rain occurs at a colder temperature, this depletion occurs more rapidly. Evaporation of water from the land back into the atmosphere can add water vapor back into the atmosphere. However, this water will be isotopically lighter because of fractionation related to evaporation.

These ideas are embodied in the concepts of a *continental effect*, an *elevation effect*, and a *latitude* effect. The continental effect describes the progressive depletion in the isotopic composition of water vapor moving from oceans onto continents. Because the rain is preferentially enriched in the heavier isotopes (that is, D and ^{18}O), the reservoir of heavier isotopes vapor is depleted. Each new rain is isotopically lighter than the previous one. This effect has been studied extensively around the world. Ingraham and Taylor (1991) provide a particularly good example of the decline in δD as vapor from the Pacific Ocean moves onto southern California. Notice in Figure 20.3 how δD decreases as a function of the distance that the air masses move from the Pacific coast.

The elevation effect describes the acceleration in the depletion of δD and δ^{18}O in water vapor as it moves over mountains. As water vapor is carried to higher elevations, there is a tendency for more rain at higher elevations, which increases the rate of depletion. Also, at higher elevations, the temperature of condensation is lower, which increases the fractionation and intensifies the rate of depletion. Figure 20.3 also illustrates this effect, with δD values declining by 25‰ over a very short distance at about 250 km from the coast. There, air masses rise up to pass over the Sierra Nevadas. In addition to the increase in elevation, there is an increase in precipitation. Once into the Great Basin (approximately 350 km), there is no change in the isotopic composition (Figure 20.3).

Observations from around the world have shown that precipitation at higher latitudes is relatively depleted in D and ^{18}O (that is, more negative) as compared to samples from lower latitudes. This latitude effect is explained by higher precipitation rates and colder temperatures at higher latitudes. For example, moving northward along the western coast of the United States to Canada, values of δD decrease from approximately -22 to -101‰ and values of δ^{18}O decrease from -4 to -14 (Gat, 1980).

▶ 20.3 VARIABILITY IN ^{18}O AND D ALONG THE METEORIC WATER LINE IN GROUND WATER

This section describes how ^{18}O and D are used in ground-water applications. It examines situations where the isotopic composition of ground water falls along the meteoric water line. With this class of problem, shallow ground water has an isotopic composition that

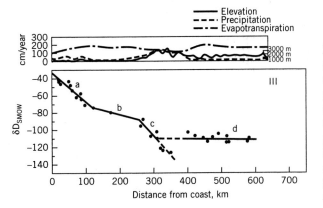

Figure 20.3 Plot of δD versus distance from the Pacific Ocean along traverse III. For reference purposes, information on elevation, precipitation, and evapotranspiration is also provided. (Modified from Ingraham and Taylor, *Water Resour. Res.*, v. 27, no.1, p. 77–90. Copyright by American Geophysical Union.)

is similar to the precipitation. For this simplest case, it is assumed that no processes are operating to change the isotopic composition of the precipitation once it recharges the ground water. The next section deals with the more complicated cases where isotopic processes produce deviations from the meteoric water line.

A ground-water system is more complicated than atmospheric systems examined so far. The reason is that a flow system can contain meteoric water that recharged the ground water thousands and even tens of thousand years ago. In effect, the ground-water system can be an archive for information on the isotopic composition of precipitation through time. Also, low-permeability units could contain the original water of formation when the unit was deposited, with a climate much different than the present one (Bradbury, 1984; Desaulniers et al., 1981; Hendry and Wassenaar, 1999; Remenda et al., 1994).

On the continents, several different factors explain why the isotopic composition of rainfall at a location might change with time. Continental glaciations that occur at intervals through geologic time are the most obvious. The most recent glaciation, which occurred 10,000 to 25,000 years ago, resulted in local rainfall that was isotopically much lighter than present-day precipitation. In effect, the temperature on a continentwide basis in the Northern Hemisphere was colder than today, resulting in an enhanced loss of heavier isotopes as air masses made their way across the continents.

Another not so obvious cause of long-term changes in the isotopic character of rainfall is a tectonic one. Over millions of years, mountain ranges can rise and change the pattern of vapor transport across the continent. A case in point occurs in the Great Basin of the United States (see Figure 20.3) where the continuing uplift of the Sierra Nevada has increased the rainout and gradually decreased the δD of rain in the Great Basin.

Isotopic Records of the Last Glacial Advance

^{18}O and D data from ground water often plot along the meteoric water line over quite a range. This pattern is commonly explained as a long-term climate change with time, with old water related to the last glaciation being isotopically lighter (that is, more negative) than present-day recharge. For example, across the northern United States and Canada, fine-grained glacial deposits (e.g., glacial tills) have often been found to contain water that formed when the deposit was laid down thousands of years ago. The hydraulic conductivity of these deposits is so low that 10,000+ years of recharge have not been able to displace the water of formation. Case Study 20-1 (from Hendry and Wassenaar, 1999) describes this situation as it developed in a thick aquitard system in Saskatchewan, Canada.

Love et al. (1994) describe a slight variation on this theme. They examined ground-water data from the Otway Basin in southwestern Australia. The deuterium levels in water collected along a flow line, approximately 100 km long, varied from an outcrop value of about $-19‰$ in δD to $-35‰$, far downgradient in the aquifer. The water at the down-gradient end of the flow system is estimated to be about 30,000 years old. During glacial maximum, sea levels fell by 150 m, moving the shoreline some 200 to 300 km westward. The continental effect then would cause meteoric water recharging the aquifer to be depleted because of the longer travel distance.

Long-Term Continental Records of Climate

Questions as to whether humans are impacting Earth's climate are being addressed on many different fronts. Geologists are attempting to place the present climate in perspective by developing paleoclimate records from the recent geologic past using ice cores and sediments. Long records, spanning hundreds of thousands of years and several glacial cycles,

ISOTOPIC INVESTIGATION OF AN AQUITARD SYSTEM, SASKATCHEWAN, CANADA

Hendry and Wassenaar examined the variation in ^{18}O and D within a thick aquitard. The aquitard consists of Quaternary clay-rich till approximately 80 m thick underlain by a Cretaceous marine clay approximately 76 m thick. The glacier depositing the till retreated from this part of Saskatchewan approximately 12,000 years ago.

Water samples were collected on a number of occasions from 21 piezometers installed across the till unit and in the upper 16 m of the Cretaceous clay. The δ^{18}O and δD values for the ground water occupy a large segment of the meteoric water line (Figure 20.4). Almost every sample is isotopically lighter than the present-day mean annual precipitation. The isotopic data can also be displayed as a function of depth (Figure 20.5). From the top of the unoxidized till (3 to 4 m deep) to a depth of about 30 m, the δD values decreased from $-136‰$ to $-178‰$. In the range from 30 m to 46 m, the minimum δD value was $-178‰$.

Below this depth, δD values gradually increased to -144‰ across the till-clay interface.

The $-178‰$ δD value is interpreted as glacial meltwater that was introduced to the till at the time of its deposition. These values are similar to values reported for similar deposits in Manitoba (Remenda et al., 1994). The hydraulic conductivity of the deposit is sufficiently low that present-day meteoric water has not made much headway in displacing this water from the till. The smooth pattern of change has likely developed as a consequence of diffusion rather than advection. The deeper ground water is characterized by a δD of $-144‰$ and is interpreted as being even older water that was recharged in a previous interglacial period when the climate was warmer. Similar patterns of isotopic variability have been observed in aquifers in Nevada and in glacial till in Wisconsin and southwestern Ontario, Canada.

Hendry and Wassenaar, 1999

are difficult to come by and were nonexistent in North America. However, in the late 1980s, a unique climate record was found to be preserved in calcite deposited along a fault zone below the water table (Winnograd et al., 1992). Devils Hole is the submerged end of a carbonate aquifer system that drains a large area of south-central Nevada. As water moves into the discharge area, it becomes supersaturated and deposits a small layer of calcite along the walls of the fault zone carrying the flow. Every year, a small layer of calcite is deposited, and through time a layer > 0. 3 m thick gradually built up. A 36-cm-long core of this calcite was collected by divers and analyzed in great detail. Every 1.26 mm of the core was analyzed to provide 285 δ^{18}O measurements. Time was estimated along the core using uranium-series dating.

Figure 20.6 displays the δ^{18}O-time curve. The record shows obvious fluctuations that Winnograd et al. (1992) interpret as changes in the isotopic composition of rainfall recharging the aquifer. The record appears to encompass several glacial cycles in the middle to late

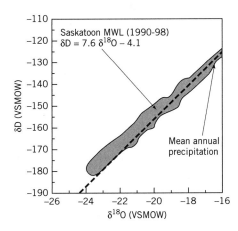

Figure 20.4 Showing of δ^{18}O and δD data for pure water samples collected from piezometers at the study site in Saskatchewan, Canada. The dashed line represents the local meteoric water line at Saskatoon. For reference, the isotopic composition of the mean annual precipitation is shown. (Modified from Hendry and Wassenaar, 1999.)

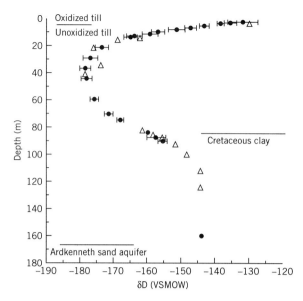

Figure 20.5 Variation in δ D value as a function of depth through the aquitard system. (Modified from Henry and Wassenaar, *Water Resour. Res.*, v.35, p.1751–1760, 1999. Copyright American Geophysical Union.)

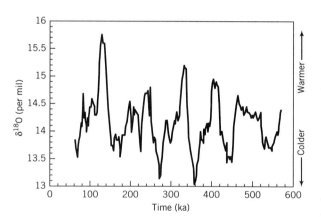

Figure 20.6 Variations in $\delta^{18}O$ values along 36-cm-long core DH-11. Ages were assigned to the $\delta^{18}O$ data by interpolating between 21 MS uranium-series-dated intervals. $\delta^{18}O$ data is reported relative to VSMOW. (Reprinted with permission from Winnograd et al., Continuous 500,000-year climate record from vein calcite in Devils Hole, Nevada. *Science*, v. 258, October 1992.) Copyright ©1992 American Association for the Advancement of Science.

Pleistocene. The record obtained at Devils Hole is amazingly similar to paleotemperature results obtained from Antarctica ice cores and ocean sediments. This example provides yet another way in which ground-water data are helping to sort out information on the Earth's climate in the recent past.

Isotopes as Tracers

^{18}O and D are also useful as tracers. One application is in sorting out the origin of inflow to streams during storm flow events. On occasion, the $\delta^{18}O$ and δ D composition of a rainstorm can be much different from that of the ground water or soil water that already exists in the basin. By careful monitoring of the isotopic composition of a stream through the storm event, one can determine what proportion of the water in the stream came from the present storm and what proportion already existed in the watershed.

This problem can be cast more formally as a two-component hydrograph separation (Sklash et al., 1976; Sklash and Farvolden, 1979). One component of the streamflow is

^{18}O IN THE STUDY OF RAINFALL-RUNOFF PROCESSES

Sklash and coworkers used ^{18}O as a tracer to study hydrologic processes in Big Creek and Big Otter Creek, two tributaries of Lake Erie in Southern Ontario. Each watershed has an areal extent of about 700 mi^2 (1812 km^2) and main channel lengths of about 85 mi (137 km). The watersheds rise in a till upland and flow across the Norfolk Sand Plain, which covers about 75% of the two watersheds.

Sixty ground-water samples had a mean δ^{18}O value of -10.1‰ with a standard deviation of 0.69. The measured δ^{18}O for baseflow in the stream was similar, though slightly heavier because of evaporation. The storm used for the analysis dropped between 2.1 cm and 2.8 cm of rain in a one-day period. The δ^{18}O content of the rain was -4.0‰, which provided a significant contrast with preexisting water in the basin.

Sklash et al., 1976

Figure 20.7 shows how δ^{18}O varied at the Vienna Gage, which is close to the outflow of the Big Otter Creek watershed. As expected, the δ^{18}O of the stream became heavier at the peak flow, which reflected the contribution from the isotopically heavier precipitation. However, the stream water is isotopically more like ground water than rain water. Using a quantitative mass balance approach, Sklash et al. determined that the peak discharge at Vienna was comprised of 70% prestorm water and 30% storm water. The results were similar at other sampling sites in the mid- to lower reaches of the channel system. In the upstream reaches, the peak flow was a 50–50 mix of storm and prestorm water. They interpreted the prestorm component of the hydrograph as ground water. Recharge near the creeks locally increased the water table elevation and the inflow to the creek.

the *event water*, or water that was delivered by the storm and has the isotopic composition of the precipitation. The other component is the *pre-event water*, or vadose zone water that already existed in the basin before the storm. For this approach to work, the isotopic composition of the event water should be different from the pre-event water. In addition, the isotopic composition of the precipitation needs to stay constant through the storm.

The simplest interpretation of these results is that the pre-event water is mostly ground water whose rate of inflow to the stream increases because of recharge due to the rainfall. The event runoff is interpreted as rain falling on the stream and overland flow, which are generated as the storm commences. Several researchers have explained that this simple interpretation doesn't always hold. Any assessment of the origin of streamflow needs to take account of both physical hydrologic measurements and the isotopic data.

What has been surprising is how much of the streamflow during a storm event is actually comprised of pre-event water. Sklash et al. (1976) and Sklash and Farvolden (1979), studying sites in Ontario and Quebec, Canada, found that about 50% to 80% of the streamflow at peak discharge (i.e., peak flow) was pre-event water. Studies conducted in Europe and New Zealand had similar results. Case Study 20-2 illustrates some of these findings in more detail.

▶ 20.4 DEVIATIONS IN ^{18}O AND D AWAY FROM THE METEORIC WATER LINE

Plotting δ^{18}O and δD data for water samples with reference to the meteoric water line provides the framework for identifying the isotopic processes that are at work. So far, we have looked at situations where essentially no processes have operated to change the isotopic composition of precipitation recharging aquifer systems. Now we will look at processes that cause data to plot away from the meteoric water line.

When water resides in a lake, pond, or even river for a long time under evaporative conditions, the isotopic composition of the water can change. Overall, there is enrichment in the heavier isotopes because the lake or pond is a finite reservoir. The dynamics of

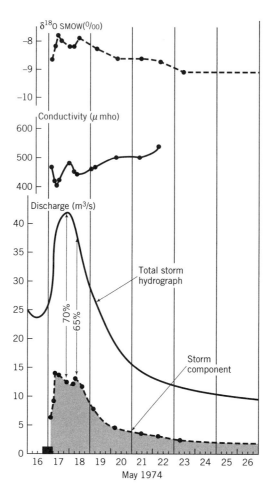

Figure 20.7 Variation in $\delta^{18}O$ and conductivity in Big Otter Creek at Vienna, Ontario, Canada. The resulting hydrograph separation shows that 70% if the peak flow is prestorm water. (Modified from Sklash et al., 1976).

the evaporation process causes the isotopic composition of the lake or pond to follow an *evaporation line*, with a slope ranging from one to five depending on the local rates of evaporation. The point of intersection of the evaporation line with the meteoric water line usually is taken as the isotopic composition of unaltered precipitation. In general, the farther along the evaporation line the data points lie, the greater the evaporation. Case Study 20-3 illustrates this process.

Isotopic exchange between minerals and ground water is important in deep, basinal flow systems or in geothermal systems. The relatively high temperatures enable oxygen and hydrogen to exchange between phases and to achieve an equilibrium distribution. This equilibrium is described by the fractionation factor for the mineral. Savin (1980) summarized fractionation factors for a variety of different minerals and described how they change as a function of temperature.

Exchange between meteoric water and minerals containing oxygen results in the kind of deviation depicted in Figure 20.2. Because only ^{18}O is involved in exchange, the samples typically fall along a horizontal line, reflecting an *oxygen shift* away from the meteoric water line. The size of the oxygen shift is proportional to the difference in original $\delta^{18}O$ between the water and rock, the temperature, and the time of contact, and inversely proportional to the water/rock ratio (Truesdell and Hulston, 1980). Yellowstone Park, Steamboat Springs, Wairaki, and Salton Sea are well-known examples.

EVAPORATION OF WATER FROM CLOSED LAKES

This study traces the changing isotopic composition of water moving through the hydrologic cycle in Surprise Valley, California. Winter snows, melting from the mountains, infiltrated the subsurface and flowed into creeks. The creeks drained into playa lakes, where much of it evaporated through the summer months. The isotopic composition of these waters is shown in Figure 20.8. As expected, the $\delta^{18}O$ and δD values for snow samples fall close to the meteoric water line. Ground water and water from the creeks are isotopically heavier, but also lie along the meteoric water line. The relatively consistent isotopic composition of the creeks reflects the important contribution of ground water to the flow.

The $\delta^{18}O$ and δD values for water from one of the playa lakes deviate from the meteoric water line (Figure 20.8). They lie along an evaporation line having a slope of 5.2. In general, the distribution sample points are explained by the gradual evaporation of up to 95% of the total volume of lake water by the end of the hot summer. The lake-water data plotting closest to the meteoric water line are from the April to June time period. Those farthest along the evaporation are from August to October.

Ingraham and Taylor, 1989

► 20.5 RADIOACTIVE AGE DATING OF GROUND WATER

Two different approaches may be used for dating ground water. The direct approach involves interpreting the concentration distribution of radioactive elements that allow a continuous age determination. The most common environmental tracers used for this purpose, tritium, carbon-14, and chlorine-36, have a worldwide distribution.

The indirect dating method relies on knowledge of how particular isotopes or organic chemicals change as a function of time in recharge. One good example is the family of chlorofluorocarbons, which have been increasing in the atmosphere since they began to be manufactured in the 1930s. The measured concentration of these man-made compounds in ground water often provides an estimate of the time when this water was recharged to the aquifer. We will examine these approaches in more detail in the following section.

Direct dating techniques interpret the distribution of a radioactive species in terms of a first-order kinetic rate law for decay. The residence time of mass in the system or the ground-water age (t) is described mathematically as

$$t = \frac{t_{1/2}}{\ln 2}\ln(A_0/A_{\text{obs}}) \tag{20.4}$$

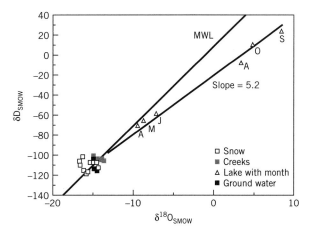

Figure 20.8 Plot of $\delta^{18}O$–δD data for samples from Surprise Valley, California (from Ingraham and Taylor, 1989). Reprinted from *J. Hydrol.*, v. 106, The effect of snowmelt on the hydrogen isotope ratios of creek discharge in Surprise Valley, California. p. 233–244). Copyright 1989, with permission from Elsevier Science.

where $t_{1/2}$ is the half-life for decay, A_0 is the activity assuming no decay occurs, and A_{obs} is the observed or measured activity of the sample.

Tritium (^{3}H, $t_{1/2} = 12.35$ yr) and carbon-14 (^{14}C, $t_{1/2} = 5730$ yr) are commonly used for age dating. However, both suffer from limitations. The relatively short half-life for ^{3}H makes it useful only for dating water less than about 40 years old. In addition, the cessation of nuclear testing in the atmosphere has eliminated the global source of new tritium. Within a few more decades, the tritium levels in precipitation will decline to an extent that tritium will be less useful as a tracer. ^{14}C, with a much longer half-life, has the potential to date water up to about 40,000 years old. However, interpreting ^{14}C data is difficult because of the need to account for other processes besides radioactive decay that influence the measured ^{14}C activity.

Philips et al. (1986) have demonstrated the potential of using chlorine-36 (^{36}Cl) in dating water. A particularly attractive aspect of this radionuclide is its long half-life ($t_{1/2} = 3.01 \times 10^5$ years) and the smaller number of potential reactions (as compared to ^{14}C) that need to be accounted for in calculating an age date (Bentley et al., 1986). One limitation associated with this radioisotope is the need for tandem accelerator mass spectrometry for measurements, which at present limits the availability of the approach.

Tritium

Tritium concentrations are reported in terms of tritium units (TU), with 1 TU corresponding to one atom of ^{3}H in 10^{18} atoms of ^{1}H (Fontes, 1980). Tritium occurs naturally in the atmosphere, with concentrations in precipitation usually less than 20 TU. However, tritium generated by thermonuclear testing in the atmosphere between about 1952 and 1963 swamped the natural production of tritium. The long-term record of ^{3}H in precipitation at Ottawa, Canada, shows that during the period of nuclear testing concentrations were often greater than 1000 TU (Figure 20.9). Tritium levels declined once weapons testing stopped in 1963, but present-day levels remain above the natural background. Gat (1980) and Fontes (1980) present details on the seasonal variation and latitude dependence of ^{3}H levels in precipitation.

Ideally, by knowing the concentration of ^{3}H in precipitation (the source) and its distribution in ground water, one should be able to date the water. In most cases, however, tritium cannot be used in such a quantitative way. The main problems stem from the uncertainty and complexity of atmospheric loading. For example, most places yield insufficient data to establish the historical pattern of ^{3}H loading. This limitation can sometimes be overcome by correlating partial local records to stations like Ottawa, Canada, with long-term records.

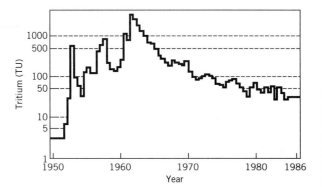

Figure 20.9 Tritium levels in precipitation at Ottawa, Ontario. (Modified from Robertson and Cherry, *Water Resources Res.*, v. 25, p. 1097–1109, 1989. Copyright ©by American Geophysical Union.)

The complex loading function also provides a problem in interpreting ^{3}H data. Without a great deal of information about tritium distributions in the ground water, it is difficult to determine whether a sample with 30 TU is a late 1950s water that has decayed through three half-lives or a 1970s water originally with 75 TU that has decayed through one half-life.

Tritium data by themselves are most commonly used in an indirect manner. Rather than looking for an actual date, tritium is used to differentiate pre-1952 water from younger water. The logic is that assuming pre–1952 water had an original ^{3}H concentration of 5 TU, the concentration in 1988 would be at maximum 0.6 TU, which is close to the detection limit even using enrichment techniques of analysis. Thus, any detectable tritium in a sample implies that the water contains some component of more recent or post–1952 water (Fontes, 1980).

Some of the limitations of tritium dating can be overcome by adding measurements of helium-3 (^{3}He), the stable daughter of tritium decay. The isotopic decay reaction is written as

$$^3\text{H} \rightarrow {}^3\text{He}^* + \beta^-$$

where the * indicates a tritiogenic source of ^{3}He (Solomon et al., 1995). Other sources of ^{3}He exist that must be appropriately accounted for. When both ^{3}H and ^{3}He* measurements are available, summing ^{3}H (TU) and ^{3}He (as TU) eliminates the decay of ^{3}H (Solomon et al., 1995). Thus, one could determine, for example, the peak concentration coinciding with 1960s bomb testing. In addition, one could determine the ^{3}H/^{3}He* age of the water. The ^{3}H/^{3}He* age of a ground-water sample can be calculated as

$$t = \frac{t_{1/2}}{\ln 2} \ln \left(\frac{^3\text{He}^*}{^3\text{H}} + 1 \right) \tag{20.5}$$

where $t_{1/2}$ is the half-life for decay of ^{3}H, and ^{3}H and ^{3}He* are tritium concentrations expressed in tritium units.

► **EXAMPLE 20.2**   A ground water was found to contain 14.8 TU of tritium and 31.6 TU of tritiogenic helium (^{3}He*). Calculate the age of the ground water. The half-life for tritium decay is 12.3 yr.

SOLUTION   The age of the water can be determined by substituting values into Eq. (20.5):

$$t = \frac{t_{1/2}}{\ln 2} \ln \left(\frac{^3\text{He}^*}{^3\text{H}} + 1 \right) = \frac{12.3}{0.693} \ln \left(\frac{31.6}{14.8} + 1 \right) = 17.7 \ln 3.14 = 20.2 \text{ yr}$$

Tritium is used in ground-water studies in a variety of ways. The most obvious applications fall in the area of age dating water less than 50 years old. The following Case Study 20-4 from Solomon et al. (1995) demonstrates ^{3}H/^{3}He* age dating at a site on Cape Cod.

Tritium is also used to ensure that old water, based on δD and δ^{18}O data, has not been contaminated during the sampling process. Old water should have undetectable tritium concentrations.

Carbon-14

Measurements of ^{14}C are reported as percent modern ^{14}C (pmc), determined as the ratio of the sample activity to that of the international standard expressed as a percentage. ^{14}C originates naturally in the upper atmosphere through a reaction involving nitrogen and neutrons. Like ^{3}H, weapons testing in the atmosphere has affected its concentration in recent years. However, except for young waters, this increase does not affect the interpretation.

[3]H/[3]He* AGE DATING AT THE FS-12 FIELD SITE, CAPE COD, MASSACHUSETTS

This study used tritium to help explain patterns of flow and recharge at a contaminated site at Cape Cod, Massachusetts. Attention was focused on the shallow unconfined glacial aquifer, which is approximately 65 m thick. Nests of five to eight piezometers were available at three locations (ITW-1, ITW-2, ITW-3) along a flow line.

The procedure for collecting water samples for ^{3}He* analysis is complicated. Readers interested in learning more about this topic and the analytical procedures should refer to Solomon et al. (1995). Ages were calculated using the ^{3}H alone and ^{3}H plus ^{3}He* data. Figure 20.10 shows how the ages change vertically and as a function of travel distance.

Solomon et al., 1995

Generally, the water ages increase as a function of depth to about 25 m. Samples from greater depths produced younger ages. Solomon et al. (1995) felt that the ages from the deeper samples were anomalously low because of dispersion effects or analytical/sampling errors. There were no obvious hydraulic reasons to contribute to these younger ages. Based on these age measurements, it was determined that the contaminant release likely occurred in 1975. Recharge rates are in the range of 70 to 115 cm/yr.

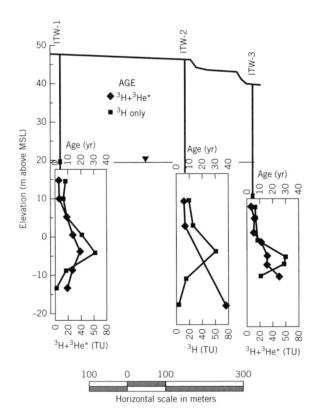

Figure 20.10 Profiles of ^{3}H/^{3}He age and ^{3}H+^{3}He*. The cross section from ITW-1 to ITW-3 is oriented along a ground-water flow line. For ITW-2 only ^{3}H is plotted because a complete profile was not obtained (from Solomon et al., 1995). Reprinted by permission of *Ground Water*. Copyright ©1995. All rights reserved.

One source of ^{14}C in ground water is the solution of $CO_2(g)$ in the soil zone. The activity of ^{14}C in CO_2 gas is approximately 100 pmc (Fritz and Fontes, 1980) and slightly higher in the ions coming from the dissolution of CO_2. When CO_2 is dissolved in the ground water, it is found mostly as HCO_3^- and CO_3^{2-}. The dating method works because, once carbonate species move below the water table, ^{14}C begins to decay and there are no additional sources.

The main problem in applying this method is that some reactive minerals (e.g., calcite) contain inorganic carbon, and carbon is transferred in and out of the ground water. These interactions can reduce the ^{14}C activity in the water and thus need to be accounted for in estimating the age. In terms of Eq. (20.4), the value of A_0 (the ^{14}C activity assuming no decay) would be lower than 100 pmc, reflecting the fact that other processes besides radioactive decay influence the ^{14}C activity of the sample. Any age calculation is meaningful as long as A_0 and A_{obs} differ only because of the effects of radioactive decay.

Mook (1980), Reardon and Fritz (1978), and Wigley et al. (1978) state that the following processes can alter the ^{14}C activity of ground water:

1. The congruent dissolution of carbonate minerals, which adds "dead carbon" or carbon without ^{14}C activity to the ground water. This process lowers the ^{14}C activity measured for the sample.

2. The incongruent dissolution of carbonate or other Ca-containing minerals, accompanied by the precipitation of calcite. This process will remove ^{14}C as calcite precipitates, and if dolomite is the mineral, dissolving additional dead carbon is added through (1) above. This process could occur in the zone of saturation following the rapid solution of calcite to equilibrium, with subsequent precipitation as dolomite slowly dissolved.

3. The addition of dead carbon from other sources such as the oxidation of old organic matter, sulfate reduction, and methanogenesis. Again these reduce the ^{14}C activity of the sample.

Two approaches can be used to estimate ground-water age (Kimball, 1984). One way is to interpret age on the basis of the ion and isotopic data for a single sample without information from other samples. The second approach involves using the ion or isotopic data from many samples in an integrated way and mass balance modeling to sort out all the major inputs and outputs of carbon.

A comprehensive review of approaches available to correct ^{14}C data is beyond the scope of this book. Some simple correction procedures are included, however, to give readers an appreciation of what is involved in working with data for a single sample. These approaches only account for the most important process affecting ^{14}C activity, which is the congruent dissolution of calcite. The reaction between water containing CO_2 and calcite is

$$CO_2 + H_2O + CaCO_3(s) = Ca^{2+} + 2HCO_3^- \tag{20.6}$$

At equilibrium according to this reaction, half the bicarbonate would be generated from a source containing ^{14}C (CO_2) and the other half would be generated from a dead source (calcite). Assuming that the activity of the CO_2 is 100 pmc and the calcite is 0 pmc, A_0 would be 50 pmc, reflecting the equal contribution of carbon from both sources.

In many cases, the reaction would not likely be at equilibrium due to the lack of carbonate minerals or kinetic effects (Pearson and Hanshaw, 1970). The following form of the reaction describes this more realistic situation (Mook, 1980):

$$(a + 0.5b)\,CO_2 + 0.5b\,CaCO_3 + H_2O = 0.5b\,Ca^{2+} + b\,HCO_3^- + a\,CO_2 \tag{20.7}$$

Excess CO_2 in this reaction results in ^{14}C activity greater than 50 pmc and perhaps close to 85 pmc.

One technique for correcting A_0 values is based on this idea. For example, one simply assumes A_0 has a value of $85 +/- 5$ pmc. Measurements of dissolved carbon in the ground water of northwestern Europe suggested that this value was in fact representative of soil water and shallow ground water in temperate climates (Vogel, 1967, 1970). Another less empirical approach is based on Eq. (20.6) and provides an A_0 value that is the weighted contribution of CO_2 and calcite. Mathematically, this correction is a simple mixing equation, or

$$(a + b)A_0 = (a + 0.5b)A_{CO2} + 0.5b\,A_c \qquad (20.8)$$

where A_{CO2} is the estimated activity of CO_2 in the soil zone (usually 100 pmc), A_c is the estimated activity of calcite (usually 0 pmc), and $(a + b)$ is the total moles of carbon in the water (C_T). Rearranging Eq. (20.8) and assuming A_c is 0 provides the desired correction equation for A_0

$$A_0 = (a + 0.5b)\,A_{CO2}/(a + b) \qquad (20.9)$$

or written in terms of C_T

$$A_0 = (C_T - 0.5b)\,A_{CO2}/C_T \qquad (20.10)$$

Equation (20.10) is known as the Tamers equation (Tamers, 1967, 1975). The first term on the right side of Eq. (20.10) assumes that for water with a normal pH the total carbon is found as CO_2 (or $H_2CO_3^*$, depending on conventions) and HCO_3^-. C_T is determined using the major ion data together with a speciation program like WATEQF. The molar concentration of HCO_3^- (i.e., b) is known from the water analysis, and A_{CO2} is taken as 100 pmc.

These and similar approaches are useful to some extent. However, they do not account for many of the processes affecting the ^{14}C content of the water (Pearson and Hanshaw, 1970). Most applications today interpret ^{14}C data using computer codes that account for the variety of processes that influence carbon mass balances. For example, the NETPATH model (Plummer et al., 1991, 1994) is commonly used for interpreting ^{14}C data when both isotopic and chemical data are available.

Aravena et al. (1995) used ^{14}C to date water from the Alliston aquifer in Ontario, Canada. Their NETPATH analysis showed that the key reactions influencing the chemical evolution of ground water in the aquifer are incongruent dissolution of dolomite, ion exchange, methanogenesis, and oxidation of sedimentary organic matter. The study illustrates the extent of the detailed geochemical evaluation that must accompany approaches for age dating. Similarly, it shows that simple correction approaches are not feasible in systems with complex geochemical processes. The early hope of collecting a single sample of water and extracting a date seems to have faded in light of the effort required for process identification.

Chlorine-36

There is strong evidence (Bentley et al., 1986; Phillips et al., 1986) that ^{36}Cl will emerge as a useful radioisotope for dating water up to two million years old. Again, the most important source of this isotope is fallout from the atmosphere, although small quantities are produced in the subsurface. The results of analyses are reported as the ratio of $^{36}Cl/Cl$, with a typical value for meteoric water lying in a range from 100 to 500×10^{-15}. Because comparatively few processes apparently affect the $^{36}Cl/Cl$ ratio, the dating equations are straightforward modifications of Eq. 20.4. The simple geochemistry of Cl in ground water avoids the complexity inherent with ^{14}C. About the only processes that contribute dead Cl to ground water are the dissolution of mixing of older water containing Cl$^-$. To date, ^{36}Cl has been used to date ground water in the Great Artesian Basin of Australia (Bentley et al., 1986) and the Milk River aquifer in Alberta, Canada (Phillips et al., 1986).

Like tritium, levels of ^{36}Cl have been elevated up to two or three orders of magnitude as a consequence of global fallout from high-yield nuclear weapons tests in the 1950s. The presence of so-called bomb-pulse ^{36}Cl in a water sample provides a clear indication of the relatively young age of the sample. To date, it has been useful in sorting out the patterns of complex unsaturated zone recharge (Fabryka-Martin et al., 1993).

► 20.6 INDIRECT APPROACHES TO AGE DATING

Indirect methods depend on interpreting systematic changes in the chemical composition of indicator species or isotopes along ground-water flow paths. Unlike the radiogenic isotopes, some chemical or isotopic pattern is interpreted in light of events or other independent data, which provide a basis for establishing times. This section reviews the use of two promising approaches that involve environmental isotopes (δ^{18}O and δD) and anthropogenic contaminants with a known history of use (chlorofluorocarbons). The indirect approaches work because the isotopic composition of some tracer changes with time in recharge. When the time variation is known, one can directly infer the age of a sample from its measured abundance in ground water.

Isotopically Light Glacial Recharge

Several examples presented earlier in this chapter illustrated how a time-varying signal in the δ^{18}O and δD levels in precipitation can be preserved in large aquifer systems or in units having extremely low hydraulic conductivity. In regions like western Canada or northern Wisconsin, where the isotopic composition of glacial meltwater is reasonably well known, isotopic data provide indirect evidence of ground water tens of thousands of years old. Thus, by carefully documenting the isotopic character of samples from low-permeability sites, one might be able to date the ground water indirectly. Water that is isotopically much lighter than present-day meteoric water is likely relatively old, originating during colder glacial climates.

This approach to indirect dating has promise in selecting sites for the long-term containment of hazardous wastes (Desaulniers et al., 1981; Hendry and Wassenaar, 1999). In effect, the isotopic indicators can be used to verify hydraulic evidence that flow velocities through these kinds of low-permeability deposits are extremely small—on the order of 1 m per 10,000 years.

Chlorofluorocarbons

Chlorofluorocarbons (CFCs or Freons) are a family of organic compounds used widely as propellants in aerosol cans and refrigerants. They are not isotopes but are generally lumped together in the family of environmental tracers, which includes mostly isotopes. Their use is being phased out because of concerns that they are destroying ozone in the Earth's upper atmosphere. Two of the CFCs in particular, dichlorodifluoromethane (CCl_2F_2 or CFC-12) and trichlorofluoromethane (CCl_3F or CFC-11), account for the greatest commercial use (Busenberg and Plummer, 1992). CFC-12 and CFC-11 have been manufactured since the 1930s and 1940s, respectively.

The potential use of environmental CFCs to date ground water has been recognized since the mid-1970s (Thompson, 1976; Thompson and Hayes, 1979). These compounds have also been used in tracer experiments (Randall and Schultz, 1976). Interest in CFC dating has been renewed owing mainly to the efforts of scientists at the U.S. Geological Survey (Busenberg and Plummer, 1992; Dunkle et al., 1993).

Based on the history of production of these compounds and direct atmospheric measurements since 1977, it is possible to develop relatively accurate estimates of expected CFC-12 and CFC-11 concentrations in the atmosphere. Figure 20.11, taken from Dunkle et al. (1993), shows how the atmospheric concentration of these compounds has changed as a function of time. Busenberg and Plummer (1992) discuss in detail how these atmospheric concentration values were reconstructed.

The dating method assumes that the concentration of CFCs in precipitation recharging the ground water is in equilibrium with the atmospheric gas-phase concentrations according to Henry's Law. The equilibrium constant describing the partitioning of gases into water is temperature dependent. Thus, the actual concentrations of CFC-11 and CFC-12 found in the ground water depend on the ground-water temperature. Figure 20.11 depicts the expected concentration of the two main CFCs in ground water, assuming equilibrium with the atmospheric concentrations at 9°C.

To employ CFC-11 and CFC-12 in dating, one would collect a sample of ground water, determine the concentrations of CFCs, and utilize the observed concentrations together with Figure 20.11 to obtain a date of recharge. All of these steps are complicated. To ensure that samples are completely isolated from the air, Busenberg and Plummer (1992) sealed samples in glass ampules by heat fusing immediately following their collection. The laboratory measurement of CFCs is also difficult to perform because it is imperative that samples (especially old samples) not be contaminated with modern air, which contains high CFC concentrations. Readers interested in a detailed discussion of the analytical techniques can refer to Busenberg and Plummer (1992).

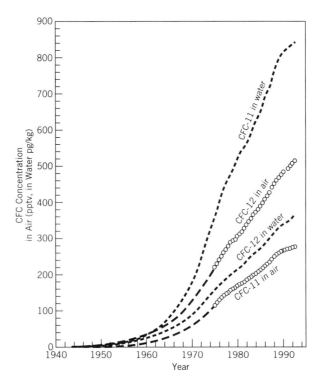

Figure 20.11 Reconstruction of atmospheric concentrations of CFC-11 and CFC-12 since 1940 in parts per trillion volume per volume air (after Busenberg et al., 1993; Elkins et al., 1993) and calculated corresponding solubilities of CFC-11 and CFC-12 in water at 9 °C in picograms per kilogram water (from Dunkle et al., *Water Resources Res.*, v. 29, p. 3837–3860, 1993). Copyright by American Geophysical union.

An example of the application of CFC dating is the study of the shallow ground water of the Delmarva Peninsula, which is located along the eastern side of Chesapeake Bay (Dunkle et al., 1993). Table 20.3 summarizes data for a nest of wells along a cross section (F1–F1′) completed in a shallow surficial aquifer near the town of Fairmont. The concentrations of CFC-11 and CFC-12 (pg/kg) are used with the relationship presented in Figure 20.11 to provide the estimated age in years. Usually, the oldest pair of dates (CFC-11 and CFC-12) is selected as the best estimate of the age. The distribution of ages (Figure 20.12) agrees with relative ages based on hydrologic arguments (Dunkle et al., 1993). In other words, samples that are furthest along the flow system are the oldest.

Like many other age dating techniques, there are limitations. For example, CFC concentrations in anaerobic environments may be reduced through microbial degradation. Processes like dispersion and sorption can also influence CFC concentrations (Dunkle et al., 1993). In spite of limitations, the approach has potential for dating relatively young waters.

▶ **EXERCISES**

20.1. A large lake has $\delta^{18}O$ and δD values of $-14.0‰$ and $-104‰$, respectively. Calculate the isotopic composition of vapor in equilibrium with this water at 10 °C.

20.2. Explain why the $\delta^{18}O$ and δD composition of rainfall changes as air masses move inland on the continents.

20.3. Shown in Figure 20.13 are isotopic data collected from the confined aquifer at Alliston, Ontario, Canada (Aravena et al., 1995). The rectangle shows the isotopic composition of present-day meteoric water at the site.

    a. Is there evidence that ground water has been evaporated in the past? Explain your answer.

    b. Given that the study area is located in the southern part of Canada, suggest one reason why some ground-water samples are isotopically different from others.

20.4. Tritium can be applied in ground-water studies as both a direct and an indirect way of dating ground waters. Explain the difference in these approaches.

20.5. An analysis of ground water found ^3H and $^3He*$ contents of 27.3 TU and 39.1 TU, respectively. Calculate the age of the water sample.

20.6. The concentration of CFC-11 in four ground-water samples was measured as 1.9, 76.9, 176.1, and 430.4 pg/kg (data from Dunkle et al., 1993). Estimate the year in which recharge occurred, assuming a recharge temperature of 9° C.

TABLE 20.3 Sample data for ground water from the Delmarva Peninsula (from Dunkle and others, 1993)

Well	Depth to Screen (top) m	CFC-11 Concentration pg/kg	CFC-11 Age year	CFC-12 Concentration pg/kg	CFC-12 Age Year
57	6.1	661.3	1985.0	281.9	1985.8
58	12.2	582.5	1982.3	260.4	1984.0
59	18.3	409.1	1976.0	216.0	1980.0
60	24.4	232.3	1971.5	152.4	1974.8
61	29.6	207.1	1970.5	95.3	1970.3

Source: Dumkle et al., 1993.

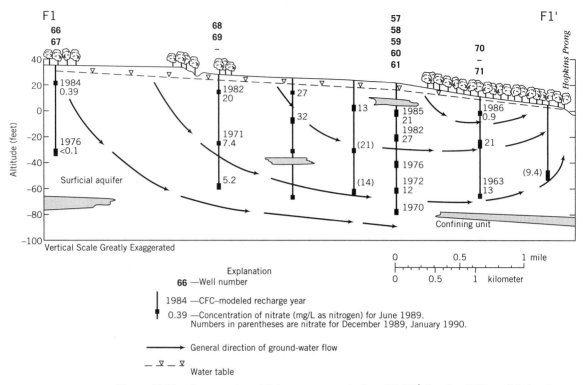

Figure 20.12 Cross section of Fairmount network along F1–F1′ showing CFC-modeled recharge years, nitrate concentrations (in milligrams per liter as N), and generalized grounw-water flow paths (from Dunkle et al., *Water Resources Res.*, v. 29, p. 3837–3860, 1993). Copyright by American Geophysical Union.

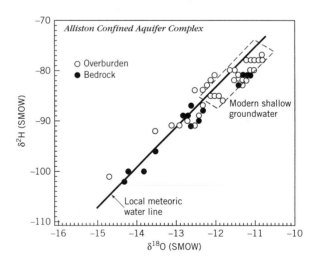

Figure 20.13 Plot of $\delta^{18}O$ versus $\delta^{18}H$ composition in the Alliston groundwater (from Aravena et al., *Water Resour. Res.*, v. 31, no. 9, 2307–2317). Copyright by American Geophysical Union.

▶ **CHAPTER HIGHLIGHTS**

1. Isotopes can be divided broadly into stable and radiogenic varieties. The stable isotopes are used mainly for flow system tracing and climate reconstruction; the radiogenic isotopes are used to date ground water. Isotopes are atoms of the same element that differ in terms of their mass. For example, hydrogen with an atomic number of 1 has three isotopes, ^1_1H, ^2_1H, and ^3_1H, with mass numbers (superscripts) of 1, 2, and 3, respectively.

2. Abundances of stable isotopes are reported as positive or negative deviations of isotope ratios from a standard. For example, the sulfur-isotope ratio is

$$\delta^{34}S_{sample} = \frac{\left(^{34}S/^{32}S\right)_{sample} - \left(^{34}S/^{32}S\right)_{standard}}{\left(^{34}S/^{32}S\right)_{standard}} \times 1000$$

where δ, reported as permil (‰), represents the deviation from the standard. A $\delta^{34}S$ value of $-20‰$ means that the sample is depleted in ^{34}S by 20‰, or 2‰ relative to the standard.

3. The decay of radioactive isotopes follows a first-order rate law, characterized by a radioactive half-life ($t_{1/2}$). Tritium [T] is measured in terms of its relative abundance in the water molecule as compared to hydrogen of mass 1 [H]. One tritium unit (TU) corresponds to one atom of ^3H in 10^{18} atoms of ^1H. Measurements of ^{14}C are reported as percent modern carbon (pmc).

4. Water circulating from oceans into the atmosphere and falling as rain has $\delta^{18}O$ and δD values that are correlated. When measured δD and $\delta^{18}O$ values for samples of meteoric water are plotted together, they lie along a straight line known as the meteoric water line. The equation for this line is approximately $\delta D = 8\delta^{18}O + 10‰$.

5. The isotopic composition of rainfall at a location can change with time owing, for example, to continental glaciations and tectonics. The most recent glaciation produced local rainfall that was isotopically much lighter than present-day precipitation. The colder temperatures enhanced the loss of heavier isotopes as air masses moved across the continents. Over millions of years, mountain ranges can rise and change the pattern of rainfall and vapor transport across the continent.

6. Plotting $\delta^{18}O$ and δD data with reference to the meteoric water line helps to identify isotopic processes. With surface water under evaporative conditions, the isotopic composition of the water follows an evaporation line that deviates from the meteoric water line with a slope ranging from one to five.

7. There are two different approaches for dating ground water. The direct approach involves interpreting the concentration distribution of radioactive isotopes (for example, tritium, carbon-14, and chlorine-36) to provide an age determination. The indirect dating method relies on knowledge of how particular isotopes or organic chemicals change as a function of time in recharge.

8. Tritium generated by thermonuclear testing in the atmosphere between about 1952 and 1963 swamped the natural production of tritium. During the period of nuclear testing, concentrations were often greater than 1000 TU. Tritium levels declined once weapons testing stopped in 1963. Any detectable tritium in a sample implies that the water contains some component of more recent or post–1952 water.

9. The key source of ^{14}C in ground water is the solution of $CO_2(g)$ in the soil zone. When CO_2 is dissolved in the ground water, it is found mostly as HCO_3^- and CO_3^{2-}. The dating method works because once carbonate species move below the water table, ^{14}C begins to decay and there are no additional sources. Other sources of carbon (mineral dissolution, oxidation of old organic matter) can alter the ^{14}C content of the water. An elaborate correction procedure is necessary to extract a date.

10. Indirect methods depend on interpreting systematic changes in the chemical composition of indicator species or isotopes along ground-water flow paths. Some chemical pattern is interpreted in light of events or other independent data, which establish the timing. Environmental isotopes ($\delta^{18}O$ and δD) and chlorofluorocarbons with a known concentration history in precipitation are useful in dating water.

21

GEOCHEMISTRY OF NATURAL WATER SYSTEMS

The mass transport and chemical concepts presented in the last few chapters provide the framework for understanding the chemical behavior of ground water. This general framework applies equally well to core hydrogeological problems such as the chemical evolution of ground water along flow systems or the behavior of ground-water contaminants. Moreover, this process approach lets us address other important problems, including the diagenesis of carbonate rocks, karst formation, the origin of some kinds of ore deposits, and the secondary migration of petroleum, which historically have resided in other areas of geology. More specifically, this chapter examines how mass transport and mass transfer processes work together to determine the chemical evolution of ground water. The examples presented are not designed to comprehensively describe all the different types of water found in natural settings. Instead, they provide illustrative examples of a process-oriented approach to understand the controls on the geochemical evolution of ground water.

▶ 21.1 MIXING AS AN AGENT FOR CHEMICAL CHANGE

Solutes and stable isotopes are affected by advection and dispersion. These processes become apparent along a flow system when water of one chemical composition displaces or simply ends up next to water with a different composition. Mechanical dispersion and diffusion create a zone of mixing between the two waters. Accordingly, phenomena most often thought of in relation to laboratory columns operate on a much larger scale and strongly influence the chemistry of natural ground waters.

Mixing of Meteoric and Original Formation Waters

Ground-water mixing can occur for several reasons. A common situation is related to large-scale structural changes in a sedimentary basin, when, following uplift, meteoric water has the opportunity to flush units that previously contained connate water or formation water. *Connate water* is water that is trapped in the pores of sediment during deposition. For marine deposits, this suggests that connate water has the same composition as seawater. Whether or not this composition is retained up to the beginning of burial depends strongly on the depositional environment as well as numerous other geologic factors. For example, Land and Prezbindowski (1981) point out that the Edwards Formation was massively altered by meteoric diagenesis prior to burial, where the original connate water was first mixed with several pore volumes of meteoric water. The term *formation water* is used when the origin of saline water in a unit is undefined.

Whatever its chemical composition at the time of burial, the displacement of formation water will generally begin when the following two conditions are satisfied: (1) the permeable unit is uplifted so that its outcrop occupies a position that permits recharge by meteoric water, and (2) the downdip portions of the unit have outlets through which the formation water can be displaced. When permeable beds pinch out with depth, definite outlets are not readily available and the water must escape through overlying, less permeable formations.

In the typical pattern, fresh, unaltered meteoric water is found in and near regions of outcrop, with progressive increases in salinity down the flow system. Freshwater or "noses" of relatively freshwater grading into saline water have often been noted. Some examples include the Madison Formation west and south of the Williston Basin (Downey, 1984), the Inyan Kara Formation in southwestern North Dakota (Butler, 1984), and the Dakota-Newcastle Formation of the Dakota Sandstone in western South Dakota (Peter, 1984). These formations have limited recharge areas that coincide with the uplifted portions of the aquifers.

Lahm et al. (1998) provide a field example in the development of a freshwater nose in a regionally extensive aquifer in the midwestern United States. The Silurian-Devonian Carbonate aquifer crops out or subcrops in the western part of Ohio (Figures 21.1 and 21.2). The aquifer dips gently eastward into the Appalachian Basin. As Figure 21.2 implies, there is a dramatic increase in salinity in the aquifer, moving 50 km from areas of outcrop eastward into the basin. Total dissolved solids contents vary from 300 mg/L in the updip part of the carbonate aquifer to greater than 300,000 mg/L in the downdip portion of the basin. Here,

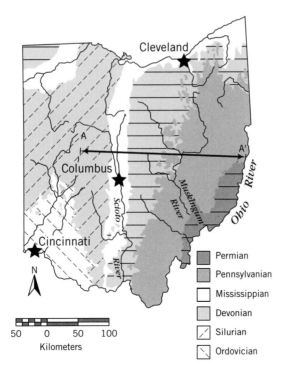

Figure 21.1 Geologic map of Ohio showing the cross-section location and major rivers (from Lahm et al., *Water Resour. Res.*, vol 34, no. 6, p. 1469–1480, 1998). Copyright by American Geophysical Union.

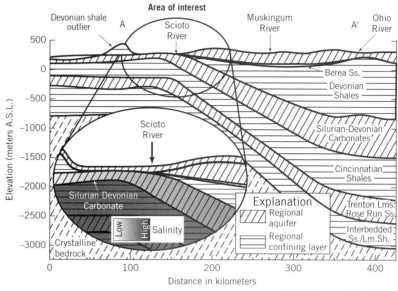

Figure 21.2 Regional geologic cross section parallel with the groundwater flow direction into the Appalachian Basin within the Silurian-Devonian carbonate aquifer. The model area (A-A′) and the area of interest are shown (from Lahm et al., *Water Resour. Res.*, v. 34, no. 6, p. 1469–1480, 1998). Copyright by American Geophysical Union.

then, is an example of how chemical mixing of waters on a large scale controls the variation in major ion chemistry in the aquifer. Lahm et al. (1998) explain that the chemical patterns have persisted for long periods because the flushing process is so inefficient.

Another type of mixing occurs whenever brines are forced upward in a compaction-driven flow system. An example of this mixing occurs in the Gulf of Mexico Basin (Bethke et al., 1988), where abnormal fluid pressure is driving pore fluids out of the basin. In the last two million years, when the effects of overpressuring have been most pronounced, brine migration has apparently pushed back an apparently much more extensive freshwater incursion into the basin (Bethke et al., 1988).

The implication of this process in terms of ground-water chemistry is that brine can move into parts of a basin where it otherwise may not have formed. An example is the Na-Ca-Cl brine found in Lower Cretaceous shelf carbonates of the Edwards Group in south-central Texas (Land and Prezbindowski, 1981). This brine may have formed in the overpressured portion of the Gulf of Mexico basin from evaporites known to occur there and subsequently migrated into shallower, updip parts of the Edwards Group.

Not all mixing in ground water is related to the large-scale tectonic deformation of sedimentary basins. In units having low hydraulic conductivity, the pore water can often be relatively old and chemically different than modern-day recharge. Examples of such materials are the clay-rich tills found in the northern part of the United States and in Canada (Bradbury, 1984; Desaulniers et al., 1981). These deposits often have hydraulic conductivities of 10^{-8} cm/s and lower. Because of the relatively short time since deposition, they can contain water incorporated when the till was deposited. Even where this is not the case, the low permeability of till means that any long-term changes in the chemistry of recharge will be reflected in the chemical patterns along the flow system because of the age differences of the water. Bradbury (1984) describes a site in Wisconsin where changing values of δD and $\delta^{18}O$ document significant differences in the isotopic composition of modern-day recharge and original pore water that was possibly present when the glacial till was deposited about 10,000 years ago.

Diffusion in Sedimentary Basins

Diffusion is the transport of mass in response to concentration gradients. Because the diffusive flux is proportional to the concentration gradient, this process can cause significant mass redistribution whenever large concentration gradients occur and can be maintained over geologic time. Thus, this process should be most effective where zones of hypersalinity underlie brackish water. Diffusion acts to eliminate the concentration gradient and thus becomes less significant as this equalization occurs.

The occurrence of brines in sedimentary basins has been attributed to three causes: dissolution of original evaporites, retention of original formation waters buried with the sediment, or membrane filtration. Whatever the mechanism, the resulting concentration gradients can give rise to significant mass redistribution over geologic time. The common pattern that emerges is a consistent increase in pore-water salinity (mainly Na^+ and Cl^-) with depth (Figure 21.3). The data from north Louisiana, in particular, show a linear increase in NaCl over 2700 m to a depth at which salt is known to occur. An analysis by Ranganathan and Hanor (1987) concluded that diffusion over several tens of millions of years could explain salinity-depth relationships similar to those observed in sedimentary basins.

Manheim and Bischoff (1969) examined the pore-water composition from six holes on the continental slope of the Gulf of Mexico and showed how the presence or absence of salt diapirs affected pore-water salinity. Samples from holes drilled near salt plugs showed systematic increases in salinity, with depth that was interpreted as being caused by salt diffusion (Figure 21.4a). The development of the diffusion profile in Figure 21.4a required about 400,000 years. Samples from holes drilled away from diapiric structures showed little change in chemistry as compared to seawater except for minor diagenetic changes

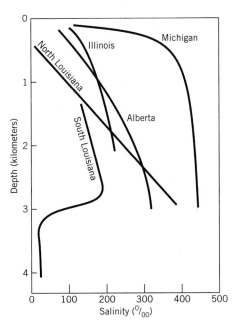

Figure 21.3 Variation in maximum pore-water salinity as a function of depth in the Alberta, Illinois, and Michigan basins as well as the Louisiana Gulf Coast (from Ranganathan and Hanor, 1987). Reprinted from *J. Hydrol.*, v. 92, A numerical model for the formation of saline waters due to diffusion of dissolved NaCl in subsiding sedimentary basins with evaporates, p. 97–120. Copyright 1987, with permission from Elsevier Science.

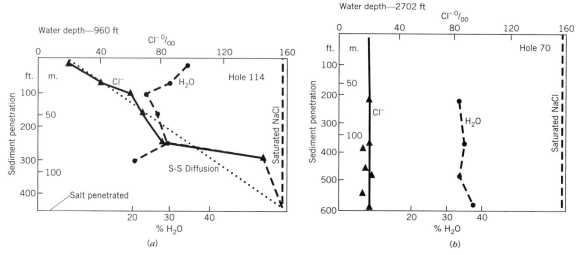

Figure 21.4   Plots of depth versus chlorinity and water content in shallow sediments of the Gulf of Mexico. (*a*) Rock salt is encountered at 433 ft. The dashed line represents the distribution of Cl^- at steady state due to diffusion between two boundaries of constant Cl^-. (*b*) Away from diapiric structures there is no significant variation in Cl^- concentration with depth (from Manheim and Bischoff, 1969). Reprinted from *Chem. Geol.*, v. 4, Geochemistry of pore waters from Shell Oil Company drill holes on the continental slope of the northern Gulf of Mexico, p. 63–82. Copyright 1969, with permission from Elsevier Science.

and a loss of SO_4^{2-} (Figure 21.4*b*). Similar studies along the southeastern Atlantic coast of the United States showed the same result over a greater range in depth (Manheim and Paull, 1981).

▶ 21.2 CHEMICAL PROCESSES AND THEIR IMPACT ON WATER CHEMISTRY

Chemical reactions in the shallow subsurface generally play the dominant role in controlling the chemistry of natural ground waters. Rainfall or snowmelt that ultimately ends up recharging most ground-water systems has little dissolved mineral matter. Yet, once this water disappears into the ground, it quickly accumulates a relatively large load of dissolved mineral matter. Armed with knowledge of how various chemical processes impact water chemistry and features of the geologic setting, it is possible to understand why the chemistry has evolved as it has or to predict what the water chemistry is likely to be for some new area. The important chemical reactions begin in the unsaturated zone as water percolates through the soil zone and moves deeper. Key reactions in this respect include (1) the dissolution of soil gases, like O_2 and CO_2, (2) the dissolution/precipitation of various minerals, like calcite, dolomite, or feldpsar, (3) cation exchange reactions, and (4) the dissolution/utilization of organic compounds. The dissolution reactions are particularly important because water is so efficient at dissolving gases and solids found in the vadose zone. Thus, by the time water gets to the water table, it has acquired a significant load of dissolved mineral matter.

The chemistry of ground water not only depends on the processes in the vadose zone but also on the reactions operating along the saturated flow system. Most of the same processes affecting ion concentrations in the unsaturated zone are also operative in the saturated zone, including the dissolution and precipitation of various minerals and cation exchange. Redox reactions within the aqueous phase become increasingly important in transferring mass among the various aqueous species.

As a general rule, the most rapidly dissolving minerals have the greatest impact on the chemistry of water. Thus, as water moves through the vadose zone, minerals like calcite ($CaCO_3$), dolomite ($CaMg[CO_3]_2$) and gypsum ($CaSO_4.2H_2O$) dissolve to saturation if they are present. Minerals like quartz and feldspar dissolve much more slowly at low temperatures and will only reach equilibrium in water having a long residence time in the subsurface. Thus, this notion of the water/rock system being a partial-equilibrium system, in equilibrium with some minerals, and out of equilibrium with respect to others is a good way to conceptualize the evolution of ground water along a flow system.

Gas Dissolution and Redistribution

The dissolution and redistribution of $CO_2(g)$ are important soil-zone processes and have a profound influence on the chemistry of ground water. Rainwater or melted snow contains relatively small quantities of mass, is somewhat acidic, and has a P_{CO_2} of about $10^{-3.5}$ atm. As this water moves downward, it rapidly dissolves CO_2, which occurs in soil at partial pressures larger than the atmospheric value. Elevated CO_2 pressures are due primarily to root and microbial respiration and to a lesser extent to the oxidation of organic matter (Palmer and Cherry, 1984). Values for soils generally range from $10^{-3.5}$ atm to more than 500 times larger (Palmer and Cherry, 1984). CO_2 dissolved in water is further redistributed among the weak acids of the carbonate system according to the following set of reactions:

$$CO_2(g) + H_2O = H_2CO_3^*$$

$$H_2CO_3^* = HCO_3^- + H^+$$

$$HCO_3^- = CO_3^{2-} + H^+$$

One direct result of dissolving $CO_2(g)$ in water is a rapid increase in the total carbonate content of the water and a decrease in pH. For a pH range of 4.5 to 5.5, $H_2CO_3^*$ is the dominant carbonate species, with HCO_3^- and H^+ the most dominant anion and cation, respectively. In general, because the partial pressures of CO_2 gas are high in the soil zone, the water becomes quite acidic and rich in dissolved carbonate species.

Another important soil-zone/atmospheric process is the dissolution of $O_2(g)$. The resulting levels of dissolved oxygen are sufficiently large, at least initially, to control the redox chemistry in shallow ground water.

Mineral Dissolution/Precipitation

CO_2-charged ground water is effective in dissolving minerals. The most common reactions involve the weak acids of the carbonate and silicate systems and the strong bases from the dissolution of carbonate, silicate, and alumino-silicate minerals. As we showed in Section 18.1, this process causes the weak acids to dissociate. In the carbonate system, the relative abundance of HCO_3^- and CO_3^{2-} increases at the expense of $H_2CO_3^*$. Overall, both the alkalinity and cation concentrations increase. Following are both generic and specific examples of weak acid, strong base reactions:

carbonate minerals $+ H^+ =$ cations $+ HCO_3^-$

for example, calcite: $CaCO_3(s) + H^+ = Ca^{2+} + HCO_3^-$

silicate minerals $+ H^+ =$ cations $+ H_2SiO_3$

for example, enstatite: $MgSiO_3(s) + 2H^+ = Mg^{2+} + H_2SiO_3$

$$\text{alumino-silicate minerals} + H^+ = \text{cations} + H_2SiO_3 + \text{secondary minerals}$$

for example, anorthite: $CaAl_2Si_3O_8(s) + 2H^+ + H_2O = \text{kaolinite} + Ca^{2+}$

albite: $2NaAlSi_3O_8(s) + 2H^+ + 5H_2O = \text{Kaolinite} + 4H_2SiO_3 + 2Na^+$

Soluble salts commonly are disseminated within particular geologic units (for example, near-surface deposits in arid areas) or occur as thick evaporite deposits in sedimentary basins. Mineral salts are extremely soluble and when present can dissolve to produce saline waters and even brines. The actual composition of the water depends on the particular minerals present (for example, halite, anhydrite, gypsum, carnalite, kieserite, and sylvite). Here are examples of salts and resulting ion species.

$$\text{halite}: NaCl(s) = Na^+ + Cl^-$$

$$\text{anhydrite}: CaSO_4(s) = Ca^{2+} + SO_4^{2-}$$

$$\text{gypsum}: CaSO_4 \cdot 2H_2O(s) = Ca^{2+} + SO_4^{2-} + 2H_2O$$

$$\text{carnalite}: KCl \cdot MgCl_2 \cdot 6H_2O(s) = K^+ + Mg^{2+} + 3Cl^- + 6H_2O$$

$$\text{kieserite}: MgSO_4 \cdot H_2O(s) = Mg^{2+} + SO_4^{2-} + H_2O$$

$$\text{sylvite}: KCl(s) = K^+ + Cl^-$$

Sulfide oxidation is one of the important redox reactions within the unsaturated zone. Minerals like pyrite or marcasite are dissolved in oxidation reactions to produce $Fe(OH)_3(s)$, SO_4^{2-} and H^+. In fact, pyrite oxidation is one of the most important acid-producing reactions in geological systems (Moran et al., 1978). In coal mining areas like the Appalachians, this reaction can be the cause of serious problems of acid-mine drainage (Moran et al., 1978). Following is an example of sulfide mineral oxidation:

$$4FeS_2(s) + 1SO_2 + 14H_2O = 4Fe(OH)_3 + 16H^+ + 8SO_4^{2-}$$

Precipitation reactions operate to remove mass from solution within the soil zone and along the ground-water flow system. Although such mass losses are important, they are difficult to notice in chemical data because they tend to be swamped by mass additions. It is these precipitation processes that give rise to the formation of cements and authigenic pore-filling minerals.

Cation Exchange Reactions

The most important exchange reactions are the water-softening reactions where Ca^{2+} and Mg^{2+} in the water exchange with sorbed Na^+ as ground water moves through clayey material, for example,

$$\begin{matrix} Ca^{2+} & & Ca^{2+} \\ Mg^{2+} & + \text{Na-clay} = 2Na^+ & + \text{Mg-clay} \\ Fe^{2+} & & Fe^{2+} \end{matrix}$$

where Na-clay is Na adsorbed onto a clay mineral.

Exchange reactions can involve clay minerals within the unsaturated zone. In large flow systems, cation exchange is usually related to clay minerals deposited in a marine environment, which contain a large reservoir of exchangeable Na. One does not have to go far in the United States or Canada to find clays or shales capable of ion exchange.

Dissolution/Utilization of Organic Compounds

Works by Wallis et al. (1981), Thurman (1985), and Hendry et al. (1986) have identified a variety of important organic reactions such as dissolution of organic litter at the ground surface, complexation of Fe and Al, sorption of organic compounds, and oxidation of organic compounds.

The dissolution of organic litter at or close to the ground surface is the major source of dissolved organic carbon (DOC) in soil water and shallow ground water. DOC concentrations typically fall in a range from 10 to 50 mg/L in the upper soil horizons and less than 5 mg/L deeper in the unsaturated zone (Thurman, 1983). Concentrations are highest at their source and decline with depth through sorption and oxidation. The most important fraction of the DOC is a group of humic substances, consisting mainly of humic and fulvic acids. Tannins and lignins, amino acids, and phenolic compounds are often present in smaller concentrations (Wallis et al., 1981).

The complexation of Fe and Al with organic matter is an important process facilitating the transport of these poorly soluble metals from the A horizon of the soil to the B horizon (Thurman, 1983). This is one key feature of the soil-forming process called podzolization (Thurman, 1983). In terms of ground-water systems, this reaction is not particularly important because most of these complexed metals sorb in the B horizon.

Dissolved organic compounds originating in the upper part of the soil horizon are not particularly mobile because of sorption. Many of the sorption models we discussed in Chapter 18 operate within the soil zone. Hydrophobic sorption occurs because of the relatively large quantities of solid organic matter present in the upper part of many soil horizons (Thurman, 1983). Similarly, the abundance of metal oxides, hydroxides, and clay minerals leads to surface complexation reactions and electrostatic interactions. Again in terms of the chemistry of shallow ground water, this is one of the processes that keeps the quantity of DOC in recharge at relatively low concentrations.

Oxidation reactions involving organic matter can influence the chemistry of shallow ground water. For example, the oxidation of dissolved organic matter (represented as CH_2O) provides a source of CO_2 gas within the unsaturated zone, which is readily dissolved in soil water. A second reaction, involving the oxidation of a sulfur-containing compound (represented by the amino acid cysteine), is thought to play a major role in the accumulation of gypsum in shallow soils (Hendry et al., 1986). This reaction is the organic counterpart to the pyrite oxidation reaction. In arid areas, this process can also produce recharge with large SO_4^{2-} concentrations (Hendry et al., 1986).

Redox Reactions

Redox conditions are most important along a ground-water flow system, where they play an important role in controlling (a) the chemistry of metal ions and solids (for example, Fe^{2+}, Mn^{2+}, and Fe_2O_3), (b) species or solids containing sulfur (for example, SO_4^{2-}, H_2S, and FeS_2), and (c) dissolved gases containing carbon (for example, CO_2, CH_4). Here are examples of key half-reactions important in aquifers:

$$\tfrac{1}{4}O_2(g) + H^+ + e^- = \tfrac{1}{2}H_2O$$

$$\tfrac{1}{2}Fe_2O_3(s) + 3H^+ + e^- = Fe^{2+} + \tfrac{3}{2}H_2O$$

$$\tfrac{1}{2}MnO_2(s) + 2H^+ + e^- = \tfrac{1}{2}Mn^{2+} + H_2O$$

$$\tfrac{1}{8}SO_4^{2-} + (9/8)H^+ + e^- = \tfrac{1}{8}HS^- + \tfrac{1}{2}H_2O$$

$$\tfrac{1}{8}CO_2(g) + H^+ + e^- = \tfrac{1}{8}CH_4(g) + \tfrac{1}{4}H_2O$$

$$\tfrac{1}{4}CO_2(g) + H^+ + e^- = \tfrac{1}{4}CH_2O + \tfrac{1}{4}H_2O$$

In some flow systems it is possible to define redox zones. These zones are parts of an aquifer in which pe is controlled by a dominant redox couple. Field studies (for example, Champ et al., 1979; Jackson and Patterson, 1982) have shown that oxygen, iron-manganese, and sulfide zones will often be present. The probable half reactions controlling pe in these zones are the reduction of oxygen to water, the reduction of iron or manganese oxides, and the reduction of sulfate to HS^- or H_2S. In a few cases, a methane zone can form from the reduction of CO_2.

Reduced sulfide and methane zones tend to develop in confined flow systems containing excess oxidizable DOC and a lack of recharge containing oxygen downgradient in the system. Thus, redox zones will be observed in an extensive artesian aquifer that receives recharge from a limited area of outcrop or in an aquifer with confining units.

Figure 21.5 summarizes the common changes in redox chemistry from zone to zone (Jackson and Inch, 1980). In zone I, oxygen initially present in recharge will decline through reduction by organic carbon

$$CH_2O + O_2 = CO_2 + H_2O \tag{21.1}$$

The concentrations of Fe^{2+} and Mn^{2+} in zone II increase because the Fe(III) and Mn(IV) minerals, which are oxidized solids, are not stable in the more reducing environment. These reactions are

$$CH_2O + 8H^+ + 4Fe(OH)_3(s) = 4Fe^{2+} + 11H_2O + CO_2 \tag{21.2}$$

$$CH_2O + 4H^+ + 2MnO_2(s) = 2Mn^{2+} + 3H_2O + CO_2 \tag{21.3}$$

Once the pe is sufficiently reduced, sulfide species appear from the reduction of SO_4^{2-}, or

$$2CH_2O + SO_4{}^{2-} + H^+ = HS^- + 2H_2O + 2CO_2 \tag{21.4}$$

When the rate of sulfide production exceeds the dissolution rate of iron and manganese oxides, Fe^{2+} and Mn^{2+} concentrations fall (Figure 21.5) as metal sulfides precipitate as the stable phase.

21.3 EXAMPLES OF HOW REACTIONS AFFECT WATER CHEMISTRY

This section provides examples of how the chemical processes presented in the previous section work to control the chemistry of ground water. For each example, we highlight the most important reaction and describe how the water evolves chemically. For most of these examples, the mineralogy of the hydrogeologic unit and climate play major roles in determining how the water evolves. Thus, with information about the geologic setting and the climate, one can begin to anticipate how water might evolve chemically.

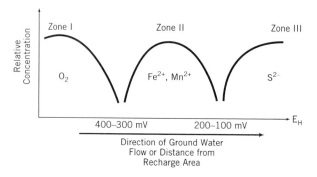

Figure 21.5 Different redox zones existing along a confined aquifer (from Jackson and Inch, 1980). Reproduced with the permission of the Minister of Public Works and Government Services, Canada, 2001 and courtesy of Environment Canada.

Chemical Evolution of Ground Water in Carbonate Terrains

> - dissolution of CO_2 gas
> - dissolution of one or two carbonate minerals

This simple example shows how water evolves as a consequence of two simple chemical processes: the dissolution of CO_2 gas in the shallow soil zone and the dissolution of calcite and/or dolomite. Rain water enters the subsurface and rapidly dissolves CO_2 gas. At the same time, it begins to dissolve carbonate that it encounters. The aggressiveness of the water (reflected by pH) depends on the partial pressure of CO_2 gas initially present in the water. Langmuir (1971) produced a graph that shows how HCO_3^- and pH change as the water approaches equilibrium with respect to calcite and dolomite. For a system that is open with respect to CO_2, water evolves to equilibrium with calcite along lines of constant P_{CO_2} (Figure 21.6). For example, moving along the P_{CO_2} line of $10^{-2.5}$ atm, the pH of a solution in equilibrium with calcite is 7.65, and the HCO_3^- concentration is about 175 ppm. The second solubility curve labeled "dolomite" illustrates the pH and HCO_3^- concentration for soil water in equilibrium with dolomite alone or calcite and dolomite together. At the temperature of calculation, the solubility curves for dolomite and dolomite plus calcite are the same (Langmuir, 1971).

Another feature of this diagram is the series of paths labeled "no CO_2 added." These lines describe how the concentration of HCO_3^- and pH change under closed-system conditions. In calculating these pathways, the P_{CO_2} is not replenished as CO_2 is depleted by the dissolution of carbonates. Whether the system is open or closed with respect to CO_2 can have an important bearing on the pore-water chemistry. For example, an open system $P_{CO_2} = 10^{-2}$ atm has a HCO_3^- concentration of about 260 ppm and a pH of about 7.34 at calcite saturation (Figure 21.6). Under closed-system assumptions, these values are 170 ppm and 7.70, respectively.

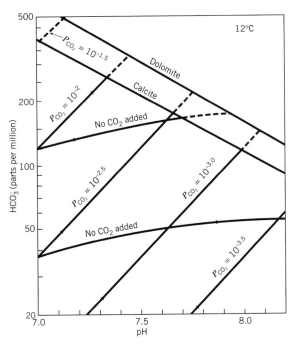

Figure 21.6 Possible approaches to equilibrium of ground water in contact with calcite and (or) dolomite at 12°C. The solubility curves of dolomite and calcite plus dolomite are the same at this temperature (from Langmuir, 1971). Reprinted from *Geochim. Cosmochim. Acta*, v. 35, The geochemistry of some carbonate ground waters in central Pennsylvania, p. 1023–1045. Copyright 1971, with permission from Elsevier Science.

A carbonate terrain provides a field situation in which the dissolution of CO_2 gas and calcite and/or dolomite describes the chemistry of soil water and ground water. Table 21.1 compares the chemistry of rain water with water from surface springs and wells in a limestone in central Pennsylvania (Langmuir, 1971). The spring water reflects relatively short travel distances and probably approximates the composition of water from the unsaturated zone. The most abundant ions are Ca^{2+} and HCO_3^-, with relatively small concentrations of Na^+ and Cl^-. The calculated P_{CO_2} for a single spring sample is $10^{-2.08}$ atm.

Generally, all of the spring water is undersaturated with respect to calcite and dolomite. Presumably, transport through fractures has been so rapid that calcite equilibrium was not achieved. Farther along the flow system, calcite equilibrium finally constrains increases in Ca^{2+} and HCO_3^- concentrations.

Once in the zone of saturation, the ground water approaches saturation with respect to both calcite and dolomite because of the longer residence time (Figure 21.7). Also, water from a predominantly limestone source rock more closely approaches saturation with respect to calcite than dolomite. This fact is illustrated in Figure 21.7 by the abundance of data points for limestone source rocks (triangles) lying above the dashed line ($SI_c = SI_d$). Even in this simple carbonate system, there are differences in the pathways of evolution, apparently depending on what carbonate minerals were encountered and in what order.

The example of evolving composition in a predominantly carbonate terrain (Langmuir, 1971) illustrates how the composition of the starting rain water simply proceeds toward equilibrium with respect to those minerals available for dissolution (for example, calcite and dolomite).

TABLE 21.1 **Chemical composition of rain water and ground water from a carbonate terrain**

Water Type	Ca^{2+}	Mg^{2+}	Na^+	K^+	HCO_3^-	SO_4^{2-}	Cl^-	pH	SI_c	SI_d
Rain[a]	4.1	0.4	1.0	0.2	9.0	2.2	1.5	6.46	–	–
Spring Water	47.5	13.9	3.8	1.6	183	22.0	8.2	7.37	−0.41	−0.62
Well Water	55.0	29.6	3.3	1.5	265	22.0	13.4	7.46	−0.15	−0.18

[a]Rain sample from Southern Canada.
Concentrations in mg/L, pH in pH units.
Source: Modified from Langmuir, 1971.

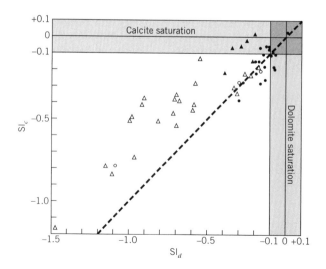

Figure 21.7 Saturation index for calcite versus the saturation index for dolomite. Spring waters are designated by open symbols; well waters by solid symbols. Triangles denote a limestone sources rock; circles denote a dolomitic source rock. Cross-hatched area shows the limits of uncertainty of SI_c and SI_d (from Langmuir, 1971). Reprinted from *Geochim. Cosmochim. Acta*, v. 35, The geochemistry of some carbonate ground waters in central Pennsylvania, p. 1023–1045. Copyright 1971, with permission from Elsevier Science.

Shallow Brines in Western Oklahoma

> • dissolution of halite and gypsum/anhydrite

This example is similar to the previous one in that the presence of a few rapidly dissolving minerals controls the chemical evolution of the ground water. In the shallow ground water of western Oklahoma and the southeastern part of the Texas Panhandle, the minerals involved are so soluble that brines are generated. Freshwater recharged through permeable units moves downward until it encounters salts at depths ranging from 10 to 250 m (Figure 21.8). The dissolving salt produces cavities at the updip limit or the top of the salt (Johnson, 1981). Periodically, the rocks overlying the cavities collapse. The process is apparently self-perpetuating because the collapse and fracturing of overlying units provide improved access to the salt for freshwater.

The evaporites, mainly halite and gypsum/anhydrite, are interbedded with a thick sequence of red-beds. Given the particular salts involved, high concentrations of Na^+ and Cl^- are not surprising. Clearly, when ground water encounters large quantities of soluble salts in the subsurface, the impact on the chemistry is considerable.

Chemistry of Ground Water in an Igneous Terrain

> • CO_2 dissolution and redistribution
> • dissolution of alumino-silicate minerals

Igneous and metamorphic rocks contain silicates and alumino-silicate minerals that are slow to react even when under attack by acidic waters. Thus, the mass dissolved in pore water will be relatively small when only these minerals are present. Garrels and MacKenzie (1967) looked at a case where waters rich in CO_2 reacted with plagioclase, K-feldspar, and biotite in igneous rock. Table 21.2 documents the evolution of water from rain through ground water with a short residence time (ephemeral springs) to ground water with a longer residence time (perennial springs). In ephemeral water, the concentration of all ions is low because only plagioclase feldspar, biotite, and K-feldspar dissolve. In fact, the concentration of most ions is so low that the initial composition of precipitation had to

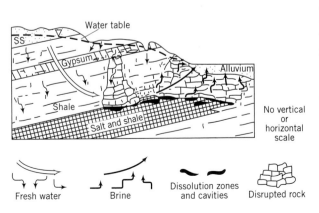

Figure 21.8 Schematic representation of the circulation of freshwater and brine in areas of salt dissolution in western Oklahoma. There is no scale, but the length of the section could range from 1 to 15 km and the thickness from 30 to 300 m (from Johnson, 1981). Reprinted *J. Hydrol.* Special Issue: Symposium on Geochemistry of Groundwater, v. 54, Dissolution of salt on the east flank of the Permian Basin in the south-western United States; p. 75–94. Copyright 1981, with permission from Elsevier Science.

TABLE 21.2   Chemical composition of rain water and ground water from a granitic terrain of the Sierra Nevadas

Water Type	Ca^{2+}	Mg^{2+}	Na^+	K^+	HCO_3^-	SO_4^{2-}	Cl^-	pH
Rain								
Ephemeral Springs	3.11	0.70	3.03	1.09	20.0	1.00	0.50	6.2
Perennial Springs	10.4	1.70	5.95	1.57	54.6	2.38	1.06	6.8

Concentrations in mg/L, pH in pH units.
Source: Modified from Garrels and MacKenzie, 1967.

be considered when Garrels and MacKenzie worked with the data. Dissolution apparently takes place under closed-system conditions with about half of the CO_2 consumed.

With longer residence time, the quantity of mass dissolved in the ground water increases. The concentrations of cations increase due to the continued reaction of biotite and plagioclase, as does the alkalinity (reflected in the HCO_3^- concentration and pH). An important source of Ca^{2+} is the dissolution of small quantities of carbonate minerals. In the deeper parts of the system, montmorillonite occurs as a weathering product of plagioclase in addition to kaolinite.

Evolution of Shallow Ground Water in an Arid Prairie Setting

> - CO_2 dissolution and redistribution
> - O_2 dissolution
> - dissolution/precipitation of calcite, pyrite, and gypsum
> - cation exchange

A study of shallow ground water from a study area in North Dakota (Moran et al., 1978) illustrates how many of the processes that we have considered work together. The system response to CO_2 dissolution and calcite distribution is the same as in the previous examples. Of greatest importance in this system is the aridity that results in abundant gypsum in the soil and ultimately large solute loads in the ground water.

Over much of the plains region of the United States and Canada, annual potential evaporation exceeds annual precipitation by a considerable amount. Thus, water that infiltrates in normal precipitation years evaporates and deposits a small quantity of gypsum. With repeated rain or snowmelt, gypsum accumulates in the upper part of the soil horizon (Moran et al., 1978). Exceptional recharge can dissolve some of this soluble material and move it down and into the ground-water system (Figure 21.9). In some arid areas, recharge water could have SO_4^{2-} concentrations in excess of 5000 mg/L (Hendry et al., 1986).

The chemical data in Table 21.3 represent mean values for 39 water samples collected from various near-surface drift units. Cation exchange operates to replace Ca^{2+}, which results from gypsum dissolution with Na^+. Thus, the abundance of Na^+ and SO_4^{2-} in the shallow ground water is attributed mainly to gypsum dissolution and cation exchange. Reaction of pyrite with oxygen also adds SO_4^{2-} to the ground water.

Redox Reactions in a Small-Scale Flow System

> - removal of O_2 by reaction with organic matter
> - reduction of Fe(III) and Mn(IV) minerals
> - reduction of SO_4^{2-}

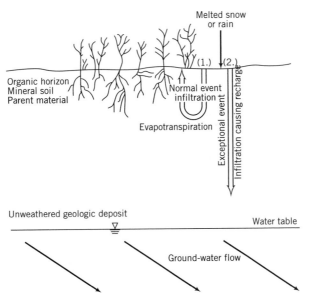

Figure 21.9 A conceptual model of subsurface flow in the plains region of the United States and Canada. (1) Annual potential evpotranspiration greatly exceeds annual precipitation, and most infiltration is lost by evapotranspiration. (2) Exceptional precipitation events produce recharge (from Moran et al., 1978).

TABLE 21.3 **Average composition of ground water collected from shallow drift units in North Dakota**

Water Type	Ca^{2+}	Mg^{2+}	Na^+	K^+	HCO_3^-	SO_4^{2-}	Cl^-	pH
Shallow Ground Water	62	25	469	7.9	748	577	4.7	8.0

Concentrations in mg/L, pH in pH units
Source: Modified from Moran et al., 1979.

A study at Chalk River Nuclear Laboratories near Ottawa, Canada, demonstrates how redox conditions can change along a small flow system. By careful measurements of pH, E_H, dissolved oxygen (DO), and the total concentration of sulfides (S_T^{2-}), as well as other key chemical parameters, Jackson and Patterson (1982) defined three redox zones. The main geologic units were two fluvial sand aquifers separated by a thin layer of interbedded clay about 1 m thick (Figure 21.10). Ground water flows southward from an upland area toward Perch Lake. Inflow to the aquifers is near the area labeled "disposal area" (Figure 21.10), with none further downgradient.

The lower aquifer in the recharge area contains dissolved oxygen and has an E_H of about 0.55 V (Figure 21.11), which is quite close to the theoretical range of 0.71 to 0.83 V. Both Fe^{2+} and Mn^{2+} concentrations are low, and sulfide is undetectable. The declining E_H along the deep flow system coincides with the reduction of oxygen. Increasing concentrations of iron and manganese are evident once the oxygen is depleted (Figure 21.11). Zone II is probably not extensive because sulfide is becoming relatively abundant at E_H values below 0.2 V. The downstream end of the flow system is a sulfide zone, with E_H values approaching the theoretical range for the sulfate-sulfide couple (i.e., −0.21 to −0.32 V). Concentrations of Fe^{2+}, Mn^{2+}, and SO_4^{2-} decline in zone III from the precipitation of ferrous sulfides (Jackson and Patterson, 1982).

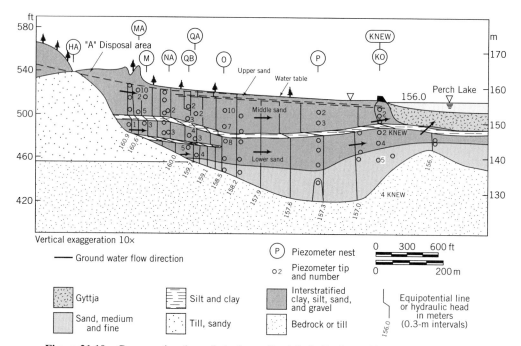

Figure 21.10 Cross section through the lower Perch Lake Basin aquifer at Chalk River Nuclear Laboratories, Canada, showing equipotential lines, piezometer tips, and the position of the water table. Hydraulic-head measurements averaged over the period of 1973–1975 (from Jackson and Patterson, *Water Resour. Res.*, v. 18, p. 1255–1268, 1982). Copyright © by American Geophysical Union.

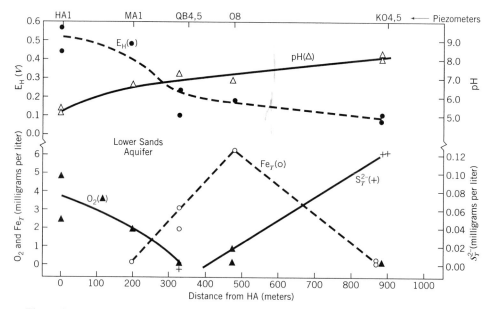

Figure 21.11 The pH, E_H, O_2, and total dissolved iron and sulfide values for the deep, confined Lower Sands aquifer. The piezometer numbers are shown on the top abscissa (from Jackson and Patterson, *Water Resour. Res.*, v. 18, p. 1255–1268, 1982). Copyright by American Geophysical Union.

Sodium-Bicarbonate (Na-HCO$_3^-$) Ground Waters

> - continuing production and redistribution of CO$_2$
> - redox reactions
> - cation exchange

In North America and elsewhere, sodium-bicarbonate (Na-HCO$_3^-$)-type waters can evolve. This water might have HCO$_3^-$ concentrations in excess of 1000 mg/L, along with relatively large concentrations of uncommon species like F$^-$. The key feature contributing to the evolution of this ground water is a continuing source of CO$_2$ all along the flow system. Interestingly, many of the redox reactions involve the oxidation organic matter (see Section 21.1), which produce CO$_2$ as a product. This CO$_2$ is redistributed among H$_2$CO$_3^*$, HCO$_3^-$, and CO$_3^{2-}$. In many of these aquifers, Ca^{2+} and Mg^{2+} are exchanged on clay minerals for Na$^+$, which provides the possibility for carbonate dissolution and even higher HCO$_3^-$ concentrations. With CO$_2$ generated by redox reactions, ion exchange, and carbonate dissolution, the water will evolve chemically to a sodium-bicarbonate type like that from the Atlantic coastal plain (Foster, 1950), the Eocene aquifers of the east Texas Basin (Fogg and Kreitler, 1982), and the Milk River aquifer (Hendry and Schwartz, 1990).

Chapelle et al. (1987) investigated the source of CO$_2$ in ground water from the Atlantic Coastal Plain in Maryland. The units of interest included the Magothy-Upper Patapsco aquifer and the Lower Patapsco aquifer. These aquifers crop out between Washington and Baltimore (Figure 21.12) and dip toward the east. Recharge occurs along the outcrop areas, with flow toward the Potomac River and under Chesapeake Bay (Figure 21.12). Although these units contain no calcareous minerals, HCO$_3^-$ concentrations increase significantly in the direction of flow. Where the Patapsco aquifer crops out, HCO$_3^-$ concentrations range from 0 to 50 mg/L. Approximately 30 km downgradient, they have increased to about 150–200 mg/L, and, finally, near Cambridge, Maryland (Figure 21.12), HCO$_3^-$ concentrations

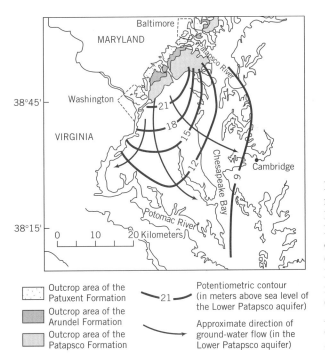

Figure 21.12 Map showing the regional outcrop area for the Patuxent, Arundel, and Patapsco Formations on the Atlantic coastal plain in Maryland. Ground-water flow in the Lower Patapsco Formation is away from the outcrop area (from Chapelle et al., *Water Resources Res.*, v. 23. p. 1625–1632, 1987). Copyright by American Geophysical Union.

range from 400 to 500 mg/L (Chapelle et al., 1987). This pattern is similar to that observed by Foster (1950) farther south in Virginia.

The potential sources of CO_2 are the bacterially mediated oxidation of solid organic matter or the abiotic decarboxylation of these materials. The presence of both sulfate-reducing and methanogenic bacteria in cores, however, points to the generation of CO_2 in redox reactions. Although sulfate-reducing bacteria are present, the relatively small quantities of SO_4^{2-} present make the contribution of CO_2 from sulfate reduction small relative to that from methanogenesis. The bacteria facilitating these reactions occur together because the theoretical *pe* of a system is nearly the same with either reaction.

▶ 21.4 CASE STUDY OF THE MILK RIVER AQUIFER

The Milk River aquifer provides an interesting study because interpretation of chemical patterns requires integration of concepts of mass transport and mass transfer. Several studies have documented the geologic and hydrogeologic setting, including Meyboom (1960), Schwartz and Muehlenbachs (1979), and Hendry and Schwartz (1988, 1990). The aquifer is part of a thick sequence of Cretaceous rocks (Figure 21.13). Units of interest from oldest to youngest are the shale of the Colorado Group, the Milk River Formation whose lower member is the aquifer, and the shale of the Pakowki Formation (Figure 21.13). The aquifer pinches out to the north and east as the result of a facies change.

Ground-water recharge occurs where the aquifer crops out in northern Montana and along an area of outcrop-subcrop in southern Alberta. Flow is mainly northward or downdip in the aquifer. Leakage from the aquifer moves upward across the Pakowki Formation and overlying units. Before extensive development of the ground-water resource, flowing wells were common.

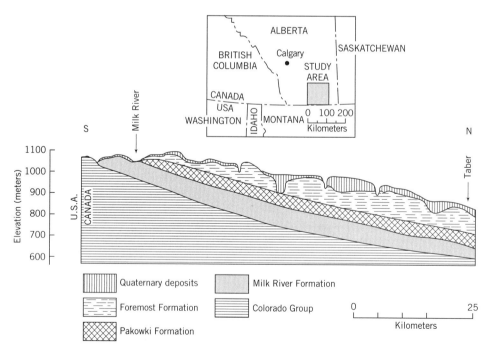

Figure 21.13 Study area for the Milk River aquifer investigation. The geological cross section illustrates the most important drift and bedrock units in southern Alberta, Canada (from Hendry and Schwartz, 1990). Reprinted by permission of *Ground Water.* Copyright © 1990. All rights reserved.

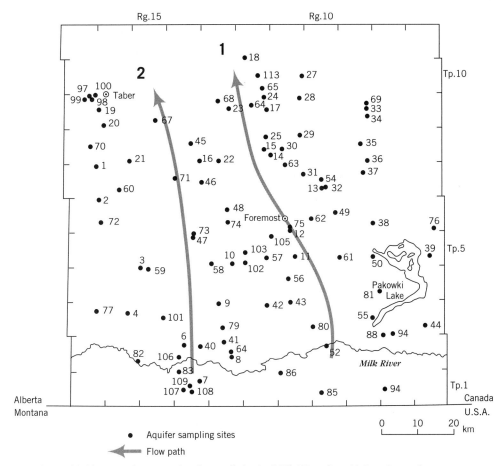

Figure 21.14 Location map showing wells in the Milk River for which major analyses are available (from Hendry et al., 1991). Reproduced from *Applied geochemistry, hydrogeology and hydrogeochemistry of the Milk River aquifer system*, v. 6, p. 369–380. Copyright 1991, with permission from Elsevier Science.

The first step in interpreting patterns of water chemistry is to examine the basic data. All the wells located on Figure 21.14 have major ion and some isotopic data available for them. All the chemical data (concentrations as mM) are included as the Excel file Milk River.xls, on the website for you to investigate.

Let's begin by looking at some of the basic data. We will concentrate on data for a few of the wells along the flow paths 1 and 2 (Figure 21.14). Simple inspection of the data in Table 21.4 suggests a significant variability along the two flow paths. In particular, there is a marked increase in Na^+, HCO_3^-, and Cl^- and a decrease in SO_4^{2-}. Ca^{2+} and Mg^{2+} are generally small and don't change much. By converting the data in Table 21.4 to % meq/L, we could display this information in a Piper diagram to depict this pattern of chemical evolution.

Given the obvious pattern of change in water chemistry from the recharge area down the flow system, there is a good chance that mapping the ion distributions would be useful. The data (Milk River.xls) are in a format that can be used with Surfer®, the industry-standard package for plotting and contouring irregularly spaced data (two space coordinates and a value). Let's illustrate this approach and make a map showing the distribution of Cl^-.

TABLE 21.4   Change in major ion chemistry in the Milk River aquifer along two flow paths

Well number	Ca		Mg		Na		Tot meq	HCO₃+CO₃		Cl		SO₄		Tot meq
	mM	%meq	mM	%meq	mM	%meq		mM	%meq	mM	%meq	mM	%meq	
52	0.04	0.39	0.02	0.02	20.4	99.4	20.5	12.6	63.8	0.3	1.3	3.44	34.8	19.7
80	0.03	0.31	0.01	0.01	19.5	99.6	19.6	11.9	66.0	0.4	2.2	2.87	31.8	18.0
11	0.02	0.23	0.01	0.01	17	99.6	17.0	14.6	91.7	1.3	8.2	0.01	0.12	15.9
75	0.01	0.12	0.01	0.01	16.6	99.8	16.7	14.8	88.9	1.8	11.0	0.01	0.12	16.7
23	0.03	0.25	0.02	0.02	23.9	99.6	24.0	20.3	83.8	3.8	15.9	0.04	0.33	24.2
106	0.63	5.43	0.53	0.46	20.9	90.0	23.2	13.2	59.2	1.3	5.7	3.9	35	22.3
6	0.15	1.61	0.15	0.16	18	96.8	18.6	9.9	54.7	1.5	8.3	3.35	37	18.1
73	0.01	0.08	0.02	0.02	24.7	99.8	24.8	20.3	81.8	3.2	12.8	0.67	5.3	24.8
71	0.02	0.11	0.04	0.02	34.9	99.7	35.0	23.6	67.6	11.3	32.3	0.01	0.05	34.9
67	0.04	0.17	0.07	0.03	48.1	99.5	48.3	23.5	48.4	25.0	51.5	0.03	0.12	48.5

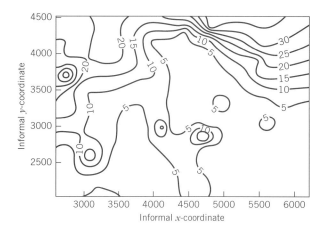

Figure 21.15 Variation in Cl⁻ concentration (mM) in the Milk River aquifer.

Before copying the data from the Excel spreadsheet, it is necessary to remove the less than signs (<) they are present. After this is completed, the data can be copied to the Surfer worksheet. Once the data are gridded, using a kriging approach, the map shown in Figure 21.15 can be prepared. It shows that Cl^- concentrations increase from less than 5 mM in the south to more than 30 mM in the north or the downgradient part of the flow system.

Na^+ and HCO_3^- concentrations increase in the direction of flow (Figure 21.16). SO_4^{2-} behaves oppositely, with the highest concentrations in the recharge area and a systematic decrease downgradient in the confined part of the aquifer (Figure 21.16). Not shown are maps for Ca^{2+} and Mg^{2+}. Except for a few areas where the Milk River crops out, concentrations of these ions are less than a few milligrams per liter.

Several different studies have been conducted to explain these patterns, such as Schwartz and Muehlenbachs (1979), Domenico and Robbins (1985a), and Phillips et al. (1986). A reexamination of these ideas prompted the development of a model that involves dispersion and advection in the aquifer and diffusion in the aquitard (Hendry and Schwartz, 1988). The change in water chemistry is explained in terms of both mass transport processes and geologic processes.

Toth and Corbet (1986) showed how the configuration of the land surface changed with time. As recently as five million years ago, the land surface was probably 700 m higher than now. Even in the early Pleistocene, the land surface was about 200 m higher than at present

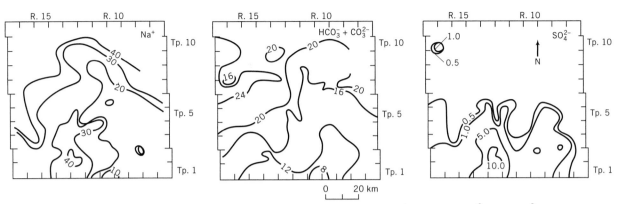

Figure 21.16 Concentration distribution (mM) of Na$^+$, HCO$_3^-$, + CO$_3^{2-}$, and SO$_4^{2-}$ in waters of the Milk River aquifer (modified from Hendry and Schwartz, 1990). Reprinted by permission of *Ground Water*. Copyright ©1990. All rights reserved.

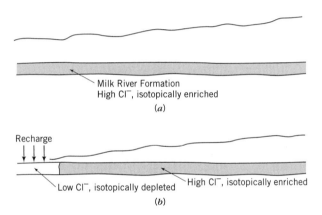

Figure 21.17 The effect of geologic changes on the chemistry of ground water in the Milk River aquifer. In (*a*) the aquifer is deeply buried. With time, erosion uncovers the upstream end of the aquifer, causing inflow of meteoric water (modified from Hendry and Schwartz, 1990). Reprinted by permission of *Ground Water*. Copyright ©1990. All rights reserved.

(Toth and Corbet, 1986). When the aquifer was deeply buried (Figure 21.17*a*), ground water probably had a mean Cl$^-$ concentration of about 85 mM (Hendry and Schwartz, 1988). Continuing erosion eventually exposed a relatively large recharge area, providing a new source of water to the aquifer. Proximity to the surface would result in recharge with a low Cl$^-$ concentration (Figure 21.17*b*). This change in the flow dynamics of the aquifer probably occurred about 1×10^6 years ago.

With time, this chemically different water has displaced the original formation water. However, in so doing, it has created diffusion gradients between the low-chlorinity water in the aquifer and the original water remaining in the shale of the Colorado Group. Thus, water moving down the aquifer experienced an increase in Cl$^-$ concentration due to aquitard diffusion. This simple model of advection and dispersion in the aquifer and diffusion in the aquitard was tested mathematically by Hendry and Schwartz (1988).

The classical interpretation of these data would point to sulfate reduction and ion exchange as the key processes responsible for generating these patterns. However, other processes are probably more important. The aquitard diffusion model, explaining the distribution of Cl$^-$, also constrains the behavior of the cations. Diffusion of only Cl$^-$ would produce a charge imbalance. Thus, counter ions must diffuse from the shale to maintain electroneutrality. Na^+ would be the likely species, given its abundance in formation waters in the basin (Hendry and Schwartz, 1990). If the Na^+ distribution is in fact related to Cl^-, a relationship should exist between Na^+ and Cl^- concentrations. The lines in Figure 21.18 are hypothetical pathways of chemical evolution generated by adding equivalent amounts of Na^+ and Cl^- to water from the recharge area. The lines fit the observed trends in the data extremely well, suggesting that the process controlling the Na^+ ion concentration is also the one affecting Cl^-.

Sulfate reduction cannot explain the increase in HCO_3^- and the decline in SO_4^{2-}. There is a lack of measurable sulfide in either ionic or gaseous forms. The most reasonable explanation of the SO_4^{2-} distribution is a geologic one. Approximately 35,000 years ago, glaciation continued to open up the recharge area of the aquifer and deposited till. The sulfate chemistry of the aquifer in areas of subcrop simply reflects unique processes that operate within the till to produce high SO_4^{2-} (Hendry et al., 1986). Thus, high SO_4^{2-} water is limited to the vicinity of the subcrop because time has not been sufficient for this chemically distinctive water to move further. On average, the transport of a tracer through the confined part of the aquifer requires about 3×10^5 yr (Hendry and Schwartz, 1988).

Mass transfer calculations downgradient of the SO_4^{2-} bulge suggest that minor cation exchange and sulfate reduction is occurring but not nearly to the extent suggested initially by the major ion distributions. The increasing HCO_3^- concentrations and large concentrations of methane gas in the downdip end of the aquifer come from the reduction of organic matter. It is not certain whether the reaction is occurring in the aquifer or whether both CH_4 and HCO_3^- are diffusing from the Colorado shale. Given that a major gas field occurs at the downdip end of the aquifer, it is most likely that the shale unit is the source rock for the gas.

More recently, the Milk River aquifer was the focus of a large multidisciplinary study designed to test isotopic and geochemical techniques for testing old ground water (Ivanovich et al., 1991). Beyond standard major ion and environmental isotope analysis (that is, tritium, D, ^{18}O), samples were measured for ^{14}C, ^{36}Cl, various noble gases (He, Ne, Ar, Kr, Xe, and ^{222}Ra), Ar isotopes (^{37}Ar, ^{39}Ar), uranium and thorium isotopes, and various isotopes of krypton.

An important result of this work is yet another model to explain the origin of salinity patterns within the Milk River aquifer. Based on the geochemistry of halides (Cl, Br, and I) and their radioactive isotopes ^{36}Cl and ^{129}I, Fabryka-Martin et al. (1991) suggest that the major influence on the major ion chemistry is the diffusion of modified seawater from low-permeability lenses or units within the Milk River aquifer. Thus, although diffusion remains a dominant process, this model looks to sources within the aquifer to provide dissolved mass as opposed to sources outside the aquifer (Hendry and Schwartz, 1988).

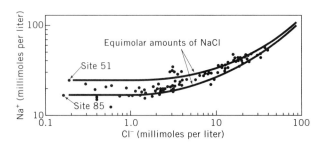

Figure 21.18 Scatter diagram illustrating how Na^+ concentration varies in relation to Cl^-. The lines depict how water from sites 51 and 85 would evolve by adding equivalent amounts of Na^+ and Cl^- (from Hendry and Schwartz, 1990). Reprinted by permission of *Ground Water*. Copyright ©1990. All rights reserved.

▶ EXERCISES

21.1. Explain why the isotopic composition of ground water found in glacial till in Canada and parts of the United States may be depleted in $\delta^{18}O$ and δD relative to precipitation sampled at the same locations.

21.2. Over the plains region of the northern United States and Canada, carbonate-rich till overlies marine shale and sandstone. Recharge from snowmelt has the following chemical composition as it moves downward through till and into shale bedrock:

Concentration (milligrams per liter)

Unit	Ca^{2+}	Mg^{2+}	Na^+	HCO_3^-	SO_4^{2-}	Cl^-	pH
till	79.0	5 0.0	210.0	436.0	61.0	14.0	7.80
shale	5.0	0.5	450.0	1044.0	6.0	53.0	8.10

Interpret the chemical evolution of the water in terms of the most likely mass transport processes.

21.3. Explain why the order in which ground water encounters minerals can be important in determining how the major ion chemistry evolves.

21.4. In carbonate rocks subject to recharge, the most dramatic changes in major ion chemistry occur over a relatively short distance as infiltration first enters the unit. Later changes are often almost insignificant by comparison. Explain why, using arguments related to kinetics and mineral equilibrium.

21.5. Using the concept of redox zones, explain why H_2S gas is rarely found in ground water close to the water table in recharge areas.

▶ CHAPTER
   HIGHLIGHTS

1. In sedimentary basins, the mixing of formation water and meteoric water that is recharging in zones of uplift is important in controlling water chemistry. Commonly, fresh, unaltered meteoric water is found near regions of outcrop, with progressive increases in salinity down the flow system. Freshwater or "noses" of relatively freshwater grading into saline water have been in the Madison Formation, the Inyan Kara Formation in southwestern North Dakota, and the Dakota-Newcastle Formation of the Dakota Sandstone.

2. Given sufficient time, diffusion can be important in moving dissolved mass in sedimentary basins. The common pattern is a consistent increase in pore-water salinity (mainly Na^+ and Cl^-) with depth. Data from north Louisiana show a linear increase in NaCl over 2700 m to a depth at which salt is known to occur.

3. Chemical reactions in the shallow subsurface generally play the dominant role in controlling the chemistry of natural ground waters. Rainfall or snowmelt has little dissolved mineral matter. Yet, once this water disappears into the ground, it quickly accumulates a relatively large load of dissolved mineral matter. The important chemical reactions begin in the unsaturated zone and continue as the water moves deeper. Key reactions in this respect include (a) the dissolution of soil gases, like O_2 and CO_2; (b) the dissolution/precipitation of various minerals, like calcite, dolomite, or feldpsar; (c) cation exchange reactions; and (d) the dissolution/utilization of organic compounds. The dissolution reactions are important because water is so efficient at dissolving gases and solids found in the vadose zone.

4. Most of the same processes affecting ion concentrations in the unsaturated zone are also operative in the saturated zone, including (a) and (c). Redox reactions (e) within the aqueous phase become increasingly important as well.

5. Here are examples of the five key reactions listed in (3) and (4):
   • *CO_2 gas dissolution and redistribution*

$$CO_2(g) + H_2O = H_2CO_3^*$$

$$H_2CO_3^* = HCO_3^- + H^+$$

$$HCO_3^- = CO_3^{2-} + H^+$$

- *Dissolution of carbonates, silicates, and salts*

$$\text{carbonate minerals} + H^+ = \text{cations} + HCO_3^-$$
$$\text{silicate minerals} + H^+ = \text{cations} + H_2SiO_3$$
$$\text{alumino-silicate minerals} + H^+ = \text{cations} + H_2SiO_3 + \text{secondary minerals}$$
$$\text{halite: } NaCl(s) = Na^+ + Cl^-$$
$$\text{anhydrite: } CaSO_4(s) = Ca^{2+} + SO_4^{2-}$$

- *Cation exchange reactions*

$$
\begin{matrix}
Ca^{2+} & & & Ca^{2+} \\
Mg^{2+} & + \text{ Na-clay} & = 2Na^+ + & Mg\text{-clay} \\
Fe^{2+} & & & Fe^{2+}
\end{matrix}
$$

- *Dissolution/utilization of organic compounds*
- *Redox reactions*

$$CH_2O + O_2 = CO_2 + H_2O$$
$$CH_2O + 8H^+ + 4Fe(OH)_3(s) = 4Fe^{2+} + 11H_2O + CO_2$$
$$CH_2O + 4H^+ + 2MnO_2(s) = 2Mn^{2+} + 3H_2O + CO_2$$
$$2CH_2O + SO_4^{2-} + H^+ = HS^- + 2H_2O + 2CO_2$$

6. The most rapidly dissolving minerals have the greatest impact on the chemistry of water. Thus, in the vadose zone, minerals like calcite, dolomite, and gypsum dissolve to saturation if they are present. Quartz and feldspar dissolve much more slowly and reach equilibrium only after long residence times. Thus, the concept of a partial-equilibrium system, water in equilibrium with some minerals, and out of equilibrium with respect to others explains the evolution of ground water.

7. Assembling these groups of mass transport and geochemical processes in different ways can explain the ground-water chemistry in particular locations. This chapter presents examples from a variety of settings to illustrate this point. In particular, the Milk River aquifer shows that one can interpret the chemical and isotopic evolution of ground waters along a flow path.

22

INTRODUCTION TO CONTAMINANT HYDROGEOLOGY

Three important attributes distinguish sources of ground-water contamination: (1) their degree of localization, (2) their loading history, and (3) the kinds of contaminants present. Given the large number of ways ground water can be contaminated, there is a spectrum of source sizes ranging from an individual well to areas of 100 km^2 or more. In practice, the terms *point* or *nonpoint* describe the degree of localization of the source. A *point source* is characterized by the presence of an identifiable, small-scale source, such as a leaking storage tank, one or more disposal ponds, or a sanitary landfill. Usually, this source produces a reasonably well-defined plume. A *nonpoint* problem refers to larger-scale relatively diffuse contamination originating from many smaller sources, whose locations are often poorly defined. Examples of nonpoint contaminants could include herbicides or pesticides that are used in farming, nitrates that originate as effluents from household disposal systems, salt derived from highways in winter, and acid rain. Typically, there are no well-defined plumes in these cases but a large enclave of contamination with extremely variable concentrations.

The loading history describes how the concentration of a contaminant or its rate of production varies as a function of time at the source. A spill is an example of pulse loading, where the source produces contaminants at a fixed concentration for a relatively short time (Figure 22.1a). This loading could occur from a one-time release of contaminants from a storage tank or storage pond. Long-term leakage from a source is termed *continuous source loading*. Figure 22.1b is one type of continuous loading in which the concentration remains constant with time. This loading might occur, for example, when small quantities of contaminants are leached from a volumetrically large source over a long time. Light nonaqueous phase liquids (LNAPLs) or dense nonaqueous phase liquids (DNAPLs) present at a source can dissolve at a slow rate over many decades to provide this kind of source loading.

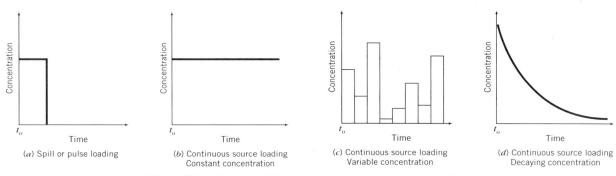

Figure 22.1 Examples of functions used to characterize contaminant loading from (a) a spill or (b, c, and d) long-term leakage.

Most sources of long-term leakage cannot be described in terms of a constant loading function. For example, the concentration of chemical wastes added to a storage pond at an industrial site can vary with time owing to changes in a manufacturing process, seasonal or economic factors, or the addition of other reactive wastes (for example, Figure 22.1c). Leaching rates for solid wastes at a sanitary landfill site could be controlled by seasonal factors related to recharge or by a decline in source strength as components of the waste (for example, organics) biodegrade. This latter source behavior could result in the loading history shown on Figure 22.1d. Typically, increasing complexity in the loading function translates directly into increasingly variable concentration distributions.

▶ 22.1 TYPES OF PROBLEMS AND CONTAMINANTS

An amazingly large list of potential activities can result in ground-water contamination (Table 22.1). As a consequence of these many different industrial, agricultural, and domestic activities, the list of potential contaminants can number in the thousands or tens of thousands of compounds. The question of how to organize this list for study is a difficult one. One approach has been to concentrate on a subset of this list and to pay special attention to contaminants that commonly occur in effluents and drinking water, produce adverse health effects, or persist within the food chain. An example is the U.S. Environmental Protection Agency's list of 129 priority pollutants, containing 114 organic compounds and 15 inorganic species, mainly trace metals.

We organize the contaminants by grouping them according to reaction type and mode of occurrence. Thus, dissolved compounds that are affected similarly by chemical, nuclear, or biological processes are grouped together. Contaminants that occur as a nonaqueous phase liquid (NAPL) are a separate subgroup. The major groups of contaminants include: (1) radionuclides, (2) trace elements, (3) nutrients, (4) othecr inorganic species, (5) organic contaminants, and (6) microbial contaminants. The NAPLs that are included in other inorganic species naturally subdivide into LNAPLs and DNAPLs.

All of these contaminants have the potential to produce health problems. In moving through the groups, we will point out the most serious ones. As a broad generalization, too much of anything in water can produce health problems in humans. For some contaminants, particularly the radionuclides, increasing exposure results in increasing health consequences. Thus, any exposure above background levels can be of concern. For other contaminants, such as the major ions (for example, Na^+ or Cl^-), there is often a threshold below which no serious health effects will occur.

TABLE 22.1 **Sources of ground-water contamination**

CATEGORY I—SOURCES DESIGNED TO DISCHARGE
    SUBSTANCES
Subsurface percolation (e.g., septic tanks and cesspools)
Injection wells
    Hazardous waste
    Nonhazardous waste (e.g., brine disposal and drainage)
    Nonwaste (e.g., enhanced recovery, artificial recharge,
      solution mining, and in situ mining)
Land application
    Wastewater (e.g., spray irrigation)
    Wastewater byproducts (e.g., sludge)
    Hazardous waste
    Nonhazardous waste

CATEGORY II—SOURCES DESIGNED TO STORE, TREAT,
    AND/OR DISPOSE OF SUBSTANCES; DISCHARGE
    THROUGH UNPLANNED RELEASE
Landfills
    Industrial hazardous waste
    Industrial nonhazardous waste
    Municipal sanitary
Open dumps, including illegal dumping (waste)
Residential (or local) disposal (waste)
Surface impoundments
    Hazardous waste
    Nonhazardous waste
Waste tailings
Waste piles
    Hazardous waste
    Nonhazardous waste
Materials stockpiles (nonwaste)
Graveyards
Animal burial
Above-ground storage tanks
    Hazardous waste
    Nonhazardous waste
    Nonwaste
Underground storage tanks
    Hazardous waste
    Nonhazardous waste
    Nonwaste
Containers
    Hazardous waste
    Nonhazardous waste
    Nonwaste

Open burning and detonation sites
Radioactive disposal sites

CATEGORY III—SOURCES DESIGNED TO RETAIN
    SUBSTANCES DURING TRANSPORT OR
    TRANSMISSION
Pipelines
    Hazardous waste
    Nonhazardous waste
    Nonwaste
Materials transport and transfer operations
    Hazardous waste
    Nonhazardous waste
    Nonwaste

CATEGORY IV—SOURCES DISCHARGING SUBSTANCES
    AS CONSEQUENCE OF OTHER PLANNED ACTIVITIES
Irrigation practices (e.g., return flow)
Pesticide applications
Fertilizer applications
Animal feeding operations
De-icing salts applications
Urban runoff
Percolation of atmospheric pollutants
Mining and mine drainage
    Surface mine-related
    Underground mine-related

CATEGORY V—SOURCES PROVIDING CONDUIT OR
    INDUCING DISCHARGE THROUGH ALTERED FLOW
    PATTERNS
Production wells
    Oil (and gas) wells
    Geothermal and heat-recovery wells
    Water-supply wells
Other wells (nonwaste)
    Monitoring wells
    Exploration wells
Construction excavation

CATEGORY VI—NATURALLY OCCURRING SOURCES
    WHOSE DISCHARGE IS CREATED AND/OR
    EXACERBATED BY HUMAN ACTIVITY
Ground water-surface water interactions
Natural leaching
Saltwater intrusion/brackish water upconing (or intrusion of
    other poor-quality natural water)

Source: Office of Technology Assessment (1984)

Radioactive Contaminants

The nuclear industry is the main generator of radioactive contaminants. Potential sources occur throughout the nuclear fuel cycle, which involves the mining and milling of uranium, uranium enrichment and fuel fabrication, power plant operation, fuel reprocessing, and waste disposal. The kinds of contaminants depend on the type of reactor and the extent of spent-fuel reprocessing. For example in the United States, Japan, France, and Germany, large numbers of light-water reactors (LWR) utilize enriched uranium (^{235}U) as the predominant fuel source, and possibly ^{239}Pu and ^{233}U. In Canada, heavy-water reactors (HWRs) use natural or enriched uranium as a fuel and heavy water (D_2O) as a coolant and moderator.

During mining when the raw ore is processed, ^{238}U, ^{230}Th, ^{226}Ra, and ^{222}Rn(gas) are potential contaminants along with nonradioactive contaminants that include trace constituents and major ions such as SO_4^{2-} or Cl^-. Areas where this contamination occurs are Colorado, New Mexico, Texas, Utah, Wyoming, northern Saskatchewan, and Ontario. The enrichment and fuel fabrication step treats raw uranium concentrate to increase the concentration of ^{235}U in the fuel relative to the more abundant ^{238}U and actually produces UO_2, which is the actual fuel. The most common contaminants are ^{238}U, ^{235}U, ^{90}Sr, and ^{137}Cs.

Many radionuclides are generated as fission products from the decay of ^{235}U or ^{239}Pu and from neutron activation of stable elements in the coolant or metallic components of the reactor. Fission is the power-generating part of a nuclear reaction whereby the heavy nucleus is split into nuclei of lighter elements and neutrons. Neutron activation is a process wherein neutrons are added to the nucleus of a stable isotope to produce a radioactive one. These processes produce radionuclides like ^{137}Cs, ^{134}Cs, ^{58}Co, ^{51}Cr, ^{54}Mn, ^{55}Fe, ^3H, and ^{131}I. Overall, in excess of 75 radionuclides could potentially be produced. Fortunately, most of these radioisotopes, except for ^3H, remain in the spent fuel or the reactor and are not a serious source of contamination at this stage in the fuel cycle. Reprocessing treats spent fuel to remove ^{235}U and ^{239}Pu. However, the other radionuclides remain to produce the high-level waste problems common to many countries of the world.

The health hazards associated with ionizing radiation are well known. An exposed individual can be affected by cancer or genetic defects that could affect his or her offspring. These risks are difficult to assess at low levels of exposure. However, most risk assessments consider that the probability of inducing cancer or genetic damage increases as a function of dose without a threshold. Other health problems associated with exposure to radiation include, for example, cataracts, nonmalignant skin damage, depletion of bone marrow, and infertility.

Trace Metals

The next major group of contaminants are the trace metals (Table 22.2). As a group, trace metals contain the largest proportion of elements found on the periodic table. The most common sources of contamination include (1) effluents from mining, (2) industrial wastewater, (3) runoff, solid wastes, or wastewater contributed from urban areas, (4) agricultural wastes and fertilizers, and (5) fossil fuels. For an excellent survey of sources and background concentrations, readers can refer to Forstner and Wittman (1981).

Trace metals can be toxic and even lethal to humans even at relatively low concentrations because of their tendency to accumulate in the body. Some studies have found positive correlations between the concentration of trace metals in water (for example, Be, Cd, Pb, and Ni) and death rates from some cancers. Bioaccumulation of trace metals in the food chain has produced the most well-known cases of metal poisonings (for example, Minamata, Japan). Organisms higher up the food chain progressively accumulate metals. Eventually, humans at the top of the chain can experience severe health problems.

Nutrients

This group of potential contaminants includes those ions or organic compounds containing nitrogen or phosphorus. By far, the dominant nitrogen species in ground water is nitrate

TABLE 22.2 Examples of trace metals occurring in ground water

Aluminum	Gold	Silver
Antimony	Iron	Strontium
Arsenic	Lead	Thallium
Barium	Lithium	Tin
Beryllium		
Boron	Manganese	Titanium
Cadmium	Mercury	Uranium
	Molybdenum	Vanadium
Chromium	Nickel	Zinc
Cobalt	Selenium	
Copper		

(NO_3^-), and of lesser importance is ammonium ion (NH_4^+). Agricultural practices including the use of fertilizers containing nitrogen, cattle feeding operations, and the cultivation of virgin soils (leading to the oxidation of large quantities of nitrogen existing in organic matter in the soil) are important sources of contamination.

Other sources are sewage that could enter ground water from septic tank systems or irrigated wastewater. According to Bouwer (1985), effluent from sewage treatment plants in the United States contains 30 mg/L of nitrogen, mainly as NH_4^+, and organic nitrogen in the form of nitrobenzenes or nitrotoluenes.

The main health effects related to contamination by nitrogen compounds are (1) methemoglobinemia, a type of blood disorder in which oxygen transport in young babies or unborn fetuses is impaired or (2) the possibility of forming cancer-causing compounds (for example, nitrosamines) after drinking contaminated water. Typically, phosphorus contamination is considered together with nitrogen. However, it is less important because of the low solubility of phosphorus compounds in ground water, the limited mobility of phosphorus due to its tendency to sorb on solids, and the lack of proven health problems. The major sources of phosphorus are again soil-applied fertilizers and wastewater.

Other Inorganic Species

This miscellaneous group includes metals present in nontrace quantities such as Ca, Mg, and Na, plus nonmetals such as ions containing carbon and sulfur (for example, HCO_3^-, HS^-, CO_3^{2-}, SO_4^{2-}, and H_2CO_3) or other species such as Cl^- and F^-. Many of these ions are major contributors to the overall salinity of ground water. Extremely high concentrations of these species make water unfit for human consumption and for many industrial uses. The health-related problems are not as serious as those caused by the other contaminant groups. However, high concentrations of even relatively nontoxic salts, particularly Na^+, can disrupt cell or blood chemistry with serious consequences. At lower concentrations, an excessive intake of Na^+ may cause less serious health effects such as hypertension.

The potential sources of major ion salinity include: (1) saline brine that is produced with oil, (2) leachate from mine tailings, mine spoil, or sanitary landfills, and (3) industrial wastewater that often has large concentrations of common ions in addition to heavy metals or organic compounds. Fluoride is probably the best example of a trace nonmetal occurring as a contaminant. Like the metals, trace quantities of these contaminants can produce health problems at relatively low concentrations. With fluoride, an increase in concentration to as little as seven or eight times the levels for combating tooth decay can cause skeletal fluorosis.

Organic Contaminants

Contamination of ground water by organic compounds is a logical consequence of the large quantities of unrefined petroleum products and man-made organic compounds used today. Of the list of sources we considered previously, almost every one either is known to contribute or has the potential to contribute organic contaminants to ground water.

Table 22.3 lists the most important families of organic contaminants. The following sections examine these constituents in detail.

Petroleum Hydrocarbons and Derivatives

Petroleum hydrocarbons and *derivatives* are made up of carbon and hydrogen that are derived from crude oil, natural gas, and coal. The organic compounds in crude oil can be divided into three main groups (Zemo et al., 1993). The alkanes or paraffins are present in most petroleum products. Common alkanes (n-alkanes) have the general formula C_nH_{2n+2} and include n-butane, n-pentane, n-hexane, and so on. The cycloalkanes have carbon atoms arranged in a circle containing either five or six carbon atoms. Examples include cyclopentane and cyclohexane. The second major group, the alkenes or olefins, are not constituents of crude oil but are formed during the refining process. These molecules have the general formula C_nH_{2n} and include compounds like ethene and propene.

The third major group of compounds in crude oil are the aromatic hydrocarbons. These compounds contain at least one *benzene ring* (that is, C_6H_6). Examples of these compounds include *be*nzene (C_6H_6), *t*oluene ($C_6H_5CH_3$), *e*thylbenzene (C_8H_{11}), and *x*ylene ($C_6H_4(CH_3)_2$). These so-called BTEX compounds are both extremely soluble in water and toxic.

The *polynuclear aromatic hydrocarbons* (PAHs) are also components of concern in petroleum hydrocarbons. These compounds derive from a series of benzene rings. Examples include anthracene and phenanthrene.

The fuels are produced from crude oil through a refining process. Distillation separates crude oil into different fractions according to the temperature of boiling. Zemo et al., (1993) provide the following general descriptions of the chemistry of these fractions:

TABLE 22.3 **Important families of organic contaminants found in ground water**

Chemical Family	Examples of Compounds
Hydrocarbons and Derivatives	
Fuels	Benzene, toluene, o-xylene, butane, phenol
PAHs	Anthracene, phenanthrene
Alcohols	Methanol, glycerol
Creosote	m-cresol, o-cresol
Ketones	Acetone
Halogenated Aliphatics	Tetrachloroethene, trichloroethene, dichloromethane
Halogenated Aromatics	Chlorobenzene, dichlorobenzene
Polychlorinated Biphenyls	2,4′-PCB, 4,4′-PCB

- Gasolines (low boiling fractions) C_4 to C_{12} alkanes, C_4 to C_7 alkenes, aromatic compounds including the BTEX compounds, C_3 benzenes, and C_4 benzenes
- Middle distillates (kerosene, diesel, home heating oil, jet fuel): C_{10} to C_{24} alkanes, slightly soluble aromatic compounds including C_3 to C_5 benzenes, C_0 to C_8 napthalenes, and C_0 to C_5 anthracenes, and
- Residual products (diesel No. 4 and 6, Bunker C, motor oil): C_{20} to C_{78} alkanes, nonsoluble aromatics (predominately PAHs)

Overall, the compositional differences among the fuels have important implications for monitoring and cleanup. For example, the low boiling fractions like gasoline contain many highly volatile components. The oils are much less volatile. The various additives designed to improve combustion and to remove combustion deposits from engines also complicate the chemistry of fuels.

Halogenated Aliphatic Compounds

These compounds are formed from chains of carbon and hydrogen atoms where certain of the hydrogens may be replaced by chlorine, fluorine, or bromine atoms. Examples of these compounds include tetrachloroethene (PCE), trichloroethene (TCE), and carbon tetrachloride (CT). Historically, the use of these solvents or their improper disposal has given rise to many of the most serious problems of contamination encountered in hydrogeological practice. Most of these compounds have specific gravities greater than 1 and thus can be found as DNAPLs.

Halogenated Aromatic Compounds

These compounds are formed from benzene rings with substituted halogens. Examples include chlorobenzene and dichlorobenzene (DCB), which are used in various industrial and agricultural applications. Again, by virtue of their relatively large specific gravity, these compounds occur as DNAPLs in ground water.

Polychlorinated Biphenyls

Polychlorinated biphenyls were widely used in the 1960s and 1970s in transformers and capacitors. Their environmental persistence toxicity has made them an important contaminant, even though their production has been curtailed. Chemically, they consist of chlorine-substituted benzene rings joined together.

Health Effects

Certain important health effects are related to drinking water contaminated by organic compounds. However, as Craun (1984) points out, it is difficult to establish which compounds are most toxic because not all have been tested, and health risks are inferred from studies of laboratory animals, poisonings or accidental ingestion, and occupational exposures. Furthermore, there is a serious lack of information on the health effects related to the combined effect of several compounds, and on the epidemiology of populations consuming contaminated water. Organic contamination may cause cancer in humans and animals, and a host of other problems, including liver damage, impairment of cardiovascular function, depression of the nervous system, brain disorders, and various kinds of lesions. More detailed information on health effects related to organic materials can be obtained from Craun (1984).

Biological Contaminants

The important biological contaminants include *pathogenic bacteria, viruses,* or *parasites.* It does not take a degree in medicine to be aware of the serious health problems associated with typhoid fever, cholera, polio, and hepatitis. Other less serious abdominal disorders are often too well known by travelers to countries with poor sanitation. As a group, these health effects are some of the most significant related to the contamination of ground water.

The main source of biological contamination is from human and animal sewage, or wastewater. Ground waters become contaminated owing to (1) land disposal of sewage from centralized treatment facilities or septic tank systems, (2) leachates from sanitary landfills, and (3) various agricultural practices such as the improper disposal of wastes from feedlots.

Particulate contaminants are less mobile than those dissolved in water. For this reason, most reported problems are of a local nature, related, for example, to (1) poor well construction, which enables surface runoff or sewage to enter the well, and (2) sewer line breaks or septic tank fields located close to a well.

▶ 22.2 NONPOINT-SOURCE CONTAMINATION

Nonpoint-source contamination originates from many diffuse sources, which are typically widespread and hard to identify. Examples of nonpoint sources include (a) rain that is contaminated by nutrients and pesticides being carried in the atmosphere, (b) animal wastes of all kinds, and (c) fertilizers and agrochemicals applied to farmland. Of this list, problems of agricultural contamination have received the most attention because of the vast areas affected and the sheer quantities of chemicals being applied to the land. The USGS reports that in the United States approximately 12 million tons of nitrogen and 2 million tons of phosphorus are now applied annually as chemical fertilizers. The rate of application increased dramatically from the 1950s to 1970s (Figure 22.2). The application of manure adds another 7 million tons of nitrogen and 2 millions tons of phosphorus (USGS, 1999). Pesticide applications in the United States (agriculture and urban) have been constant at about 1 billion pounds per year. Of this total approximately 60% are herbicides. Insecticide use is significant but relatively much smaller (Figure 22.3).

The problems of nonpoint agricultural contamination are manifested in both ground and surface waters. Typically, surface waters are most affected because streams receive contributions of nutrients and pesticides from both ground water and overland flows. The extent to which ground waters are impacted usually depends on the near-surface hydrogeologic conditions. With shallow unconfined aquifers, contaminants have easy access to the aquifer. Deeper, confined aquifers are less vulnerable because of low-permeability confining beds.

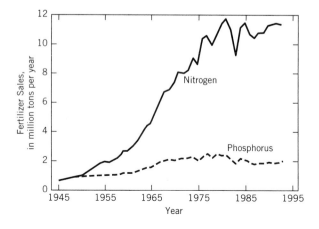

Figure 22.2 Use of nitrogen and phosphorous fertilizers in the United States (from USGS, 1999).

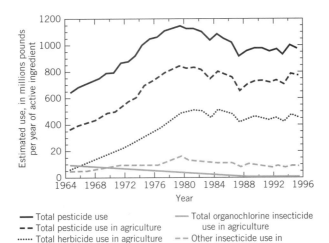

Figure 22.3 Use of agricultural pesticides in the United States (from USGS, 1999).

A case in point is the White River Basin of southern Indiana (USGS, 1999). Approximately 60% of the total watershed area is covered by cropland. Corn is grown on about half of this acreage, with nitrogen fertilization rates often in excess of 200 pounds per acre. Interestingly, even with the high fertilization rates, nitrates and pesticides do not contaminate deep aquifers because the clay-till deposits provide a barrier to migration (Figure 22.4). In addition, tile drains that are used to control water tables effectively route waters containing nutrients and pesticides to surface drains and streams. Thus, during spring months, rainstorms are very efficient in moving nutrients and pesticides to the surface-water systems. Commonly, NO_3^- concentrations exceed regulatory limits of 10 mg/L in streams, which in the Midwest often serve as the water supplies for towns and cities.

Where low-permeability units do not protect aquifers, nitrates can impact ground-water supplies. In the White River Basin, glacial outwash aquifers along the major streams are prone to NO_3^- contamination (Figure 22.4). In heavily fertilized areas, the following factors promote the contamination of ground water by nitrates (USGS, 1999):

1. High rainfall, snowmelt, and (or) excessive irrigation especially following recent fertilizer application.

2. Well-drained and permeable soils that are underlain by sand and gravel or karst, which enable rapid downward movement of water.

3. Areas where crop-management practices slow runoff and allow more time for water to infiltrate the ground.

4. Low organic matter content and high levels of dissolved oxygen, which can minimize the transformation of nitrate to other forms.

These vulnerabilities and high NO_3^- loading commonly occur at sites around the world. In the United States, the problem of nitrate contamination is particularly serious in Nebraska, Delaware, and California.

The contamination of ground waters by phosphorous compounds in commercial fertilizer is less of a problem than nitrogen. In general, PO_4^{3-} is sorbed to soil solids and is not particularly mobile in ground water. Erosion of soil particles, however, can produce significant phosphorous loading to streams. Although PO_4^{3-} concentrations are usually lower than NO_3^- in streams, this loading commonly contributes to excessive plant growth in lakes or reservoirs.

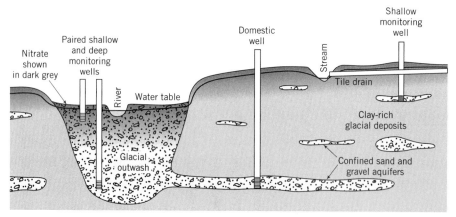

Figure 22.4 Conceptual model of nonpoint-source contamination by nitrate in the White River Basin. Seventeen percent of wells in glacial outwash had nitrate concentrations that exceeded the drinking water standard of 10 mg/L (from USGS, 1999).

In the United States, issues of nonpoint contamination are being addressed as part of the comprehensive National Water-Quality Assessment Program (NAWQA) of the U.S. Geological Survey. A major impetus of this program is to elucidate water-quality problems and their impact on biological systems. Studies have addressed problems in more than 50 of the largest basins in the United States. Syntheses of these studies (for example, USGS, 1999) and reports for individual study areas provide a wealth of information related to problems of nonpoint problems. The following Case Study (22-1) looks at the water quality in the San Joaquin-Tulare Basins Study Unit. This intensively farmed area in California is experiencing water-quality problems in ground water and surface waters alike.

Strategies for Studying Nonpoint Problems

Because of their diffuse and regional character, nonpoint problems are difficult to study. As the National Water-Quality Assessment (NAWQA) study of the San Joaquin-Tulare Basins shows, many different contaminants can be involved with temporally and spatially variable patterns of loading. Invariably, studies require an integrated approach that focuses on ground waters, surface waters, and ecosystems.

The NAWQA program now underway in the United States provides extremely useful guidance on how these studies should be carried out. Readers can find much of this material online at the USGS website. Detailed chemical monitoring at stream-sampling sites needs to be combined with regional monitoring of ground water. Experience has shown that special attention needs to be given to storm events, when most agrochemicals are flushed off the land and into the surface waters and ground waters. To provide a context for these assessments, contaminant levels in sediments and aquatic organisms need to be considered along with information on (1) land use, (2) soil and aquifer properties, (3) fertilizer-use rates, and (4) climate. The USGS has found it useful to conduct a few site-scale investigations to look at more specific problems in detail and to confirm conclusions from the regional evaluation.

As will be evident, the repertoire of hydrogeological techniques use for the investigation of a nonpoint problem is much different from that for a point-source problem. Large-scale tools like satellite imagery, regional geologic mapping, assessment of driller's

WATER QUALITY IN THE SAN JOAQUIN-TULARE BASINS, CALIFORNIA

The San Joaquin-Tulare Basins Study unit is located in central California. As shown in Figure 22.5, this area includes the San Joaquin Valley and basins draining the Coast Ranges and the Sierra Nevada. The valley is almost completely agricultural land, which produces about 5% of the agricultural products of the United States. Approximately 38% of the surface water is imported to the watershed for irrigation purposes. The remaining water comes from the Sierra Nevada.

The intensive agriculture has a major impact on the quality of ground water and surface water. Irrigation water returns to surface water as direct runoff and/or from drainage systems designed to control the water table. This water contains elevated concentrations of dissolved solids, nutrients, and pesticides. Excess irrigation water also recharges the shallow, regional aquifer, impacting the water quality.

Rainfall is generally seasonal, with most rain due to winter storms. The greatest precipitation amounts are recorded at high elevations on the Sierra Nevada as rain and snow. The San Joaquin Valley itself is quite dry, with rainfall amounts varying from 5 to 15 inches.

Nitrates are a problem in both ground water and surface water. Surveys conducted by the USGS in 1993–95 found that 21 of 88 domestic wells completed in the regional aquifer in the eastern San Joaquin Valley exceeded the drinking-water standard of 10 mg/L. Seventy-seven percent had NO_3^- concentrations > 2 mg/L, which is considered to be background. In cases where the well data could be related to a particular land use (Figure 22.6), it was found that the highest ground-water NO_3^- levels were associated with almond growing, which has the highest fertilizer applications. Generally, all three agricultural land uses contributed to elevated nitrate levels in ground water.

The San Joaquin River is being impacted by nutrient loading from agricultural drainage, wastewater treatment plants, and runoff from dairies. Generally, NO_3^- concentrations do not exceed the drinking water standard, but concentrations are of concern and increasing with time (Figure 22.7). The main cause of the increase in nitrates from the 1970s on is increasing

drainage from agricultural lands. It is likely that the leaching of native soil is more important in this respect than fertilizer applications.

Pesticides are found everywhere at low concentrations in ground waters and surface waters. Because many different pesticides (more than 50) are used for a variety of purposes, the story is complicated. In ground water, 61 of the 88 domestic wells contained pesticides at low concentrations (generally, less than 0.1 μg/L). The most common pesticides included simazine, 1, 2-dibromo-3-chloropropane (DBCP), atrazine, desethylatrazine (a breakdown product of atrazine), and diuron. DBCP concentrations exceeded the EPA drinking-water standard of 0.2/μg/L in 20% of the domestic water supply wells sampled. Most surface waters tested contained mixtures of pesticides at low concentrations (that is, less than 1 μg/L). In no cases did concentrations exceed available drinking-water standards, although concentrations of several pesticides exceeded criteria related to the protection of aquatic life. The most common pesticides were the herbicides simazine, dacthal, metolachlor, and EPTC, and the insecticides diazinon and chlorpyrifos. Long-banned insecticides like DDT, toxaphene, and chlorodane were still being detected as a consequence of erosion of contaminated agricultural fields.

Pesticide concentrations varied seasonally in the San Joaquin River, depending on climatic conditions and use in the watershed. An example is diazinon, which is an organophosphate insecticide used mainly to control wood-boring insects in almond orchards. Normally, this insecticide is applied from December to February, when the almond orchards are dormant. The timing of this application coincides with the winter storms. Typically, farmers apply this insecticide in short dry periods that might occur in a sequence of winter storms. When applications coincide with storms, diazinon concentrations in the San Joaquin River commonly exceed 0.35/μg/L (Figure 22.8). This concentration of diazinon is acutely toxic to water fleas. During the summer dry period, diazinon concentrations in the river is relatively small.

Dubrovsky et al., 1998

---

logs, and Geographical Information Systems (GIS) are used in preference to piezometer installation, borehole sampling, and geophysical surveys, which are more commonly used in the investigation of contaminated sites.

► 22.3 POINT-SOURCE CONTAMINATION: PLUMES OF DISSOLVED CONTAMINANTS

Point-source contamination emanates from one or several small-scale sources that can be reasonably well identified and described. To an important extent, the complexity of a point-

Figure 22.5 Map showing the location of the San Joaquin-Tulare Basins Study Unit in Central California (from Dubrovsky et al., 1998).

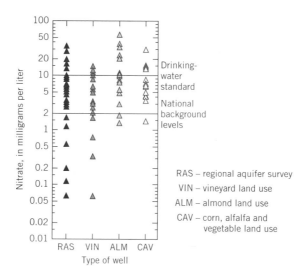

Figure 22.6 Summary of nitrate data from a survey of domestic wells (from Dubrovsky et al., 1998).

source problem is determined by whether or not *nonaqueous phase liquids* (NAPLs) are present. NAPLs are classified according to whether they are more or less dense than water. *Dense nonaqueous phase liquids* (DNAPLs) have a specific gravity greater than one and sink through water. LNAPLs have a specific gravity less than one and float on water.

Figure 22.9*a* is an example of the simplest case of point-source contamination where there is only a plume of dissolved contaminants and no NAPLs. Figure 22.9*b* illustrates a more complicated point-source problem developed due to a DNAPL spill. Contamination

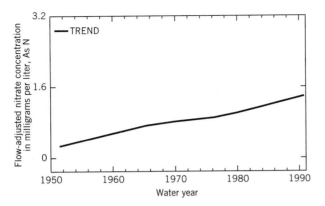

Figure 22.7 The flow-adjusted nitrate concentration in the San Joaquin River has increased from about 0.3 to 1.4 milligrams per liter (mg/L) over the last four decades (from Dubrovsky et al., 1998).

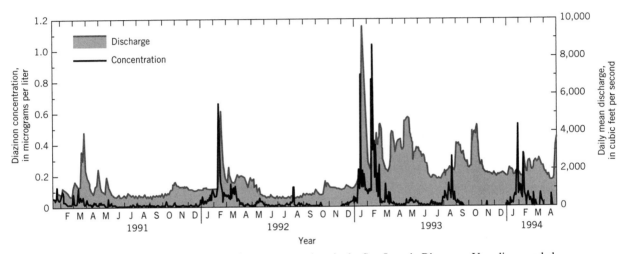

Figure 22.8 Diazinon concentrations in the San Joaquin River near Vernalis exceeded concentrations toxic to aquatic life (0.35 microgram per liter) during storm runoff periods (from Dubrovsky et al., 1998).

by NAPLs raises the level of complexity because contaminants can migrate as a separate liquid phase, a dissolved phase, and a vapor phase. The figure shows how DNAPL can occur as residual fluid within the pores and/or "ponded" on low-permeability layers. This DNAPL slowly dissolves into the flowing ground water to create a large plume. Within the unsaturated zone, volatilization promotes the spreading of DNAPL as a vapor phase.

Plume Formation

Plumes develop as dissolved contaminants are transported down the flow system. If the spill is old and the contaminant is nonreactive, plumes can be miles in length. Generally, a separate plume develops for each dissolved contaminant. Thus, at complex sites, one could draw 20 or 30 different plumes. The flow velocity, the extent to which dispersion occurs, and the types of chemical interactions (for example, sorption or biodegradation) determine their size and shape.

The largest plumes developed at any site are due to chemically nonreactive species like Cl^-, which are mobile because they are advected at the linear ground-water velocity.

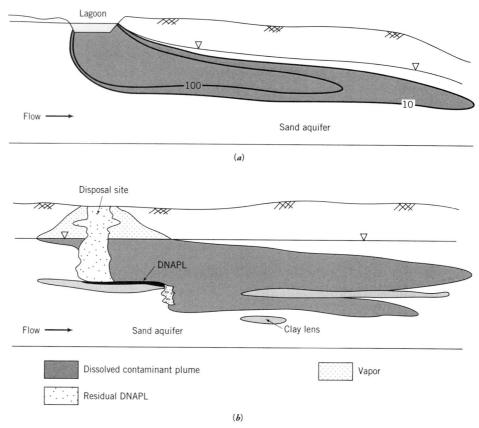

Figure 22.9 Conceptualization of contamination problems of varying complexity. In Panel (*a*), waste dissolved in water leaking from a lagoon creates a large plume of dissolved contaminants. In Panel (*b*), the presence of DNAPLs results in a much more complex problem, where the contamination occurs as a pure organic liquid, a vapor in the soil zone, and a dissolved phase in the ground water (from Domenico and Schwartz, 1998. *Physical and chemical hydrogeology*). Copyright ©1998 by John Wiley & Sons, Inc. Reprinted by permission of John Wiley & Sons. Inc.

Typically, dispersion is the only other process that influences the concentration distributions of such species. It spreads the plume in the direction of flow and transverse to the direction of flow. When contaminants react, they are generally less mobile and end up moving at a velocity that is less than linear ground-water velocity. Contaminants that are strongly sorbed or biodegrade with short half-lives only spread a short distance from the source. Case Study 22-2 illustrates features of a nonreactive plume in a relatively simple hydrogeological setting.

Reactions and Their Impact on Plume Development

Historically, there has not been much interest in understanding the details of chemical processes and their impact on contaminant mobility. The state-of-practice usually involved carefully mapping the distribution of a few key contaminant species and inferring what processes where operative. In some cases, it is relatively easy to identify transport processes and their impact on transport. A case in point is sorption of contaminants onto aquifer solids.

BABYLON, NEW YORK, CASE STUDY

At Babylon, New York (see insert Figure 22.10), Cl^-, nitrogen compounds, trace metals, and various organic compounds originating from a landfill have contaminated shallow ground water (Kimmel and Braids, 1980). Landfilling at the site began in 1947 with urban refuse, incinerated garbage, cesspool waste, and industrial refuse. The refuse is 18 to 24 m thick and is often placed below the water table. The cesspool wastes were treated to some extent before being discharged into lagoons on the north and west parts of the property.

The surficial sand aquifer is approximately 27.5 m thick and has a hydraulic conductivity of 1.7×10^{-3} m/s. Flow in the aquifer is generally to the south-southeast (Figure 22.10). Contaminants include major ions Ca^{2+}, Mg^{2+}, Na^+, K^+, HCO_3^-, SO_4^{2-}, and Cl^-; nitrogen species such as NH_4^+ and NO_3^-; heavy metals, particularly iron and manganese; and organic compounds.

Shown in Figure 22.11 is the Cl^- plume between 9.1 and 12.1 m below the water table. Because Cl^- does not react, the plume is a manifestation of only mass transport processes. Of these, the most important is advection. Longitudinal dispersion is relatively significant, but transverse dispersion is negligible. A three-dimensional analysis of the main portion of the Cl^- plume by Kelly (1985) determined an advective velocity of 2.9×10^{-4} cm/s, which is similar to the velocity reported by Kimmel and Braids (1980). Values of dispersivity in the longitudinal and two transverse directions (y and z) are estimated to be 18.6 m, 3.1 m, and 0.6 m, respectively. The tendency for α_x to be about six times larger than α_y and for α_z to be much smaller than either of the others is in line with expectations from carefully run tracer tests. The relatively smooth decline in Cl^- concentration away from the source suggests a relatively constant rate of loading. However, with the given information there is no way to establish whether the estimate of α_x contains a component due to variable source loading that might inflate the dispersivity value.

Kimmel and Braids (1980)

For example, organic compounds with a large octanol/water partition coefficient tend to be strongly sorbed and form a small plume, whereas a compound with a small K_{ow} should form a large plume.

Jackson et al. (1985) describe the Gloucester site near Ottawa, Canada, where a variety of organic compounds were disposed in a shallow trench. These contaminants included diethyl ether, tetrahydrofuran, 1.4-dioxane, carbon tetrachloride, benzene, and 1,2-dichloroethane. Plumes maps for three of these compounds (Figure 22.12) indicate variable rates of spreading. However, the extent of spreading is inversely related to the hydrophobicity or "water hating" character of the compound as measured by the octanol/water partition coefficient. Thus, 1,2-dichloroethane with a high K_{ow} migrates at a much slower velocity than 1,4-dioxane, which is much less strongly sorbed by the aquifer (small K_{ow}). This behavior suggests that sorption of the organic compounds is the most important process controlling the spread of contaminants.

With the advent of a remedial action scheme called natural attenuation, there has been much more interest in systematically documenting the variety of reactions that occur. *Natural attenuation* is the capability of natural systems to destroy contaminants before they impact downstream receptors. To demonstrate natural attenuation of contaminants requires that details of the chemical processes be absolutely and unambiguously defined. Thus, in recent years, there has been a significant leap forward in understanding complex chemical and biological processes in ground water and the pathways for chemical transformation.

Most studies simply focus on the particular contaminants of interest being transported in the ground-water system. However, new contaminant species can form as the product of microbial transformations. For example, in some settings, contaminants added at the source (e.g., PCE, TCE) biodegrade in some settings to produce hazardous products like

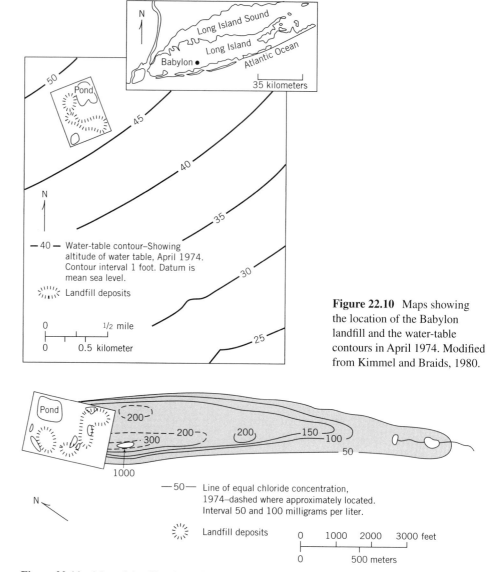

Figure 22.10 Maps showing the location of the Babylon landfill and the water-table contours in April 1974. Modified from Kimmel and Braids, 1980.

Figure 22.11 Map of the Cl⁻ plume in 1974 at depths below the water table ranging between 9.1 and 12.1 m. (Modified from Kimmel and Braids, 1980.)

DCE and VC. These species "grow in" as transport occurs and may reach their maximum concentrations away from the source. In addition, when contaminant concentrations are high, these reactions can impact the concentration of other "natural" constituents in the ground water that are caught up in the redox reactions. For example, redox processes can change the concentrations of common electron acceptors like iron, sulfate, and oxygen. The reality is that contaminants added to ground water produce complexity because of (1) the variety of species involved in the spill, (2) the possibilities for creating many new species via transformation reactions, and (3) natural species which may be caught up in the package of reactions.

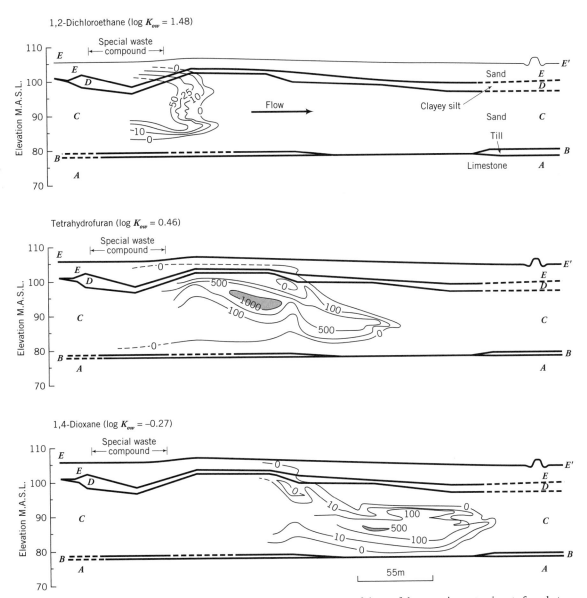

Figure 22.12 Sections illustrating the occurrence of three of the organic contaminants found at the Gloucester sanitary landfill near Ottawa, Ontario (from Jackson et al., 1985). Reproduced with the permission of the Minister of Public Works and Government Services, Canada, 2001 and courtesy of Environment Canada.

▶ 22.4 NONAQUEOUS PHASE LIQUIDS IN THE SUBSURFACE

One of the first considerations in dealing with LNAPLs or DNAPLs is defining some measure of abundance. *Saturation* represents the abundance of NAPL in a porous medium as the volume of the i_{th} fluid per unit void volume. For a representative elemental volume,

$$S_i = \frac{V_i}{V_{\text{voids}}} \qquad (22.1)$$

where V_i is the volume of the i_{th} fluid and V_{voids} is the volume of the voids. In a multicomponent system, the sum of all the saturations is equal to 1.

In most NAPL/water systems, it is difficult for the NAPL to fill the entire pore or to drain out completely. For example, NAPL entering a porous medium cannot displace all the water; some remaining water saturation is likely. Similarly, when NAPL drains from a porous medium, some NAPL is left behind. When saturations of one of the fluid components (e.g., water or NAPL) get low, that fluid phase becomes disconnected and essentially can't flow. The term *residual saturation* defines the saturation at which a fluid component becomes unable to flow. The following section explores this idea in more detail.

Features of NAPL Spreading

Consider a spill of LNAPLs or DNAPLs on the ground surface. With time, free product percolates downward through the unsaturated zone toward the water table. The most important process influencing downward movement of the free product is gravity-driven flow. The NAPL does not displace the water as it goes, but moves around it from pore to pore, once saturation exceeds the residual saturation.

Several important factors control the flow of NAPLs. In the case of a noncontinuous source (Figure 22.13), the volume of free product gradually decreases because some of the downward-moving NAPL is trapped in each pore at residual saturation. Thus, if the spill is relatively small, downward percolation in the unsaturated zone will stop once the total spill volume is at residual saturation. Another pulse of NAPLs is necessary to move the product downward. The main threat to ground water from such small spills is the opportunity for continuing dissolution of the NAPL by infiltration or vapor phase migration in the vadose zone.

The NAPL in Figure 22.13 also tends to spread horizontally as it moves downward. This spreading is due to capillary forces, which operate together with gravity forces to control migration. The presence of layers of varying hydraulic conductivity also influences NAPL spread (Farmer, 1983). Even a relatively thin, low-permeability unit (Figure 22.14) will inhibit downward percolation, and force the free product to move laterally. If such a layer is continuous, the NAPL will spread only within the unsaturated zone. If the layer is discontinuous, the NAPL will eventually spill over and continue to move downward toward the water table.

The flow of NAPLs also depends on the way the spill occurs (Farmer, 1983). Releasing a relatively large volume of contaminants over a relatively short time causes rapid downward and lateral migration (Figure 22.15a). Spreading is maximized, and a relatively large

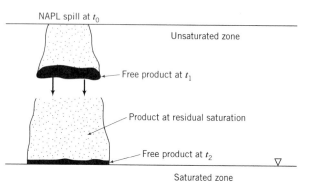

Figure 22.13 Downward percolation of a NAPL in the unsaturated zone. As the contaminant moves downward, the quantity of mobile fluid decreases and an increasing quantity is trapped as residual saturation (from Domenico and Schwartz, 1998. *Physical and chemical hydrogeology*). Copyright ©1998 by John Wiley & Sons, Inc. Reprinted by permission of John Wiley & sons, Inc.

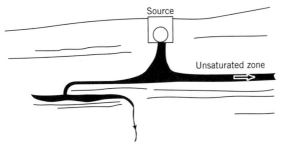

Figure 22.14 Presence of zones of low-hydraulic conductivity within the unsaturated zone can cause NAPLs to mound and spread laterally. In this case, some of the contaminant is able to move around the ends of discontinuous units. (Modified from Farmer, 1983.)

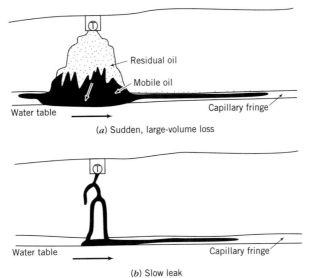

Figure 22.15 Pattern of flow is determined by the nature of the spill. In (a), a sudden, large-volume loss results in maximum spreading, large residual saturation, and collapse of the capillary fringe. In (b) a slow leak casues the product to follow a set of channels, a large volume of product to reach the water table, and minimal disturbance of the capillary fringe. (Modified from Farmer, 1983.)

volume of residual contamination remains in what Farmer (1983) refers to as the *descent cone* (Figure 22.15a). With slow leakage over a long time, the contaminant moves along the most permeable pathways (Figure 22.15b). These pathways can be a single channel or a more complex array of smaller channels, arranged in a dendritic pattern. Both the extent of lateral spreading and the volume of product held at residual saturation are considerably less than with a large-volume spill. Overall, more of the mobile liquid will reach the water table from slow leakage.

The way LNAPLs interact with the capillary fringe depends on the rate at which product is supplied (Farmer, 1983). A large volume of fluid, reaching the capillary fringe over a relatively short period of time, collapses the fringe and depresses the water table (Figure 22.15a). The extent of depression depends on the quantity of product and its density. Alternatively, a slow rate of supply has little effect on the capillary fringe or the configuration of the water table (Figure 22.15b). With this loading, the fluid occurs mainly within the capillary fringe. For spills that reach the capillary fringe, spreading continues until the total spill is at residual saturation.

Another feature of a ground-water system that influences LNAPL distributions is water-table fluctuation. The complexities in LNAPL saturation outlined above and the lack of mobility may conspire to trap LNAPLs below a rapidly rising water table. Over the long term, continuing water-table fluctuations will smear the free product above and below the water table.

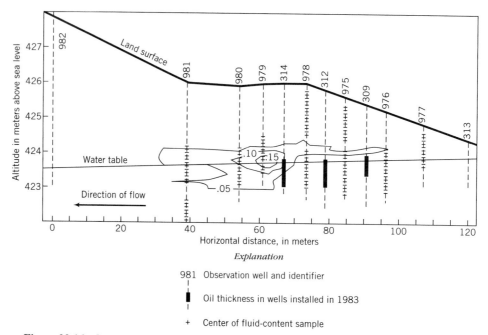

Figure 22.16 Longitudinal transect through the southern oil pool at the Bemidji site. Shown on the figure is the location of core samples, the oil thickness in wells, and the oil content as cm³/cm³ (from Hess et al., 1992). Reprinted from *J. Contam. Hydrol.*, v. 10, Determination of subsurface fluid contents at a crude-oil spill, p. 75–96. Copyright 1992, with permission from Elsevier Science.

The possibility for LNAPL to occur below the water table is illustrated by detailed saturation data developed for a single transect at a crude-oil spill near Bemidji, Minnesota (Hess et al., 1992). The distribution of LNAPL at this site appears to be influenced by sediment heterogeneities and water-table fluctuations. In particular, Figure 22.16 illustrates how LNAPL may be trapped due to water-table fluctuations.

It is actually difficult to measure how much LNAPL is present near the water table. Standard practice is to install a water-table observation well (screened for some distance above and below the water table) and to measure the thickness of free product in the well (Figure 22.17) using an interface probe. The thickness measured in the well is known as the *apparent thickness*. The measurement is crude because it really cannot establish the complexities of LNAPL saturation near the water table. Experience shows that problems are probably not as bad as indicated by measurements of free product measurements in observation wells.

When a NAPL is present at a site, its distribution is usually described by using an isopach map that shows the thickness and lateral extent. These kinds of maps are most commonly presented for LNAPLs because the distribution of LNAPLs is much more "regular" in comparison to DNAPLs that can be spread extensively through an aquifer.

Typically, the thickness of LNAPL is measured in water-table observation wells across the site using an interface probe. The measured thickness of LNAPL is plotted on a map of the site next to the location of each well. Lines of equal NAPL thickness are contoured to provide the desired map. Recall that the thickness of LNAPL shown on such a map is an *apparent thickness*, which is greater than the actual NAPL thickness. Figure 22.18 is an example of an isopach map showing the distribution of LNAPL at a site.

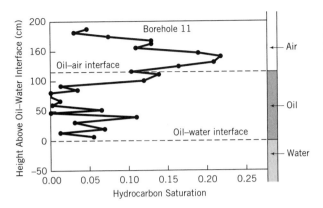

Figure 22.17 Measured hydrocarbon saturation profile compared to LNAPL distribution obtained from an observation well. Modified from Huntley et al., 1992. Reprinted by permission of Conference on Petroleum Hydrocarbons and Organic Chemicals in Ground Water-Prevention, Detection, and Restoration. Copyright ©1992. all rights reserved.

Occurrence of DNAPLs in the Saturated Zone

Unlike the case of LNAPLs that move along the capillary fringe, DNAPLs can move downward to the base of the aquifer. Downward moving DNAPLS displace water because they have a specific gravity much greater than water. DNAPL accumulating on low-permeability units will move downhill following the topography of the boundary (Figure 22.19). This flow in many cases will be in a direction that is different from the ground water. Spreading continues until the spill is at residual saturation. Within both the saturated and the unsaturated flow systems, these zones of residual saturation are a source of dissolved contamination as long as DNAPL remains. As was the case with LNAPLs, this simple conceptual model disguises the complex patterns of saturation in zones with *free product*.

Not surprisingly, few sites are as simple as that depicted in Figure 22.19. Heterogeneities of all kinds may occur within the saturated ground-water system to produce complex patterns of NAPL distribution. Figure 22.20 illustrates how subtle variations in permeability and the attitude of heterogeneities can lead to complex distributions of DNAPL "pools." (In pools, DNAPL saturations are close to the maximum.) The figure also illustrates complexities added by fracturing. Thus, in complex geologic settings, predicting migration pathways for DNAPLs can be relatively difficult.

Secondary Contamination Due to NAPLs

NAPLs in the subsurface can serve as important sources of *secondary contamination*. Problems develop when organic contaminants present as the free or residually saturated products partition into the soil gas through volatilization or into the ground water through dissolution. Figure 22.21 illustrates how even a small-volume spill of a volatile organic liquid in the unsaturated zone can produce a plume of dissolved contaminants more significant than the original spill (Mendoza and McAlary, 1990). The spreading of volatiles laterally and downward away from the source can partition into available soil moisture. Through continued infiltration, this water ultimately contaminates the aquifer.

Mendoza and Frind (1990 a, b) simulated the vapor transport of volatile organic solvents in the unsaturated zone. Their model considered (1) transport due to diffusion, and advection related to density gradients and vapor mass generation at the source and (2) attenuation related to dissolution into soil moisture. Of particular importance is density-driven advection caused by density gradients, which may develop with dense (that is, relative to soil gas) chlorinated solvent vapors moving downward and away from a spill (Mendoza and Frind, 1990a). What is surprising about this transport is the speed and extent to which

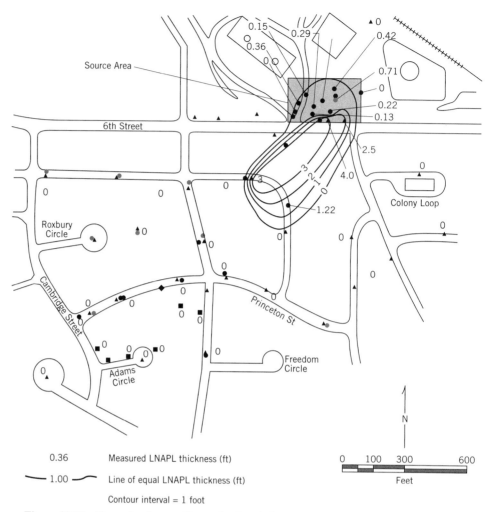

0.36   Measured LNAPL thickness (ft)

——— 1.00 ——   Line of equal LNAPL thickness (ft)

Contour interval = 1 foot

Figure 22.18 Example of a map illustrating how information on apparent LNAPL thickness can be presented. In the vicinity of the source, the maximum thickness is 3 ft and thicker. Modified from Wiedemeir et al., 1996.

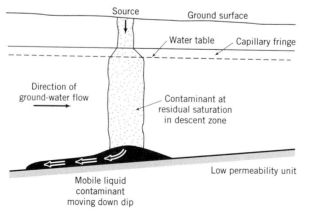

Figure 22.19 Pattern of DNAPL flow. In this case, the liquid moves along the bottom of the aquifer in a direction that is opposite to that of the ground water (from Domenico and Schwartz, 1998. *Physical and chemical hydrogeology*). Copyright ©1998 by John Wiley & Sons, Inc. Reprinted by permission of John Wiley & Sons, Inc.

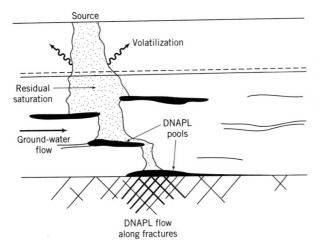

Figure 22.20 Complex distribution of DNAPLs in the vicinity of a surface source. Heterogeneities in hydraulic conductivity cause pooling. At depth, DNAPL flows into the bedrock along fractures (from Domenico and Schwartz, 1998. *Physical and chemical hydrogeology*). Copyright ©1998 by John Wiley & Sons, Inc. Reprinted by permission of John Wiley & Sons, Inc.

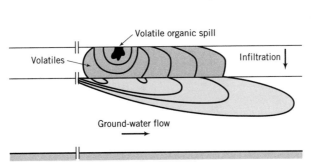

Figure 22.21 NAPLs in the vadose zone as a source of ground-water contamination. Volatilization of organic compounds leads to vapor phase migration and dissolution into the ground water (from Domenico and Schwartz, 1998. *Physical and chemical hydrogeology*). Copyright ©1998 by John Wiley & Sons, Inc. Reprinted by permission of John Wiley & Sons, Inc.

volatiles are able to spread through the soil-gas system. Simulation results showed vapor migration of up to several tens of meters in a few weeks. This tendency for volatile NAPLs to contaminate soil gases will be enhanced as the spill grows in size and spreads along the capillary fringe.

NAPLs in the subsurface also have the opportunity to dissolve into mobile water that moves through the zone of contamination. Figure 22.22 illustrates the plume of dissolved contaminants that could develop in conjunction with a free LNAPL moving along the capillary fringe. Figure 22.23 shows the plume from DNAPLs at residual saturation within the cone of descent and from the free product. Experimental studies by Anderson et al. (1987), together with the work they review, show that water moving through DNAPLs at residual saturation requires only 10 cm or so of contact distance before saturation is reached in the ground water. Although the solubilities of volatile organic compounds (for example, benzene, carbon tetrachloride, and trichloroethane) in water can be relatively low, the quantity of dissolved contaminant in water is often several orders of magnitude greater than the current standards permit (Anderson et al., 1987). Thus, these secondary sources of contamination are important in a regulatory sense.

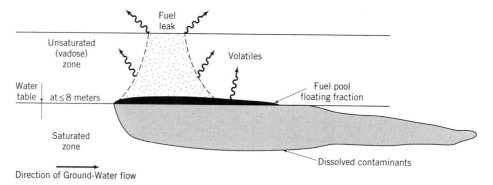

Figure 22.22 LNAPL dissolves into flowing ground water to form a contaminant plume (from Walther et al., 1986). Reprinted by permission of Conference on Petroleum Hydrocarbons and Organic Chemical in Ground Water-Prevention, Detection, and Restoration. Copyright ©1986. All rights reserved.

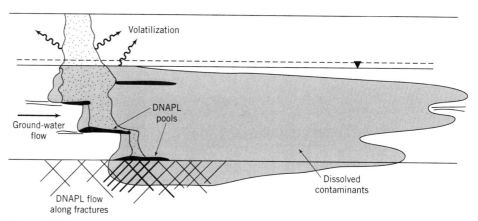

Figure 22.23 DNAPL dissolves into flowing ground water from residually saturated free product and pools to form a contaminant plume (from Domenico and Schwartz, 1998. *Physical and chemical hydrogeology*). Copyright ©1998 by John Wiley & Sons, Inc. Reprinted by permission of John Wiley & Sons Inc.

▶ 22.5 TECHNIQUES IN PROBLEM INVESTIGATION

Source Identification

Often, information on where wastes have been disposed, where tanks may have been abandoned, or where piping has leaked is not available for sites. Fortunately, a few approaches have proven useful for identifying potential sources, using preexisting data. One possibility is an analysis of historical aerial photographs. Thus, even though a former disposal area may now be buried, it might have been captured in aerial photographs taken, say, 30 years ago. Often the historical photos capture the layout of old landfills, trenches, drum disposal areas, and disposal ponds. They also might contain information on the location of old facilities (e.g., buildings, storage tanks), which can provide useful clues as to where contamination is likely. Another commonly used approach involves interviewing retired or long-time employees at the facility who can recall disposal practices.

Once the preexisting information has been used, you can turn to rapid investigative approaches that involve geophysics. Geophysical methods are useful in investigating old and abandoned sites, where former landfill sites, waste-storage ponds, or drum-disposal areas are now hidden as a result of cosmetic changes to the ground surface. Of the array of geophysical approaches discussed in Chapter 7, electrical and radar-based approaches are the most useful. For example, in Chapter 7, we presented a case study from Jordan et al. (1991) to show how high-resolution electromagnetic methods could be used to map the distribution of wastes in the near surface. Typically, equipment like a Geonics EM31 terrain conductivity meter can be used for this purpose. Jordan and Costantini (1995) describe another site where high-resolution electromagnetic surveying helped to pinpoint buried drum-disposal sites, buried bulk wastes, and slag in just a few days. Ground-penetrating radar (GPR) has also proven to be capable of finding underground targets like storage tanks or drum-storage areas.

Preliminary Assessments of Contamination

Typically, a site investigation seeks to determine where contaminants are located and the factors that influence patterns of spreading. The key investigative approach is to utilize wells to provide samples for chemical analysis. Early in an investigation, the wells of residents can often be sampled to provide a preliminary indication of contaminant distributions. This kind of preliminary information along with the preexisting hydrogeologic data is useful in selecting sites for the installation of monitoring wells.

With large sites (for example, refineries, military bases, airports, or railroad yards), there is sometimes little pre-existing information to guide the installation of permanent monitoring wells. In these situations, reconnaissance investigations are often used to provide the needed information. An increasing number of reconnaissance-type tools can be utilized to establish contaminant distributions. Some of these approaches, like cone penetrometry or detailed solids sampling using a soil probing machine, have already been discussed in earlier chapters. The main advantage of these approaches is their ability to investigate large areas at relatively modest costs. The main disadvantage is that they provide a "snapshot" of contaminant distributions at the time the survey is undertaken. Permanent wells are usually required to provide samples over the life of the investigation and site cleanup.

Other approaches have been developed to deal especially with contaminated sites. The most useful of these alternative approaches is soil-gas analysis, which we examine in the following section.

Soil-Gas Characterization

Characterizing the composition of soil gases has emerged as an industry-standard technique for tracing volatile organic compounds in ground water. The approach involves defining zones of ground-water contamination based on the presence of volatile components in the soil gas. With time, volatiles at the capillary fringe partition into the soil gas and gradually diffuse upward to the ground surface. The presence of volatiles is commonly established by collecting soil gas from some fixed depth and analyzing the sample with a gas chromatograph. However, this approach does not detect all organic contaminants because not all are volatile. In addition, the organic compound should not be too soluble. Highly soluble volatiles moving through the unsaturated zone will dissolve into any water present. Reisinger et al. (1987) illustrate the typical components of gasoline, which might be successfully detected in a soil-gas survey (Figure 22.24).

A metal probe with a perforated tip is driven from 2 to 4 m into the unsaturated zone, and a small sample of soil gas is extracted by pumping. A variety of commercially available equipment is available to assist with placing the sampling probes. Gas samples are

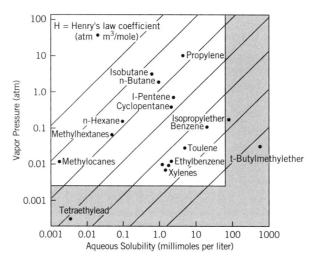

Figure 22.24 The unshaded area shows the range in vapor pressure and aqueous solubility for the common constituents in gasoline for which soil-gas surveying will be successful (from Reisinger et al., 1987). Reprinted by permission of First Outdoor Action Conference on Aquifer Restoration, Ground Water Monitoring and Geophysical Methods. Copyright ©1987. All right reserved.

usually analyzed on-site with a gas chromatograph (GC). The GC measurement provides a quantitative estimate of the mass of a particular volatile compound per volume of soil gas (for example, μg/L). When plotted on a map, these data can be used to infer zones of contamination in the unsaturated and saturated zones. By relating contaminant concentrations in the soil gas to measured concentrations in the ground water, the soil-gas data can be transformed to provide a quantitative estimate of concentrations in ground water.

Soil-gas sampling provides a rapid and economical way of surveying large sites for contamination. Thus, it is attractive for reconnaissance studies aimed at discovering what volatile contaminants are present and where they are located (see Case Study 22-3). This information often assists in designing a conventional sampling program. In some studies, the information provided by a soil-gas survey may be all that is required to establish whether or not a contamination problem exists or to identify a source.

These approaches may not work well under some circumstances. A low-permeability layer in the unsaturated zone can inhibit the upward diffusion of vapor and promote extensive

---

CASE STUDY 22-3

VERONA WELL FIELD

Wittmann (1985) describes a case study that demonstrates the potential of soil-gas sampling for delineating contaminated ground water. The case considered the origin of contamination in the Verona Well Field at Battle Creek, Michigan. Conventional ground-water monitoring pointed to the presence of an additional but unidentified source. Soil-gas surveying was utilized to explore a large rail yard. The sampling approach involved driving a metal probe approximately 1 m into the soil and extracting the soil gas with a hand pump. Sample analyses were performed on site using a gas chromatograph. Figure 22.25 shows PCE

results for 43 samples. The survey detected three areas of elevated soil gas in the rail yard. The most concentrated of the three was a small solvent disposal area that had gone undetected in previous investigations. Subsequently, two wells confirmed the PCE source. Overall, the approach worked well, except that it did not reveal the full extent of the PCE plume. Migrating away from the source, the plume moved deeper and lost touch with the water table. With this condition, volatiles were unable to partition into the soil gas.

Wittmann, 1985

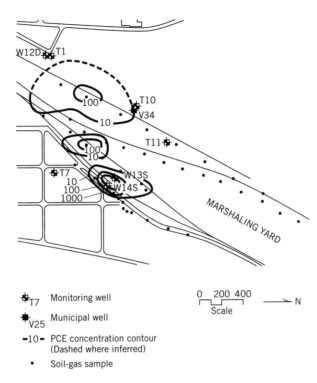

Figure 22.25 Results of a soil-gas survey (July 1984) for PCE in a rail yard in Battle Creek, Michigan. Modified from Wittmann, 1985. Reprinted by permission of Conference on Petroleum Hydrocarbons and Organic Chemicals in Ground Water—Prevention, Detection, and Restoration. Copyright ©1985. All rights reserved.

horizontal spreading. This situation will produce an estimated distribution of contaminants that is much larger than the actual one. Other problems relate to how the contaminant occurs. With LNAPLs, vapor concentrations in the unsaturated zone will be larger and easier to detect than if the contaminant is dissolved in the ground water (Reisinger et al., 1987). Furthermore, when contaminants are present in localized fracture zones (for example, in sandstone or limestone), the rates of diffusion away from the fractures may be so slow that vapor phase transport is limited in extent (Reisinger et al., 1987). Thus, care must be exercised in interpreting the results, and where necessary, conclusions should be confirmed using an independent approach.

Distribution of Dissolved Contaminants

The most definitive assessment of contaminant distributions requires the installation of monitoring wells and the collection/analysis of water samples. Monitoring usually continues over a period of several years and requires several sampling rounds. A water analysis would typically include the major cation and anions (routine analysis) and packages of organic and/or metal analyses, depending on the type of contamination involved. Detailed information on the installation of piezometers and water-table wells is presented in Chapter 7. Protocols for developing the wells and sampling them are presented in Chapter 16, along with information on quality assurance/quality control.

The information collected from this surveying provides the basis for depicting the distribution of contaminants in the subsurface. One way is to represent a contaminant plume as a fully three-dimensional shape where the shape represents an isosurface. An *isosurface* is the three-dimensional equivalent of a contour line and is the volume of the plume defined by some specified concentration value. These plots are often not very practical because they require significant data and specialty software, and have problems with the "noise" commonly associated with real chemical data. A simpler and more common approach is to produce a two-dimensional map-view, often called a *plume map*. This map is prepared by plotting the location of monitoring wells on a site map along with the measured concentration for the contaminant of interest. When sampling wells are nested at the same location (that is, piezometer nests), one selects the largest concentration value at the point of interest regardless of depth. Thus, a plume map depicts the worst case of contamination but provides no information on depths (Figure 22.26). There can be deeper and/or shallower zones where there may be no contamination (Figure 22.26).

If the aquifer is thick, an alternative is to map a contaminant plume in different depth zones. This approach provides some capability of visualizing the distribution of contaminants as a function of depth. Plume maps constructed for the Islip landfill, Long Island, New York (Kimmel and Braids, 1980) illustrate this point well. The Islip landfill began operation in about 1933 in an old sand pit. Through the years, it grew to a maximum size of 17 acres. Because the landfill has no liner or surface seal, water infiltrates and reacts with the various industrial and urban wastes to contaminate the shallow unconfined aquifer at the site. Because the landfill leachate is slightly denser than the ambient ground water, it sinks through the entire 170 ft of the aquifer. Slice maps for specific conductance turned out to be useful in visualizing how the leachate (represented by specific conductance) has spread both laterally and vertically (Figure 22.27). In this case, the plume is "boot" shaped in three dimensions. Most spreading is evident in Figure 22.27*d* at a depth of 94–114 ft below the water table. The largest specific conductance values (that is, greatest concentrations of contaminants) are found near the base of the aquifer (see Figures 22.27*c*, *d*). This pattern in specific conductance is indicative of variable-density flow.

Another way of presenting contamination data is to plot concentration data along cross sections. Commonly, such sections are oriented along the midline of the plume. The combination of cross sections and a plume map describes the contaminant variation in space.

▶ 22.6 FIELD EXAMPLE OF AN LNAPL PROBLEM

This case study (from Wiedemeier et al., 1996) examines the complex pattern of chemical changes that accompany the migration of contaminants away from a former fire training area (Site FT-002) at Plattsburg Air Force Base (AFB). Plattsburg AFB is located 167 miles north of Albany, New York, near the border with Canada. Figure 22.28 is a map showing the fire training area and the ground-water contamination related to training activities.

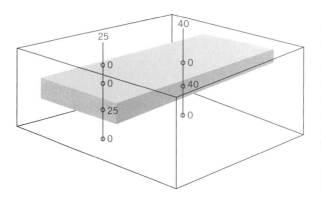

Figure 22.26 A plume map is constructed by using the maximum concentration observed at a monitoring site. In this example, values of 25 and 40 would be plotted (from Domenico and Schwartz, 1998. *Physical and chemical hydrogeology*). Copyright ©1998 by John Wiley & Sons, Inc. Reprinted by permission of John Wiley & Sons, Inc.

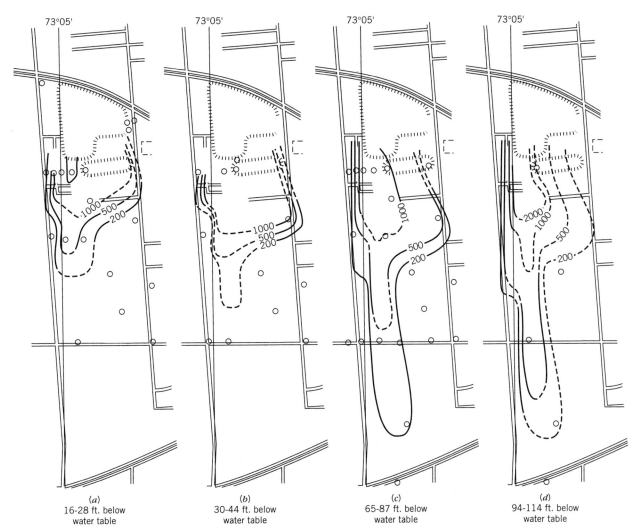

Figure 22.27 Series of slice maps through the plume at Islip, New York. Specific conductance values are highest at depth, suggesting that variable-density flow is operative (from Kimmel and Braids, 1980).

Contamination was caused as jet fuel containing PCE and TCE was spilled during the training exercises. There is a zone at the source (Figure 22.28) that contains LNAPL floating on the water table.

Figure 22.29 shows the geology of the site along the cross section. A well-sorted fine to medium sand comprises the shallow, unconfined aquifer impacted by the contamination. This unit is approximately 50 ft thick and has a mean hydraulic conductivity of 11.6 ft/d. The linear ground-water velocity in this unit is about 0.39 ft/d. Underlying the sand are clay and glacial till.

Contaminants present in the LNAPL have dissolved over time to create various plumes. As expected, the most important contaminants dissolved from the fuel are benzene, toluene, ethylbenzene, and xylene (the so-called BTEX compounds). The BTEX compounds, along with TCE, form plumes of contamination (Figure 22.30a,b). The BTEX plume extends for

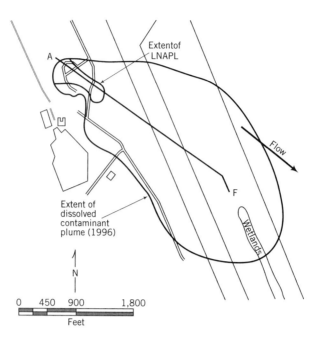

Figure 22.28 Map of the site at Plattsburg Air Force Base, New York. There is a small zone where NAPL is found. Plumes of dissolved contaminants have spread downgradient away from the source (modified from Wiedemeir et al., 1996).

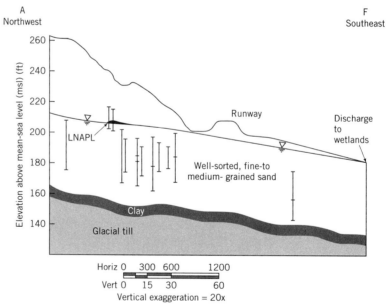

Figure 22.29 Cross section along *A-F* showing the key geologic units, sampling wells, and the location of the water table (modified from Wiedemeir et al., 1996).

about 2000 ft downgradient from the source. In the source area, BTEX compounds are observed at 17 mg/L. The BTEX plume is shown at steady state and effectively was not spreading further at that time. The TCE plume was generally larger than the BTEX plume and was also shown to be at steady state.

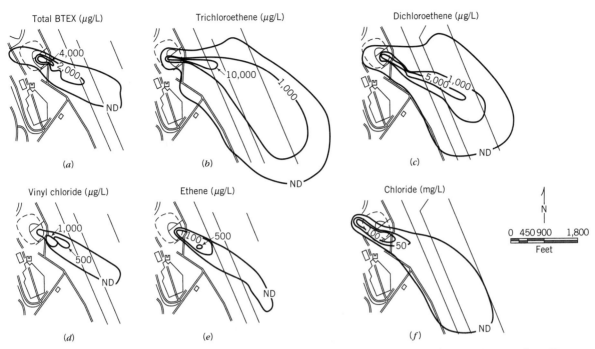

Figure 22.30 Plume maps for the major contaminants and their breakdown products (from Wiedemeir et al., 1996).

Interestingly, nearly all of the organic compounds present in the system are being biodegraded. TCE is biotransformed through the process of reductive dechlorination with the formation of DCE, vinyl chloride (VC), and ethane as Cl atoms are replaced with H. These compounds are being cometabolized with the BTEX compounds, which serve as the primary substrate for microbial population and promote reducing conditions in the aquifer. Thus, even though DCE, VC, ethane, and Cl^- are not present at the source as original contaminants, plumes develop as these compounds form from TCE (Figure 22.30c, d, e, f). Once the BTEX compounds are effectively removed via biodegradation (at about 1500 ft to 2000 ft downgradient from the source), the reductive dechlrination reactions cease as well.

The mineralization of BTEX compounds requires an electron acceptor. The concentration distribution of oxygen, nitrate, and sulfate (Figure 22.31) as compared to that of BTEX indicates that they function as electron acceptors. This case study shows that by careful study of contaminants, their breakdown products, and their natural constituents, one can interpret likely patterns of chemical interaction.

► EXERCISES

22.1. Shown in the figure for this problem (Figure 22.32) is a series of plumes from a sanitary landfill. Examine these plumes in detail and answer the following questions.

a. Qualitatively evaluate the extent to which advection and dispersion are important in controlling contaminant spread at the site.

b. Given the type of source and the resulting plume shapes, what can you say about the type of source loading?

c. Suggest what processes could be operating to cause the increasing pH away from the source.

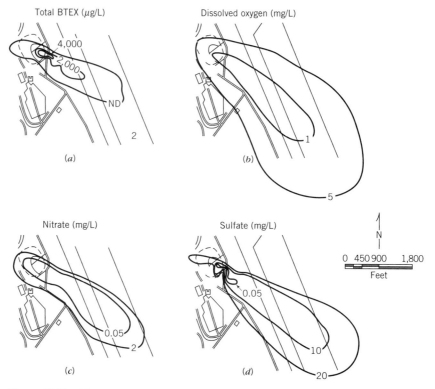

Figure 22.31 Plume maps for BTEX and the major electron acceptors (from Wiedemeir et al., 1996).

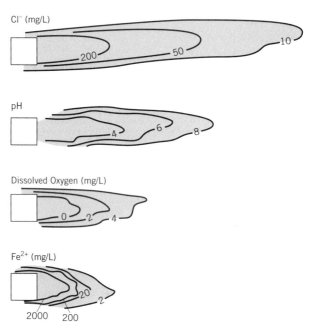

Figure 22.32 Plume maps for the major contaminants and their breakdown products.

d. Metals, for example Fe, tend to be relatively abundant in landfill leachates. However, Fe^{2+} tends to be strongly attenuated relative to mobile species like Cl^-. Explain why iron behaves in this way.

22.2. There are 100 g/m^3 of some organic constituent in an aquifer. The solubility of the constituent is 1000 μ g/L, and the drinking water standard is 5 μg/L. Calculate the potential volume of water that can be rendered undrinkable.

22.3. On diagrams (a) and (b) shown in Figure 22.33, illustrate the contaminant distribution for the organic compounds with the specified chemical properties. Consider all important spreading mechanisms.

22.4. Water in "Local Creek" was being contaminated by inflows of shallow ground water. In an attempt to discover the source of contamination, three shallow wells (MW2A, MW3, MW5) were installed (see Figure 22.34) Although slight contamination was evident in some of these wells, this information did not help to pinpoint the source (Tillman et al., 1988). A soil-gas survey was undertaken to help find the source of contamination. The map (Figure 22.34) shows total volatiles in the soil gas (μg/L).

a. Contour the soil-gas data and determine the source of ground-water contamination.

b. Explain why the three wells were not particularly successful in pinpointing the source of contamination.

22.5. This problem involves ground-water contamination associated with the sewage disposal system that formerly operated at Otis Air Base on Cape Cod, Massachusetts. The location of sampling wells is depicted in Figure 22.35 Some of the chemical information relating to these wells (1983) is compiled in Table 22.4.

a. Recall that many different contaminants are present at Otis. Creating a plume map of elevated specific conductance helps to define generally where the ground-water contamination is located. Given the information in the table, construct the specific conductance plume map.

b. For some of the wells, organic contamination, represented as the total concentration of volatile compounds, has been analyzed. Construct a plume map illustrating the distribution of volatile organic compounds (VOCs).

c. Explain some of the reasons why the VOC plume is different from that for specific conductance.

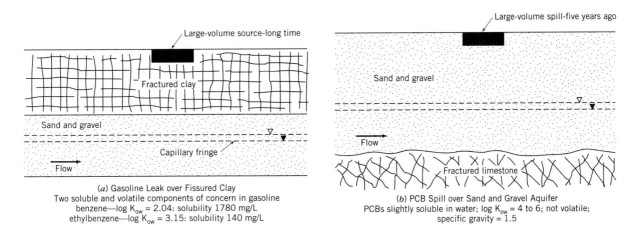

(a) Gasoline Leak over Fissured Clay
Two soluble and volatile components of concern in gasoline
benzene—log K_{ow} = 2.04: solubility 1780 mg/L
ethylbenzene—log K_{ow} = 3.15: solubility 140 mg/L

(b) PCB Spill over Sand and Gravel Aquifer
PCBs slightly soluble in water; log K_{ow} = 4 to 6; not volatile;
specific gravity = 1.5

Figure 22.33 Two typical hydrogeological settings.

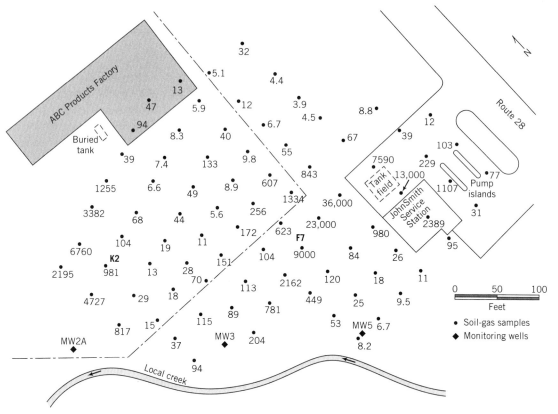

Figure 22.34 Plume maps for the major contaminants and their breakdown products (from Tillman et al., 1988).

1. Three attributes distinguish sources of ground-water contamination: (a) their degree of localization, (b) their loading history, and (c) the kinds of contaminants present. Given the large number of ways of contaminating ground water, there is a spectrum of source sizes ranging from an individual well to areas of 100 km^2 or more. The terms *point* or *nonpoint* describe the degree of localization of the source. A point source is an identifiable, small-scale source, such as a leaking storage tank, one or more disposal ponds, or a sanitary landfill. Usually, this source produces a reasonably well-defined plume. A nonpoint problem refers to larger-scale, relatively diffuse contamination originating from many smaller sources, whose locations are often poorly defined.

2. Examples of nonpoint sources include (a) rain that is contaminated by nutrients and pesticides being carried in the atmosphere, (b) animal wastes of all kinds, and (c) fertilizers and agrochemicals applied to farmland. Agricultural contamination is important because of the vast areas affected and the sheer quantities of chemicals being applied to the land. In the United States, about 12 million tons of nitrogen and 2 million tons of phosphorus are being applied to farmlands.

3. The loading history describes how the concentration of a contaminant or its rate of production varies as a function of time at the source. There are two different types of source behaviors, including: (a) pulse loading, where the source produces contaminants at a fixed concentration for a relatively short time, or (b) continuous source loading, where the source is active for a long time to produce loading at constant or variable concentrations.

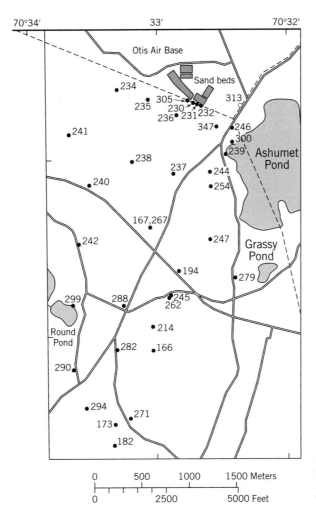

Figure 22.35 Plume maps for the major contaminants and their breakdown products (from Thurman et al., 1994).

4. The major groups of contaminants include: (a) radionuclides, (b) trace elements, (c) nutrients, (d) other inorganic species, (e) organic contaminants, and (f) microbial contaminants.

5. Most contaminants have the potential to cause health problems. For some contaminants, particularly the radionuclides, increasing exposure results in increasing health consequences. Thus, any exposure above background levels can be of concern. For other contaminants, there is often a threshold before health effects become evident.

6. The complexity of problems involving organic contaminants is determined by whether or not nonaqueous phase liquids (NAPLs) are present. NAPLs are classified depending on whether they are more or less dense than water. Dense nonaqueous phase liquids (DNAPLs), for example, the chlorinated solvents commonly used as degreasers, have a specific gravity greater than 1 and sink in an aquifer. LNAPLs have a specific gravity less than 1 and float on the water table. LNAPLs include the common fuels like gasoline and diesel, and crude oil.

7. The abundance of NAPL in a porous medium is reported as saturation. For a representative elemental volume,

$$S_i = \frac{V_i}{V_{\text{voids}}}$$

where V_i is the volume of the ith fluid.

TABLE 22.4 Summary of some of the chemical data collected for ground water at Otis Air Base

Well No.	Depth (feet)	Specific Cond. (μmhos/cm)	Total VOCs (μg/L)	Well No.	Depth (feet)	Specific Cond. (μmhos/cm)	Total VOCs (μg/L)
166	67	100	0.5	254	54	220	8.1
167	55	48	na	254	26	70	nd
173	69	122	na	262	159	125	na
182	69	80	nd	262	85	255	46.7
194	57	145	1.1	262	69	200	3.7
214	60	83	na	262	41	90	na
230	48	115	na	267	155	122	na
231	57	145	na	267	136	95	na
232	58	153	0.8	267	111	120	na
234	99	122	na	267	88	195	9.6
235	94	81	na	271	165	125	nd
236	106	126	1.0	271	141	132	nd
237	88	128	0.9	271	85	150	nd
238	106	95	na	271	41	55	nd
239	64	190	268.4	279	86	76	na
240	95	57	na	279	61	73	na
241	98	62	na	282	123	143	nd
242	77	51	nd	282	94	208	5.9
244	90	230	390.6	282	70	215	nd
245	25	200	nd	282	49	100	na
246	35	152	na	288	97	142	30.8
247	70	120	na	290	91	90	na
254	216	59	1.0	294	89	139	0.5
254	168	115	1.3	299	20	76	nd
254	140	175	4.5	300	30	410	3.9
254	107	235	93.7	300	10	138	na

na—not analyzed; nd—not detected.
Source: Thurman et al., 1984.

8. It is difficult for NAPL to fill an entire pore or to drain completely. NAPL entering a porous medium cannot displace all the water. Similarly, when NAPL drains from a medium, some is left behind. NAPL saturations can become so small that the NAPL phase is disconnected and unable to flow. This saturation is termed *residual saturation.*

9. NAPLs in the subsurface can serve as important sources of secondary contamination. Problems develop when organic contaminants present as the free or residually saturated products partition into the soil gas through volatilization or into the ground water through dissolution. Volatiles originating from NAPL in the unsaturated zone can be transported to the water table and produce a significant plume of dissolved contaminants. As ground water moves through source zones with LNAPLs or DNAPLs, they can dissolve soluble components and create long plumes.

10. The site investigation determines where contaminants are located and the factors that influence patterns of spreading. Wells are required to provide samples for chemical analysis. Early on, samples from the wells of nearby residents might provide a preliminary indication of contaminant distributions. This kind of preliminary information along with the preexisting hydrogeologic data is useful in selecting sites for the installation of monitoring wells.

11. Eventually, wells will need to be installed. Reconnaissance investigations are often used to help site these wells. Rapid investigations are possible with cone penetrometry, detailed solids sampling using a soil probing machine, and/or soil-gas analyses. The main advantage of these approaches is the

ability to investigate large areas at relatively modest costs. However, they do not replace permanent wells that yield samples throughout the investigation and cleanup.

12. With soil-gas sampling, a metal probe with a perforated tip is driven into the unsaturated zone, and a small sample of soil gas is extracted. Gas samples can be analyzed on-site with a gas chromatograph to provide quantitative estimates of organic vapor concentrations (mg/L). When plotted on a map, these data infer zones of subsurface contamination. In a number of circumstances this approach may not work well.

13. Chemical data collected from monitoring wells can be plotted in several different ways. The most common approach is to produce a plume map by plotting the location of monitoring wells on a site map along with the measured concentration for the contaminant of interest. When sampling wells are nested at the same location, select the largest concentration value at the point of interest regardless of depth. Thus, a plume map depicts the worst case of contamination but provides no information on depths.

TABLE OF ATOMIC WEIGHTS

Element	Symbol	Atomic Number	Atomic Weight	Element	Symbol	Atomic Number	Atomic Weight
Actinium	Ac	89		Gold	Au	79	196.97
Aluminum	Al	13	26.98	Hafnium	Hf	72	178.49
Americium	Am	95		Helium	He	2	4.00
Antimony	Sb	51	121.75	Holmium	Ho	67	164.93
Argon	Ar	18	39.95	Hydrogen	H	1	1.01
Arsenic	As	33	74.92	Indium	In	49	114.82
Astatine	At	85		Iodine	I	53	126.90
Barium	Ba	56	137.34	Iridium	Ir	77	192.20
Berkelium	Bk	97		Iron	Fe	26	55.85
Beryllium	Be	4	9.01	Krypton	Kr	36	83.80
Bismuth	Bi	83	208.98	Lanthanum	La	57	138.91
Boron	B	5	10.81	Lead	Pb	82	207.19
Bromine	Br	35	79.91	Lithium	Li	3	6.94
Cadmium	Cd	48	112.30	Lutetium	Lu	71	174.97
Calcium	Ca	20	40.08	Magnesium	Mg	12	24.31
Californium	Cf	98		Manganese	Mn	25	54.94
Carbon	C	6	12.01	Mendelevium	Md	101	
Cerium	Ce	58	140.12	Mercury	Hg	80	200.59
Cesium	Cs	55	132.90	Molybdenum	Mo	42	95.94
Chlorine	Cl	17	35.45	Neodymium	Nd	60	144.24
Chromium	Cr	24	52.00	Neon	Ne	10	20.18
Cobalt	Co	27	58.93	Neptunium	Np	93	
Copper	Cu	29	63.54	Nickel	Ni	28	58.71
Curium	Cm	96		Niobium	Nb	41	92.91
Dysprosium	Dy	66	162.50	Nitrogen	N	7	14.01
Einsteinium	Es	99		Nobelium	No	102	
Erbium	Er	68	167.26	Osmium	Os	76	190.20
Europium	Eu	63	151.96	Oxygen	O	8	16.00
Fermium	Fm	100		Palladium	Pd	46	106.40
Fluorine	F	9	19.00	Phosphorus	P	15	30.97
Francium	Fr	87		Platinum	Pt	78	195.09
Gadolinium	Gd	64	157.25	Plutonium	Pu	94	
Gallium	Ga	31	69.72	Polonium	Po	84	
Germanium	Ge	32	72.59	Potassium	K	19	39.10

Element	Symbol	Atomic Number	Atomic Weight	Element	Symbol	Atomic Number	Atomic Weight
Praseodymium	Pr	59	140.91	Tantalum	Ta	73	180.95
Promethium	Pm	61		Technetium	Tc	43	
Protactinium	Pa	91		Tellurium	Te	52	127.60
Radium	Ra	88		Terbium	Tb	65	158.92
Radon	Rn	86		Thallium	Tl	81	204.37
Rhenium	Re	75	186.20	Thorium	Th	90	232.04
Rhodium	Rh	45	102.90	Thulium	Tm	69	168.93
Rubidium	Rb	37	85.47	Tin	Sn	50	118.69
Ruthenium	Ru	44	101.07	Titanium	Ti	22	47.90
Samarium	Sm	62	150.35	Tungsten	W	74	183.85
Scandium	Sc	21	44.96	Uranium	U	92	238.03
Selenium	Se	34	78.96	Vanadium	V	23	50.94
Silicon	Si	14	28.09	Xenon	Xe	54	131.30
Silver	Ag	47	107.87	Ytterbium	Yb	70	173.04
Sodium	Na	11	22.99	Yttrium	Y	39	88.90
Strontium	Sr	38	87.62	Zinc	Zn	30	65.37
Sulfur	S	16	32.06	Zirconium	Zr	40	91.22

▶ REFERENCES

ABEELE, W. V. 1984. Hydraulic testing of rushed Bandelier Tuff. Report No. LA-10037-MS, Los Alamos National Laboratory, Los Alamos, New Mexico.

ADAMS, E. E., and L. W. GELHAR. 1992. Field study of dispersion in a heterogeneous aquifer. 2. Spatial moments analysis. Water Resour. Res., v. 28, no. 12, p. 3293–3307.

ALLER, L., et al. 1991. Handbook of suggested practices for the design and installation of ground-water monitoring wells. EPA/600/4-89/034, 221 p.

ALLEY, W. M., T. E. REILLY, and O. L. FRANKE. 1999. Sustainability of ground-water resources. U.S. Geol. Surv. Circular 1186.

ALLISON, G. B., C. J. BARNES, M. W. HUGHES, and F. W. J. LEANEY. 1983. The effect of climate and vegetation on oxygen-18 and deuterium profiles in soils. Proc. 1983 Int. Sym. Iso. Hydrol. Water Resource Dev., IAEA, Vienna, No. IAEA-SM-270/20.

ALLISON, G. B., and M. W. HUGHES. 1978. The use of environmental chloride and tritium to estimate total recharge to an unconfined aquifer. Aus. J. Soil Res., 16, p. 181–195.

ANDERSON, M. P., and W. W. WOESSNER. 1992. Applied Groundwater Modeling. Academic Press, San Diego, 381 p.

ANDERSON, M. R., R. L. JOHNSON, and J. F. PANKOW. 1987. The dissolution of dense non-aqueous phase liquid (DNAPL) from a saturated porous medium. Proc. of Petroleum Hydrocarbons and Organic Chemicals in Ground Water: Prevention, Detection and Restoration, National Ground Water Assoc., Columbus, Ohio, p. 409–428.

ARAVENA, R., L. I. WASSENAAR, and L. N. PLUMMER. 1995. Estimating ^{14}C groundwater ages in a methanogenic aquifer. Water Resour. Res., v. 31, no. 9, p. 2307–2317.

Arizona Department of Water Resources. 1994. Riparian Protection Program Legislative Report. Phoenix, Arizona, p. 209–280.

Arizona Department of Water Resources. 1997. Preliminary Determination Report on the Safe-Yield Status of the Prescott Active Management Area, Phoenix, Arizona.

ARMSTRONG, J. M., L. E. LEACH, R. M. POWELL, and S. V. VANDERGRIFT, and J. T. WILSON. 1980. Bioremediation of a fuel spill: Evaluation of techniques for preliminary site characterization. NWWA/API Conference Petroleum Hydrocarbons and Organic Chemicals in Ground Water: Prevention, Detection and Restoration: Dublin, Ohio, National Water Well Assoc., p. 931–944.

ASTM. 1992. ASTM Standards on ground water and vadose zone investigations. ASTM Publication Code Number (PCN): 03-418192-38, Philadelphia, Pennsylvania, 166 p.

BACK, W. 1961. Techniques for mapping of hydrochemical facies. U.S. Geol. Surv. Prof. Paper 424-D, p. 380–382.

BALKWILL, D. L., J. K. FREDRICKSON, and J. M. THOMAS. 1989. Vertical and horizontal variations in the physiological diversity of the aerobic chemoheterotrophic bacteria microflora in deep southeast Coastal Plain subsurface sediments. Appl. Environ. Microbiol., v. 55, p. 1058–1065.

BANKS, S. 1993. Exploratory modeling for policy analysis. Oper. Res., v. 41, 435.

BARCELONA, M. J., J. P. GIBB, J. A. HELFRICH, and E. E. GARSKE. 1985. Practical Guide for Ground-Water Sampling. Illinois State Water Survey, ISWS Contract Report 374, 94 p.

BEAR, J. 1972. Dynamics of Fluids in Porous Media. Elsevier, New York, 764 p.

BEAR, J. 1979. Hydraulics of groundwater. McGraw-Hill, New York, 569 p.

BENNETT, R. R. 1962, Flow net analysis. In J. G. FERRIS, D. B. KNOWLES, R. H. BROWN, and R. W. STALLMAN (eds.), Theory of aquifer tests. U.S. Geol. Surv. Prof. Paper 708, 70 p.

BENNETT, R. R., and R. R. MEYER. 1952. Geology and ground water resources of the Baltimore area [Maryland] Maryland Dept. of Geology, Mines and Water Resour. Bull. 4, 573 p.

BENTLEY, H. W., F. M. PHILLIPS, S. N. DAVIS, M. A. HABERMEHL, P. L. AIREY, G. E. CALF, D. ELMORE, H. E. GROVE, and T. TORGERSEN. 1986. Chlorine 36 dating of very old groundwater 1. The great artesian basin, Australia. Water Resour. Res., v. 22, no. 13, p. 1991–2001.

BETHKE, C. M., W. J. HARRISON, C. UPTON, and S. P. ALTANER. 1988. Supercomputer analysis of sedimentary basins. Science, v. 239, p. 261–267.

BIRSOY, Y. K., and W. K. SUMMERS. 1980. Determination of aquifer parameters from step tests and intermittent pumping data. Ground Water, v. 18, p. 137–146.

BOGGS, J. M. et al. 1992. Field study of dispersion in a heterogeneous aquifer. 1, Overview and site description. Water Resour. Res., v. 28, no.12, p. 3281–3291.

BOGGS, S. 1987. Principles of Sedimentology and Stratigraphy. Merrill Pub. Co., Columbus, Ohio.

BOHLKE, J. K., and J. M. DENVER. 1995. Combined use of groundwater dating, chemical and isotopic analyses to resolve the history and fate of nitrate contamination in two agricultural watersheds, Atlantic coastal plain, Maryland. Water Resour. Res., v. 31, no. 9, 2319–2339.

BOLT, G. H., and P. H. GROENEVELT. 1969. Coupling phenomena as a possible cause for non-Darcian behavior of water in soil. Bull. Intern. Assoc. Sci. Hydrol. 14, p. 17–26.

BOULDING, J. R., and M. J. BARCELONA. 1991. Geochemical sampling of subsurface solids and ground water. In Site Characterization for Subsurface Remediation. EPA/625/4-91/026, p. 123–154.

BOULTON, N. S. 1954. The drawdown of the water table under nonsteady conditions near a pumped well in an unconfined formation. Proc. Inst. Civil Engrs., v. 3, p. 564–579.

BOULTON, N. S. 1955. Unsteady radial flow to a pumped well allowing for delayed yield from storage. Proc. Gen. Assembly, Rome Intern. Assoc. Sci. Hydrol. Publ., v. 37, p. 472–477.

BOULTON, N. S. 1963. Analysis of data from nonequilibrium pumping test allowing for delayed yield from storage. Proc. Inst. Civil Engrs., v. 26, p. 469–482.

BOULTON, N. S. 1970. Analysis of data from pumping test in unconfined anisotropic aquifers. J. Hydrol., v. 10, 369.

BOUWER, H. 1985. Renovating waste water with groundwater recharge. In E. T. SMERDON, and W. R. JORDAN (eds.), Issues in Groundwater Management. Texas A&M Univ., College Station, p. 331–346.

BOUWER, H. 1989. The Bouwer and Rice slug test—An update. Ground Water, v. 27, p. 304–309.

BOUWER, H., and R. C. RICE. 1976. A slug test for determining hydraulic conductivity of unconfined aquifers with completely or partially penetrating wells. Water Resour. Res., v. 12, p. 423–428.

BOYD, F. M., and C. W. KREITLER. 1986. Hydro-geology of a Gypsum Playa. Northern Salt Basin, Texas. Bureau of Economic Geology, Univ. of Texas at Austin, Report of Investigations No. 158, 37 p.

BRADBURY, K. R. 1984. Major ion and isotope geochemistry of groundwater in clayey till, northwestern Wisconsin, U.S.A. First Canadian/American Conference on Hydrogeology, B. HITCHON and E. I. WALLICK (eds.). National Water Well Assoc., Dublin, Ohio, p. 284–289.

BRAHANA, J. V., D. MULDERINK, J. MACY, and M. W. BRADLEY. 1986. Preliminary delineation and description of the regional aquifers of Tennessee—The East Tennessee aquifer system. U.S. Geol. Surv. Water-Resour. Investigation Report 82-4091, 30 p.

BRAHANA, J. V., J. THRAILKILL, T. FREEMAN, and W. C. WARD. 1988. Carbonate rocks. In W. BACK, J. S. ROSENSHEIN, and P. R. SEABER (eds.), Hydrogeology, The Geology of North America, v. O-2. Geol. Soc. Am., Boulder, Colorado, p. 333–352.

BREDEHOEFT, J. D., and L. F. KONIKOW. 1993. Editorial—Ground-water models: Validate or invalidate. Ground Water, v. 31, no. 2, p. 178–179.

BREDEHOEFT, J. D., C. E. NEUZIL, and P. C. D. MILLY. 1983. Regional flow in the Dakota Aquifer, a study of the role of confining layers. U.S. Geol. Surv. Water-Supply Paper 2237, 45 p.

BREDEHOEFT, J. D., E. G. REICHARD, and S. M. GORELICK. 1995. If it works, don't fix it; benefits from regional groundwater management. In Groundwater Models for Resources Analysis and Management. 1994 Pacific Northwest/Oceania Conference, Aly I. El-Kadi, Honolulu, HI, United States, March 21–23. 1994, p. 101–121.

BREEN, K. J., A. L. KONTIS, G. L. ROWE, and R. J. HAEFNER. 1995. Simulated ground-water flow and sources of water in the Killbuck Creek valley near Wooster, Wayne County, Ohio. U.S. Geol. Surv., Water-Resour. Investigations Report 94-4131.

BROOKS, R. H., and A. T. COREY. 1966. Properties of porous media affecting fluid flow. J. Irrig. Drainage Div., ASCE Pro. 72(IR2), p. 61–88.

BROWN, R. H. 1963. Drawdown resulting from cyclic intervals of discharge. U.S. Geol. Surv. Water-Supply Paper 1536-I, p. 324–330.

Bureau of Reclamation, U.S. Department of Interior. 1995. Ground water manual. A Water Resour. Technical Publication, 2nd edition, 661 p.

BURNETT, R. D., and E. O. FRIND. 1987. An alternative direction Galerkin technique for simulation of ground-water contamination transport in three dimensions, 2 dimensionality effects. Water Resour. Res., v. 23, no. 4, p. 695–705.

BUSENBERG, E., and L. N. PLUMMER. 1992. Use of chlorofluoromethanes (CCl_3F and CCl_2F_2) as hydrologic tracers and age-dating tools: Example, The alluvium and terrace system of Central Oklahoma. Water Resour. Res., v. 28, p. 2257–2283.

BUTLER, R. D. 1984. Hydrogeology of the Dakota aquifer system. Williston Basin, North Dakota. In D. G. JORGENSEN and D. C. SIGNOR (eds.), Proc. of the Geohydrology Dakota Aquifer Symposium. Water Well Journal Pub. Co., Worthington, Ohio, p. 12–23.

BUURSINK, M. L., and J. W. LANE Jr. 1999. Characterizing fractures in a bedrock outcrop using ground-penetrating radar at Mirror Lake, Grafton County, New Hampshire. USGS Toxic Substances Hydrology Meeting, Charleston, South Carolina, March 8–12.

CAMPBELL, M. D., and J. H. LEHR. 1973. Water Well Technology. McGraw Hill Book Company, New York, 681 p.

CANT, D. J. 1982. Fluvial facies models and their application. In P. A. SCHOLLE and D. SPEARING (eds.), Sandstone Depositional Environments. Am. Assoc. Petrol. Geol. Mem. 31, p. 115–138.

CARTWRIGHT, K., and M. R. McCOMAS. 1968. Geophysical surveys in the vicinity of sanitary landfills in Northern Illinois. Ground Water, v. 16, no. 5, p. 23–30.

CASAGRANDE, A. 1937. Seepage through dams. Harvard Graduate School Eng. Pub. 209.

CHAMP, D. R., J. GULENS, and R. E. JACKSON. 1979. Oxidation-reduction sequences in ground-water flow systems. Can. J. Earth Sci., v. 16, no. 1, p. 12–23.

CHAPELLE, F. H. 1993. Ground-Water Microbiology and Geochemistry. John Wiley & Sons, New York, 424 p.

CHAPELLE, F. H., J. L. ZELIBOR Jr., D. J. GRIMES, and L. L. KNOBEL. 1987. Bacteria in deep coastal plain sediments of Maryland: A possible source of CO_2 to groundwater. Water Resour. Res., v. 23, no. 8, p. 1625–1632.

CHERRY, J. A. 1983. Piezometers and other permanently-installed devices for groundwater quality monitoring. Proc. Seminar on Groundwater and Petroleum Hydrocarbons Protection, Detection, Restoration. Petroleum Assoc. for Conservation of the Canadian Environment, Ottawa, p. IV-1–IV-39.

CHIANG, C. Y., K. R. LOOS, R. A. KLOPP, and M. C. BELTZ. 1989. A real-time determination of geological/chemical properties of an aquifer by penetration testing. NWWA/API Conference Petroleum Hydrocarbons and Organic Chemicals in Ground Water—Prevention, Detection, and Restoration, Proc. National Water Well Association, Dublin, Ohio, p. 175–189.

CLARK, L. 1977. The analysis and planning of step-drawdown tests. Quart. J. Eng. Geol. v. 10, no. 2, p. 125–143.

COCHRAN, J. R., and K. E. DALTON. 1995. Using high-density magnetic and electromagnetic data for waste site characterization. Proc. of the Symposium on the Application of Geophysics to Engineering and Environmental Problems. Environmental and Engineering Geophysical Society, Englewood, Colorado, p. 117–127.

COHEN, P., O. L. FRANKE, and B. L. FOXWORTHY. 1968. An atlas of Long Island's water resources. New York Water Resources Commission Bull. 62, 117 p.

CONKLING, H. 1946. Utilization of groundwater storage in stream system development. Trans. Am. Soc. Civil Engrs., v. 111, p. 275–305.

COOPER, H. H., J. D. BREDEHOEFT, and I. S. PAPADOPULOS. 1967. Response of a finite diameter well to an instantaneous charge of water. Water Resour. Res., v. 3, p. 263–269.

COOPER, H. H., and C. E. JACOB. 1946. A generalized graphical method for evaluating formation constants and summarizing well field history. Trans. Am. Geophys. Union, v. 27, p. 526–534.

COTTER, R. D., R. D. HUTCHINSON, E. L. SKINNER, and D. A. WENTZ. 1969. Water Resources of Wisconsin, Rock-Fox Basin: Geological Survey Water-Resour. Investigations Atlas HA-360. 4 sheets.

COZZARELI, I. M., R. P. EAGANHOUSE, and M. J. BAEDECKER. 1988. The fate and effects of crude oil in a shallow aquifer: II. Evidence of anaerobic degradation of monoaromatic hydrocarbons. U.S. Geol. Surv. Toxic Waste Ground-Water Contamination Program, U.S. Geol. Surv. Water-Resources Investigations Report 88-4220, p. 21–33.

CRAIG, H. 1961. Isotopic variations in meteoric water. Science, v. 133, p. 1702–1703.

CRAUN, G. F. 1984. Health aspects of groundwater pollution. In G. BITTON and C. P. GERBA (eds.), Groundwater Pollution Microbiology. John Wiley & Sons, New York, p. 135–179.

CRWMS M&O. 1998. Chapter 8, Saturated Zone Flow and Transport. Total System Performance Assessment—Viability Assessment (TSPA-VA): Analyses Technical Basis Document. B00000000-01717-4301-00008 Rev 01. Las Vegas, Nevada, CRWMS M&O.

DAGAN, G. 1967. A method of determining the permeability and effective porosity of unconfined anisotropic aquifers. Water Resour. Res., v. 3, p. 1059–1071.

DAGAN, G. 1982. Stochastic modeling of groundwater flow by unconditional and conditional probabilities. 2. The solute transport. Water Resour. Res., v. 18, no. 4, p. 835–848.

DAGAN, G., and J. BEAR. 1968. Solving the problem of local interface upconing in a coastal aquifer by the method of small perturbations. J. Hydrol. Res., v. 6, p. 15–44.

DANIELS, J. J. 1989. Technical review: Ground penetrating radar. In Proc. (2nd) Symposium on the Application to Engineering and Environmental Problems. Soc. Engineering and Mineral Exploration Geophysicists, Golden, Colorado, p. 62–142.

DANSGAARD, W. 1964. Stable isotopes in precipitation. Tellus, v. 16, p. 436–438.

DARCY, H.P.G. 1856. Les fontaines publiques de la Ville de Dijon. Victor Dalmont, Paris.

DARTON, N. H. 1909. Geology and underground waters of South Dakota. U.S. Geol. Surv. Water-Supply Paper 227.

DAVIS, S. N. 1969. Porosity and permeability of natural materials. In R.J.M. DEWIEST (ed.), Flow through Porous Materials. Academic Press, New York, p. 54–89.

DAVIS, S. N. 1988. Sandstones and shales. In W. BACK, J. S. ROSENSHEIN, and P. R. SEABER (eds.), Hydrogeology, The Geology of North America, v. O-2, Geol. Soc. Am., Boulder, Colorado, p. 323–332.

DAVIS, S. N., and R.J.M. DE WIEST. 1966. Hydrogeology. John Wiley & Sons, New York, 463 p.

DESAULNIERS, D. E., J. A. CHERRY, and P. FRITZ. 1981. Origin, age, and movement of pore water in argillareous Quaternary deposits at four sites in southwestern Ontario. J. Hydrol., v. 50, p. 231–257.

DOMENICO, P. A. 1972. Concepts and Models in Groundwater Hydrology. McGraw-Hill, New York, 405 p.

DOMENICO, P. A. 1987. An analytical model for multidimensional transport of a decaying contaminant species. J. Hydrol., v. 91, p. 49–58.

DOMENICO, P. A., and M. D. MIFFLIN. 1965. Water from low permeability sediments and land subsidence. Water Resour. Res., v. 4, p. 563–576.

DOMENICO, P. A., and G. A. ROBBINS. 1985a. The displacement of connate water from aquifers. Geol. Soc. Am. Bull. 96, p. 328–335.

DOMENICO, P. A., and G. A. ROBBINS. 1985b. A new method of contaminant plume analysis. Ground Water, v. 23, p. 476–485.

DOMENICO, P. A., and F. W. SCHWARTZ. 1998. Physical and Chemical Hydrogeology. John Wiley & Sons, New York, 506 p.

DOUGLAS, B. J., and R. S. OLSEN. 1981. Soil classification using electric cone penetrometer. Cone Penetrometer Testing and Experience, ASCE, p. 209–227.

DOWNEY, J. S. 1984. Geology and hydrology of the Madison limestone and associated rocks in parts of Montana, Nebraska, North Dakota, South Dakota, and Wyoming. U.S. Geol. Surv. Prof. Paper 1273-G, 152 p.

DUBROVSKY, N. M., C. R. KRATZER, L. R. BROWN, J. M. GRONBERG, and K. R. BUROW. 1998. Water quality in the San Joaquin-Tulare Basins, California, 1992-95. U.S. Geol. Surv. Circular 1159, 38 p.

DUNKLE, S. A., L. N. PLUMMER, E. BUSENBERG, P. J. PHILLIPS, J. M. DENVER, P. A. HAMILTON, R. L. MICHEL, and T. B. COPLEN. 1993. Chlorofluorohydrocarbons (CCl_3F and CCl_2F_2) as dating tools and hydrologic tracers in shallow groundwater of the Delmarva Peninsula, Atlantic Coastal Plain, United States Water Resour. Res. v. 29, p. 3837–3860.

DYCK, J. H., W. S. KEYS, and W. A. MENELEY. 1972. Application of geophysical logging to groundwater studies in southern Saskatchewan. Can. J. Earth Sci., v. 9, no. 1, p. 78–94.

EAKIN, T. A., et al. 1976. Summary appraisals of the nation's groundwater resources: Great Basin. U.S. Geol. Surv. Prof. Papers 813-G.

EAGLESON, P. S. 1978. Climate, soil, and vegetation. 3. A simplified model of soil moisture movement in the liquid phase, Water Resour. Res., v. 14, no.5, p. 722–730.

ELKINS, J. W., T. M. THOMPSON, T. H. SWANSON, J. H. BUTLER, B. D. HALL, S. O. CUMMINGS, D. A. FISHER, and A. G. RAFFO. 1993. Decrease in growth rates of atmospheric chlorofluorocarbons 11 and 12. Nature, v. 364, p. 780–783.

FABRYKA-MARTIN, J., D. O. WHITTEMORE, S. N. DAVIS, P. W. KUBIK, and P. SHARAMA. 1991. Geochemistry of halogens in the Milk River aquifer, Alberta, Canada. Applied Geochemistry, v. 6, p. 447–464.

FABRYKA-MARTIN, J., S. J. WIGHTMAN, W. J. MURPHY, M. P. WICKHAM, M. W. CAFEE, G. J. NIMZ, J. R. SOUTHON, and P. SHARMA. 1993. Distribution of chlorine-36 in the unsaturated zone at Yucca Mountain: An indicator of fast transport paths. Proceedings Focus '93: Site Characterization and Model Validation. American Nuclear Society, La Grange Park, Illinois, p. 58–68.

FARMER, V. E. 1983. Behaviour of petroleum contaminants in an underground environment. Proc. Seminar on Groundwater and Petroleum Hydrocarbons, Protection, Detection, Restoration. Petroleum Assoc. for Conservation of the Canadian Environment, Ottawa, II-1–II-16.

FERRIS, J. G., and D. B. KNOWLES. 1963. The slug-injection test for estimating the coefficient of transmissibility of an aquifer. U. S. Geol. Surv. Water-Supply Paper 1563-I.

FERRIS, J. G., D. B. KNOWLES, R. H. BROWN, and R. W. STALLMAN. 1962. Theory of aquifer tests. U.S. Geol. Surv. Prof. Paper 708, 70 p.

FOGG, G. E., and C. W. KREITLER. 1982. Ground-water Hydraulics and Hydrochemical Facies in Eocene Aquifers of the East Texas Basin. Bureau of Economic Geology, Univ. of Texas at Austin, Report of Investigations No. 127, 75 p.

FONTES, J. C. 1980. Chapter 3. Environmental isotopes in groundwater hydrology. In P. FRITZ and J. C. FONTES (eds.), Handbook of Environmental Isotope Geochemistry, v. 1. Elsevier, Amsterdam, p. 75–140.

FORD, D. C. 1980. Thresholds and limit effects in karst geomorphology. In D. R. COATES and J. D. VITEK (eds.), Thresholds in Geomorphology. Allen and Unwin, Boston, p. 345–362.

FORD, D. C. 1983. Karstic interpretation of the Winnipeg Aquifer (Manitoba, Canada). J. Hydrol., v. 61, p. 177–180.

FORSTNER, U., and T. W. WITTMAN. 1981. Metal Pollution in the Aquatic Environment. Springer-Verlag, Berlin Heidelberg, Germany, 486 p.

FOSTER, M. D. 1950. The origin of high sodium bicarbonate waters in the Atlantic and Gulf Coastal Plains. Geochim. Cosmochim. Acta, v. 1, p. 33–48.

FRANCIS, R. M. 1981. Hydrogeological properties of a fractured porous aquifer, Winter River Basin, Prince Edward Island. Unpublished M.Sc. Thesis, University of Waterloo, 153 p.

FRANKE, O. L., T. E. REILLY, D. W. POLLOCK, and J. W. LaBAUGH. 1998. Estimating areas contributing recharge to wells, Lesson from previous studies. U.S. Geol. Surv. Circular 1174.

FRANKS, B. J. (ed.). 1987. U.S. Geol. Surv. Program on Toxic Waste—Ground-Water Contamination: Proc. of the Third Technical Meeting, Pensacola, Florida, March 23–27, 1987. U.S. Geol. Surv. Open-File Report 87-109.

FREDRICKSON, J. K., F. J. BROCKMAN, R. J. HICKS, and B. A. DENOVAN. 1990. Biodegradation of nitrogen-containing aromatic compounds in deep subsurface sediments. Proceedings of the First International Symposium on Microbiology of the Deep Subsurface, (eds.) C. B. FLIERMANS and T. C. HAZEN. Westinghouse Savannah River Company, p. 6-27–6-44.

FREEZE, R. A. 1969a. Theoretical analysis of regional groundwater flow. Department of Energy Mines and Resources, Inland Waters Branch, Scientific Series No. 3, 147 p.

FREEZE, R. A. 1969b. Regional groundwater flow—Old Wives Lake Drainage basin, Saskatchewan. Department of Energy Mines and Resources, Inland Waters Branch, Scientific Series No. 5, 243 p.

FREEZE, R. A., and P. A. WITHERSPOON. 1966. Theoretical analysis of regional groundwater flow. I: Analytical and numerical solutions to the mathematical model. Water Resour. Res., v. 2, p. 641–656.

FREEZE, R. A., and P. A. WITHERSPOON. 1967. Theoretical analysis of regional groundwater flow. II: Effects of water table configuration and subsurface permeability variations. Water Resour. Res., v. 3, p. 623–634.

FREYBERG, D. L. 1986. A natural gradient experiment on solute transport in a sand aquifer 2. Spatial moments and the advection and dispersion of nonreactive tracers. Water Resour. Res., v. 22, no. 13, p. 2031–2046.

FREYBERG, D. L. 1988. An exercise in ground-water model calibration and prediction. Ground Water, v. 26, no. 3, p. 350–360.

FRITZ, P., and J. C. FONTES. 1980. Introduction. In P. FRITZ and J.C. FONTES (eds.), Handbook of Environmental Isotope Geochemistry, v. 1. Elsevier, Amsterdam, p. 1–19.

FUSILLO, T. V., T. A. EHLKE, M. MARTIN, and B. P. SARGENT. 1987. Movement and fate of chlorinated solvents in ground water: Preliminary results and future research plans. In U.S. Geol. Surv. Program on Toxic Waste— Ground-Water Contamination, Proceedings of the Third Technical Meeting. U.S. Geol. Surv. Open-File Report 87-109, p. D-5–D-12.

GALE, J. E. 1982. Assessing the permeability characteristics of fractured rock. In T. N. NARASIMHAN (ed.), Recent Trends in Hydrogeology. Geol. Soc. Am. Special Paper 189, p. 163–181.

GARABEDIAN, S. P., and D. R. LeBLANC. 1991. Overview of research at the Cape Cod Site: Field and laboratory studies of hydrologic, chemical and microbial processes affecting transport in a sewage-contaminated sand and gravel aquifer. U.S. Geol. Surv., Toxic Waste Ground-Water Contamination Program, Water-Resour. Investigations Report 91-4034, p. 1–9.

GARDNER, W. H. 1986. Water content in methods of soil analysis, A. KLUTE (ed.), Amer. Soc. of Agron., Madison, Wisconsin, p. 493–544.

GARRELS, R. M., and F. T. MacKENZIE. 1967. Origin of the chemical compositions of some springs and lakes. In R. F. GOULD (ed.), Equilibrium Concepts in Natural Water Systems. Am. Chem. Soc., Adv. Chem. Ser. 67, Washington, DC, p. 222–242.

GAT, J. R. 1980. Chapter 1. The isotopes of hydrogen and oxygen in precipitation. In P. FRITZ and J. C. FONTES (eds.), Handbook of Environmental Isotope Geochemistry, v. 1. Elsevier, Amsterdam, p. 21–47.

GELHAR, L. W., and C. L. AXNESS. 1983. Three-dimensional stochastic analysis of macrodispersion in aquifers. Water Resour. Res., v. 19, no. 1, p. 161–180.

GELHAR, L. W., A. MANTOGLOU, C. WELTY, and K. R. REHFELDT. 1985. A review of field-scale physical solute transport processes in saturated and unsaturated porous media. Electric Power Research Instit. EPRI EA-4190 Project 2485-5, 116 p.

GELHAR, L. W., C. WELTY, and K. R. REHFELDT. 1992. A critical review of data on field-scale dispersion in aquifers. Water Resour. Res., v. 28, no. 7, p. 1955–1974.

GHIORSE, W. C., and J. T. WILSON. 1988. Microbial ecology of the terrestrial subsurface. Advances in Applied Microbiology, v. 33, p. 107–172.

GHYBEN, W. B. 1899. Notes in verband met Voorgenomen Put boring Nabji Amsterdam, Tijdschr. Koninhitk, Inst. Ingrs., The Hague.

GLOVE, R. E. 1964. Ground water movement. U.S. Bureau of Reclamation Engineering Monograph 31, p. 31–34.

GOODE, D. J., and C. A. APPEL. 1992. Finite-difference interblock transmissivity for unconfined aquifers and for aquifers having smoothly varying transmissivity. U.S. Geol. Surv. Water-Resour. Investigations Report 92-4124, 79 p.

GRAY et al. 1970. Handbook on the Principles of Hydrology. Canadian National Committee for the International Hydrological Decade, National Research Council of Canada.

GREEN, W. H., and C. A. AMPT. 1911. Studies on soil physics. I. The flow of air and water through soils. Journal of Agricultural Sciences, IV (Part I), p. 1–24.

GREENHOUSE, J. P., and D. D. SLAINE. 1983. The use of reconnaissance electrical methods to map contaminant migration. Ground Water Mon. Rev., v. 3, p. 47–49.

GRIFFIN, R. A., and W. R. ROY. 1985. Interaction of Organic Solvents with Saturated Soil-Water Systems. Environmental Institute for Waste Management Studies. Univ. of Alabama, Open File Report No. 3, 86 p.

GRISAK, G. E., and R. E. JACKSON. 1978. An appraisal of the hydrogeological processes involved in shallow subsurface radioactive waste management in Canadian terrain. Environment Canada, Scientific Series No. 84, 194 p.

GRISAK, G. E., and J. F. PICKENS. 1980. Solute transport through fractured media, 1. The effect of matrix diffusion. Water Resour. Res., v. 16, no. 4, p. 719–730.

GROVE, D. B. 1971. U.S. Geological Survey tracer study, Amargosa Desert, Nye County, Nevada, II, An analysis of the flow field of a discharging-recharging pair of wells. U.S. Geol. Surv. Rep. 474-99, 56 p.

GROVE, D. B., and W. A. BEETEM. 1971. Porosity and dispersion constant calculations for a fractured carbonate aquifer using the two well tracer method. Water Resour. Res., v. 7, p. 128–134.

GÜVEN, O., R. W. Falta, F. J. Molz, and J. G. Melville. 1986. A simplified analysis of two-well tracer tests in stratified aquifers. Ground Water, v. 24, no. 1, p. 63–71.

HABERMEHL, M. A. 1980. The Great Artesian Basin, Australia. BMR J. Aust. Geol. Geophys., 5, p. 9–38.

HAMBROOK, J. A., G. F. KOLTUN, B. B. PALCSAK, and J. S. TERTULIANI. 1997. Hydrologic disturbance and response of aquatic biota in Big Darby Creek, Ohio. U.S. Geol. Surv. Water Resour. Investigations Report 96-4315, 82 p.

HANTUSH, M. S. 1956. Analysis of data from pumping tests in leaky aquifers. Trans. Am. Geophys. Union, v. 37, p. 702–714.

HANTUSH, M. S. 1960. Modification of the theory of leaky aquifers. J. Geophys. Res., v. 65, p. 3713–3725.

HANTUSH, M. S. 1961a. Tables of the function $W(u, \beta) = \int_u^\infty \dfrac{e^{-z-\frac{\beta^2}{4y}}}{y}\, dy$, New Mexico Inst. Mining and Technology Prof. Paper 104, 13 p.

HANTUSH, M. S. 1961b. Tables of the function $H(u, \beta) = \int_u^\infty \dfrac{e^{-y}}{y}\, erfc\left(\dfrac{\beta\sqrt{u}}{\sqrt{y(y-u)}}\right) dy$, New Mexico Inst. Mining and Technology Prof. Paper 103, 12 p.

HANTUSH, M. S. 1961c. Aquifer tests on partially penetrating wells. Proc. Am. Soc. Civ. Engrs., v. 87, p. 171–195.

HANTUSH, M. S. 1964. Hydraulics of wells. In V. T. Chow (ed.), Advances in Hydroscience. Academic Press, New York, p. 281–432.

HANTUSH, M. S., and C. E. JACOB. 1955. Nonsteady radial flow in an infinite leaky aquifer. Trans. Am. Geophys. Union, v. 36, p. 95–100.

HARBAUGH, A. W. 1992. A generalized finite-difference formulation for the U.S. Geological Survey modular three-dimensional finite-difference ground-water flow model. U.S. Geol. Surv. Open-File Report 91-494, 60 p.

HARBAUGH, A. W. 1995. Direct solution package based on alternating diagonal ordering for the U.S. Geological Survey modular finite-difference ground-water flow model. U.S. Geol. Surv. Open-File Report 95-288, 46 p.

HARBAUGH, A. W., and M. G. MCDONALD. 1996a. Programmer's documentation for MODFLOW-96, an update to the U.S. Geol. Surv. modular finite-difference ground-water flow model. U.S. Geol. Surv. Open-File Report 96-486, 220 p.

HARBAUGH, A. W., and M. G. MCDONALD. 1996b. User's documentation for MODFLOW-96, an update to the U.S. Geol. Surv. modular finite-difference ground-water flow model. U.S. Geol. Surv. Open-File Report 96-485, 56 p.

HARRIL, J. R. 1971. Determining transmissivity from water-level recovery of a step-drawdown test, U.S. Geol. Surv. Prof. Paper 700-C, C212–C213.

HARVEY, R. W., R. L. SMITH, and L. GEORGE. 1984. Microbial distribution and heterotrophic uptake in a sewage plume. Movement and Fate of Solutes in a Plume of Sewage-Contaminated Ground Water, Cape Cod, Massachusetts. U.S. Geol. Surv. Toxic Waste Ground-Water Contamination Program, U.S. Geol. Surv. Open-File Report 84-475, p. 139–152.

HASSETT, J. J., W. L. BANWART, and R. A. GRIFFIN. 1983. Chapter 15. Correlation of compound properties with sorption characteristics of nonpolar compounds by soils and sediments: Concepts and limitations. In C. W. FRANCIS and S. I. AUERBACH (eds.), Environment and Solid Wastes: Characterization, Treatment and Disposal. Butterworth Publishers, p. 161–178.

HAVERKAMP, R., M. VAUCLIN, J. TOVINA, P. J. WIERENGA, and G. VACHAUD. 1977. A comparison of numerical simulation models for one-dimensional infiltration. Soil Science Society of America Proceedings, v. 41, p. 285–294.

HEALY, R. W. 1990. Simulation of solute transport in variably saturated porous media with supplemental information on modification to the U.S. Geol. Survey's computer program VS2D. U.S. Geol. Surv. Water-Resour. Investigations Report 90-4025, 125 p.

HEATH, R. C. 1983. Basic ground-water hydrology. U.S. Geol. Surv. Water-Supply Paper 2220, 84 p.

HEATH, R. C. 1988. Hydrogeologic setting of regions. In W. BACK, J. S. ROSENSHEIN, and P. R. SEABER (eds.), Hydrogeology, The Geology of North America, v. O-2. Geol. Soc. Am., Boulder, Colorado, p. 15–23.

HENDRY, M. J., J. A. CHERRY, and E. I. WALLICK. 1986. Origin and distribution of sulfate in a fractured till in southern Alberta, Canada. Water Resour. Res., v. 22, no. 1, p. 45–61.

HENDRY, M. J., and F. W. SCHWARTZ. 1988. An alternative view on the origin of chemical and isotopic patterns in groundwater from the Milk River aquifer. Water Resour. Res., v. 24, no. 10, p. 1747–1764.

HENDRY, M. J., and F. W. SCHWARTZ. 1990. The chemical evolution of ground water in the Milk River aquifer, Canada. Ground Water, v. 28, no. 2, p. 253–261.

HENDRY, M. J., F. W. SCHWARTZ, and C. Robertson. 1991. Hydrogeology and hydrogeochemistry of the Milk River aquifer system. Applied Geochemistry, v. 6, p. 369–380.

HENDRY, M. J., and L. I. WASSENAAR. 1999. Implication of the transport of δD in pore waters for groundwater flow and the timing of geologic events in a thick aquitard system. Water Resour. Res., v. 35, p. 1751–1760.

HERZBERG, B. 1901. Die Wasserversovgung einiger Nordseebaser. J. Gasbeleucht and Wasserversorv, v. 44, p. 815–819.

HESS, K. M., W. N. HERLKELRATH, and H. I. ESSAID. 1992. Determination of subsurface fluid contents at a crude-oil spill site. J. Contam. Hydrol., v. 10, p. 75–96.

HIGGINS, C. G., and D. R. COATES. 1990. Groundwater geomorphology. Geol. Soc. Am. Special Paper 252, 368 p.

HILL, M. C. 1990. Preconditioned Conjugate-Gradient 2 (PCG2), A computer program for solving ground-water flow equations, U.S. Geol. Surv. Water-Resour. Investigations Report 90-4048, 43 p.

HILL, M. C. 1992. A computer program (MODFLOWP) for estimating parameters of a transient, three-dimensional ground-water flow model using nonlinear regression. U.S. Geol. Surv. Open-File Report 91-484.

HORTON, R. E. 1933. The role of infiltration in the hydrologic cycle, Trans. Am. Geophys. Union, v. 14, p. 446–460.

HORTON, R. E. 1940. Approach toward physical interpretation of infiltration capacity, Soil Sci. Soc. Am. Proc., v. 5, p. 339–417.

HSIEH, P. A., and J. R. FRECKLETON. 1993. Documentation of a computer program to simulate horizontal-flow barriers using the U.S. Geol. Surv.'s modular three-dimensional finite-difference ground-water flow model. U.S. Geol. Surv. Open-File Report 92-477, 32 p.

HUBBERT, M. K. 1940. The theory of ground water motion. J. Geology, v. 48, no. 8, pt. 1, p. 785–944.

HUNT, C. D., C. J. EWART, and I. V. CLIFFORD. 1988. Hawaii Islands. In W. BACK, J. S. ROSENSHEIN, and P. R. SEABER (eds.), Hydrogeology, The Geology of North America, v. O-2. Geol. Soc. Am., v. O-2. Boulder, Colorado, p. 255–262.

HUNTER, J. A., S. E. PULLAN, R. A. BURNS, R. M. GAGNE, and R. L. GOOD. 1984. Shallow seismic reflection mapping of the overburden-bedrock interface with the engineering seismograph—some simple techniques. Geophysics, v. 49, p. 1381–1385.

HUNTER, J. A., S. E. PULLAN, R. A. BURNS, R. M. GAGNE, and R. L. GOOD. 1988. Applications of a shallow seismic reflection method to groundwater and engineering studies. Proc. of Exploration '87: Third Decennial

International Conference on Geophysical and Geochemical Exploration for Minerals and Groundwater. Ontario Geological Survey, Special Volume, p. 704–715.

HUNTLEY, D., R. N. HAWK, and H. P. CORLEY. 1992. Non-aqueous phase hydrocarbon saturations and mobility in a fine-grained, poorly consolidated sandstone. Proc. of the 1992 Petroleum Hydrocarbons and Organic Chemicals in Ground Water: Prevention, Detection, and Restoration. National Ground Water Association, Columbus, Ohio, p. 223–237.

HUYAKORN, P. S., P. F. ANDERSEN, O. GÜVEN, and F. J. MOLZ, 1986. A curvilinear finite element model for simulating two-well tracer tests and transport in stratified aquifers. Water Resour. Res., v. 22, no. 5, p. 663–678.

HUYAKORN, P. S., and G. F. PINDER. 1983. Computational Methods in Subsurface Flow. Academic Press, New York, 473 p.

INGRAHAM, N. L., and B. E. TAYLOR. 1989. The effect of snowmelt on the hydrogen isotope ratios of creek discharge in Surprise Valley, California. J. Hydrol., v. 106, p. 233–244.

INGRAHAM, N. L., and B. E. TAYLOR. 1991. Light stable isotope systematics of large-scale hydrologic regimes in California and Nevada. Water Resour. Res., v. 27, no. 1, p. 77–90.

IVANOVICH, M., K. FROHLICH, and M. J. HENDRY. 1991. Dating very old groundwater, Milk River Aquifer, Alberta, Canada. Applied Geochemistry, v. 6, no. 4, p. 367–472.

JACKSON, R. E., and R. J. PATTERSON. 1982. Interpretation of pH and Eh trends in a fluvial-sand aquifer system. Water Resour. Res., v. 18, no. 4, p. 1255–1268.

JACKSON, R. E., R. J. PATTERSON, B. W. GRAHAM, J. BAHR, D. BELANGER, J. LOCKWOOD, and M. PRIDDLE. 1985. Contaminant Hydrogeology of Toxic Organic Chemicals at a Disposal Site, Gloucester, Ontario: 1. Chemical Concepts and Site Assessment. Environment Canada, National Hydrol. Research Instit., Paper No. 23, Ottawa, 114 p.

JACOB, C. E. 1940. On the flow of water in an elastic artesian aquifer. Trans. Am. Geophys. Union, v. 22, p. 574–586.

JACOB, C. E. 1946. Radial flow in a leaky artesian aquifer. Trans. Am. Geophys. Union, v. 27, no. 2, p. 198–205.

JACOB, C. E. 1947. Drawdown test to determine effective radius of artesian well. Transactions, ASCE, v. 112, Paper 2321, p. 1047–1070.

JACOB, C. E. 1963. Determine the permeability of water-table aquifers. In Ray Bentall (comp.), Methods of Determining Permeability, Transmissivity, and Drawdown. U.S. Geol. Surv. Water-Supply Paper 1536-I, p. 245–271.

JOHNSON, K. S. 1981. Dissolution of salt on the east flank of the Permian Basin in the southwestern U.S.A. J. Hydrol., Special Issue, Symposium on Geochemistry of Groundwater, W. BACK and R. LÉTOLLE (eds.), 54, p. 75–94.

JOHNSTON, R. H., and J. A. MILLER. 1988. Region 24, Southeastern, United States. In W. BACK, J. S. ROSENSHEIN, and P. sR. SEABER (eds.), Hydrogeology, The Geology of North America, v. O-2. Geol. Soc. Am., Boulder, Colorado, p. 229–236.

JORDAN, T. E., and D. J. CONSTANTI. 1995. The use of non-invasive electromagnetic (EM) techniques for focusing environmental investigations. The Professional Geologist, June, p. 4–9.

JORDAN, T. E., D. G. LEASK, D. SLAIN, I. MacLEOD, and T. M. DOBUSH. 1991. The use of high resolution electromagnetic methods for reconnaissance mapping of buried waste. Proc. of the Fifth National Outdoor Action Conference on Aquifer Restoration: Ground Water Monitoring and Geophysical Methods, p. 849–862.

JORGENSEN, D. G., J. O. HELGESEN, and J. L. IMES. 1993. Regional aquifers in Kansas, Nebraska, and parts of Arkansas, Colorado, Missouri, New Mexico, Oklahoma, South Dakota, Texas, and Wyoming—Geohydrologic framework. U.S. Geol. Surv. Prof. Paper 1414-B, 72 p.

KAIDA, Y., M. MATSUBARA, R. GHOSE, and T. KANEMORI. 1995. Very shallow seismic reflection profiling using portable vibrator. Proc. of the Symposium on the Application of Geophysics to Engineering and Environmental Problems. Environ. and Engr. Geophys. Soc., Englewood, Colorado, p. 601–607.

KALYONCU, R. S., W. G. COPPINS, D. T. HOOIE, and M. J. SNYDER. 1978. Characterization and analysis of Devonian shales: 1. Physical characterization. U.S. Dept. Energy, MERC/SP-7715, p. 255–258.

KARICKHOFF, S. W. 1981. Semi-empirical estimation of sorption of hydrophobic pollutants on natural sediments and soils. Chemosphere, v. 10, no. 8, p. 833–846.

KARICKHOFF, S. W. 1984. Organic pollutant sorption in aquatic systems. J. Hydraulic Engr. (ASCE), v. 110, no. 6, p. 707–735.

KARICKHOFF, S. W., D. S. BROWN, and T. A. SCOTT. 1979. Sorption of hydrophobic pollutants on natural sediments. Water Res., v. 13, p. 241–248.

KELLY, F. X., K. J. DAPSIS, and D. A. LAUFFENBURGER. 1988. Effect of bacterial chemotaxis on dynamics of microbial competition, Microb. Ecol., v. 16, p. 115–131.

KELLY, V. 1985. Field determination of dispersivity of comingling plumes. Unpublished M.S. Thesis, Texas A&M University, College Station.

KENNEDY, E. J. 1984. Discharge ratings at gaging stations. U.S. Geol. Surv. Techniques of Water-Resour. Investigations, Book 3, Chap. A10, 59 p.

KEYS, W. S. 1990. Borehole geophysics applied to ground-water investigations. U.S. Geol. Surv. Techniques of Water-Resour. Investigations, Chap. E-2, 150 p.

KHARAKA, Y. K., and I. BARNES. 1973. SOLMNEQ: Solution-Mineral Equilibrium Computations. National Tech. Infor. Serv. Tech. Report PB214-899, 82 p.

KILLEY, R. W. D., J. O. MCHUGH, D. R. CHAMP, E. L. COOPER, and J. L. YOUNG, 1984. Subsurface Cobalt-60 migration from a low-level waste disposal site. Environ. Sci. Technol., v. 18, p. 148–157.

KIMBALL, B. A. 1984. Ground water age determinations, Piceance Creek Basin, Colorado. First Canadian/American Conference on Hydrogeology, B. HITCHON and E. I. WALLICK (eds.). National Water Well Assoc., Dublin, Ohio, p. 267–283.

KIMMEL, G. E., and O. C. BRAIDS. 1980. Leachate Plumes in Ground Water from Babylon and Islip Landfills, Long Island, New York. U.S. Geol. Surv. Prof. Paper 1085, 38 p.

KIPP, K. L., Jr. 1997. Guide to the revised heat and solute transport simulator, HST3D–Version 2. U.S. Geol. Surv. Water-Resour. Investigations Report 97-4157, 149 p.

KONIKOW, L. F., and J. D. BREDEHOEFT. 1978. Computer model of two-dimensional solute transport and dispersion in ground water. U.S. Geol. Surv. TWRI, Book 7, 1 Chap. C2, Reston, Virginia, 40 p.

KUNIANSKY, E. L. 1990. Potentiometric surface of the Edwards–Trinity aquifer system and contiguous hydraulically connected units, west-central Texas, winter 1974–1975. U.S. Geol. Surv. Water-Resour. Investigations Report 89-4208 Scale: 1:750,000, 2 sheets.

LAHM, Terry, D., E. S. BAIR, and J. VANDERKWAAK. 1998. Role of salinity-derived variable-density flow in the displacement of brine from a shallow, regionally extensive aquifer. Water Resour. Res., v. 34, no. 6, p. 1469–1480.

LAMOREAUX, P. E., B. A. MEMON, and H. IDRIS. 1985. Groundwater development, Kharga Oases, Western Desert of Egypt: A long-term environmental concern. Environ. Geol. Water Sci., v. 7, no. 3, p. 129–149.

LAND, L. S., and D. PREZBINDOWSKI. 1981. The origin and evolution of saline formation waters; Lower Cretaceous carbonates, south-central Texas, U.S.A. J. Hydrol., v. 54, p. 51–74.

LANGMUIR, D. 1971. The geochemistry of some carbonate ground waters in central Pennsylvania. Geochim. Cosmochim. Acta, v. 35, p. 1023–1045.

LANGMUIR, D. 1978. Uranium solution-mineral equilibria at low temperatures with applications to sedimentary ore deposits. Geochim. Cosmochim. Acta, v. 42, p. 547–569.

LANGMUIR, D., and J. MAHONEY. 1984. Chemical equilibrium and kinetics of geochemical processes in ground water studies. First Canadian/American Conference on Hydrogeology., B. HITCHON and E. I. WALLICK (eds.). National Water Well Assoc., Dublin, Ohio, p. 69–95.

LAPPALA, E. G. 1978. Quantitative Hydrogeology of the Upper Republican Natural Resources District, Southwest Nebraska. U.S. Geol. Surv. Water Resour. Investigations Report 78–38.

LAPPALA, E. G., R. W. HEALY, and E. P. WEEKS. 1987. Documentation of computer program VS2D to solve the equations of fluid flow in variably saturated porous media, U.S. Geol. Surv. Water-Resour. Investigations Report 83-4099, 184 p.

LAPPIN, A. R. 1988. Summary of site-characterization studies conducted from 1983 through 1987 at the Waste Isolation Pilot Plant (WIPP) site, Southeastern New Mexico. SAND88-0157, Sandia National Laboratories, Albuquerque, New Mexico.

LATTMAN, L. A., and R. R. PARIZEK. 1964. Relationship between fracture traces and the occurrence of groundwater in carbonate rocks. J. Hydrol., v. 2, p. 73–91.

LEAKE, S. A., P. P. LEAHY, and A. S. NAVOY. 1994. Documentation of a computer program to simulate leakage from confining units using the modular finite-difference ground-water flow model. U.S. Geol. Surv. Open-File Report 94-59, 70 p.

LEAKE, S. A., and D. E. PRUDIC. 1988. Documentation of a computer program to simulate aquifer-system compaction using the modular finite-difference ground-water flow model. U.S. Geol. Surv. Techniques of Water-Resour. Investigations Report 06-A2, 68 p.

LEBLANC, D. R. (ed.). 1984a. Movement and Fate of Solutes in a Plume of Sewage-Contaminated Ground Water, Cape Cod, Massachusetts. U.S. Geol. Surv. Toxic Waste Ground-Water Contamination Program. U.S. Geol. Surv. Open-File Report 84-475, 175 p.

LEBLANC, D. R. 1984b. Digital modeling of solute transport in a plume of sewage-contaminated ground water. Movement and Fate of Solutes in a Plume of Sewage-Contaminated Ground Water, Cape Cod, Massachusetts. U.S. Geol. Surv. Toxic Waste Ground-Water Contamination Program. U.S. Geol. Surv. Open-File Report p. 11–45.

LEBLANC, D. R., S. P. GARABEDIAN, K. M. HESS, L. W. GELHAR, R. D. QUADRI, K. G. STOLLENWERK, and W. W. WOOD. 1991. Large-scale natural gradient tracer test in sand and gravel, Cape Cod, Massachusetts 1. Experimental design and observed tracer movement. Water Resour. Res., v. 27, no. 5, p. 895–910.

LeBlanc, D. R., K. M. Hess, D. B. Kent, R. L. Smith, L. B. Barber, K. G. Stollenwerk, and K. W. Campo. 1999. Natural restoration of a sewage plume in a sand and gravel aquifer, Cape Cod, Massachusetts. D.W. Morganwalp and H.T. Buxton (eds.), U.S. Geol. Surv. Water Resour. Investigations Report 99-4018C, p. 245–259.

Lefkoff, L. J., and S. M. Gorelick. 1990a. Simulating physical processes and economic behavior in saline, irrigated agriculture – model development. Water Resour. Res., v. 26, no. 7, p. 1359–1369.

Lefkoff, L. J., and S. M. Gorelick. 1990b. Benefits of an irrigation water rental market in a saline stream-aquifer system. Water Resour. Res., v. 26, no. 7, p. 1371–1381.

LeGrand, H. E. 1954. Geology and groundwater in the Statesville area, North Carolina. North Carolina Dept. Conser. Dev., Div. Mineral Resources, Bull. 68.

Lennox, D. H. 1966. Analysis and application of step-drawdown test. ASCE, Proc. Sep. v. 92, no. 4967, November.

Leonards, G. A. 1962. Engineering properties of soil. In G. A. Leonards (ed.), Foundation Engineering. McGraw-Hill, New York, p. 66–240.

Levine, S. N., and W. C. Ghiorse. 1990. Analysis of environmental factors affecting abundance and distribution of bacteria, fungi and protozoa in subsurface sediments of the Upper Atlantic Coastal Plain, USA. Proceedings of the First International Symposium on Microbiology of the deep subsurface, (eds.) C. B. Fliermans and T. C. Hazen. Westinghouse Savannah River Co., p. 5-31–5-45.

Lindberg, R. D., and D. D. Runnells. 1984. Ground water redox reactions: An analysis of equilibrium state applied to Eh measurements and geochemical modeling. Science, v. 225, p. 925–927.

Lindstrom, F. T., R. Hague, V. H. Freed, and L. Boersma. 1967. Theory on the movement of some herbicides in soils—linear diffusion and convection of chemicals in soils. Journal of Environmental Sciences and Technology, v. 1, p. 561–565.

Linsley, R. K., M. A. Kohler, and J. I. H. Paulhus. 1958. Hydrology for Engineers. McGraw-Hill, New York.

Lohman, S. W. 1972. Ground water hydraulics. U.S. Geol. Surv. Prof. Paper 708, 70 p.

Loucks, R. G., M. M. Dodge, and W. E. Galloway. 1984. Regional controls on diagenesis and reservoir quality in Lower Tertiary sandstones along the Texas Gulf Coast. In D.A. McDonald and R.C. Surdam (eds.), Clastic Diagenesis. Amer Assoc. Petrol. Geol. Mem. 37, p. 15–45.

Love, A. J., A. C. Herczeg, F. W. Leaney, M. F. Stadter, J. C. Dighton, and D. Armstrong. 1994. Groundwater residence time and paleohydrology in the Otway Basin, South Australia: ^{2}H, ^{18}O, and ^{14}C data. J. Hydrol., v. 153, p. 157–187.

Lyngkilde, J., and T. H. Christensen. 1992. Fate of organic contaminants in the redox zones of a landfill leachate pollution plume (Vejen, Denmark). J. Contam. Hydrol., v. 10, p. 291–307.

Mackay, D. M., D. L. Freyberg, and P. V. Roberts. 1986. A natural gradient experiment on solute transport in a sand aquifer 1. approach and overview of plume movement. Water Resour. Res., v. 22, no. 13, p. 2017–2029.

Mackay, D. M., and T. M. Vogel. 1985. Ground water contamination by organic chemicals: Uncertainties in assessing impact. Second Canadian/American Conference on Hydrogeology, eds. B. Hitchon and M. Trudell. National Water Well Assoc., Dublin, Ohio, p. 50–59.

Mackay, L. D., J. A. Cherry, and R. W. Gillham. 1993. Field experiments in a fractured clay till 1. Hydraulic conductivity and fracture aperture. Water Resour. Res., v. 29, no. 4, p. 1149–1162.

Macumber, P. G. 1984. Hydrochemical processes in the regional ground water discharge zones of the Murray Basin, Southeastern Australia. First Canadian/American Conference on Hydrogeology, eds. B. Hitchon and E. I. Wallick. National Water Well Assoc., Dublin, Ohio, p. 47–63.

Manheim, F. T., and J. F. Bischoff. 1969. Geochemistry of pore waters from Shell Oil Company drill holes on the continental slope of the northern Gulf of Mexico. Chem. Geol., Special Issue, Geochemistry of Subsurface Brines, E. E. Angino and G. K. Billings (eds.), v. 4, p. 63–82.

Manheim, F. T., and C. K. Paull. 1981. Patterns of groundwater salinity changes in a deep continental-oceanic transect off the southeastern Atlantic coast of the U.S.A. J. Hydrol. Special Issue, Symposium on Geochemistry of Groundwater, W. Back and R. LéTolle (eds.), v. 54, p. 95–106.

Manning, J. C. 1987. Applied Principle of Hydrology. Merrill Publishing Co., Columbus, Ohio.

Marquardt, D. W. 1963. An algorithm for least-squares estimation of non-linear parameters, J. Soc. Ind. Appl. Math., 11, p. 431–441.

Masterson, J. P., D. A. Walker, and J. Savoie. 1997. Use of particle tracking to improve model calibration and to analyze ground-water flow and contaminant migration, Massachusetts Military Reservation, Western Cape Cod Massachusetts. U.S. Geol. Surv. Water-Supply Paper 2482, 50 p.

Matheron, G. 1967. Eléments pour une théorie des milieux poreux. Masson et Cie, Paris.

McCarty, P. L., B. E. Rittman, and E. J. Bouwer. 1984. Micro-biological processes affecting chemical transformations in groundwater. In G. Bitton and C. P. Gerba (eds.), Groundwater Pollution Microbiology. John Wiley & Sons, New York, p. 89–115.

McDonald, M. G., and A. W. Harbaugh. 1984. A modular three-dimensional finite-difference ground-water flow model. U.S. Geol. Surv. Open-File Report 83–875, 528 p.

McDonald, M. G., and A. W. Harbaugh. 1988. A modular three-dimensional finite-difference ground-water flow model. U.S. Geol. Surv. Techniques of Water-Resour. Investigations, book 6, Chap. A1, 586 p.

McNabb, J. F., and G. Mallard. 1984. Microbial sampling in the assessment of groundwater pollution. In G. Bitton and C. P. Gerba (eds.), Pollution Microbiology. John Wiley & Sons, New York, p. 235–260.

Meinzer, O. E. 1923. Outlines of groundwater in hydrology with definitions. U.S. Geol. Surv. Water-Supply Paper, v. 494.

Meinzer, O. E. 1927. Plants as indicators of groundwater. U.S. Geol. Surv. Water-Supply Paper, v. 577.

Mendoza, C. A., and E. O. Frind. 1990a. Advective-dispersive transport of dense organic vapors in the unsaturated zone, 1, Model development. Water Resour. Res., v. 26, no. 3, p. 379–387.

Mendoza, C. A., and E. O. Frind. 1990b. Advective-dispersive transport of dense organic vapors in the unsaturated zone, 2. Sensitivity analysis. Water Resour. Res., v. 26, no. 3, p. 388–398.

Mendoza, C. A., and T. A. McAlary. 1990. Modeling of ground-water contamination caused by organic solvent vapors. Ground Water, v. 28, no. 2, p. 199–206.

Meyboom, P. 1960. Geology and groundwater resources of the Milk River sandstone in southern Alberta. Research Council of Alberta Memoir 2, Edmonton.

Meyboom, P. 1961. Estimating ground water recharge from stream hydrographs, J. Geophys. Res., v. 66, p. 1203–1214.

Meyboom, P. 1966. Groundwater studies in the Assiniboine River drainage basin, pt. 1. The evaluation of a flow system in southern central Saskatchewan. Can. Dept. Mines, Tech. Surv. Geol. Surv. Can. Bull. 139.

Meyboom, P. 1967. Mass transfer studies to determine the groundwater regime of permanent lakes in hummocky moraine of western Canada. J. Hydrol., v. 5, p. 117–142.

Mikels, F. C. 1952. Report on hydrogeological survey for city of Zion, Illinois. Ranney Method Water Supplies, Inc., Columbus, Ohio.

Miller, J. A. 1990. Ground water atlas of the United States, Segment 6, Alabama, Florida, Georgia, South Carolina, Hydrologic Investigations Atlas 730-G. U.S. Geol. Surv., Reston, Virginia.

Miller, J. A. Ground Water Atlas of the United States, Introduction and National Summary, U.S. Geol Surv., http://capp.water.usgs.gov.gwa/ch_a/index.html.

Miller, J. A., and C. L. Appel. 1997. Ground water atlas of the United States, Segment 3, Kansas, Missouri, and Nebraska, Hydrologic Investigations Atlas 730-D. U.S. Geol. Surv., Reston, Virginia.

Miller, R. D., and J. Xia. 1997. High Resolution Seismic Reflection Survey to Map Bedrock and Glacial/Fluvial Layers at the U.S. Navy Northern Ordnance Plant (NIROP) in Fridley, Minnesota, Open-File Report No. 97-12, Kansas Geological Survey, University of Kansas, Lawrence, Kansas.

Millington, R. J., and J. M. Quirk. 1961. Permeability of porous solids. Trans. Faraday Soc., 57, 27, p. 324–332.

Moench, A. F. 1993. Computation of type curves for flow to partially penetrating wells in water-table aquifers. Ground Water, v. 31, p. 966–971.

Moench, A. F. 1995. Combining the Neuman and Boulton models for flow to a well in an unconfined aquifer. Ground Water, v. 33, p. 378–384.

Moench, A. F. 1996. Flow to a well in a water-table aquifer: an improved Laplace transform solution. Ground Water, v. 34, p. 593–596.

Moench, Allen, and Ogata, Akio. 1984. Analysis of constant discharge wells by numerical inversion of Laplace transform solutions, In J. S. Rosenshein and G. D. Bennett (eds.), Groundwater Hydraulics: Water Resources Monograph 9, American Geophysical Union, Washington, DC, p. 146–170.

Moody, J. B. 1982. Radionuclide migration/retardation: Research and development technology status report. Office of Nuclear Waste Isolation, Battelle Memorial Institute, ONWI-3231, 61 p.

Mook, W. G. 1980. Chapter 2. Carbon-14 in hydrogeological studies. In P. Fritz and J. C. Fontes (eds.), Handbook of Environmental Isotope Geochemistry, v. 1, Elsevier, Amsterdam, p. 49–74.

Moran, S. R., J. A. Cherry, P. Fritz, W. M. Peterson, M. H. Somerville, S. A. Stancel, and J. H. Ulmer. 1978a. Geology, Groundwater Hydrology, and Hydrochemistry of a Proposed Surface Mine and Lignite Gasification Plant Site Near Dunn Center, North Dakota. North Dakota Geological Survey, Report of Investigation No. 61, 263 p.

Moran, S. R., G. H. Groenwold, and J. A. Cherry. 1978b. Geologic, Hydrologic, and Geochemical Concepts and Techniques in Overburden Characterization for Mined-land Reclamation. North Dakota Geological Survey, Report of Investigation No. 63, 152 p.

Morel, F.M.M., and J. G. Hering. 1993. Principles and Applications of Aquatic Chemistry. John Wiley & Sons, New York, 588 p.

Mualem, Y. 1976. A new model for predicting the hydraulic conductivity of unsaturated porous media. Water Resour. Res., v. 12, no. 3, p. 513–522.

Muckel, D. C. 1959. Replenishment of groundwater supplies by artificial means. U.S. Dept. Agric. Res. Service, Tech. Bull. 1195.

MUSKAT, M. 1937. The Flow of Homogeneous Fluids through Porous Media. McGraw-Hill, New York.

National Research Council. 1993. In Situ Bioremediation, When Does It Work? National Academy Press, Washington, DC, 207 p.

NELDER, J. A., and R. MEAD. 1965. A simplex method for function minimization. Computer Journal, v. 7, p. 308–313.

NERETNIEKS, I. 1985. Transport in fractured rocks. Proc. of Hydrogeology of Rocks of Low Permeability. International Assoc. of Hydrogeologists, Tucson, Arizona, p. 306.

NEUMAN, S. P. 1972. Theory of flow in unconfined aquifers considering delayed response of the water table. Water Resour. Res., v. 8, p. 1031–1045.

NEUMAN, S. P. 1973. Supplementary comments on "Theory of flow in unconfined aquifers considering delayed response of the water table." Water Resour. Res., v. 9, p. 1102–1103.

NEUMAN, S. P. 1974. Effect of partial penetration on flow in unconfined aquifers considering delayed response of the water table. Water Resour. Res., v. 9, p. 1102–1103.

NEUMAN, S. P. 1975. Analysis of pumping test data from anisotropic unconfined aquifers considering delayed gravity response. Water Resour. Res., v. 11, p. 329–342.

NIELSEN, D. M. 1996. Design and construction of contaminant monitoring wells to improve performance and cut monitoring costs. Workshop Notebook, Tenth National Outdoor Action Conference and Exposition. National Ground Water Association, Worthington, Ohio, p. 34–42.

NIELSEN, P. H., and T. H. CHRISTENSEN. 1994a. Variability of biological degradation of aromatic hydrocarbons in an aerobic aquifer determined by laboratory batch experiments. J. Contam. Hydrol., v. 15, p. 305–320.

NIELSEN, P. H., and T. H. CHRISTENSEN. 1994b. Variability of biological degradation of phenolic hydrocarbons in an aerobic aquifer determined by laboratory batch experiments. J. Contam. Hydrol., v. 17, p. 55–67.

NORRIS, S. E., and R. E. FIDLER. 1973. Availability of water from limestone and dolomite aquifers in southwest Ohio and the relation of water quality to the regional flow system. U.S. Geol. Surv. Water-Resour. Investigations Report 17-73, 42 p.

NORTON, D., and R. KNAPP. 1977. Transport phenomena in hydrothermal systems: Nature of porosity. Am. J. Sci., v. 27, p. 913–936.

Office of Technology Assessment. 1984. Protecting the nation's groundwater from contamination. Office of Technology Assessment, OTA-0-233, Washington, D.C., 244 p.

Ogata, Akio, and R. B. BANKS. 1961. A solution of the differential equation of longitudinal dispersion in porous media. U.S. Geol. Surv. Prof. Paper 411-A, p. A1–A9.

OKI, D. S., G. W. TRIBBLE, W. R. SOUZA, and E. L. BOLKE. 1999. Ground-water resources in Kaloko-Honokohau National Historical Park, Island of Hawaii, and numerical simulation of the effects of ground-water withdrawals. U.S. Geol. Surv. Water-Resour. Investigations Report 99-4070, p. 49.

OLCOTT, P. 1992. Ground water atlas of the United States, Segment 9, Iowa, Michigan, Minnesota, and Wisconsin, Hydrologic Investigations Atlas 730-J. U.S. Geol. Surv., Reston, Virginia.

ORESKES, N., K. SHRADER-FRECHETTE, and K. BELITZ. 1994. Verification, validation, and confirmation of numerical models in earth sciences. Science, v. 263., p. 641–646.

PALMER, A. N. 1990. Groundwater processes in karst terranes. In C. G. HIGGINS and D. R. COATES (eds.), Groundwater Geomorphology: The Role of Subsurface Water in Earth-Surface Processes and Landforms. Geol. Soc. Am. Special Paper 252, p. 177–209.

PALMER, C. D., and J. A. CHERRY. 1984. Geochemical evolution of groundwater in sequences of sedimentary rocks. J. Hydrol., v. 75, p. 27–65.

PANNO, S. V., K. C. HACKLEY, K. CARTWRIGHT, and C. L. LIU. 1984. Geochemistry of the Mahomet Bedrock Valley aquifer, East-Central Illinois: Indicators of recharge and ground-water flow. Ground Water, v. 32, no. 4, p. 591–604.

PAPADOPULAS, I. S., J. D. BREDEHOEFT, and H. H. COOPER. 1973. On the analysis of slug test data. Water Resour. Res., v. 9, p. 1087–1089.

PAPADOPULOS, I. S., and H. H. COOPER. 1967. Drawdown in a well of large diameter. Water Resour. Res., v. 3, p. 241–244.

PARKHURST, D. L., L. M. PLUMMER, and D. C. THORSTENSON. 1982. BALANCE—A Computer Program for Calculation of Chemical Mass Balance. U.S. Geol. Surv. Water-Resour. Investigations Report 82-14. NTIS Tech. Report PB82-255902, Springfield, Virginia, 33 p.

PARKHURST, D. L., D. C. THORSTENSON, and L. N. PLUMMER. 1980. PHREEQE — A computer program for geochemical calculation. U.S. Geol. Surv. Water-Resources Investigations Report 80–96, 210 p.

PEARSON, F. J., Jr., and B. B. HANSHAW. 1970. Sources of dissolved carbonate species in groundwater and their effects on carbon-14 dating. Isotope Hydrology. Int. Atomic Energy Agency, Vienna, p. 271–286.

PETER, K. L. 1984. Hydrochemistry of lower Cretaceous sandstone aquifers, Northern Great Plains. In D. G. JORGENSON and D. C. SIGNOR (eds.), Proc. of the Geohydrology Dakota Aquifer Symposium, Water Well Journal Publ., Worthington, Ohio, p. 163–174.

PETERS, J. G. 1987. Description and comparison of selected models for hydrologic analysis of ground-water flow, St. Joseph River Basin, Indiana, U.S. Geol. Surv. Water-Resour. Investigations Report 86-4199, p. 125.

PETERSON, D. M., and J. L. WILSON. 1988. Field study of ephemeral stream infiltration and recharge. New Mexico Water Resour. Res. Inst. Technical Completion Report No. 228, New Mexico State University, New Mexico.

PHILIP, J. R. 1957. The theory of infiltration. 4. Sorptivity and algebraic infiltration equations. Soil Sciences, v. 84, p. 257–264.

PHILLIPS, F. M., H. W. BENTLEY, S. N. DAVIS, D. ELMORE, and G. SWANICK. 1986. Chlorine 36 dating of very old groundwater 2. Milk River Aquifer, Alberta. Water Resour. Res., v. 22, no. 13, p. 2003–2016.

PHILLIPS, F. M., J. L. MATTICK, T. A. DUVAL, D. ELMORE, and P. W. KUBIK, 1988. Chlorine-36 and Tritium from nuclear weapons fallouts as tracer for long-term liquid and vapor movement in desert soils. Water Resour. Res., v. 24, p. 1877–1891.

PICKENS, J. F., R. E. JACKSON, K. J. INCH, and W. F. MERRITT. 1981. Measurement of distribution coefficients using a radial injection dual-tracer test. Water Resour. Res., v. 17, no. 3, p. 529–544.

PITZER, K. S., and J. J. KIM. 1974. Thermodynamics of electrolytes: 4. activity and osmotic coefficients for mixed electrolytes. J. Amer. Chem. Soc., v. 96, p. 5701–5707.

PLANERT, M., and J. S. WILLIAMS. 1995. Ground water atlas of the United States: Segment 1, California and Nevada, Hydrologic Investigations Atlas 730-B. U.S. Geol. Surv., Reston, Virginia.

PLUMMER, L. N., B. F. JONES, and A. H. TRUESDELL. 1976. WATEQF—A Fortran IV version of WATEQ, A Computer Program for Calculating Chemical Equilibrium of Natural Waters. U.S. Geol. Surv. Water Resour. Investigations Report 76-13, 61 p.

PLUMMER, L. N., E. C. PRESTEMON, and D. L. PARKHURST. 1991. An interactive code (NETPATH) for modeling NET geochemical reactions along a flow path. U.S. Geol. Surv. Water Resour. Investigations Report 91-4078, 227 p.

PLUMMER, L. N., E. C. PRESTEMON, and D. L. PARKHURST. 1994. An interactive code (NETPATH) for modeling NET geochemical reactions along a flow PATH—Version 2.0. U.S. Geol. Surv. Water-Resour. Investigations Report 94-4169, 130 p.

POHLMANN, K. F., R. P. BLEGEN, and J. W. HESS. 1990. Field comparisons of ground-water sampling devices for hazardous waste sites: An evaluation using volatile organic compounds. EPA/600/4-90/028 (NTIS PB91-181776), 102 p.

POLLOCK, D. W. 1989. Documentation of computer programs to compute and display pathlines using results from the U.S. Geol. Surv. modular three-dimensional finite-difference ground-water flow model. U.S. Geol. Surv. Open-File Report 89-381, 188 p.

POLLOCK, D. W. 1994. User's Guide for MODPATH/MODPATH-PLOT, Version 3: A particle tracking post-processing package for MODFLOW, the U.S. Geol. Surv. finite-difference ground-water flow model. U.S. Geol. Surv. Open-File Report 94-464.

PRESS, W. H., B. P. FLANNERY, S. A. TEUKOLSKY, and W. T. VETTERLING. 1987. Numerical Recipes. Cambridge, Cambridge University Press, 818 p.

PRICKETT, T. A. 1965. Type curve solution to aquifer tests under water table conditions. Ground Water, v. 3, p. 5–14.

PRUDIC, D. E. 1989. Documentation of a computer program to simulate stream-aquifer relations using a modular, finite-difference, ground-water flow model. U.S. Geol. Surv. Open-File Report 88-729, 113 p.

PRUESS, K. 1987. TOUGH user's guide, LBL-20700, NUREG/CR-4645. Lawrence Berkeley Laboratory, Berkeley, California.

PYTKOWICZ, R. P. 1983. Equilibria, Nonequilibria & Natural Waters, Volume II. John Wiley & Sons, New York, 353 p.

QUINLAN, J. F., and R. O. EWERS. 1985. Ground water flow in limestone terranes: Strategy, rationale and procedure for reliable, efficient monitoring of ground water quality in Karst areas. Proc. Fifth National Symposium and Exposition on Aquifer Restoration and Ground Water Monitoring. National Water Well Assoc., Dublin, Ohio, p. 197–234.

RAGONE, S. E. (ed.). 1988. U.S. Geological Survey Program on Toxic Waste—Ground Water Contamination. Proc. of the Second Technical Meeting, Cape Cod, Massachusetts, October 21–25, 1985. U.S. Geol. Surv. Open-File Report 86-481.

RANDAL, M. A. 1978. Hydrogeology of the Southeastern Georgina Basin and Environs, Queensland and Northern Territory. Geological Survey of Queensland, Publication 366.

RANDALL, J. H., and T. R. SCHULTZ. 1976. Chlorofluorocarbons as hydrologic tracers: A new technology. Hydrol., Water Res. Ariz. Southwest, v. 6, p. 189–195.

RANGANATHAN, V., and J. S. HANOR. 1987. A numerical model for the formation of saline waters due to diffusion of dissolved NaCl in subsiding sedimentary basins with evaporites. J. Hydrol., v. 92, p. 97–120.

RAVI, V., and J. R. WILLIAMS. 1998. Estimation of infiltration rate in the vadose zone. V. I: Compilation of simple mathematical models. EPA/600/R-97/128a, 26 p.

REA, B. A., D. B. KENT, L.C.D. ANDERSON, J. A. DAVIS, and D. R. LEBLANC. 1994. The transport of inorganic contaminants in a sewage plume in the Cape Cod Aquifer, Massachusetts. U.S. Geol. Surv. Water-Resour. Investigations Report 94-4015, p. 191–198.

REARDON, E. J., and P. FRITZ. 1978. Computer modeling of ground water ^{13}C and ^{14}C isotope compositions. J. Hydrol., v. 36, p. 201–224.

REED, J. E. 1980. Type curves for selected problems of flow to wells in confined aquifers. U.S. Geol. Surv. Techniques of Water-Resour. Investigations, Chap. B3, 106 p.

REEVES, M., D. S. WARD, N. D. JOHNS, and R. M. CRANWELL. 1986. Theory and implementation for SWIFT II, The Sandia Waste-Isolation Flow and Transport Model (SWIIFT) Release 4.81, NUREG/CR-2324 and SAND81-2516. Sandia National Laboratories, Albuquerque, New Mexico.

REHFELDT, K. R., J. M. BOGGS, and L. W. GELHAR. 1992. Field study of dispersion in a heterogeneous aquifer, 3, Geostatistical analysis of hydraulic conductivity. Water Resour. Res., v. 28, no. 12, p. 3309–3324.

REHM, B. W., T. R. STOLZENBURG, and D. G. NICHOLS. 1985. Field measurement methods for hydrologic investigations: A critical review of the literature. Electric Power Research Institute, EPRI EA-4301, Palo Alto, California.

REISINGER, H. J., D. R. BURRIS, L. R. CESSAR, and G. D. MCCLEARY. 1987. Factors affecting the utility of soil vapor assessment data. Proc. First Outdoor Action Conference on Aquifer Restoration, Ground Water Monitoring and Geophysical Methods. National Water Well Assoc., Dublin, Ohio, p. 425–435.

REMENDA, V. H., J. A. CHERRY, T. W. D. EDWARDS. 1994. Isotopic composition of old ground-water from Lake Agassiz – Implications for Late Pleistocene climate. Science, v. 266: (5193) p. 1975–1978.

RENDER, F. W. 1970. Geohydrology of the metropolitan Winnipeg area as related to groundwater supply and construction. Can. Geotech. J., v. 7, p. 243–274.

RIGGS, H. C. 1989. Frequency curves. Techniques of water-resource investigations of the United States Geological Survey, Book 4, Chap. A2, 15 p.

RITTMANN, B. E. 1993. The significance of biofilms in porous media: Water Resour. Res., v. 29, no. 7, p. 2195–2202.

ROBBINS, G. A. 1983. Determining Dispersion Parameters to Predict Groundwater Contamination. Ph.D. dissertation, Texas A&M Univ., College Station, Texas, 226 p.

ROBERTS, P. V., L. SEMPRI, G. D. HOPKINS, D. GRBIC-GALIC, P. L. MCCARTY, and M. REINHARD. 1989. In-Situ Restoration of Chlorinated Aliphatics by Methanogenic Bacteria. EPA/600/2-89/033, 214 p.

ROBERTS, W. J., and J. B. STALL. 1967. Lake evaporation in Illinois. Illinois State Water Survey Report of Investigation No. 57.

ROBSON, S. G., and E. R. BANTA. 1995. Ground water atlas of the United States, Segment 2: Arizona, Colorado, New Mexico, Utah., Hydrologic Investigations Atlas 730-C. U.S. Geol. Surv., Reston, Virginia, 32 p.

ROMM, E. S. 1966. Flow characteristics of fractured rocks (in Russian.) Nedra, Moscow.

RORABAUGH, M. I. 1953. Graphical and theoretical analysis of step-drawdown test of artesian well. ASCE, v. 79, Proceedings Separate No. 362, p. 1–23.

ROSENSHEIN, J. S. 1988. Region 18, Alluvial valleys. In W. BACK, J. S. ROSENSHEIN, and P. R. SEABER (eds.), Hydrogeology, The Geology of North America, v. O-2. Geol. Soc. Am., Boulder, Colorado, p. 165–175.

RUST, B. R., and E. H. KOSTER. 1984. Coarse alluvial deposits. In R. G. WALKER (ed.), Facies Models, Geoscience Canada Reprint Series 1.

RUTLEDGE, A. T., and C. C. DANIEL III. 1994. Testing and automated method to estimate ground-water recharge from streamflow records. Ground Water, v. 32, p. 180–189.

RYDER, P. D. 1996. Ground water atlas of the United States, Segment 4, Oklahoma, Texas, Hydrologic Investigations Atlas 730-E. U.S. Geol. Surv., Reston, Virginia.

SALVUCCI, G. D., and D. ENTEKHABI. 1994. Explicit expressions for Green-Ampt (delta function diffusivity) infiltration rate and cumulative storage. Water Resour. Res., v. 30, no. 9, p. 2661–2663.

SAVIN, S. M. 1980. Chapter 8. Oxygen and hydrogen isotope effects in low-temperature mineral-water interactions. In P. FRITZ and J. C. FONTES (eds.), Handbook of Environmental Isotope Geochemistry, v. 1. Elsevier, New York, p. 283–327.

SCALF, M. R., J. F. MCNABB, W. J. DUNLAP, R. L. COSBY, and J. FRYBERGER. 1981. Manual of Ground Water Sampling Procedures. EPA 660/2-81-160, 93 p.

SCHEIDEGGER, A. E. 1961. General theory of dispersion in porous media: Jour. Geophys. Research, v. 66, no. 10, p. 3273–3278.

SCHELESINGER, W. H. 1991. Biogeochemistry—an analysis of global change. Academic Press, San Diego.

SCHENKER, A. R., D. C. GUERIN, T. H. ROBEY, C. A. RAUTMAN, and R. W. BARNARD. 1995. Stochastic hydrogeologic units and hydrogeologic properties development for total-system performance assessments. SAND94-0244. Sandia National Laboratory, Albuquerque, New Mexico.

SCHMORAK, S., and A. MERCADO. 1969. Upconing of freshwater-seawater interface below pumping wells. Water Resour. Res., v. 5, p. 1290–1311.

SCHNOEBELEN, D. J., and N. C. KROTHE. 1999. Reef and nonreef aquifers – a comparison of hydrogeology and geochemistry, northwestern Indiana. Ground Water, v. 37, no. 2, p. 194–203.

SCHWARTZ, F. W., and F. J. LONGSTAFFE. 1988. Ground water and clastic diagenesis. In W. BACK, J. S. ROSENSHEIN, and P. R. SEABER (eds.), Hydrogeology, The Geology of North America, v. O-2, Geol. Soc. Am., Boulder, Colorado, p. 413–434.

SCHWARTZ, F. W., and K. MUEHLENBACHS. 1979. Isotope and ion geochemistry of groundwaters in the Milk River aquifer, Alberta. Water Resour. Res., v. 15, no. 2, p. 259–268.

SCHWARZENBACH, R. P., and W. GIGER. 1985. Behavior and fate of halogenated hydrocarbons in ground water. In Ground Water Quality. C. H. WARD, W. GIGER, and P. L. McCARTY (eds.), Wiley-Interscience, New York, p. 446–471.

SCHWARZENBACH, R. P., and J. WESTALL. 1981. Transport of nonpolar organic compounds from surface water to groundwater. Laboratory studies. Environ. Sci. Technol., v. 15, p. 1300–1367.

SCOTT, R. B., and J. BONK. 1984. Preliminary geologic map of Yucca Mountain, Nye County, Nevada, with geologic sections. U.S. Geol. Surv. Report OFR-84-494.

SEIDEL, G. E. 1980. Application of the GABHYD groundwater model of the Great Artesian Basin, Australia, BMR. J. Aust. Geol. Geophys., p. 39–45.

SHARMA, M. L., and M. W. HUGHES. 1985. Groundwater recharge estimation using chloride, deuterium and oxygen-8 profiles in the deep coastal sands of western Australia. J. Hydrol., v. 8, p. 93–109.

SHARP, J. M. 1988. Alluvial aquifers along major rivers. In W. BACK, J. S. ROSENSHEIN, and P. R. SEABER (eds.), Hydrogeology, The Geology of North America, v. O-2. Geol. Soc. Am., Boulder, Colorado, p. 273–282.

SIBUL, U., and A. V. CHOO-YING. 1971. Water resources of the Upper Nottawasaga River Basin Ontario Water Resources Comm., Water Resour. Report 3, 128 p.

SIKORA, R. F., D. L. CAMPBELL, and R. P. KUCKS. 1995. Aeromagnetic surveys across Crater Flat and part of Yucca Mountain, Nevada. U.S. Geol. Surv. Open-File Report 95-812, 13 p. Denver Federal Center, Denver, Colorado.

SIMUNEK, J., K. HUANG, and M. Th. VAN GENUCHTEN. 1995. The SWMS_3D code for simulating water flow and solute transport in three-dimensional variable-saturated media, Version 1.0, Research Report No. 139. U.S. Salinity Laboratory, Riverside California.

SKLASH, M. G., and R. N. FARVOLDEN. 1979. The role of groundwater in storm runoff. J. Hydrol., v. 43, p. 45–65.

SKLASH, M. G., R. N. FARVOLDEN, and P. FRITZ. 1976. A conceptual model of watershed response to rainfall, developed through the use of oxygen-18 as a natural tracer. Can. J. Earth Sci., v. 13, p. 271–283.

SMOLENSKI, W. J., and J. M. SULFLITA. 1987. Biodegradation of cresol isomers in anoxic aquifers. Applied and Environmental Microbiology, v. 53, p. 710–716.

SMOLLEY, M., and J. C. KAPPMEYER. 1991. Cone penetrometer tests and Hydropunch sampling: A screening technique for plume definition. Ground Water Mon. Rev., v. 11, no. 2, p. 101–106.

SNOW, D. T. 1968. Rock fracture spacings, openings, and porosity. J. Soil Mech., Found. Div. Proc. Am. Soc. Civil Engrs., v. 94, p. 73–91.

SOLOMON, D. K., R. J. POREDA, P. G. COOK, and A. HUNT. 1995. Site characterization using $^3H/^3He$ ground-water ages, Cape Cod, Massachusetts. Ground Water, v. 33, no. 6, p. 988–996.

SOPHOCLEOUS, M. 1997. Managing water resources systems—Why "safe yield" is not sustainable. Ground Water, v. 35, p. 561.

SPRINGER, A. E. 1990. An evaluation of wellfield-protection area delineation methods as applied to municipal wells in the stratified-drift aquifer at Wooster, Ohio. M.S. Thesis, The Ohio State University, 167 p.

SPRINGER, A. E., and E. S. BAIR. 1992. Comparison of methods used to delineate capture zones of wells: 2. Stratified-drift buried-valley aquifer. Ground Water, v. 30, no. 6, p. 908–917.

STALLMAN, R. W. 1952. Nonequilbrium type curves for two well systems. U.S. Geol. Surv. Groundwater Notes. U.S. Geol. Surv. Open-file Report 3.

STANNARD, D. I. 1986. Theory, construction and operation of simple tensiometers. Ground Water Monitoring and Remediation, p. 70–78.

STEFEST, H. 1970. Numerical inversion of Laplace transforms. Commun. ACM, v. 13, p. 47–49.

STEIN, R. 1987. A hydrogeological investigation of the origin of saline soils at Blackspring Ridge, southern Alberta. Unpublished M.S. thesis, Univ. of Alberta, Edmonton, 272 p.

STEIN, R., and F. W. SCHWARTZ. 1990. On the origin of saline soils at Blackspring Ridge, Alberta, Canada. J. Hydrol., v. 117, p. 99–131.

STEPHENSON, D. A., A. H. FLEMMING, and D. M. MICKELSON. 1988. Glacial deposits. In W. BACK, J. S. ROSENSHEIN, and P. R. SEABER (eds.), Hydrogeology, The Geology of North America, v. O-2. Geol. Soc. Am., Boulder, Colorado, p. 301–314.

STEWART, M. T., and M. C. GAY. 1986. Evaluation of transient electromagnetic soundings for deep detection of conductive fluids. Ground Water, v. 24, p. 351–356.

STOLLENWERK, K. G. 1991. Simulation of molybdate sorption with diffuse layer surface-complexation model. U.S. Geol. Surv. Water-Resour. Investigations Report 91-4034, p. 47–52.

STOODLEY, P., D. DE BEER, and Z. LEWANDOWSKI. 1994. Liquid flow in biofilm systems. Applied and Environmental Microbiology, August, p. 2711–2716.

STRELTSOVA, T. D. 1972. Unsteady radial flow in an unconfined aquifer. Water Resour. Res., v. 8, p. 1059–1066.

STRELTSOVA, T. D., and K. RUSHTON. 1973. Water table drawdown due to a pumped well. Water Resour. Res., v. 9, p. 236–242.

STRINGFIELD, V. T., and H. E. LEGRANDE. 1966. Hydrology of limestone terranes. Geol. Soc. Am. Spec. Paper 3.

STRUTYNSKY, A. I., and T. J. SAINEY. 1990. Use of cone penetrometer testing and penetrometer groundwater sampling for volatile organic contaminant plume detection. NWWA/API Conference Petroleum Hydrocarbons and Organic Chemicals in Ground Water—Prevention, Detection, and Restoration, Proc. National Water Well Association, Dublin, Ohio, p. 71–84.

STUMM, W., and J. J. MORGAN. 1981. Aquatic Chemistry, 2nd edition. John Wiley & Sons, New York, 780 p.

SUDICKY, E. A. 1986. A natural gradient experiment on solute transport in a sand aquifer: Spatial variability of hydraulic conductivity and its role in the dispersion process. Water Resour. Res., v. 22, no. 13, p. 2069–2082.

SUDICKY, E. A., and E. O. FRIND. 1982. Contaminant transport in fractured porous media: analytical solutions for s system of parallel fractures. Water Resour. Res., v. 18, no. 3., 1634–1642.

SWENSON, F. A. 1968. New theory of recharge to the artesian basin of the Dakotas. Geol. Soc. Am. Bull. 79, p. 163–182.

TAKASAKI, K. J., and S. VALENCIANO. 1969. Water in the Kahuku area, Oahu, Hawaii. U.S. Geol. Surv. Water-Supply Paper 1874, 59 p.

TAMERS, M. A. 1967. Surface-water infiltration and groundwater movement in arid zones of Venezuela. Isotopes in Hydrology. Intl. Atomic Energy Agency, Vienna, p. 339–351.

TAMERS, M. A. 1975. Variability of radiocarbon dates on groundwater. Geophys. Survey, v. 2, p. 217–239.

TANG, D. H., E. O. FRIND, and E. A. SUDICKY. 1981. Contaminant transport in fractured porous media: Analytical solution for single fracture. Water Resour. Res., v. 17, no. 3, p. 555–564.

THEIS, C. V. 1935. The relation between the lowering of the piezometric surface and rate and duration of discharge of a well using groundwater storage. Trans. Am. Geophys. Union, v. 2, p. 519–524.

THEIS, C. V. 1963. Drawdowns resulting from cyclic rates of discharge. In Ray Bentall (comp.), Methods of Determining Permeability, Transmissivity, and Drawdown. U.S. Geol. Surv. Water-Supply Paper 1536-I, p. 319–329.

THIEM, G. 1906. Hydrologische Methode. Gebhardt, Leipzig.

THOMAS, G. B., Jr. 1972. Calculus and analytic geometry, 3rd edition. Addison-Wesley Publishing Company, Reading, Massachusetts.

THOMPSON, C., G. MCMECHAN, R. SZERBIAK, and N. GAYNOR. 1995. 3-D GPR imaging of complex stratigraphy within the Ferron Sandstone, Castle Valley, Utah. Proc. of the Symposium on the Application of Geophysics to Engineering and Environmental Problems. Environmental and Engineering Geophys. Soc., Englewood, Colorado, p. 435–443.

THOMPSON, D. B. 1987. A microcomputer program for interpreting time-lag permeability test. Ground Water, v. 25, 212–218.

THOMPSON, G. M. 1976. Trichloromethane: A new hydrologic tool for tracing and dating groundwater. Ph.D. Dissertation, Department of Geology, Indiana University, Bloomington, 93 p.

THOMPSON, G. M., and J. M. HAYES. 1979. Trichlorofluoromethane in groundwater: A possible tracer and indicator of groundwater age. Water Resour. Res., v. 15, p. 546–554.

THORNTHWAITE, C. W. 1948. An approach toward a rational classification of climate. Geograph. Rev., v. 38., p. 55–94.

THURMAN, E. M. 1985. Organic Geochemistry of Natural Waters. Martinus Nijhoff/Dr. W. Junk Publishers, Dordrecht, 497 p.

THURMAN, E. M., L. B. BARBER, M. L. CEAZAN, R. L. SMITH, M. G. BROOKS, M. P. SCHROEDER, R. J. KECK, A. J. DRISCOLL, D. R. LEBLANC, and W. J. NICHOLS Jr. 1984. Sewage contaminants in ground water. U.S. Geol. Surv. Open-File Report 84-475, p. 47–87.

TILLMAN, N., K. RANLET, and N. TILLMAN. 1988. Advanced soil gas surveys and their application to hazardous waste management. Proceedings of HAZTECH '88, Cleveland, p. 301–311.

TODD, D. K. 1959. Groundwater Hydrology, 1st edition. John Wiley & Sons, New York.

TODD, D. K. 1980. Groundwater Hydrology, 2nd edition. John Wiley & Sons, New York, 535 p.

TÓTH, J. 1962. A theory of groundwater motion in small drainage basins in Central Alberta. J. Geophys. Res., v. 67, p. 4375–4387.

TÓTH, J. 1963. A theoretical analysis of groundwater flow in small drainage basins. J. Geophys. Res., v. 68, p. 4795–4812.

TÓTH J., and T. CORBET. 1986. Post-Paleocene evolution of regional groundwater flow systems and their relation to petroleum accumulations, Taber area, southern Alberta. Bull. Canadian Petroleum Geol., v. 34, no. 3, p. 339–363.

TRAINER, F. W. 1988. Plutonic and metamorphic rocks. In W. BACK, J. S. ROSENSHEIN, and P. R. SEABER, (eds.), Hydrogeology, The Geology of North America, v. O-2. Geol. Soc. Am., Boulder, Colorado, p. 367–380.

TRAPP, H., Jr., and M. A. HORN. 1997. Ground water atlas of the United States, Segment 11: Delaware, Maryland, New Jersey, North Carolina, Pennsylvania, Virginia, West Virginia. U.S. Geol. Surv., Reston, Virginia, 24 p.

TRUESDELL, A. H., and J. R. HULSTON. 1980. Chapter 5. Isotopic evidence on environments of geothermal systems. In P. FRITZ and J. C. FONTES (eds.), Handbook of Environmental Isotope Geochemistry, v. 1. Elsevier, Amsterdam, p. 179–226.

ULARA Watermaster. 1994. Watermaster service in the upper Los Angeles River Area, Los Angeles County. ULARA Watermaster Report, May.

U.S. Department of Agriculture, Soil Conservation Service. 1972. National Engineering Handbook, Hydrology Section 4. Washington, DC.

U.S. Department of the Navy, Naval Facilities Engineering Command. 1992. Soil Mechanics. NAVFAC Design Manual 7.1.

U.S. EPA. 1987. Compendium of superfund field operations methods, Part 2. EPA/540/P-87/001 (NTIS PB88-181557), 644 p.

U.S. EPA. 1987. Handbook of Ground Water. EPA/625/6-87/016, 212 p.

U.S. EPA. 1990. Handbook, Ground Water, v.1, Ground Water and Contamination. USEPA Office of Research and Development, EPA 625/6-90/016a, 144 p.

U.S. Geological Survey (USGS). 1999. The quality of our nation's waters—Nutrients and pesticides. U.S. Geol. Surv. Circular 1225, 82 p.

VAN DEN BERG, J. 1980. Petroleum Industry in Illinois, 1978. Illinois Geological Survey, Illinois Petroleum 116, 132 p.

VAN GENUCHTEN, M. TH. 1980. A closed-form equation for predicting the hydraulic conductivity of unsaturated soils. Soil Science of American Proceedings, v. 44, no. 5, p. 892–898.

VAN GENUCHTEN, M. TH., and W. J. ALVES. 1982. Analytical solutions of the one-dimensional convective-dispersive solute transport equation. U.S. Department of Agriculture, Technical Bulletin No. 1661, 151 p.

Virginia Soil and Water Conservation Commission. 1980. Virginia Erosion and Sediment Control Handbook. Richmond, Virginia, 600 p.

VOGEL, J. C. 1967. Investigation of groundwater flow with radiocarbon, Isotopes in Hydrology. Intl. Atomic Energy Agency, Vienna, p. 355–368.

VOGEL, J. C. 1970. Carbon-14 dating of groundwater, Isotopes in Hydrology. Intl. Atomic Energy Agency, Vienna, p. 235–237.

VOGEL, T., K. HUANG, R. ZHANG, and M. TH. VAN GENUCHTEN. 1996. The HYDRUS code for simulating one-dimensional water flow, solute transport, and heat movement in variable-saturated media. Version 5.0, Research Report No. 140, U.S. Salinity Laboratory, Riverside, California.

VOSS, C. I. 1984. A finite-element simulation model for saturated-unsaturated, fluid-density-dependent groundwater flow with energy transport or chemically-reactive single-species solute transport. U.S. Geol. Surv. Water-Resour. Investigations Report 84-4369, 409 p.

WALKER, R. G., and D. J. CANT. 1984. Sandy fluvial systems. In R. WALKER (ed.), Facies Models. Geoscience Canada Reprint Series 1, p. 71–89.

WALLIS, P. M., H. B. N. HYNES, and S. A. TELANG. 1981. The importance of groundwater in the transportation of allochthonous dissolved organic matter to the streams draining a small mountain basin. Hydrobiologia, v. 79, p. 77–90.

WALTER, D. A., D. R. LeBLANC, K.G. STOLLENWERK, and K. W. CAMPO. 1999. Phosphorous transport in sewage-contaminated ground water, Massachusetts Military Reservation, Cape Cod, Massachusetts, D. W. MORGANWALP and H. T. BUXTON (eds.). Water Resour. Investigations Report 99-4018C, p. 305–315.

WALTHER, E. G., A. M. PITCHFORD, and G. R. OLHOEFT. 1986. A strategy for detecting subsurface organic contaminants. Proc. Petroleum Hydrocarbons and Organic Chemicals in Ground Water: Prevention, Detection and Restoration. National Water Well Assoc., Dublin, Ohio, p. 357–381.

WALTON, W. C. 1962. Selected analytical methods for well and aquifer evaluation. Illinois State Water Survey, Bull. 49, 81 p.

WEEKS, E. P. 1969. Determining the ratio of horizontal to vertical permeability by aquifer test analysis. Water Resour. Res., v. 5, p. 196–214.

WENZEL, L. K. 1936. The Thiem method for determining permeability of water-bearing materials and its application to the determination of specific yield, results of investigations in the Platte River Valley, Nebraska. U.S. Geol. Surv. Water-Supply Paper 679-A, 57 p.

WHITE, W. B. 1990. Surface and near-surface Karst landforms. In C. G. HIGGINS and D. R. COATES (eds.), Ground-Water Geomorphology, The Role of Subsurface Water in Earth-Surface Processes and Landforms. Geol. Soc. Am. Special Paper, 252, p. 157–175.

WHITE, W. B., and E. L. WHITE. 1987. Ordered and stochastic arrangements within regional sinkhole populations. In B. F. BECK and W. L. WILSON (eds.), Karst Hydrogeology: Engineering and Environmental Applications, A.A. Balkema Publishers, Accord, Massachusetts, p. 85–90.

WHITEHEAD, R. L. 1996. Ground water atlas of the United States, Segment 8: Montana, North Dakota, South Dakota, Wyoming. Hydrologic Investigations Atlas 730-I. U.S. Geol. Surv., Reston, Virginia, 24 p.

WIEDEMEIR, T. H., M. A SWANSON, D. A. MOUTOUX, E. K. GORDON, J. T. WILSON, B. H. WILSON, D. H. KAMBELL, J. E. HANSEN, P. HAAS, and F. H. CHAPELLE. 1996. Technical protocol for evaluating natural attenuation of chlorinated solvents in groundwater. Air Force Center for Environmental Excellence, San Antonio, Texas.

WIERENGA, P. J., L. W. GELHAR, C. S. SIMMONS, G. W. GEE, and T. J. NICHOLSON. 1986. Validation of stochastic flow and transport models for unsaturated soil: A comprehensive field study. U.S. Nuclear Regulatory Commission, NUREG/CR-4622.

WIERENGA, P. J., M. H. YOUNG, G. W. GEE, R. G. HILLS, C. T. KINCAID, T. J. NICHOLSON, and R. E. CADY. 1993. Soil characterization methods for unsaturated low-level waste sites. U.S. Nuclear Regulatory Commission, NUREG/CR-5988 (PNL-8480).

WIGLEY, T.M.L., L. N. PLUMMER, and F. J. PEARSON Jr. 1978. Mass transfer and carbon isotope evolution in natural water systems. Geochim. Cosmochim. Acta, v. 42, p. 1117–1139.

WILLIAMS, R. A., and T. L. PRATT. 1996. Chapter 13, Detection of the base of Slumgullion landslide, Colorado, by seismic Reflection and Refraction Methods. In D. J. VARNES and W. Z. SAVAGE (eds.), The Slumgullion Earth Flow: A Large-Scale Natural Laboratory. U.S. Geol. Surv. Bull. 2130, U.S. Government Printing Office, Washington, DC.

WILLIAMS, J. R., Y. OUYANG, J. CHEN, and V. RAVI. 1997. Estimation of infiltration rate in the vadose zone. V. II: Application of selected mathematical models. EPA/600/R-97/128b, 44 p.

WINNOGRAD, I. J., T. B. COPLEN, J. M. LANDWEHR, A. C. RIGGS, K. R. LUDWIG, M. J. SZABO, P. T. KOLESAR, and K. M. REVESZ. 1992. Continuous 500,000-year climate record from vein calcite in Devils Hole, Nevada. Science, v. 258.

WINTER, T. C. 1976. Numerical simulation analysis of the interaction of lakes and groundwaters, U.S. Geol. Surv. Prof. Paper 1001.

WINTER, T. C. 1978. Numerical simulation of steady-state, three dimensional groundwater flow near lakes. Water Resour. Res., v. 14, p. 245–254.

WINTER, T. C., J. W. HARVEY, O. L. FRANKE, and W. M. ALLEY. 1998. Ground water and surface water, a single Resource. U.S. Geol. Surv. Circular 1139, Denver, Colorado.

WINTER, T. C., and H. O. PFANNKUCH. 1984. Effect of anisotropy and groundwater system geometry on seepage through lakebeds, 2. Numerical simulation analysis. J. Hydrol., v. 75, p. 239–253.

WITTMANN, S. G. 1985. Use of soil gas sampling techniques for assessment of ground water contamination. NWWA/API Conference Petroleum Hydrocarbons and Organic Chemicals in Ground Water—Prevention, Detection, and Restoration, Proc. National Water Well Association, Dublin, Ohio, p. 291–309.

WOLERY, T. J. 1979. Calculation of Chemical Equilibrium between Aqueous Solutions and Minerals: The EQ3/EQ6 Software Package. UCRL-52658, University of California, Lawrence Livermore Laboratory, Livermore, California, 41 p.

WOOD, W. W., and L. A. FERNANDEZ. 1988. Plutonic and metamorphic rocks. In W. BACK, J. S. ROSENSHEIN, and P. R. SEABER (eds.), Hydrogeology, The Geology of North America, v. O-2. Geol. Soc. Am., Boulder, Colorado, p. 353–365.

YEH, G. T. 1981. AT123D- Analytical transient one-, two-, and three-dimensional simulation of waste transport in the aquifer system. Oak Ridge, Tennessee, Oak Ridge National Laboratory, Environmental Science Division Publication No. 1439, 83 p.

YONG, R. N. 1985. Interaction of clay and industrial waste: a summary review. Second Canadian/American Conference on Hydrogeology, B. HITCHON and M. TRUDELL (eds.). National Water Well Assoc., Dublin, Ohio, p. 13–25.

ZACHARA, J. 1990. Hydrogeology in relation to microorganisms: Proceedings of the First International Symposium on Microbiology of the Deep Subsurface, C. B. FLIERMANS and T. C. HAZEN (eds.). Westinghouse Savannah River Company, p. 5-3–5-13.

ZALASIEWICZ, J. A., S. J. MATHERS, and J. D. CORNWELL. 1985. The application of ground conductivity measurements to geological mapping. Q. Journal Eng. Geol. London, v. 18, p. 139–148.

ZAPOROZEC, A. 1972. Graphical interpretation of water quality. Ground Water, v. 10, p. 32–43.

ZAPOROZEC, A., and J. C. MILLER. 2000. Ground-water pollution. UNESCO International Hydrological Program, 24 p.

ZEMO, D. A., T. E. GRAF, and J. E. BRUYA. 1993. The importance and benefit of fingerprint characterization in site investigation and remediation focusing on petroleum hydrocarbons. Proc. of Petroleum Hydrocarbons and Organic Chemicals in Ground Water: Prevention, Detection and Restoration. National Ground Water Assoc., Columbus, Ohio, p. 39–54.

ZHENG, C. 1990. MT3D: A modular three-dimensional transport model. S. S. Papadopulos and Associates, Inc., Bethesda, Maryland.

ZHENG, C., and G. D. BENNETT. 1995. Applied contaminant transport modeling. Internl. Thomson Pub. Co., New York, 440 p.

ZHENG, C., and P. P. WANG. 1999. A modular three-dimensional multispecies transport model for simulation of advection, dispersion and chemical reactions of contaminants in groundwater systems (Release DoD_3.50.A). Prepared for U.S. Army Corps of Engineers, Washington, DC.

ZIMMERMAN, U., D. ENHALT, and K. O. MUNNICH, 1966. Soil-water movement and evapotranspiration: Changes in the isotopic composition of the water, Isotopes in Hydrology. Proceedings of the IAEA Symposium 1966, Vienna, p. 567–584.

ZYVOLOSKI, G. A., B. A. ROBINSON, Z. V. DASH, and L. L. TREASE. 1997. Summary of the models and methods for FEHM Application—a finite-element heat- and mass-transfer code. NTIS, U.S. Department of Commerce, Springfield, Virginia.

INDEX

577